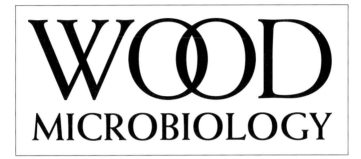

WOOD MICROBIOLOGY
Decay and Its Prevention

ROBERT A. ZABEL
Professor Emeritus
College of Environmental Science and Forestry
State University of New York
Syracuse, New York

JEFFREY J. MORRELL
Department of Forest Products
Oregon State University
Corvallis, Oregon

ACADEMIC PRESS, INC.
Harcourt Brace Jovanovich, Publishers
San Diego   New York   Boston   London   Sydney   Tokyo   Toronto

Cover photograph: The background is the surface of a hardwood board displaying the black zone lines associated with the decay of *Xylaria polymorphia* (see Figure 7-4B).

This book is printed on acid-free paper. ∞

Copyright © 1992 by ACADEMIC PRESS, INC.
All Rights Reserved.
No part of this publication may be reproduced or transmitted in any form or by any means, electronic or mechanical, including photocopy, recording, or any information storage and retrieval system, without permission in writing from the publisher.

Academic Press, Inc.
1250 Sixth Avenue, San Diego, California 92101-4311

United Kingdom Edition published by
Academic Press Limited
24–28 Oval Road, London NW1 7DX

Library of Congress Cataloging-in-Publication Data

Zabel, R. A. (Robert A.)
    Wood microbiology : decay and its prevention / Robert A. Zabel,
  Jeffrey J. Morrell.
      p. cm.
    Includes bibliographical references and indexes.
    ISBN 0-12-775210-2
    1. Wood--Deterioration.   2. Wood-decaying fungi.   3. Wood-
-Microbiology.   I. Morrell, Jeffrey J., date.   II. Title.
TA423.2.Z33   1992
620.1'22--dc20   91-41769
                                                                                           CIP

PRINTED IN THE UNITED STATES OF AMERICA
92 93 94 95 96 97   EB   9 8 7 6 5 4 3 2 1

*Dedicated
to
Theodore C. Scheffer
and
his students and colleagues
at the
Forest Products Laboratory at Madison
whose research achievements over
a half century have laid
the foundation for
the principles and practices
for the microbiology of wood*

Contents

Preface xix

Part 1

CHAPTER 1
Introduction to Wood Microbiology
   Wood—A Remarkable Material   3
   Wood Value and Uses   4
   Potential Uses of Wood   4
   Wood Disadvantages   5
   Decay Losses and Future Wood Needs   6
   Reducing Decay Losses   6
   Wood Pathology versus Wood Microbiology   6
   Historical Perspectives of Wood Pathology   8
   Concepts and Terminology in Wood Microbiology   12
   Summary   14
   Sources of Historical Information on Early Developments in Forest Pathology, Wood Pathology, and Wood Preservation   14
   Sources for Mycological, Phytopathological, and Forestry Terminology and Definitions   15
   Sources for Information on Wood Uses and Treatments   15
   References   16

Part 2

CHAPTER 2
Wood Deterioration Agents

   Abiotic Damage   22
   Biotic Damage   22
   Wood Weathering   23
   Wood Thermal Decomposition   24
   Chemical Decomposition of Wood   26
   Mechanical Wear   28
   Insect Damage to Wood   29
      Isoptera   29
      Coleoptera   34
      Hymenoptera   38
   Marine Borer Damage   40
      Shipworms   41
      Pholads   43
      Isopods   43
   Wood Decays and Discolorations Caused by Fungi   44
   Summary   45
   References   47

CHAPTER 3
Characteristics and Classification of Fungi and Bacteria

   Fungi in Relation to Other Life Forms   52
   Bacteria   54
   Fungi   55
      Macroscopic Appearances of Fungi   55
      Microscopic Features of Fungi   57
      Hyphal Wall Structure   58
      Ultrastructure of Fungi   60
      Specialized Hyphae   62
      Cultural Characteristics   64
      Reproduction   66
      Life Cycles   69
      Reproductive Capacity   71
      Variability   72
      Growth Requirements   72

A Classification of Fungi   73
   Eumycota   74
   Mastigomycotina and Zygomycotina   75
   Ascomycotina (Ascomycetes)   76
   Deuteromycotina (Deuteromycetes or Fungi Imperfecti)   77
   Basidiomycotina   77
   Hymenomycetes   78
   Aphyllophorales   80
A Classification of Bacteria   82
Roles of Fungi and Bacteria in Ecosystems and Human Affairs   84
Summary   85
References   86
   Characteristics of Fungi   86
   Classification of Fungi   88
   Characteristics and Classification of Bacteria   89

CHAPTER 4

Factors Affecting the Growth and Survival of Fungi in Wood (Fungal Ecology)

Major Growth Needs of Wood-Inhabiting Fungi   90
   Water   91
   Oxygen   96
   Temperature   102
   Substrate (Food Sources)   107
   Hydrogen Ion Concentration (pH)   108
   Chemical Growth Factors   108
   Vitamins and Minor Metals   110
   Light   110
   Miscellaneous Factors   111
Summary   111
References   111

CHAPTER 5

Fungal Metabolism in Relation to Wood Decay

Metabolism Defined   116
Energy Sources, Transfer, and Storage   117
   Oxidation–Reduction Reactions   117
   Electron Carriers and High-Energy Compounds   118

Enzymes   118
    Structure and Mode of Action   118
    Types and Classifications of Enzymes   121
Digestion and Hydrolases   122
    Cellulose   124
    Hemicelluloses   124
    Lignin   124
Absorption of Digestion Products   125
Aerobic Respiration   125
    Glycolysis   125
    Citric Acid Cycle   127
    Electron Transport Chain   128
    Pentose Shunt   128
Fermentation   129
Anaerobic Respiration   129
Enzyme Inhibitors   130
Nutrition in Relation to Fungal Growth Requisites and Decay Control   130
Summary   131
References   134

CHAPTER 6

The Decay Setting: Some Structural, Chemical, and Moisture Features of Wood in Relation to Decay Development

Wood Functions   136
Structural Features of Wood   137
    Growth Patterns and Microscopic Features   137
    Cell-Wall Ultrastructure   141
    Cell-Wall Pittings   143
Major Chemical Constituents of Wood   145
    Cellulose   146
    Hemicellulose and Other Carbohydrates   146
    Lignin   150
    Miscellaneous Cell-Wall Chemicals   151
Amounts and Distributions of Cell Wall Components   153
Distributions of the Major Chemicals in the Wood Cell Wall   154
Organizational Levels in the Cell Wall   157

Wood–Water Relationships   158
Wood Variability   163
Summary   164
References   166

CHAPTER 7
General Features, Recognition, and Anatomical Aspects of Wood Decay

The Dual Nature of Decay   168
General Features of Wood Decay   169
    Stages   169
    Macroscopic Features   170
    Decay Patterns   170
    Major Disadvantages of Decay   172
    Decay Losses   173
Recognition of Decay (Visual Evidences)   173
    Macroscopic Decay Evidences   174
    Microscopic Decay Evidences   176
Other Decay-Detection Procedures   176
Types and Classifications of Wood Decay   176
Soft Rots—A New Decay Type   178
A Classification of Wood Modifications by Microorganisms   179
    Wood Decayers   179
    Nondecaying Wood Inhabitors   180
Other Common Wood-Decay Groups   180
Some Anatomical Features of Wood Decay   182
An Early History and Major Contributions to the Anatomy of Decay   182
Principal Anatomical Features of Decay   183
    Entrance and Early Colonization   184
    Cell-Wall Penetration   184
Some Research Needs   188
Summary   190
References   191

CHAPTER 8
Chemical Changes in Wood Caused by Decay Fungi

Changes in Cell-Wall Components by Decay Type   196
Chemical Mechanisms of Wood Decay   203

Cellulose Decomposition   204
    Initial Concepts on Cellulolytic Enzymes C_1-C_x Hypothesis   204
Hemicelluloses Decomposition   208
    Xylans   208
    Mannans   209
Lignin Decomposition   209
    Lignin Determinations   210
    Chemical Modifications in Decayed Lignin   211
    Lignin-Degrading Enzymes from White-Rot Fungi   212
    Lignin Degradation by Non-White-Rot Fungi   215
Roles of Bacteria in Wood Decomposition   215
A Decay Model and Related Research Needs   217
Summary   219
References   220

CHAPTER 9
Ultrastructural Features of Wood Decay

Electron Microscopy   225
    Transmission Electron Microscopy   226
    Scanning Electron Microscopy   227
Some Wood and Fungal Ultrastructural Features   227
    Wood Ultrastructure   227
    Fungal Ultrastructure   228
Wood Ultrastructural Changes during Decay   229
    White-Rot Fungi   229
    Brown Rots   232
    Soft Rots   235
    Bacterial Erosion   238
Microbial Interactions in Decay   239
Detection and Quantification of Wood Preservatives   239
Summary   240
References   241

CHAPTER 10
Changes Caused by Decay Fungi in the Strength and Physical Properties of Wood

Wood Weight Loss (Biomass Loss)   249
    Density Loss   250

Strength (Mechanical) Properties   251
Hygroscopicity   257
Caloric Value   258
Permeability   259
Electrical Properties   260
Acoustic Properties   260
Summary   261
References   262

CHAPTER 11
Colonization and Microbial Interactions in Wood Decay
Some Ecological Concepts and Terminology   266
Types of Microbial Interactions in Wood   268
Common Wood Inhabitants during Decay   269
Microecology of Wood Decay   270
Colonization Strategies   271
   Opportunists   271
   Stress Resistors   271
   Combatants   271
Colonization Patterns of Wood by Fungi   272
   Standing Tree   272
   Standing, Dying, or Dead Trees   272
   Seasoning Lumber   273
   Untreated Wood in Ground Contact   273
   Treated Wood in Ground Contact   274
   Other Wood Conditions   274
Succession in Wood Decay   274
Research Needs on Ecology of Decay Fungi   276
Summary   276
References   277

Part 3

CHAPTER 12
Decays Originating in the Stems of Living Trees
Historical Highlights   284
Stem-Decay Types   284

Stem-Decay Origins   285
    Heart Rots   285
    Sap Rots   286
Types of Stem Wounds   286
Stem Tissue Reactions to Wounding   287
Compartmentalization and Succession   287
Rates of Decay Development   289
Recognition of Stem Decays   290
Some Common Stem Decays   291
    Major Heart Rots   293
    Common Sap Rots   299
Host-Specific Stem-Decay Fungi   300
Some Colonization Strategies of Stem Decay Fungi   300
Summary   301
References   303

CHAPTER 13
Biodeterioration of Stored Wood and Its Control

Types of Storage Loss   307
General Control Practices   307
Logs and Bolts   308
    Water Storage   310
    Water Spraying   312
    Chemical Treatments   312
    Biological Controls   313
Poles and Piling   313
Pulpwood   315
Pulpwood Chips   317
Summary   321
References   323

CHAPTER 14
Wood Stains and Discolorations

History   326
Types of Wood Discoloration   327
    Enzymatic and Chemical Stains   327
Color Changes Associated with Incipient Decay   332
Fungal Stains or Molds   332

    Molds   332
    Sapstains   333
    Stain Development   335
    Effects of Fungal Stain on Wood Properties   336
    Sapstain Control   336
Summary   339
References   339

CHAPTER 15

Decay Problems Associated with Some Major Uses of Wood Products

Decay Hazard   345
    Types of Decay Hazard   345
    Types of Wood Products—Decay Fungi   347
Decay of Wood Products   348
    Wood Buildings (Homes)   348
    Utility Poles   353
    Foundation or Marine Piling   357
    Railroad Ties, Mine Timbers, and Bridges   358
    Cooling Towers   359
    Wooden Boats   359
    Pallets and Boxes   360
    Panel Products   360
A Decay-Control Principle   360
Summary   361
References   361

CHAPTER 16

Detection of Internal Decay

Decay-Detection Difficulties   365
Basic Sampling for Decay   366
An Ideal Decay-Detection Device   366
Physical Decay Detection   367
    Sounding   367
    Boring   367
    Visual Examination of Increment Cores   368
Mechanical Decay Detection   368
    Compression Tests   368

  Penetration Resistance   369
  Drilling Torque Release   369
  Pick or Splinter Test   369
  Extensiometer   369
  Vibration   370
 Electrical Decay Detection   371
  Moisture Meters   371
  Shigometer   371
  X-ray   372
  Tomography   372
 Acoustic Decay Detection   372
  Acoustic Emissions   372
  Stress-Wave Timers   373
  Wave-Form Analysis   373
 Laboratory Decay Detection   374
  Culturing   374
  Microscopy   374
  Chemical Indicators   375
  Lectins   375
  Serological Tests   375
  Analytical Techniques   376
  Olfactory Detection   377
 Summary   378
 References   378

CHAPTER 17
Paint Mildew and Related Degradative Problems
 Types of Paint Biodeterioration   384
 Paint Types and Compositions   385
 Paint Microflora   386
  Characteristics and Growth Features of A. pullulans   387
  Factors Affecting Mildew Development   390
 General Control Practices   392
 Mildewcide Evaluations   392
 Some Research Considerations   393
 Related Degradative Problems of Microorganisms on Surfaces   393
  Slime Formation during Pulp and Paper Manufacture   394
  Molds within the Home   395

  Mold Growth on Inert Materials   395
Summary   395
References   396

CHAPTER 18
Natural Decay Resistance (Wood Durability)

 Variations in Decay Resistance   400
  Species Variations   400
  Stem Position Variations   400
 Factors Affecting Durability   402
  Lignification   402
  Growth Characteristics   403
  Miscellaneous Factors   404
 Decay Resistance and Heartwood Formation   404
  Polyphenols   405
  Terpenoids   405
  Tropolones   406
  Tannins   407
 Evaluating Natural Durability   408
 Summary   409
 References   410

CHAPTER 19
Chemical Protection of Wood (Wood Preservation)

 A Brief Early History   412
 Treatment Choice and the Biological Hazard   413
  Short-Term Wood Protection   414
  Long-Term Wood Protection   415
 Major Wood Preservatives   419
  Organic (Oilborne) Preservatives   419
  Waterborne Chemicals   422
  Miscellaneous Compounds   423
 Nonchemical Methods for Improving Wood Performance   425
 Environmental Considerations   426
 Wood Preservative Development and Testing   427
  Petri Plate Tests   427
  Soil-Block Test   427
  Decay Cellars   429

Termite Tests   429
Field Stake Tests   429
Wood Assembly Tests   430
Marine-Borer Tests   430
Preservative Resistance (Tolerance)   433
    Decomposition of Creosote   434
    Decomposition of PCP   435
    Decomposition of Inorganic Arsenicals   436
Summary   436
References   437

Part 4

CHAPTER 20
Some Trends in Wood Microbiology Research and a New Emphasis (Biotechnology)

Changes in Forests and Wood Supplies   446
New Decay Control Approaches   446
Wood as Substrate for Mushroom Production   447
Microbial Generation of Feedstocks   448
Biological Control of Fungal Stain and Decay   448
Biotechnology in Pulp and Paper   450
Biotechnology in Chemical Waste Management   451
The Future   452
Summary   453
References   453

Author Index   458

Subject Index   467

Preface

We believe a textbook on wood decay caused by fungi and related biologic deteriorations is needed to collect and summarize the rapidly expanding literature and experience in a single source. A need also exists to relate this information to the basic principles of biology.

It is our hope that this introductory text on the principles of decay and discoloration processes in wood will facilitate wiser use of wood resources and stimulate speculation and research in the area. We also hope it stimulates a fascination with fungi and their unique capabilities and important roles in the biologic world.

The textbook is based on a series of lectures on wood decay, its prevention, and control, presented over the past few decades to juniors and seniors majoring in forest biology or wood products at State University of New York and Oregon State University.

Emphasis is placed on the major fungal damages to wood which may develop during the growth, harvesting, storage, conversion, and use of wood for a range of major purposes. The characteristics and appearances, causes (etiology), detection, effects on various use properties, and preventions or controls are stressed.

In Part 1, Chapters 1 and 2 trace the origin and history of wood microbiology, discuss the major types of wood deterioration, and relate wood decay to the broader subject of biodeterioration. Chapters 3–5 review the general characteristics, types and classification, growth needs, and metabolism of fungi as the major cause of wood decays, and discolorations. An emphasis is placed on relating the growth needs of fungi to decay control principles. Chapter 6 summarizes key features of the structural aspects of wood and wood–moisture relationships central to fungal survival and growth in wood and the subsequent development of damaging decays and discolorations.

In Part 2, the emphasis is on the basic anatomical, physical, and chemical aspects of wood decay. Chapters 7-10 cover the basic aspects of wood decay, including types, appearances, and evidences; the anatomical and ultrastructural features of decay; decay effects on physical and strength properties of wood; and the chemical aspects of decay. Chapter 11 reviews the way fungi colonize wood and their interactions during decay development. It is a prelude to Part 3, which considers the major decay problems.

In Part 3, Chapters 12-15 review the principal decays and discolorations that may develop in wood in standing stems, during harvest, storage, conversion, and various major uses. Special emphasis is placed on the decay problems in buildings and utility poles. Chapter 16 discusses the methodologies and approaches to decay detection. Chapter 17 reviews mildew problems on wood-based coatings and related industrial problems caused by fungi and bacteria. Chapters 18-19 review decay control and prevention by use of naturally durable woods and wood preservatives. The characteristics and special uses of the major wood preservatives are discussed along with environmental concerns and restraints. In Part 4, a final chapter speculates on future decay prevention involving biological controls, new wood treatments, and developing more durable woods. An emphasis is placed on the role of biotechnology and future wood uses and modifications. Research trends and future career opportunities in wood microbiology are discussed.

The textbook is designed specifically for use in a two- or three-credit hour one-semester course. It assumes the student has had courses in general biology and organic chemistry.

The book is designed also to serve as a useful information source for wood processors, engineers, architects, and other professionals engaged in the practical aspects of wood who need to know more about the principal biodeterioration problems of a major wood use, why decay occurs, as well as how to recognize and prevent or minimize its development. It is also intended to provide some background on wood biomodifications for those interested in applying fungi for useful purposes in this period of expanding biotechnology.

The topics in each chapter are summarized to what we believe are the key essentials. The references and suggested further readings will lead interested readers to more detailed coverage of each topic.

It is a challenging and arduous task to write a textbook in the midst of an information explosion, new biologic insights, and mounting environmental concerns. Rapid advances in the related fields of mycology, physiology, genetics, electron microscopy, and biochemistry have required constant substantial revision of ideas. Yet it is the rapid changes and flux of ideas about wood decay that puts a new urgency on the need for a generalization and synthesis of the current literature for students.

Special thanks are extended to our many colleagues and associates for their encouragement, wise counsel, and assistance in reviews and improvements to the chapters. Particularly helpful to the senior author at State University of New York were James Worrall, Paul Manion, Chun K. Wang, and David Griffin. Colleagues who reviewed chapters are John Simeone

(Chapter 2—Insects as Biodeterioration Agents), Chun Wang (Chapter 3—Classifications of Fungi), David Griffin (Chapter 4—Growth Factors Affecting Fungi), James Nakas (Chapter 5—Metabolism), Wilfred C. Côté (Chapter 6—Wood Structure, Chemical, and Moisture Characteristics), James Worrall (Chapter 7—General Features of Decay), Tore E. Timell (Chapter 8—Chemical Aspects of Decay), Robert Hanna (Chapter 9—Ultrastructural Features of Decay), Dudley Raynal (Chapter 11—Colonization and Microbial Interactions in Wood Decay), and Paul Manion (Chapter 12—Stem Decays).

The authors also acknowledge the helpful comments provided by Jerrold E. Winandy, U.S. Forest Products Laboratory (Chapter 10—Changes in the Strength and Physical Properties of Wood Caused by Decay Fungi), Wayne Wilcox, University of California, Berkeley (Chapter 15—Decay Problems Associated with Some Major Uses of Wood Products), Theodore C. Scheffer of Oregon State University [Chapter 14—Wood Stains and Discoloration; Chapter 18—Natural Decay Resistance (Durability)], and Robert D. Graham of Oregon State University [Chapter 19—Chemical Protection of Wood (Wood Preservation)]. Their collective efforts significantly improved these chapters and are greatly appreciated.

Special thanks are due also to the many students in this course for their interest and innumerable sharp and penetrating questions, which placed many issues in clearer perspective.

The authors also acknowledge the assistance of Pat Spoon and Nadia Iwachiw in the typing and assembly of this manuscript, Susan Anagnost and George Snyder for a range of photographic assistance, Chris Biermann and Susan Smith of Oregon State University in the preparation of figures and chemical structures, as well as the patience of our families in dealing with this task.

As with any document, we expect that some portions will become outdated as new information becomes available. It is our special hope that this text stimulates others, much in the same way that our mentors encouraged and stimulated our interests in the field of wood microbiology.

R. A. Zabel
J. J. Morrell

PART 1

CHAPTER 1

Introduction to Wood Microbiology

This textbook focuses primarily on the damaging decays and discolorations that fungi may cause to wood under some conditions of use. A major emphasis is placed on recognition, causes, conditions favoring development, effects on various wood-use properties, and preventions or controls of these serious wood defects.

This introductory chapter begins with a review of the unique qualities of wood for a wide range of uses, and the benefits of its production. This is done to put the negative defect problems that may occur in some wood uses in proper perspective. It is not too early to emphasize that when wood is properly used or handled, most wood-defect problems can be avoided or greatly minimized.

Other agents that may degrade or destroy wood are briefly discussed, because occasionally their damages can be confused with decay. Wood-decay losses at various stages of production and use are reviewed to justify the level of control effort and suggest research priorities. The historical aspects of major wood-decay understandings are traced to establish its relation to other fields, and probable emergence as a specialized facet of microbiology. The fungal cause of decay has dominated much of the early information on decay in phytopathological or mycological journals. Unfortunately, this early information on decay prevention and controls failed to reach engineering and architectural sources by whom the design of wood structures was determined. Later we shall see that the design of structures in terms of water shedding or retention can be crucial to effective decay prevention. The introductory chapter ends with some basic concepts, and the definitions of some terms that will be used throughout the book for convenience and precision.

Wood—A Remarkable Material

Wood is a remarkable material of great value and importance in the world economy. It is used extensively as a structural material, fuel, or industrial raw material in many parts of the world. It is estimated that wood currently accounts for one fourth of the value of the major industrial materials in the United States (National Research Council, 1990). As a renewable natural resource, it is available in large quantities at relatively low costs. An estimated one third of the landed area of the world is in forests. As a land crop, its supplies can be increased in both volume and quality by wise forest-management practices. Wood production in the forest ecosystem is often associated with many other forest values and amenities such as soil development and enrichment, wildlife resources, moderation and extension of water run-off, provision of superb recreational settings, reduction of atmospheric pollution, and landscape aesthetics.

Wood is also unique in a plant evolutionary sense since it was the vertical development of perennial vascular tissue (wood) that led to the aerial development of land plants. The vertical stem or trunk of the tree consists of elongated cells with unusual strength, flexibility, and durable properties at both macroscopic and ultrastructural levels. These properties permit stems to bear heavy crown loads and to withstand high horizontal stress from wind loads. The long-term selection for these properties has led to many unique properties of wood.

Wood Value and Uses

As a structural material, wood has high strength per unit weight, and is easily shaped and fastened. It is a convenient energy source and a major cheap source of cellulose and its many derivatives for the chemical industry. The color patterns and textures of woods are often pleasing, leading to uses for many decorative purposes. Wood is available in a wide range of textures, colors, densities, and chemical compositions, supporting a wide range of important uses such as construction timbers and lumber, decorative paneling, plywood, piling and wharves, railroad ties, poles and posts, packaging and crates, paper and paper products, cellulose derivatives, charcoal, and thousands of specialized miscellaneous uses ranging from pipe bowls to fiddle heads.

Potential Uses of Wood

Wood looms even more significantly as a valuable raw material in the future as expanding world populations place increasing stress on natural land ecosystems as sources for food, fiber, and energy. Modern intensive forestry practices have the potential to increase substantially both the yields and quality of wood. Trees are efficient radiant-energy transducers in many regions, and their biomass might be converted to alcohol, as alternative combustion-engine energy sources. The selective decay actions of fungi and related fermentation activities may permit the utilization of wood as a cheap source of animal feed or protein. As a material, wood is readily biodegradable under certain conditions and returns natural substances to ecosystem cycling. It has low energy requirements for conversion into various use products as compared with other structural materials. As relative raw material costs and availability change, wood has the potential to replace petroleum as the base for production of a wide range of industrial chemicals and polymers and perhaps energy itself.

Wood Disadvantages

Wood has some serious disadvantages that limit its usefulness for some purposes.

1. Wood is biodegradable, primarily by fungal action; under some use conditions, it may decay and weaken or discolor, requiring replacement. Other biologic agents also attack wood. Termites are a serious threat to untreated wood in many tropical locations. Marine invertebrates chew tunnels in wood in various salt water uses and cause serious damage. These various bioagents that can degrade or destroy wood are discussed in more detail in Chapter 2. In many cases their actions can be controlled or minimized by judicious use of treated or naturally durable wood.
2. Wood combusts at low kindling temperatures and, in certain size configurations and conditions, burns readily. Chemical treatments and wood design can reduce the combustion hazard.
3. Wood is dimensionally unstable at moisture contents below the fiber saturation point (f.s.p.) and swells as it wets and shrinks as it dries. This problem is compounded by the largest dimensional changes' occurring in the tangential plane. In round stock, such as poles or piling, the differential shrinkage often leads to deep check formation in the radial plane. Chemical treatments may reduce the dimensional changes, but these treatments are expensive. Increased use of composites should minimize any swelling or shrinkage of the wood in use.
4. Wood, as a natural product, displays considerable variability in its appearance, chemical composition, and physical properties. Some differences are the result of species or growth conditions and are mitigated by specifications. High safety factors in critical design uses of wood are used to minimize this disadvantage. Conversely, the variability in color and texture of wood provides its beauty and aesthetic appeal for many home uses.
5. Wood has a large bulk per unit weight for fuel, pulping, and chemical uses.

Effective wood handling, properly designed and maintained structures, chemical treatments, and proper use of standards and specifications can dramatically minimize these disadvantages. In a long-term environmental setting, the biodegradability of wood may minimize accumulation of the solid wastes that are created when less degradable materials such as plastics are used.

Decay Losses and Future Wood Needs

Accurate estimates of decay losses are useful to justify controls and serve as research incentives. Decay losses are difficult to quantify because of the multiplicity of wood uses under a wide range of environmental conditions. Experienced guesses are that 10% of the annual timber cut in the United States is used to replace wood that decays in service, much of it primarily from improper use and care. Added to this base raw material cost would be, in many cases, those costs representing processing, fabricating, finishing, merchandising, and assembly or replacement operations. The substantial labor costs incurred by replacement and, in some cases, the inconvenience of interrupted services would have to be added. Another subtle loss source may be eventual wood replacement by more expensive, less satisfactory materials.

Large additional supplies of wood are required to meet burgeoning population needs in the next century. A substantial first step in meeting future timber goals may be simply to handle and use wood more effectively, thereby drastically reducing decay losses. Forest pathologists and entomologists recognize that control of tree diseases and forest insect pest problems also can dramatically increase future wood supplies.

Reducing Decay Losses

Properly used wood is an amazingly durable organic material. Only a few specialized microorganisms, primarily the higher fungi, have solved the biochemical riddle of its rapid digestion.

Experts agree that much is already known about effective and economical control of decay in most wood uses. The *central control problem* is that much of this information is fragmented and not readily available or generally known by wood processors, designers, merchandisers, and users. Until recently, wood has been available at low cost, and the ease of replacement has reduced the incentive to conserve. Furthermore, the central information about wood microbiology spans many disciplines and becomes inaccessible at both academic and trade levels.

Wood Pathology versus Wood Microbiology

The subject of decays and stains in wood products has generally been referred to as either wood pathology or products pathology. Traditionally, wood decay has been considered a specialized facet of forest pathology (which is an area of phytopathology), and most forest pathology textbooks have included a chapter on wood products decays and lumber stains. This was logical and natural since forest pathology emphasized stem-decay problems and the causal fungi in its early development. This approach reflected, in part, the early emphasis of foresters on forest protection and wood harvesting and the phytopathologist's concern with fungi.

Mounting afforestation problems, epidemic diseases, and shifts in phytopathology focused the interest of forest pathology more on the disease process itself—the adverse reaction of living organisms to disease agents—and began the separation of wood biodeterioration problems from tree disease problems.

Actually, the term *wood pathology* is a misnomer; it is not a pathology at all since it deals primarily with the deterioration of nonliving materials caused by biotic and abiotic agents. *Products pathology* suffers from the same *nonliving* dilemma. Products pathology is more correctly a subject matter in phytopathology that involves the diseases of stored fruits, vegetables, and grains, which are living entities.

A more logical setting for fungal-related, wood-defect problems is *applied industrial microbiology*, which studies the microorganisms that adversely affect the properties or the appearance of food products, textiles, leathers, organic materials, paper, or wood. This field also includes substrate biomodification and fermentation by microorganisms.

We might see the emerging discipline of wood microbiology in clearer perspective by listing some closely related or overlapping fields in which information on wood decays and stains and their causal agents may appear.

1. *Phytopathology* deals with the understanding and control of plant disease, although there are some similarities between decay and tissue-disintegration disease.
2. *Forest pathology* deals with tree diseases. Some decays and stains originate in the living stem and continue as problems in the wood product; some tree pathogens are also wood saprogens.
3. *Mycology* is the study of all aspects of fungi. Most wood saprogens are fungi, although a few bacteria are also important.
4. *Microbiology* includes the study of all small organisms. It includes the diseases and deterioration they cause and all facets of their use (biotechnology).
5. *Wood preservation* is an engineering subject dealing with the physical and chemical aspects of protecting wood from fire, insects, and fungi.

The reality is that such discipline boundaries are in constant flux, and shift as need and opportunity arise. *Wood microbiology* integrates wood deterioration and wood biomodifications by microorganisms. The field should focus information on wood biodeterioration as a recognized discipline.

We shall see in subsequent chapters that much recent wood-defect literature is now appearing in wood technology and microbiology journals rather than in phytopathological or forestry journals.

Historical Perspectives of Wood Pathology

Historical awareness of a subject is important because it clearly displays the ever-evolving nature of knowledge and the occasional integration of the pieces into great unifying concepts. History also demonstrates the danger of dogma and the need for constant questioning and probing in the search for better explanations of events.

In this section, the term *wood pathology* will be used for historical reasons. Wood pathology had its origin in forest pathology, in which pathologists were interested in the nature of wood decay in tree stems, building rots, and wood storage.

A concern for wood decay and pragmatic methods to reduce this damage long preceded any recorded understanding of the cause and nature of decay. The high value of the biblical cedars of Lebanon was the result of their natural durability, important in shipbuilding and temple construction. The early Greeks knew that vertical bearing beams should rest on stone and not in direct contact with soil. Pliny, the Roman historian, recorded the susceptibility of sapwood to decay, listed durable woods, and reported that soaking wood in cedar oil reduced decay. In 1832, more than 40 years before the discovery of the true fungal nature of decay, the first successful wood-preservation process, Kyanizing using mercuric chloride was introduced in Europe. At about this time in Germany, Theodor Hartig in 1833 first recorded the microscopic appearance of fungal hyphae in decayed wood. Microscopic forms of life were then assumed to arise spontaneously from decomposition products. In 1863, Schacht reported on the effects and microscopic features of hyphae on a tropical wood; nearly 100 years later, it was recognized that his descriptions and drawings were typical of a new type of decay called soft rot. He also assumed the hyphae were the result and not the cause of the cell wall decomposition.

Forest pathology originated at the time of the settling of the great spontaneous generation versus *life from life* controversy of the mid-nineteenth century. Shortly after the classic researches of Tyndall and Pasteur had destroyed the concept of spontaneous generation and established the clear role of microorganisms in causing fermentations, the stage was set in 1874 for Robert Hartig's clear resolution of the causal relationship between the presence of hyphae and subsequent decay in wood. His early insights into the nature of decay were remarkable. He clearly connected the external fruiting body to the internal hyphae, and the hyphae and their growth on the cell walls to decay. He established later that some decays were specific for a kind of fungus and subsequently attributed these differences to enzymes. His extensive publications on wood decay and tree diseases, his cadre of students and research associates, led to rapid developments in forest pathology and wood pathology in Europe and the United States (Merrill *et al.*, 1975).

Two decades later, in 1899 in the United States, the federal government established the Mississippi Valley Laboratory at St. Louis, Missouri for reconnaissance and research on forest disease problems (Hartley, 1950). Dr. Herman von Schrenk was the initial appointee and later director of a small group. He was extremely productive and within a few years published a series of pioneer papers on the stem decays of timber species, wood decay in buildings, blue stain, preservative evaluations, and wood durability. Associated later with him in the early 1900s were G. Hedgcock, P. Spaulding, and Catherine Rumbold. Hedgcock published a classic study on chromogenic fungi that discolor wood. Spaulding contributed studies on the culturing of decay fungi and slash decomposition. Catherine Rumbold studied blue stain problems and insect roles as vectors.

At this time, the expansion of the railroads and the building program associated with the rapid western expansion of the country had led to the use of many local, nondurable woods for construction, as the supplies of durable oaks, chestnut, cypress, and cedars dwindled or were not easily available. This also was a period of increased harvesting of southern and western pines, and serious sapstain problems developed with shifts to summer air seasoning. Serious decay and discoloration problems led to concerns at the national level. One outcome was the formation of the American Wood Preservers' Association in 1904 for standardization of specifications for wood preservatives and treatments. Another was that the research of von Schrenk's group shifted increasingly to study decay of structural timbers or railroad ties and began research on sapstain control and wood preservation. At this time there was also growing concern about the threat of chestnut blight and other introduced tree pathogens.

A decision was made at the federal level to separate the forest and wood pathology programs. In 1907, the Mississippi Valley Laboratory (MVL) was discontinued, and the U.S. Forest Service's Forest Products Laboratory (FPL) at Madison, Wisconsin was assigned research responsibilities for the wood pathology program. Thus, it became the first titled program of wood pathology in the United States and has played a leadership role in wood pathology and wood preservation matters. The initial FPL research responsibilities included studies of the causes and controls of decay and stains in wood products, mycology, and the physiology of wood-products fungi, and wood preservatives. The forest pathology program was transferred to the Bureau of Plant Industry as the Office of Investigations in Forest Pathology. In 1953, it was also assigned administratively to the U.S. Forest Service.

From this beginning in 1899 in the United States, the field of wood pathology expanded steadily. Principal researchers were in the FPL, some forestry colleges, military organizations responsible for supplies, the chemical industry, and allied fields such as botany and wood technology, to name a few.

Many major contributions shaped the development of wood microbiology to the present period. In 1931, Hubert published the first American textbook on forest pathology, which contained a special section titled "Wood Pathology" and reviewed the principal research contributions of this early period. Some detail is presented here to show the rapid progress and significant accomplishments of the initial handful of dedicated researchers who defined the future course of the field. The literature sources for these early contributions are available in Hubert's Textbook cited above (1931). The major research topics with selected contributions were as follows:

1. Decay effects on wood properties such as strength (Colley, 1921) and heat conductivity (Hubert, 1924);
2. The chemical properties of decayed wood (Hawley and Wise, 1926) and the use of decayed pulpwood for paper (Rue, 1924);
3. Wood durability was attributed to soluble extractives (Hawley et al. 1924), and extensive durability tests on local timbers were reported (Humphrey 1916, Schmitz and Daniels, 1921);
4. The relationship of various moisture content levels in wood to decay development (Snell, 1921) and sapstain (Colley and Rumbold, 1930);
5. Decay in buildings and lumber caused by *Meruliporia incrassata* (Humphrey, 1923);
6. Decay problems and associated causal fungi in ties, mine timbers, poles, and piling (Humphrey, 1917, 1920, 1923);
7. Taxonomic studies of sapstain fungi (Hedgcock, 1906; Rumbold, 1929) and their control (Hubert, 1929); and
8. The evaluation of preservatives (Humphrey and Flemming, 1915; Richards, 1923).

In 1940, Scheffer and Lindgren completed a detailed study on the causes of lumber stains and their control. The application of their recommendations did much to minimize the serious sapstain problems of the period. In 1946, Cartwright and Findlay published their classic *Decay of Timber and Its Prevention*, which presented a worldwide summary of wood pathology information. The book was revised in 1958 and still serves as an important reference source for the field.

Davidson *et al.* (1942) and Nobles (1948) published cultural keys that greatly facilitated the identification of decay fungi isolated from timber and wood products.

In 1954, Savory established clearly that a new and significant type of decay termed *soft rot* was caused by some microfungi and Ascomycetes. The nearly century-old dogma that only Basidiomycetes can cause decay was finally dispelled. Contributions toward understanding this rot type were made by Corbett (1965), Duncan (1960), and Levy (1978). Nilsson (1973) established that substantial numbers of microfungi may cause soft rots.

In 1961, Cowling reported on the strikingly differential effects of white- and brown-rot fungi on a range of physical and chemical properties of sweet gum sapwood. Particularly significant were insights into the nature of the enzyme systems involved, based on ultrastructural dimensional restraints.

In 1965, Duncan and Lombard provided extensive information, which has been collected at the FPL over a 50-year period, on the major decay fungi associated with various wood products and wood uses, nationwide.

Verrall (1966) contributed to the understanding of decay problems in buildings in their relation to various water sources and (1965) demonstrated the effectiveness of dip or brush preservative treatments for some above-ground wood uses.

New insights into the anatomical and ultrastructural features of decay were provided by Wilcox (1970) and Liese (1970). Koenigs (1972) postulated an intriguing nonenzymatic process to explain how drastically early in the decay process, brown rot alters cell wall components in crystalline zones inaccessible to enzymes. In a decade of study, Reese (1977) and associates provided insight into the nature and action mode of the *cellulase* enzyme complex and opened the door to current commercial wood-fermentation possibilities. Eriksson (1978) clarified identities, roles, and sequences of the cellulose-degrading enzymes for a white-rot fungus. Currently Blanchette *et al.* (1987) are providing new information on the nature of decay by white-rot fungi with the scanning electron microscope.

In long-term comprehensive studies on decay origins and organism sequences in stem decays, Shigo (1967), and Shigo and Marx (1977) have demonstrated a compartmentalizing protective system in living stems, which may have major implications in subsequent wood-product treatments and use.

Butcher (1970) and Rayner and Todd (1979) have provided new insights into the sequences and many interactions of fungi in the decay process, invalidating the old concept of one fungus—one decay. Nilsson and Holt (1983) and Nilsson and Singh (1984) have shown that bacteria can cause soft-rot cavities and unusual tunnels in the secondary cell wall.

Treatments with agricultural fumigants were developed by Graham and Corden (1980) for controlling decay in poles and piling in service. Alternatives to the protection of wood by loading the cell cavities with potent toxicants are being explored increasingly as a result of growing environmental concerns. Preston (1986) has reported on new wood preservative compounds and proposed accelerated testing procedures. Rowell and Ellis (1979) have reported on chemical ways to modify wood and enhance its decay resistance.

Recent, exciting contributions by Kirk and Farrell (1987) are clarifying the enzymes in the *ligninase* system with its significant delignifying and biopulping possibilities.

Another indication of the rapidly growing interest in wood as a valuable renewable resource and its decay problems is the appearance of three books in the past several decades. Loewus and Runeckles (1977) and Higuchi (1985) edited books covering the biosynthesis and degradation of wood. Liese (1975) edited a book on the interactions of fungi and bacteria in decay development and the enzymatic mechanisms on the decay process. Nicholas (1973) edited a book on wood deterioration and its prevention that has become the most widely used book in the field. A textbook on the microbial and enzymatic degradation of wood and wood components has been prepared by Eriksson *et al.* (1990). Several comprehensive textbooks stressing the ecological aspects of wood-inhabiting microorganisms during decay development have been prepared by Rayner and Todd (1979) and Rayner and Boddy (1988).

The historical highlights of achievements in wood microbiology indicate an initial primary concern with understanding and controlling fungal damage to wood. Current trends suggest that in the future, wood microbiology will also concern itself increasingly with the beneficial uses of fungi to enhance wood properties or transform it into new products.

Concepts and Terminology in Wood Microbiology

An important part of the preparation for any profession is learning the specific vocabulary or language of that field. Terms developed to denote specific concepts, structures, or events help to simplify the complexities of the biological world, making differences, similarities, and various interrelationships more easily understood. It must be remembered that these concepts and definitions are arbitrary, and only their usefulness justifies their existence. In some cases, concepts cannot be precisely defined to include all related phenomena. Many times a series of events occurs in an intergrading series or a continuum without easy separations. In other cases, authorities simply disagree on definitions, and their conflicting views appear in the literature. The following terms will be used in this book, and an understanding of their *meaning* will be helpful in understanding the literature of wood microbiology. The terms are contrasted with their counterparts in forest pathology where applicable.

First, a clear distinction between *deterioration* and *disease* is necessary because of the forest pathology origin and current wood pathology designation of what we now prefer to call wood microbiology.

Deterioration is the destructive change in the properties of a nonliving material caused by a wide range of chemical, physical, or biotic agents. It contrasts directly with disease. Textiles, paper products, wood, plastics, and coatings are examples of wood-related organic materials that may be degraded or destroyed by biotic agents. The weakening of fabrics by the ultraviolet portion of sunlight is an example of the use of a physical deterioration agent. Corrosion of the steel beams in a bridge by salt is an example of chemical deterioration.

Disease (plant) is a sustained abnormal physiological process (or processes) of a plant or its part that may threaten the life of the plant or its parts or reduce its economic value. Disease involves reactions and changes (symptoms as responses) in the stressed living organisms, whereas deterioration concerns only changes in nonliving organic materials. Biotic or abiotic agents may instigate the changes in either case.

Biodeterioration (biodegradation) is a subset of deterioration. It is a negative term and can be defined as any undesirable change in the properties of a nonliving material caused by the activities of living organisms. The agents involved are many and varied, including bacteria, algae, fungi, invertebrates, and vertebrates such as rodents and birds. The major processes involved are *assimilation*, including invasion and digestion of organic materials such as wood or textiles; *mechanical* damage by the rasping activities of marine borers or insects; the *corrosion* of metals by the chemical activities of bacteria; and *function impairments* such as the fouling of ships' hulls by growth and accumulation of barnacles.

Two major types of wood biodeterioration, which are the central focus of this textbook, are *decay* and *discoloration* (stain). *Decay* is changes in the chemical and physical properties of wood caused primarily by the enzymatic activities of microorganisms. *Discoloration* is changes in the normal color of wood resulting from growth of fungi on the wood or chemical changes in cells or cell contents.

Other examples of materials that may be degraded or destroyed by the activities of fungi and bacteria are electrical and optical equipment, stored foods, leathers, plastics, petroleum products, textiles, and pharmaceutical products.

Biomodification is the positive term used for the biotic processes involved in the breakdown or conversion of organic or waste materials into innocuous or useful products by microorganisms. Topics of current interest are the bioconversion of trash, garbage, toxic chemical wastes, pesticides, and sewage sludges into useful products or harmless wastes. A major chemical industry utilizes fungi or bacteria to produce a wide range of chemicals and pharmaceuticals from organic materials. Great interest has also developed in the biological delignification and conversion of wood wastes into palatable animal fodders, alcohol for fuels, or yeast as protein and vitamin sources.

Fungi and *bacteria* are important destructive agents in both disease and biodeterioration. *Bacteria* are unicellular prokaryotes that reproduce by fission. *Fungi* are filamentous eukaryotes without chlorophyll that digest various carbon compounds externally. A single filament of a fungus is known as a *hypha* (plural hyphae). Collectively a mass of hyphae is commonly called a *mycelium*. Fungi reproduce primarily by spores. *Spores* are one- to several-celled units of reproduction in the fungi. They are released from the mycelium, and each cell is capable of reproducing the fungus. Biodeterioration is recognized by changes in the appearance or properties of the organic material or the physical presence of the causal agent. Disease is recognized by signs or symptoms. *Signs* are the actual physical presence of the disease-causing organism on the diseased plant. *Symptoms* are the physiological responses or reactions of the host to the presence of the disease-causing agent or organism (e.g., tissue swelling, resinosis, dwarfing). A *saprogen* is an organism that secures its food from dead materials or carbon sources. A *parasite* is an organism dependent part or all of the time on another organism of a different species and deriving all or part of its food from this living organism. A *substratum* is a nonliving organic material that serves as food to a living organism. A *host* is a living organism serving as a source of food to a parasite. A *pathogen* is an organism capable of causing a disease. Generally pathogens are parasites, but toxin-producers can be exceptions.

As we proceed through the remaining chapters, it should become clear that deterioration involves a complex array of interacting processes involving microorganisms, insects, and the environment acting on the nonliving and, therefore, nonresponsive wood substrate. This is an important concept, since the wood can not react to protect itself, and this has major implications for wood use.

Summary

- Wood has numerous beneficial properties, but is subject to degradation.
- The active study of wood deterioration dates from the late 1800s and has gradually progressed from a simple quantification of losses to more basic studies of the nature of decay.
- The term *wood microbiology* may be useful for describing the field, since it cuts across a variety of disciplines related to the decay process.

Sources of Historical Information on Early Developments in Forest Pathology, Wood Pathology, and Wood Preservation

Graham, R. (1973). History of wood preservation. *In* "Wood Deterioration and Its Prevention by Preservative Treatments. Vol. 1, Degradation and Protection of Wood" (Darrel Nicholas, ed.), pp. 1–30. Syracuse University Press, Syracuse, New York.

Hartley, C. (1950). The Division of Forest Pathology, U.S.D.A., *Plant Disease Reporter* (Suppl.) **195**:445–462. (Detailed coverage of the early history of the Division of Forest Pathology and detailed review of principal research achievements).

Hubert, E. E. (1931). "Outline of Forest Pathology." John Wiley & Sons, Inc., New York, New York. (A brief historical treatment of forest pathology and wood pathology including the early developments in Europe).

Liese, W. (1967). History of wood pathology. *Wood Science* and *Technology* 1:169-173.
Merrill, W., D. H. Lambert, and W. Liese. (1975). "Robert Hartig 1839-1901." Phytopathological Classics No. 12. American Phytopathological Society, St. Paul, Minnesota. (The authors' introduction to their English translation of Hartig, R. 1874. Wichtige Krankheiten der Waldbäume.)
Van Groenou, H. B., H. W. L. Rischen, J. Van Den Berge. (1951). "Wood Preservation during the Last 50 Years." A. W. Sijthoff's Uitgeversmaatschappij. N. V. Leiden (Holland).

Sources for Mycological, Phytopathological, and Forestry Terminology and Definitions

Agrios, G. N. (1988). "Plant Pathology," 3rd Ed. Academic Press, New York. (A useful glossary).
Hawksworth, D. L., B. C. Sutton, and G. C. Ainsworth. (1983). "Ainsworth & Bisby's Dictionary of the Fungi," 7th Ed. Commonwealth Mycological Institute, Kew, Surrey, England.
Link, G. K. K. (1933). Etiological phytopathology. *Phytopathology* **23**: 843-862.
Society of American Foresters. (1971). "Terminology of Forest Science: Technology Practice and Products." Society of American Foresters, Washington, D.C.

Sources for Information on Wood Uses and Treatments

Anonymous (1987). "Wood Handbook: Wood as an Engineering Material." Agriculture Handbook No. 72. Forest Products Laboratory, Forest Service, U.S.D.A.
Core, H. A., W. A. Côté, and A. C. Day (1979). "Wood Structure and Identification," 2nd Ed. Syracuse University Press, Syracuse, New York.
Fleischer, H. O. (1976). Future role of wood as a material. *Al Che E Symposium Series* **71** (146):1-3.
Goldman, I. S. (1975). Potential for converting wood into plastics. *Science* (Sept. 12): 847-852.
Haygreen, J. G., and J. L. Bowyer. (1989). "Forest Products and Wood Science: An Introduction." Iowa State University Press, Ames, Iowa.
Kollmann, F. F. P., and W. A. Côté. (1968). "Principles of Wood Science and Technology," Vol. 1. Springer-Verlag, New York.
Nicholas, D. D. (1973). "Wood Deterioration and Its Prevention by Preservative Treatments. Vol. 1. Degradation and Protection of Wood. Vol. II. Preservatives & Preservative Systems." Syracuse University Press, Syracuse, New York.
Panshin, A. J., and C. deZeeuw. (1980). "Textbook of Wood Technology: Structure, Identification, Uses, and Properties of the Commercial Woods of the United States," 4th Ed. McGraw-Hill, New York.
President's Advisory Panel on Timber and the Environment (PAPTE). (1977). "Report of the President's Advisory Panel on Timber and the Environment." U.S. Government Printing Office, Washington, D.C.

References

(Articles prior to 1931 are cited in the historical sources listed above)

Blanchette, R. A., L. Otjen, and M. C. Carlson. (1987). Lignin distribution in cell walls of birch wood decayed by white rot Basidiomycetes. *Phytopathology* **77**(5): 684–690.
Butcher, J. (1970). Techniques for the analysis of fungal floras in wood. *Material und Organismen* **6**:209–232.
Cartwright, K. St. G. and W. P. K. Findlay. (1958). "Decay of Timber and Its Prevention," 2nd Ed. Her Majesty's Stationery Office, London.
Corbett, N. H. (1965). Micro-morphological studies of the degradation of lignified cell walls by Ascomycetes and Fungi Imperfecti. *Journal of the Institute of Wood Science* **14**:18–29.
Cowling, E. B. (1961). "Comparative Biochemistry of the Decay of Sweetgum Sapwood by White-Rot and Brown-Rot Fungi." U.S.D.A. Technical Bulletin No. 1258, Madison, Wisconsin.
Davidson, R. W., W. A. Campbell, and D. B. Vaughn (1942). "Fungi Causing Decay of Living Oaks in Eastern United States and Their Cultural Identification. U.S.D.A. Technical Bulletin No. 785, Washington, D.C.
Duncan, C. G. (1960). "Wood-Attacking Capacities and Physiology of Soft-Rot Fungi." U.S.D.A. Forest Products Laboratory Report 2173, Madison, Wisconsin.
Duncan, C. G., and F. F. Lombard. (1965). "Fungi Associated with Principal Decays in Wood Products in the United States." U.S. Forest Service Research Paper WO-4, U.S.D.A., Washington, D.C.
Eriksson, K. E. L. (1978). Enzyme mechanism involved in cellulose hydrolysis by the white-rot fungus, *Sporotrichum pulverulentum*. *Biotechnology and Bioengineering* **20**:317–332.
Eriksson, K. E. L., R. A. Blanchette, and P. Ander. (1990). "Microbial and Enzymatic Degradation of Wood and Wood Components." Springer-Verlag, New York.
Graham, R. D., and M. E. Corden. (1980). "Controlling Biological Deterioration of Wood with Volatile Chemicals." Electric Power Research Institute EL-1480, Palo Alto, California.
Higuchi, T. (ed.). (1985). "Biosynthesis and Biodegradation of Wood Components." Academic Press, New York.
Kirk, T. K., and R. L. Farrell. (1987). Enzymatic "combustion": The microbial degradation of lignin. *Annual Review of Microbiology* **41**:465–505.
Koenigs, J. W. (1972). Effects of hydrogen peroxide on cellulose and its susceptibility of cellulase. *Material und Organismen* **7**:133–147.
Levy, C. R. (1978). Soft rot. *Proceedings of the American Wood Preservers' Association* **74**:145–163.
Liese, W. (ed.). (1975). "Biological Transformation of Wood by Microorganisms." Springer-Verlag, New York.
Liese, W. (1970). Ultrastructural aspects of woody tissue disintegration. *Annual Review of Phytopathology* **8**:231–258.
Loewus, F. A., and V. C. Runeckles (Eds.). (1977). "The Structure, Biosynthesis, and Degradation of Wood." Plenum Press, New York.
National Research Council. (1990). "Forestry Research: A Mandate for Change." National Academy Press, Washington, D.C.

Nilsson, T. (1973). "Studies on Wood Degradation and Cellulolytic Activity of Microfungi." Studia Forestalia Suecica 104, Stockholm, Sweden.

Nilsson, T., and D. M. Holt. (1983). Bacterial attack occurring in the S2 layer of wood fibers. *Holzforschung* **37**:107-108.

Nilsson, T., and A. P. Singh. (1984). "Cavitation Bacteria." International Research Group for Wood Preservation Document IRG/WP/1235.

Nobles, M. K. (1948). Studies in forest pathology VI. Identification of cultures of wood rotting fungi. *Canadian Journal of Botany* **26**:281-431.

Preston, A. F. (1986). Development of new wood preservatives: The how and the what. *Proceedings of the Wood Pole Conference*. 78-97. Portland, Oregon.

Schacht, H. (1863). Über die Veränderungen durch Pilze in abgestorben Pflanzellen. *Jahrb. f. Wiss. Botan.* **3**:442-483.

Scheffer, T. C., and R. M. Lindgren. (1940). "Stains of Sapwood Products and Their Control." U.S.D.A., Technical Bulletin No. 714, Washington, D.C.

Shigo, A. L. (1967). Successions of organisms in discoloration and decay of wood. *International Review of Forestry Research* **2**:237-299.

Shigo, A. L., and H. Marx. (1977). "CODIT (Compartmentalization of Decay in Trees)." U.S.D.A., Information Bulletin No. 405, Washington, D.C.

Rowell, R. M., and W. D. Ellis. (1979). Chemical modification of wood. Reaction of methyl isocyanate with southern pine. *Wood Science* **12**(1):52-58.

Savory, J. G. (1954). Breakdown of timber by Ascomycetes and Fungi Imperfecti. *Annals of Applied Biology* **41**:336-347.

Schacht, H. (1863). Über die Veränderungen durch Pilze in abgestorben Pflanzellen. *Jahrb. f. Wiss. Botan.* **3**:442-483.

Scheffer, T. C., and R. M. Lindgren. (1940). "Stains of Sapwood Products and Their Control." U.S.D.A., Technical Bulletin No. 714, Washington, D.C.

Shigo, A. L., and H. Marx. (1977). "CODIT (Compartmentalization of Decay in Trees)." U.S.D.A., Information Bulletin No. 405, Washington, D.C.

Shigo, A. L. (1967). Successions of organisms in discoloration and decay of wood. *International Review of Forestry Research* **2**:237-299.

Verrall, A. F. (1966). "Building Decay associated with Rain Seepage." U.S.D.A. Forest Service Technical Bulletin 1356, Washington, D.C.

Verrall, A. F. (1965). "Preserving Wood by Brush, Dip, and Short-Soak Methods." U.S.D.A. Forest Service, Technical Bulletin 1334, Washington, D.C.

Wilcox, W. W. (1970). Anatomical changes in wood cell walls attacked by fungi and bacteria. *Botanical Review* **36**(1):1-28.

PART 2

CHAPTER

2

Wood Deterioration Agents

Though wood is readily decomposed and recycled in the forest ecosystem by various biotic and abiotic agents, it is a very durable organic material when properly used and maintained. Evidence of this durability in the northeastern United States lies in the sound condition of many wooden homes after several hundred years of use. Longer, effective uses of wood are known for wooden churches in Norway and ornate wooden temples in Japan. Samples of wood found in the tombs of the Pharaohs are reported to show no signs of deterioration.

Yet, under some conditions of exposure or use—often when it becomes wet—wood is rapidly decomposed by biotic, chemical, or physical agents acting alone or in combination.

Three major types of biotic agents are responsible for most wood destruction. *Fungi*, and to a lesser extent, *bacteria*, are believed to cause the major wood loss. *Insects* such as termites and beetles also cause major deterioration losses, particularly in tropical regions. *Marine borers* cause substantial damage to wood submerged in salt or brackish waters in temperate and tropical zones.

Combustion and weathering are the principal abiotic types of wood destruction. Because wood in natural environments can be subject to a variety of agents, combinations of these agents are often involved in the wood-decomposition process. Termites, carpenter ants, beetles, and wasps may invade wood before and in the early decay stages and accelerate the destruction. Insect adults and larvae may introduce decay and stain fungi to the wood and, in some cases, utilize them as a food source. Presumed symbiotic associations have developed between some wood insects and fungi. Wood weathering appears to result from the combined action of photodegradation, chemical oxidants, and fungi.

While this book focuses primarily on wood decay and discoloration damage caused by fungi, it is important to be aware of the other types of damage. Some may be confused with types or stages of decay. Others may be associated with decay development or occur independently on the same wood product. Mechanical damage to loading platforms or docks and chemical spills on floors may superficially resemble late stages of certain decays that are fibrous in texture. Advanced weathering damage on a utility pole surface may superficially resemble soft rot. Prolonged exposures of wood to high temperatures, such as in dry kilns, produce surface discolorations resembling some biotic stains. Enzymatic stains that develop in seasoning lumber can be confused with sapstain caused by fungi. Termite or carpenter ant damage may superficially resemble the final stages of decay in which the annual rings separate.

This chapter presents an overview of the major deterioration agents other than fungi that, under some conditions of exposure or use, may damage wood. Emphasis will be placed on defect appearances, causal agents, factors affecting development, wood property changes, prevention or controls, and any relationship to fungal defects. Since coverage of these non-fungal wood-deterioration problems is not a prime purpose of this presentation, major reliance will be placed on references and recent review articles.

The major agents and types of wood decomposition are grouped under abiotic and biotic categories and listed as follows:

Abiotic Damage

1. Weathering—primarily photodegradation by ultraviolet (UV) light and oxidation
2. Thermal decomposition—distillation or burning
   a. Low-temperature exposure (below 200°C)
   b. High-temperature exposures in absence of oxygen (above 200°C)
   c. Combustion (above 275°C)
3. Chemical decomposition—hydrolysis and oxidation
   a. Exposure to strong acids
   b. Exposure to strong bases
   c. Exposure to strong oxidizing agents and some organic solvents
4. Mechanical wear—breakage and erosion of surface fragments

Biotic Damage

5. Animal attack—mechanical disruption
   a. Boring and surface rasping by marine borers
   b. Tunneling and excavations by insects (termites, boring beetles, and Hymenoptera such as carpenter ants) and marine borers (shipworms, pholads, and isopods)
6. Decays and discolorations—penetration and digestion
   a. Cell-wall etching and tunneling by *bacteria*

b. Surface molding by *fungi*
c. Sapwood staining by *fungi*
d. Decay by *fungi* (soft rots, brown rots, and white rots)

Wood Weathering

Weathered wood often has a familiar gray color that many find aesthetically pleasing. As a result, weathered wood is highly valued for decorative and interior wall paneling because of its pleasing color and texture. When wood is unprotected by coatings and exposed to atmospheric agents and sunlight, a slow chemical and physical disintegration occurs near the surface. This damage has no appreciable effect on strength and is primarily a disfigurement problem. Freshly surfaced wood begins to change color after a few weeks of outdoor exposure. Light woods darken, slowly initially, to a brown. Dark woods first bleach and then also become brown. After several years of outdoor exposure, the wood surface gradually develops an attractive gray sheen and roughened texture.

The major events associated with weathering are photochemical damage (short-wave and long-wave UV) to wood cell-wall constituents, oxidation of the breakdown products, leaching of the soluble decomposition products, and related mechanical damage of surface elements from the constant swelling and shrinkage of the wood associated with surface wetting and drying.

The initial color change to brown is the result of lignin and extractive photo-chemical decomposition, forming free radicals, which lead to further decomposition of the structural carbohydrates and oxidation of phenolic moieties. The origins, identities, and characteristics of these free radicals have been reported by Hon *et al.* (1980). Surface leaching then removes the soluble decomposition products, exposing the more photoresistant structural carbohydrates that are also photochemically degraded and oxidized by decomposition products and atmospheric agents. Xylans are decomposed and leached more readily than cellulose or glucan-rich hemicelluloses. The residual cellulose and the surface growth of pigmented fungi such as *Aureobasidium pullulans* form the gray color. Once the weathered outer shell has developed, it greatly reduces the weathering rate and protects the surface wood from further photochemical damage. However, continual wetting and drying of the weathered surface, with its concomitant swelling and shrinking, lead to surface checking, localized mechanical failures, and slow exfoliation of the weathered surface. These weathering losses are negligible over the service life of most large wood assemblies; however, they may become significant for long-term uses of wood in implements, thin plywood surfaces, and some round stock, such as insulator pins on utility poles. There is considerable variation in the rates of weathering, probably reflecting differences in geographic location, test method, and wood species. Jamison (1937) reported that the weathering of dowels of western white pine, 5 cm in diameter and 45.7 cm in length and suspended above ground outdoors in Idaho for 10 years, caused weight losses of 7.9, 4.8, and 2.3%, respectively, in full sunlight, partial shade, and dense shade. Smaller dowels, 1.25 cm in diameter, lost 16.4% of their weight after a 7-year exposure in full sunlight. Weathering has been estimated to remove 6–7 mm of the outer wood surface per century in temperate zones (Browne, 1960; Feist, 1977; Kühne et al., 1972) and 1 mm per century for wood exposed in northern climes.

Microscopic examinations of weathered wood indicate that substantial microchecking occurs between cells and that wood substance is lost around bordered pits on radial walls. After long exposures to UV light, only remnants of the ray cells remain (Minuitti, 1967). Microchecking and splits between cells suggest wall embrittlement by photodegradation, which probably facilitates the surface-fragment exfoliation with moisture changes. The earlywood fibers are damaged more readily than latewood fibers, accounting for the rough, corrugated surface of severely weathered wood.

Coatings or films that absorb or reflect the damaging UV portion of light and reduce surface-moisture changes are the conventional methods of preventing weathering in outdoor wood exposures. Water-repellent treatments also reduce moisture fluctuations in some outdoor wood uses. Wood treatments with chemicals such as Cr_2O_3 reduce weathering and are reported to double the service life of latex and oil-based paints (Feist and Ellis, 1978). A comprehensive review of wood weathering and an extensive bibliography on the topic has been assembled by Feist (1982).

The actinic degradation of cellulose (tendering) and synthetic polymers is a related topic of interest. Here, initial goals of prevention now include attempts to accelerate photodecomposition of disposable paper products to facilitate biodegradation and minimize ecological concerns. The yellowing and embrittlement of aging paper in libraries is a serious problem that results from the oxidation of residual lignin moieties and is related again to the more general phenomenon of wood weathering.

Wood Thermal Decomposition

The thermal decomposition of wood, as with most carbon compounds, occurs readily at elevated temperatures. Initial, slow changes begin around 100°C. There are color changes, serious strength losses, a reduction in hygroscopicity, weight losses, and the evolution of gases such as CO, CO_2, CH_4, and water vapor. The changes are time dependent and increase rapidly at the higher temperatures. Combustion, with the emission of light and heat, occurs at temperatures around 275°C.

Low-Temperature Processes (Below 200°C)

Low-temperature effects on wood are particularly important because significant strength losses occur within this range. Wood thermal decomposition begins at a temperature of 100°C. The wood turns brown, the surface becomes brittle, and slow losses in weight and strength occur. This temperature effect can be readily observed on blocks left in a hot-air oven (104°C) for a few weeks or lumber that is over dried in a dry kiln. The brownish color and surface brashness of the wood resemble the early stages of some brown rots. The uniformity of the discoloration and the lack of fungal structures readily separates heat damage from early decay.

MacLean (1951) determined wood weight losses associated with various temperatures and periods of exposure, averaged for 11 commercial species, as follows:

Exposure period	Temperature (°C)	Weight loss (%)
1 yr	93	2.7
470 days	121	26.8
400 hr	149	14.8
102 hr	167	21.4

Strength is rapidly reduced by exposures to elevated temperatures (Fig. 2-1). No smoke or wood glowing is produced when wood is exposed to temperatures below 200°C. The major gases released are CO_2 and water vapor.

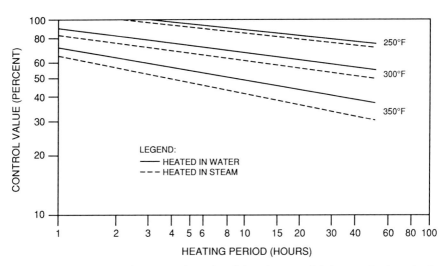

Figure 2-1 Comparison of average values of modulus of rupture (%) for Douglas fir and Sitka spruce specimens heated in water and in steam at 250°F, 300°F, and 350°F. From MacLean (1954).

High-Temperature Processes (Above 200°C)

Pyrolysis (heating in absence of O_2) is a well-known wood-distillation process during which the flammable gases evolved are CH_4 and CO. Also, many compounds such as acetic acid, methanol, formic acid, furfural, phenols, and cresols are released. The acids cause the eye sting of smoke, and the furfural products contribute to the characteristic odor of wood smoke. The remaining final product is charcoal (carbon), which has many industrial uses.

In combustion, the wood is rapidly decomposed at temperatures above 200°C in the presence of oxygen, and the flammable gases CH_4 and CO are released. At temperatures around 275°C, the gases ignite (the kindling temperature) and, thereafter, the heat released accelerates the burning or decomposition process.

The order of wood cell-wall component breakdown with increasing high temperatures is hemicelluloses, cellulose, and lignin. Hemicelluloses are least heat stable and decompose within a temperature range of 225 to 325°C. Lignin decomposes over a temperature range of 250 to 500°C, whereas cellulose decomposes at higher temperatures and in a more limited range (325–375°C) (Shafizadeh and Chin, 1977). The carbonized wood at late stages of decomposition texturally resembles a brown cubical rot, and it is of interest to note also that lignin, as in some decays, is the last wood component to be completely consumed.

Fire-Retardant Chemicals

Effective retardant chemicals are ammonium phosphates, ammonium sulphate, borax, and zinc chloride. These have been empirically determined, and their protective mechanisms are poorly understood.

A reference book has been assembled, *Chemistry and Uses of Fire Retardants* (Lyons, 1970), and a comprehensive review of the degradation and protection of wood from thermal attack has been written by Goldstein (1973).

Chemical Decomposition of Wood

As a structural material, wood displays considerable resistance to attack by most chemicals. For this reason, wood is often used to construct storage vats, tanks, cooling towers, or structures where contact with caustic chemicals may occur by condensation, aerosols, or splashing. For example, evidence of considerable wood deterioration was reported in several Kraft pulp mills owing to prolonged wood exposures to weak acids and bases at elevated temperatures and high humidities (Barton, 1982). Because wood is a major chemical raw material for the paper and cellulose-derivative industries, a wealth of information has been developed on the reactions of wood and its constituents to many chemicals. This information forms the basis of a wide range of industrial processes. Information on these topics is available in wood chemistry, cellulose chemistry, and paper-making textbooks. This section reviews only those cases in which wood contact with caustic chemicals in various structural uses causes degradation that may be confused with some types of decay, or the resulting damage has close analogies to the decay process.

Smith (1980) developed a list of timber species recommended for use in various corrosive environments such as containers for acids, exposure to acid fumes, or containers for mild corrosive liquids. Coniferous woods are generally more resistant to corrosive chemical attack than are most hardwoods. Useful criteria for the chemically resistant woods are those species high in α cellulose and lignin, and low in xylans.

Acids degrade primarily the carbohydrates in wood, and the high resistance of lignin to strong acids is the basis for its analytical determination by solubilizing wood carbohydrates with 72% H_2SO_4. The resulting filtered insoluble residue is defined as Klason lignin. Acids hydrolyze the $\beta(1\rightarrow4)$ glycosidic linkages in cellulose and hemicelluloses, resulting in drastic reductions in tensile strength. The wood in early decomposition stages turns brown and becomes brittle or brash. The depolymerization mechanisms and early reductions in some strength properties are analogous to those of brown-rot degradation.

Alkalis attack wood more severely at equivalent concentrations and time–temperature conditions than do acids. Alkalis dissolve the hemicelluloses and modify the lignin to form soluble lignin–alkali complexes. The cellulose is essentially unmodified. Many wood-pulping processes employ alkaline reactions of this type.

At high concentrations, strongly alkaline chemicals cause the wood to become fibrous and bleached in a manner similar to that of wood attacked by some white rots. The wood swells, and sharp strength reductions occur.

Selected data from Wangaard (1966) illustrate the differences between the effects of a strong acid and strong base at several concentrations and temperatures on the strength of a conifer and a hardwood (Table 2-1).

Erickson and Reese (1940) have shown that wood exposed to a series of alcohols, acetone, and benzene experienced decreased swelling and strength loss as molecular weight and structurally complexity of these organic compounds increased. Treatment of wood with ammonia temporarily causes large reductions in bending resistance and permits the wood to be bent at sharp angles into a variety of shapes without breaking.

The treatment of wood with some salts is reported to increase crushing resistance. The treatment with acid salts such as Na_2CrO_3 has been reported to reduce strength (Ross, 1956). The treatment of wood with salt-type preservatives, such as the oxides or acid salts of copper, chromium, and arsenic (CCA), does not appear to seriously affect strength (Thompson, 1982) unless the wood is subsequently dried at high temperatures (Barnes and Winandy, 1986).

Table 2-1
Effects of temperature and several concentrations of HCl and NaOH
on wood strength (modulus of rupture)[a]

	Modulus of Rupture (as a percentage of control)							
	2% HCl		10% HCl		2% NaOH		10% NaOH	
	20°C	50°C	20°C	50°C	20°C	50°C	20°C	50°C
Douglas fir	91	85	76	57	56	40	39	28
White oak	70	51	39	30	26	22	20	15

[a]Data selected from Wangaard (1966).

The prolonged contact of wood with iron causes localized embrittlement and loss in tensile strength (Baker, 1974). There are reports that wood decomposition from iron reduces nail-holding properties in cases in which the nail was driven initially into green lumber. The use of galvanized nails or dry lumber, when feasible, minimizes the problem. As iron oxidizes (rusts) to form ferric hydroxide, it catalyzes the oxidation and depolymerization of cellulose into oxycellulose. It is interesting to note that one theory on brown-rot decay mechanisms proposed a similar catalytic role for iron to initiate cellulose decomposition (Koenigs, 1974). A comprehensive study of the various effects of iron on wood properties has been reported by Marion and Wissing (1960) and Graham et al. (1976).

Mechanical Wear

The mechanical wear of wood is a minor source of wood deterioration and involves forces that rupture and detach small portions of the surface wood. They are important only in a few cases of special wood uses in which surface friction or rupturing is severe, such as stair treads, baffles in cooling towers, factory floors around heavy machinery, and spikes and plate contacts in railroad ties. Wind-driven sand particles can cause considerable mechanical damage to poles, posts, and unpainted wood in desert regions and along beaches. Another example of mechanical damage is seen frequently on loading dock or platforms. Heavy loads with sharp corners abrade and split the surface wood, which over time develops a fibrous texture similar to that of some late decay stages. Preventive methods include the selection of woods with high surface hardness, the edge grain alignment of wood in severe friction zones, and in some cases, the protection of high-damage zones with metal plates or use of polymer-hardened wood.

Insect Damage to Wood

Insects, like fungi, are also major agents in the biodeterioration of wood. Insects use wood for habitation and as a food source. For both purposes, the wood is chewed into small fragments. Collections of the chewed fragments and/or fecal material, known as *frass*, are often a useful indicator of hidden insect damage. Insect damage to wood generally consists of discrete tunnels, surface channels, or chewed zones and, in most cases, can be distinguished readily from the fungal-associated stains and decays. Insects are common vectors of stain and decay fungi, and insect and fungal damage often develop under the same conditions and are associated in many wood uses. Hence, a review of the major types of insect damage, their characteristics, prevention, and/or control is necessary.

Insects are the largest class in the Arthropoda and are characterized by segmented bodies; jointed legs (six); and a hard, chitinous exoskeleton. Of the six insect orders that cause wood damage, the Isoptera, Coleoptera, and Hymenoptera are the most important. Insects are reported to cause billions of dollars in damage to wood structures annually (Coulson and Lund, 1973; Baker, 1972; Furniss and Carolin, 1977). In fact, one insect group, the subterranean termite, causes over $1.5 billion in damage in the United States per year (Ebeling, 1968). In many cases insect infestations are also associated with fungal decay that further aggravates the damage (Amburgey, 1979).

Types of insect damage are classified according to the causal species for scientific purposes and the type of damage, the wood product, or time of attack for practical purposes. Wood can be damaged in the living tree, the freshly fallen log, the sawn or round products in storage, or in the final product during service. Considerable insect damage takes place in weakened or freshly fallen trees and stored logs, but the effects show up later in the finished wood product. Common terms for the many types of damage are powder posting, pith flecks, birdseye, pitch pockets, pinholing, honeycombing, and grub holes.

Insects typically have life cycles that begin with an egg and progress through several immature molts, climaxing with a fully functional adult that mates and produces eggs to complete the life cycle. Two types of development are *complete metamorphosis*, in which the insect passes through egg, larval, pupal, and adult stages; and *incomplete metamorphosis*, consisting of egg, nymph, and adult stages.

The stage at which insects attack wood varies, but generally the Coleoptera damage wood during the larval stages, whereas the Isoptera and Hymenoptera cause their damage in the nymphal and adult stages.

Isoptera

The wood-attacking termites are members of the order Isoptera and live in large social colonies consisting of castes that differ in appearance and function (Krishna and Weesner, 1969; Ebeling, 1968; Beal *et al.*, 1983). All members in a given colony are descendants of one original pair. Termites are often confused with ants and are commonly called *white ants*. Termites differ in many ways from ants, including morphology, food sources, and environmental requirements (Fig. 2-2).

There are over 2,000 termite species located worldwide. These species are generally confined to regions where the average annual temperature exceeds 10°C (50°F). This zone lies approximately between the latitudes of 50° south and 50° north (Fig. 2-3). Some termites may extend their ranges farther north or south of these zones in heated, man-made structures (Esenther, 1969).

Termites have specific environmental requirements, including a food source (generally wood), oxygen, and adequate moisture levels. Termites generally occupy wood internally and are negatively phototropic. They seem to require higher than ambient levels of carbon dioxide. The wood is chewed, and digested in the hindgut by enzymes released from associated symbiotic protozoa and/or bacteria. These protozoa are not present in the newly hatched nymph but are transferred by exchange of body secretions and by consuming dead or dying members of the colony (Moore, 1979). Termites utilize primarily the cellulose in the wood, and the fecal pellets contain high levels of lignin. It is speculated that more advanced termite species may produce cellulase themselves or utilize, in some cases, the cellulase enzymes released in wood by decay fungi. Some termites appear to be attracted to the chemicals produced by some decay fungi (Esenther *et al.*, 1961). Fungal termite attractants could be useful in termite detection and control procedures.

Termites vary in the amounts of water required to establish successful colonies. Drywood termites, so called because of their ability to attack dry wood (<13% water content), obtain their water needs from the wood and are highly efficient in their uses of water. Dampwood and subterranean termites require more water and invade wood that is constantly moist and usually in ground contact. Some termite species construct earthen tubes that connect wood above the ground with the soil. The humid air in these tubes contacts the wood above the ground and increases its moisture content.

Termite groups Of the six major termite families, only the dampwood, subterranean, and drywood termite families are important in the United States (Table 2-2). Of these three groups, the subterranean termites have the widest distribution and cause the most damage.

The majority of subterranean termites of economic importance in the United States are species in the genus *Reticulitermes*. As their name suggests, subterranean termites build their nests in the soil, although they can survive in extremely wet wood not in ground contact. These termites infest woody debris in soil and invade wood structures through direct soil contact. Subterranean termites also build earthen tubes over masonry or concrete foundations to reach wood above the ground. Initially the termite workers chew and digest the less dense springwood, leaving the summerwood, unless they

Figure 2-2 Termite and carpenter ant damage are often confused, though their differences are striking. (A) Carpenter ant workers with deep constrictions between the body segments. (B) Termite workers without noticeable constrictions between the body segments. (C) Carpenter ant damage is characterized by clean tunnels confined primarily to the springwood. (D) Wood damaged by termites has a "dirty" appearance, and the tunnels often contain soil and fecal pellets. (Courtesy of Dr. John Simeone.)

32 / 2. Wood Deterioration Agents

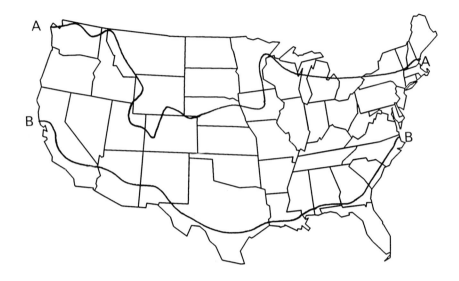

Figure 2-3 Northern limits for distribution of termites in the United States (A) subterranean (B) dry wood. From Forest Products Laboratory, U.S. Forest Service.

Table 2-2
Families of termites that can attack wood

Family	Common name	Type of damage
Rhinotermitidae	Subterranean termites	Honeycomb damp wood
Kalotermitidae	Drywood termites	Honeycomb dry wood
Termitidae	Subterranean, mound builders	Honeycomb cellulose
Termapsidae	Dampwood termites	Honeycomb damp wood
Mastotermitidae	—	Honeycomb damp/dry wood
Hodotermitidae	Harvester termites	Plant consumers

are forced to utilize this wood later (Behr et al., 1972). As they chew and tunnel through the wood, termite workers deposit small amounts of soil and fecal matter into the wood, giving the damaged wood a characteristic "dirty" appearance (Fig. 2-2B).

A recently introduced group, the Formosan termite (*Coptotermes formosanus*), has caused serious concern among wood users in the southeastern United States. Formosan termites are characterized by a rapid feeding rate, large colony size, and apparent tolerance of many commonly used wood preservatives. A major effort is underway to reduce the spread of this insect from the few areas where it has become established. Formosan termite nests can be distinguished from subterranean termites by the presence of an extremely durable secretion called *cardboard*.

Dampwood termites are confined to the Pacific Northwest, the Pacific Southwest, and southern Florida. These species are characterized by their need for very wet wood. Primarily the wood invaded is in ground contact, but reproductive swarmers can also infest very wet wood in solid lumber stacks or wood subjected to continuous wetting. In many instances, these insects are associated with wood decay due to the higher wood-moisture levels. As with subterranean termites, dampwood termites remove the softer springwood first and also preferentially remove the weaker, decayed wood. The wood damaged by dampwood termites can be recognized by the appearance of the frass, which is squeezed into pellets to recover moisture.

The drywood termites do not require soil contact or high wood-moisture content to invade wood. As a result, these insects can attack wood in roofs, rafters, and other building zones not normally considered susceptible to termite attack. Fortunately, drywood termites are confined to the Pacific Southwest and cause difficulty in only a few areas. Wood damaged by drywood termites differs from the others in that the galleries overlap the springwood/summerwood boundaries. Also, drywood termites produce distinctive barrel-shaped fecal pellets that are pushed out of the gallery through holes termed *kickholes*. These holes are immediately resealed, but the deposition of the fecal pellets outside the wood is an excellent indicator of termite attack.

Termite Prevention Entomologists have long sought effective preventive methods for limiting termite damage (Moore, 1979; Snyder, 1969). Much of this effort has been exerted to control subterranean termites, since these insects cause most of the wood damage. Termite damage can be reduced by several simple construction practices, including removing wood debris around structures, properly sealing cracks in cement floors or foundations, filling the top layer of hollow concrete blocks with cement or capping with solid, so-called *termite blocks* to prevent termite invasion, and using pressure-treated wood when wood is in soil contact (DeGroot, 1982; Moore, 1979; Beal et al., 1983). Regular inspections of crawl spaces or foundations in structures are recommended to detect infestations before substantial damage occurs.

Termite attack can also be prevented by the use of chemical soil drenches around the base of a wood structure. Chemicals formerly used for this purpose have included aldrin (0.5%), chlordane (1.0%), dieldrin (0.5%), or heptachlor (0.5%). Chloropyrifos has been used recently with some success (Moore, 1979), as have some synthetic pyrethroids. Because of many changes in insecticides permitted at the federal and state levels, checks should be made with local Extension Service offices to determine current recommended insecticides for termite control. Dampwood termites are controlled by the procedures used for the subterranean termite; however, prevention can also be effected by removing wood from direct soil contact or by removing moisture sources.

Drywood termite attack is more difficult to control, since the wood need not be wet or in ground contact for damage to occur. Infestations can be limited by screening around vents or crawl spaces, removing infested wood, and by the use of fluoridated silica aerogel dusts (Weesner, 1965).

Termite Control Once a termite colony has been detected in a structure, many of the same techniques used for prevention can be used to eliminate the infestation. In addition, application of emulsified insecticides under pressure through holes drilled in the wood can speed up colony demise. External application of these chemicals has little effect, since termites rarely venture out of the wood. Fumigations with methyl bromide or sulfuryl fluoride have been successful with drywood infestations (Moore, 1979), but these chemicals have short residual times in the wood and will not prevent reinfestation.

Coleoptera

Beetles are members of the order Coleoptera, the largest order of insects, containing nearly 40% of the known species. Nine families of Coleoptera cause wood damage; most species attack only living trees or logs in storage or seasoning lumber (Table 2-3), but they are important because the defects they cause appear later in the final wood product and may be confused with active wood infestations. Most beetle damage is caused while the insects are in the larval stages. The wood species attacked and the conditions necessary for attack vary widely among insect species.

Preharvest Beetles A number of beetles in the families Brentidae, Lymexylidae, Scolytidae, and the Platypodidae attack standing or freshly cut trees. These insects normally do not cause damage to seasoned wood, although they may continue to damage wood as it initially seasons. The first two families cause extensive damage to hardwood logs that are not removed promptly from the woods, whereas the last two families cause relatively minor

Table 2-3
Families of wood-destroying Coleoptera

Family	Common name	Damage	Product type
Anobiidae	Death-watch beetle	Powder posting	Furniture, structures
Bostrichidae	Powder-post beetle	Powder posting	Hardwood lumber
Brentidae	Timber worms	Tunneling	Hardwood logs
Buprestidae	Flat-headed borers	Tunneling	Lumber and products
Cerambycidae	Round-headed borers	Tunneling	Trees and products
Lyctidae	Powder-post beetle	Powder posting	Hardwoods
Lymexylidae	Timber worms	Pinholes	Hardwod logs
Platypodidae	Flat-footed beetles	Pinholes, stain	Logs
Scolytidae	Bark beetle	Pinholes, stain	Trees or green logs

pinhole damage to the sapwood but are effective vectors of stain fungi that decrease the wood value (Fig. 2-4). Prompt removal of the bark and drying the wood can prevent damage by these beetles. Where rapid processing is not feasible, the use of ponding or continuous spraying makes the wood too wet for insect development.

The Scolytidae generally attack living or freshly harvested trees with bark, and their damage is usually limited to the very outer sapwood. These insects often carry the spores of stain fungi that can further degrade the sapwood. This problem is particularly acute with the Ambrosia beetles that invade freshly felled logs, forming small tunnels that may penetrate deep into the sapwood and later become associated with sapstain. Ambrosia beetles infect the tunnel walls with the spores of fungi carried in special structures termed *mycangia*. They obtain nourishment from the growth of the fungi, which cause also a shallow, gray stain in the tunnel wall. Some ambrosia beetles attack living trees. The Columbia timber beetle (*Corthylus columbianus*) commonly attacks soft maples, oaks, and sycamore in the eastern United States. The unique stain pattern associated with successive beetle attacks, when seen on log ends or board surfaces, is often confused with early decay (Fig. 2-4). Ambrosia beetle attack can be limited by ponding, rapid processing, and spraying the logs with insecticides (Fisher *et al.*, 1954; Gray and Borden, 1985; McLean, 1985). The control of bark beetle attacks is presented in greater detail in the later chapter on sapstains.

Postharvest Beetles Once the wood is cut, milled, and dried, it is still susceptible to attack by members of the Anobiidae, Bostrichidae, Lyctidae, Cerambycidae, and Buprestidae. The former three groups are collectively called the powder-post beetles because of the flour-like frass that the larvae leave in their tunnels (Moore and Koehler, 1980). Powder-post beetles are reported to be the most important wood products-attacking beetles.

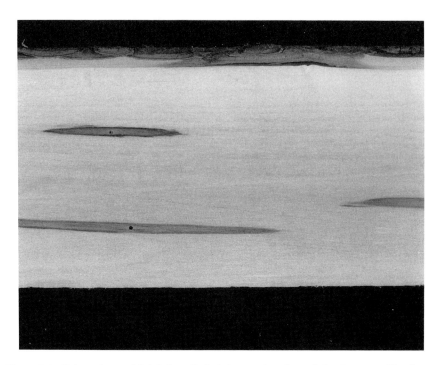

Figure 2-4 Stain and tunnel (pinhole galleries) damage in soft maple lumber caused by the attack of the Columbian timber beetle (*Corthylus columbianus* Hopkins) while the tree was living. Most ambrosia beetles invade dying trees, slash, or green roundwood. This ambrosia beetle attacks living trees, and the lumber may be riddled by old and recent galleries and is often confused with decay unless the tunnels are detected.

Anobiidae The powder-post beetles in this family are also called the death-watch or furniture beetles. The name derives from the tapping sounds adult beetles make with their heads as a mating signal. This sound is most easily heard in the walls of a quiet room such as that occupied by a person sitting with an ill person, hence the name origin. There are numerous species of Anobiidae, *Anobium punctatum* (an introduced pest), the common furniture beetle (*Xyletinus peltatus*, and *Hemicoelus carinatus*) that cause considerable wood damage. The Anobiidae are common in southern pine building timbers in the southeastern and hardwood structures in the northeastern United States. Powder-post beetles attack primarily sapwood, although heartwood is not completely immune to attack. The Anobiidae digest cellulose in the wood cell wall with the assistance of yeast cells in the digestive tract. In general, beetles in this family can attack wood at moisture contents of 15% and above, although the optimal moisture levels for development are at moisture contents greater than 30% or in decayed wood (Moore, 1979). Damaged wood has numerous small tunnels packed with frass and small exit holes on the wood surface.

Bostrichidae These beetles are called the false powder-post beetles and occur commonly in dying twigs and branches of many hardwoods. They cause significant damage in hardwood lumber. These beetles attack hardwood sapwood, utilizing the starches present in the ray cells. Wood damaged by these beetles has numerous small tunnels filled with tightly packed frass.

Lyctidae Lyctids, the true powder-post beetles, are reported to be the most important destructive agents of hardwoods (Williams, 1985). The adult beetles attack hardwood sapwood to obtain the free sugars in the ray cells. Lyctids infest wood at moisture contents ranging from 8 to 32%, with the greatest activity between 10 and 20% (Christian, 1941; Moore, 1979). Unfortunately, many seasoned wood products fall within this moisture range. Wood damaged by Lyctids is filled with small tunnels loosely packed with frass. In older infestations, adult exit bore holes may be evident, and frass will fall from these holes when the wood is jarred. Lyctid infestations can be prevented by sealing the surface to prevent access to the vessel elements, but sealing will not affect existing infestations.

Cerambycidae The round-headed borers comprise one of the largest beetle families, and many are associated with trees or wood products. These beetles have long antennae that give rise to their also being called long-horned beetles.

In general these beetles infest green or partially seasoned lumber when the bark remains attached. Long-horn beetle damage consists of round to oval tunnels tightly packed with frass throughout the wood. Often these galleries are exposed in sawn material after the damage has occurred. Whereas most long-horn beetles do not reinfest timber, several species, including the old-house borer (*Hylotrupes bajulus*), can repeatedly reinfest the same wood and cause serious structural damage. This species is believed to be an imported pest that primarily attacks seasoned coniferous wood but can also invade unseasoned wood. The old-house borer is a serious problem in buildings in coastal regions. Generally, larvae develop more rapidly in wood between 15 and 25% moisture content, but will survive for long periods at lower moisture levels. The amount of damage associated with the larval stage is confined to galleries loosely packed with frass in a few isolated boards; however, more severe damage occasionally results when reinfestation occurs in areas with moisture problems.

Buprestidae The flat-headed borers are also confined to wood and wood products where they tunnel beneath the bark, in twigs, and in the heartwood or sapwood of freshly cut logs. Most species do not attack seasoned wood, although the golden buprestid (*Buprestis aurulenta*) can survive for long periods in dry wood. Many buprestids are metallic colored, and members of this group are also called the metallic wood borers. Larvae of these species are distinguished by their flattened appearance near the mouth, and the tunnels they bore are tightly packed with frass. Following pupation the adults chew their way out of the wood, leaving characteristic D-shaped exit holes on the surface. Generally the amount of damage associated with buprestid attacks is minimal; however, heavy infestations of the golden buprestid can cause significant structural damage in log structures and utility poles. Although reinfestation is not reported, the high levels of damage in isolated structures suggest that reinfestation does occasionally occur.

Hymenoptera

Whereas termites and beetles cause the major wood damage, several members of the Hymenoptera including the Siricidae, Apidae, and Formicidae also significantly damage wood.

Siricidae The Siricidae or horntail wasps attack trees that are stressed and declining as well as fire-damaged or freshly harvested timber. The female penetrates the bark and wood with a long ovipositor and lays eggs, along with a deposit of fungal spores. As the larvae grow, they depend on the fungus mycelium for food. Typically, a larva chews a 25- to 75-cm long, C-shaped tunnel over a 2- to 3-yr period. After pupating, the adult emerges through a large circular hole. Generally, the damage associated with these insects is minimal, but the size of the exit holes and the occasional disturbing appearance of a large and noisy adult indoors can cause concern (Morgan, 1968).

Anthophoridae Members of the Anthophoridae or carpenter bees belong to the genus *Xylocarpa* and construct their nests in the wood. Carpenter bees do not use wood as a food source. They excavate a series of 12-mm diameter tunnels along the grain for 10 to 15 cm to deposit their eggs, along with nectar and pollen to provide nourishment for the developing larvae.

Generally carpenter bees attack uncoated soft wood (cedar, pine, etc.), but weathered wood of almost any species can be attacked. Carpenter bees can cause substantial wood damage if infestations go undetected for several years.

Formicidae The wood-attacking carpenter ants, belonging to the genus *Camponotus*, are social insects that have queens, winged males, and workers of varying sizes in a given colony (Simeone, 1954; Furniss, 1944). Carpenter ants are often confused with termites. Carpenter ants have constrictions between individual body segments, and the winged adults have two pairs of unequal-sized wings. Termites have two pairs of equal-sized wings, and their body segments are not constricted (Fig. 2-2). Carpenter ants remove the wood to construct galleries to raise their young, but their food resources come from outside the nest. Termites, of course, also use wood as a food source.

Carpenter ants occur throughout the Untied States but are most important in the Pacific Northwest and the Northeast, where lengthy infestations can result in considerable damage to houses. Also, ant infestations in the home are a nuisance owing to the large numbers of foraging insects. In the northeastern United States, carpenter ants commonly invade the untreated heartwood zones of cedar and Douglas fir transmission poles. Of the native carpenter ants, the black carpenter ant, *Camponotus pennsylvanicus* Degeer, has been studied most extensively. Carpenter ants are scavengers, and common food sources include aphid secretions and insects. Winged reproductives emerge and swarm in the late spring and early summer. After mating, the males die, and the females search for a suitable site. In general, females will search for moist wood or other materials. Simeone (1954) has shown that successful colonies can be established only in wood above 15% moisture content. Partially decayed wood in structures also is selected often for nesting sites (Moore, 1979). Colonies develop slowly at first but increase rapidly after the first year, ultimately approaching two to three thousand individuals when the winged reproductives are produced.

Wood damaged by carpenter ants has numerous clean tunnels primarily in the springwood (Fig. 2-2C). Generally, the nest can be detected by the presence of piles of frass and insect fragments below the entrances to the infested wood (Fig. 2-5). Carpenter ants do not cause significant wood-strength losses unless the colony is left undisturbed for long periods. Infestations by carpenter ants can be limited by keeping the wood dry, using pressure-treated wood in high-hazard areas, and eliminating wood debris

Figure 2-5 Carpenter ant invasion and internal damage in a treated Douglas fir utility pole is indicated by the accumulating pile of fresh frass (arrow) at the base of the pole.

from around structures (Furniss, 1944). Local extension agents should be contacted to determine insecticides currently recommended for ant control.

Marine Borer Damage

Marine borer is the collective term used for the many invertebrates that burrow into and damage wood exposed to ocean or brackish waters. The two major phyla involved are the Mollusca (mollusks) and Crustacea (crustaceans). These animals chew and burrow into wood primarily for protection and are not able to use it for food. It has been estimated that marine borers cause losses in marine structures of approximately $500 million annually (USN, 1965).

Whereas marine borer damage to shipping has been a long-term problem, recorded historically as early as 350 BC by Theophrastus (Turner, 1959), very little is known about the biology of the organisms, and certain stages in the life cycles of these invertebrate animals are still poorly understood. As a result, our ability to develop effective prevention and control methods, based on a thorough knowledge of vulnerable points in the marine borer life cycles, has lagged far behind the methods available for other pest problems. In general, we continue to depend on a limited number of highly toxic, broad-spectrum chemicals to protect wood in marine environments.

The wood-attacking marine borers are separated into 3 major groups (shipworms, pholads, and isopods) based on anatomy, physiology, and the nature of wood attack (Morrell et al., 1984; Helsing, 1979). The shipworms and pholads are mollusks.

Shipworms

The shipworms, or Teridinidae, are important in temperate waters, where they begin life as microscopic, free-swimming larvae that are filter feeders for several hours to a few weeks before they settle on a wood surface (Quayle, 1959). Once settled, the larvae bore into the wood using a pair of tiny, chitinous shells located near the head and become trapped for the remainder of their lives. This inability to move was exploited by captains when they sailed their shipworm-infested wooden vessels into tidal *freshwater* rivers for several months to eliminate the shipworm infestations. Once shipworm larvae bore into the wood, they undergo rapid changes in morphology. The two shells near the head calcify and develop teeth, and the body begins to elongate into the worm-like adult (Fig. 2-6C). As the shipworm grows, it burrows farther into the wood, depositing a fine layer of calcium carbonate on the surface of the tunnel (Fig. 2-6B). This layer is seen readily when the wood is X-rayed and can be used to detect shipworm infestations (Fig. 2-6A). Although shipworms grow to lengths of 0.3 to 1.5 m, the only external sign of infestation is a small hole (<0.30 cm in diameter) through which the animal exposes a pair of feather-like siphons that filter food and exchange oxygen and waste products with the surrounding water. At the slightest sign of danger, the shipworm withdraws the siphons and covers the hole with a hardened pallet, making surface detection extremely difficult. This pallet resists desiccation and allows shipworms to survive out of water for up to 10 days.

The two major shipworms are *Teredo* and *Bankia;* however, these species have different salinity and temperature requirements and generally do not occur together. For example, *Teredo navalis* is found in warmer waters, and a reduced salinity tolerance permits it to survive far upstream in many estuaries. This tolerance to low salinity resulted in losses totaling $25 million in the San Francisco Bay area during a period of low stream flow in the mid-1920s and stimulated interest in marine borer control (Hill and Kofoid, 1927).

Bankia setacea is also found in a variety of West Coast temperate waters but can not tolerate low salinities. Unlike *Teredo, Bankia* adults are much larger, eventually reaching 1.5 to 1.8 m in length. Although they can cause tremendous destruction, they are short-lived with life cycles lasting only 1 to 2 years.

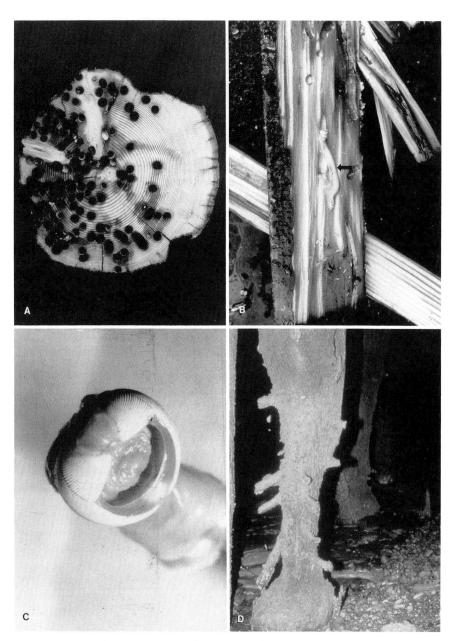

Figure 2-6 Marine borers cause severe damage to wood in salt-water exposures in some regions. (A) A cross section of a Douglas fir piling showing numerous tunnels caused by a *Teredo* sp. (B) A longisection of a split piling showing a shipworm in a tunnel (arrow), (C) The two shells in the mouth of the Teredo shipworm that are used to rasp away wood to form the tunnel (D) A piling from the intertidal zone at the point of failure from damage by *Limnoria* spp.

Whereas adults of both species are destructive to essentially all species of untreated wood, their attack can be prevented by using wood pressure-treated with creosote or inorganic salts. Attack can also be prevented by encasing the submerged wood in plastic, metal, or concrete barriers (Steiger and Horezcko, 1982). In early seafaring days, shipbuilders followed a similar practice by covering the hull with copper nails or sheathing.

Pholads

Pholads are wood- and rock-burrowing mollusks that, like shipworms, begin life as small, free-swimming larvae. Once these organisms settle on the wood, they burrow near the surface and remain entirely within their shells, which enlarge as they grow. Pholad damage weakens the wood surface to the point where waves can erode the damaged wood and expose new wood to attack. The most common wood-boring pholad, *Martesia striata*, grows to be 5 to 6.25 cm long by 2.5 cm wide and feeds through a hole 0.25 cm in diameter. Although they are small in size, pholads can completely destroy untreated piling within 1 year. Pholads are a serious problem, primarily in Hawaii and portions of southern Florida. Pholad attack can be prevented by the use of barriers or wood pressure-treated with creosote. Frequently a second marine borer resistant to creosote is associated with pholads. In these cases, the wood must be protected with both creosote and an inorganic arsenical in a process known as dual treatment.

Isopods

Unlike shipworms and pholads, the wood-boring isopods are small crustaceans that move freely about the wood surface. The *Limnoria* species, also known as gribbles, are by far the most destructive of these isopods and attack the wood surface from the mudline to the high-tide line (Kalnins, 1976; Lane, 1959; Ray, 1959). Most of this attack is concentrated in the more highly oxygenated tidal zone where the waves wear the weakened wood away to produce hourglass-shaped piling (Fig. 2-6D). Gribbles chew small tunnels that penetrate only a short distance into the wood (<2.5 cm). The *Limnoria* swim freely in and out of the burrow, and the wood may be colonized by thousands of individuals.

Limnoria are small, averaging 0.3 to 0.6 cm in length, and their taxonomy is quite complex (Menzies, 1959). There are 20 species of *Limnoria*, but only *L. lignorum*, *L. quadripunctata*, and *L. tripunctata* are of importance along the coastal United States. The latter species is the most important owing to its ability to attack creosoted wood in warm-water ports (Kalnins, 1976). The ability of this species to tolerate high levels of creosote has long perplexed scientists. This phenomenon has been attributed to selective detoxification, which renders the creosote harmless, or to the presence of symbiotic flora that degrade creosote in the midgut of the animal (Geyer, 1982; Ray, 1959). There is normally a dense microflora on the surface of submerged wood (Kohlmeyer and Kohlmeyer, 1979; Barghoorn and Linder, 1944; Boyle and

Mitchell, 1984; Kirchman and Mitchell, 1983; Cundell and Mitchell, 1977; Meyers and Reynolds, 1957), and it appears to provide some nutrition to the *Limnoria*.

Limnoria attack can be prevented by the use of barriers or by pressure-treating the wood with creosote where *L. tripunctata* resistance is absent or also with an inorganic arsenical where it is present (AWPA, 1986).

In addition to wood attack by these three groups, a number of other minor wood-boring organisms attack wood in marine environments. *Sphaeroma terebrans* is a mobile, warm-water crustacean that normally burrows in mangrove roots but has a remarkable tolerance of inorganic arsenical salt-treated wood. A shipworm *Xylophaga* sp. has been found to attack wood at depths of up to 2000 m and has caused concern where wood supports are used at great depths. This species probably plays an important role in recycling nutrients into a generally barren environment.

Whereas we know a great deal about the marine wood borers, there is still much more to be learned before we can develop control methods based on knowledge of the biology of the target organism. Such information will become increasingly important as environmental concerns are brought to bear on the broad-spectrum pesticides currently used to protect wood in marine environments. The reduction in pollution of marine waters in harbors will also favor increased borer damage.

Wood Decays and Discolorations Caused by Fungi

The decays and discolorations in wood caused by fungi, and to a lesser extent by bacteria, are major sources of loss in both timber production and wood uses. Understanding the nature of these agents and identifying prevention or control strategies are major purposes of this volume.

Decays and discolorations differ sharply in cause and nature from the other wood-deterioration agents discussed above. Microorganisms, the casual agents, are unique organisms that have evolved systems to penetrate, invade, externally digest, and absorb soluble constituents from complex substrates such as wood. The great role of fungi and bacteria in ecosystems is to decompose and steadily release carbon dioxide and other elements critical to plant photosynthesis and, thereby, continuing life. In this role, a small, specialized group of fungi are the major decomposers of wood.

Bacteria grow primarily on wood surfaces and are carried into wood by microfauna, fungi, and water menisci during wetting and drying. Their damages to wood are minor. Bacteria cause localized wall etchings and tunnels or cavities in the cell walls. When wood is stored underwater or kept wet, bacteria may destroy parenchyma cells and pit membranes, causing substantial increases in wood permeability.

Molds are fungi that grow on the surface of extremely wet wood, utilizing available simple carbon compounds. The growth and secretions of fungal hyphae on the wood surface produce colors such as black, gray, green, purples, and red; *moldlike* odors; and in some cases, the huge numbers of associated spores present potential problems as allergens. Molds can normally be removed by brushing or planing and can cause major wood-quality losses.

Stains are degrading discolorations caused by fungi that invade the sapwood of many commercial woods during log storage or lumber seasoning. The stain fungi primarily invade parenchymatous tissues in the sapwood, and the discolorations result from the masses of pigmented hyphae in the wood cells. Though staining fungi cause little damage to the prosenchyma cells in wood, several properties in addition to color, such as toughness and permeability, may be adversely affected.

Decays are the major type of damage to wood in use and are caused by fungi. Decay essentially is the result of wood digestion by fungi. The slow, progressive digestion of the wood causes a continuum of changes in its appearance and physical and chemical properties. Only a limited group of fungi possess the enzymatic capability of digesting wood. Various groups of fungi attack the wood cell-wall constituents in different ways and sequences that result in several types of decay. *Soft rots* are caused by microfungi that selectively attack the S_2 portion of the cell wall. High wood-moisture contents and soil exposures seem to favor soft-rot development. *Brown rots* are caused by a group of fungi that attack primarily the carbohydrates in the cell wall. *White rots* are caused by a group of fungi that attack both the carbohydrates and lignin in the cell wall. The white and brown fungi are in the Basidiomycotina subdivision. All decays, in the final stages, result in drastic changes in strength and other use properties or total destruction.

The decays and discolorations in wood, briefly described above, are the central theme of this book and are discussed in detail in subsequent chapters.

The characteristic appearances of most decays and discolorations and the presence of characteristic hyphae or microscopic features in the wood permit their separation from all other types of wood deterioration.

Summary

The major types of wood damage, the causal agents, and descriptions are summarized in Table 2-4.

- Fungi and insects are the major biotic agents responsible for wood destruction in use, and recycling in terrestrial ecosystems. Fungi cause white-, brown-, or soft-rot degradation that results in the loss or significant reduction in many wood-use properties. Fungi also cause surface molding and sapwood discolorations, which may develop during seasoning or when the wood becomes too wet.

Summary continues

Table 2-4
Major types of wood damage and their descriptions

Type of damage	Causal agent(s)	General descriptions	Prevention or control
Weathering	Ultraviolet light, oxidation, swelling and shrinkage, leaching, and fungi	Unprotected surfaces develop a gray color and roughened texture	Ultraviolet light-resistant coatings
Thermal decomposition	High temperature	<200°C, uniform surface brittleness >200°C, charcoal in absence of oxygen, combustion around 275°C	Fire-retardant chemicals
Chemical decomposition	Caustic chemicals	With acids wood turns brown, chars, and becomes brittle; with bases wood bleaches and defibrillates	Chemically resistant woods
Mechanical damage	Mechanical forces rupturing surface tissues	Selective surface erosion in heavy friction zones	High-specific-gravity woods, edge grain, or chemically hardened woods
Insect damage	Termites Borers Ants	Localized honeycomb cavities, wood soiled and filled with frass Tunnels, cavities, pinholes Localized honeycomb cavities, wood channels clean	Insecticides or keep wood dry
Marine borer damage	Shipworms Pholads Gribbles	Interior tunnels with lime-coated walls Large interior tunnels—near surface Surface tunneling in tidal zone	Protective surface barriers or use wood preservatives
Decay	Fungi	*White* fibrous pockets or punky texture. *Brown* fibrous pockets or cubical checking pattern. *Soft* surface embrittlement and exfoliation in small fragments	Keep wood dry or use wood preservatives
Molds	Fungi	Colored spores or mycelium on the wood surface	Dry wood or use protective chemicals
Stains	Fungi	Sapwood discolored gray, black, brown, blue and intensified in ray parenchyma	Dry wood or use protective chemicals
Ray cell and cell-wall damage	Bacteria	Soft surfaces, ray cells destroyed, microscopic tunnels in cell walls	Keep wood dry or use wood preservatives

- *Termites, beetles, and hymenopterans are the principal types of insects that damage wood in use. Termites chew the wood, forming large cavities for nests that have a honeycomb pattern.* Subterranean termites, which build their nests in soil and require wood with high moisture content, are the most economically important insects in the United States. Beetles and ants chew tunnels, channels, or cavities in wood, and many attack wood initially in the tree, as stored logs, or in the green condition.
- Insect and fungal damage in wood are often associated because of similar environmental requisites or the development of insect vectors for fungus transmission to wood. Two basic prevention or control measures for both agents in many wood uses are to keep the wood dry or to treat the wood with protective chemicals.
- Marine borers cause great damage to unprotected wood used in marine environments. Shipworms, pholads, and *Limnoria* are the major marine degradation agents. They may riddle the wood interior with tunnels or chew the surface in the intertidal zone. Principal preventions are installation of physical barriers around the wood or the use of wood treated with protective chemicals.
- Combustion and weathering are the principal abiotic types of wood destruction. Fire retardants and structural design can reduce the fire hazard in many wood uses. Weathering is the surface destruction of wood by the combined action of ultraviolet radiation, oxidation, leaching, and mechanical forces. This process results in severe aesthetic losses in some wood uses and can be controlled by maintaining protective coatings on the exposed surfaces.
- Caustic chemicals and mechanical forces degrade and damage wood in some special uses or situations, and their presence must be considered when wood is used in industrial applications.

References

Amburgey, T. L. (1979). Review and checklist of the literature on interactions between wood-inhabiting fungi and subterranean termites. 1960–1978. *Sociobiology* 4(2):279–296.

American Wood Preservers' Association. (1986). "Book of Standards." Stevensville, Maryland.

Baker, A. J. (1974). "Degradation of wood by-products of metal corrosion." U.S.D.A. Forest Service, FPL Research Paper No. FPL 229. Madison, Wisconsin.

Baker, W. L. (1972). "Eastern Forest Insects." U.S.D.A. Miscellaneous Publication 1175, Washington, D.C.

Barghoorn, E. S., and D. H. Linder. (1944). Marine fungi: Their taxonomy and biology. *Farlowia* 1:395–467.

Barnes, H. M., and J. E. Winandy. (1986). Effect of seasoning and preservatives on properties of treated wood. *Proceedings of the American Wood Preservers' Association* **82**:95–105.

Barton, G. M. (1982). A study of wood deterioration in two British Columbia Kraft pulp mills. *In* "Structural Uses of Wood in Adverse Environments" R. W. Meyer and R. M. Kellogg, eds. pp. 142–148, Society of Wood Science and Technology Van Nostrand Reinhold, New York.

Beal, R. H., J. K. Maulin, and S. C. Jones. (1983). "Subterranean Termites: Their Prevention and Control in Buildings. U.S.D.A. Forest Service Home and Garden Bulletin 64. Washington, D.C.

Behr, E. A., C. T. Behr, and L. F. Wilson. (1972). Influence of wood hardness on feeding by the eastern subterranean termite, *Reticulitermes flavipes* (Isoptera: Rhinotermidae). *Annals of the Entomological Society of America* **65**:457–460.

Boyle, P. J., and R. Mitchell. (1984). The microbial ecology of crustacean wood borers. *In* "Marine Biodeterioration: An Interdisciplinary Study." (J. D. Costlow and R. C. Tipper, eds.), pp. 17–23. Naval Institute Press, Annapolis, Maryland.

Browne, F. L. (1960). Wood siding left to weather naturally. *Southern Lumberman* 141–143.

Christian, M. B. (1941). Biology of the powder-post beetle *Lyctus planicollus* Le Conte and *Lyctus parallelopipedus* (Melsh). Part II. *Louisiana Conservation Review* **10**(1):40–42.

Coulson, R. N., and A. E. Lund. (1973). The degradation of wood by insects. *In* "Wood Deterioration and Its Prevention by Preservative Treatment" (D. D. Nicholas, ed.), pp. 277–305 Syracuse University Press, Syracuse, New York.

Cundell, A. M., and R. Mitchell. (1977). Microbial succession on a wooden surface exposed to the sea. *International Biodeterioration Bulletin* **13**:67–73.

DeGroot, R. C. (1982). Alternatives to termiticides in building protection. *In* "Termiticides in Building Protection." pp. 91–94. National Institute of Building Sciences, Washington, D.C.

Ebeling, W. (1968). "Termites: Identification, Biology, and Control of Termites Attacking Buildings. California Agricultural Experiment Station Extension Service Manual 38. 74p Berkeley, California

Erickson, H. D., and L. W. Reese. (1940). The effect of several chemicals on the swelling and crushing strength of wood. *Journal of Agricultural Research* **60**: 593–603.

Esenther, G. R. (1969). Termites in Wisconsin. *Annals of the Entomological Society of America* **62**(6):1274–1284.

Esenther, G. R., T. C. Allen, J. E. Casida, and R. D. Shenfelt. (1961). Termite attractant from fungus-infected wood. *Science* **134**:50.

Feist, W. C. (1982). Weathering of wood in structural uses. *In* "Structural Uses of Wood in Adverse Environments" (R. W. Meyer, and R. M. Kellog, eds.), pp. 156–176. Van Nostrand Reinhold, New York.

Feist, W. C. (1977). Finishing wood for exterior applications—paints, stains, and pretreatments. *In* "Wood Technology: Chemical Aspects" I. S. Goldstein, ed. pp. 294–300. American Chemical Society, Washington, D.C.

Feist, W. C., and W. D. Ellis. (1978). Fixation of hexavalent chromium on wood surfaces. *Wood Science* **11**(2):76–81.

Fisher, R. C., G. H. Thompson, and W. E. Webb. (1954). Ambrosia beetles in forest and sawmill: Their biology, economic importance, and control. *Forestry Abstracts* **15**(1):3–15.

Furniss, R. L. (1944). "Carpenter Ant Control in Oregon." Oregon Agricultural Experiment Station Circular 158. Corvallis, Oregon.
Furniss, R. L., and V. M. Carolin. (1977). "Western Forest Insects." U.S.D.A. Forest Service Miscellaneous Publication 1339, Washington, D.C.
Geyer, H. (1982). The influence of wood-inhabiting marine fungi on food selection, feeding, and reproduction of *Limnoria tripunctata* Menzies (Crustacea: Isopoda). *International Journal of Wood Preservation* 2(2):77-98.
Goldstein, I. S. (1973). Degradation and protection of wood from thermal attack. In "Wood Deterioration and Its Prevention by Preservative Treatment. Vol. 1, Degradation and Protection of Wood" (D. D. Nicholas, ed.) pp. 307-399. Syracuse University Press, Syracuse, New York.
Graham, R. D., M. M. Wilson, and A. Oteng-Amoako. (1976). "Wood-Metal Corrosion: An Annotated Survey." Research Bulletin 21. Oregon State University, Forest Research Laboratory, Corvallis, Oregon.
Gray, D. R., and J. H. Borden. (1985). Ambrosia beetle attack on logs before and after processing through a dryland sorting area. *Forestry Chronicle* (August):229-301.
Helsing, G. G. (1979). "Recognizing and Controlling Marine Borers." Bulletin SG-49. Extension Marine Advisory Program, Oregon State University, Corvallis, Oregon.
Hill, C. L., and C. A. Kofoid. (1927). "Marine Borers and Their Relation to Marine Construction on the Pacific Coast." Final Report, San Francisco Bay Piling Committee. University of California Press, San Francisco, California.
Hon, D. N-S., G. Ifju, and W. C. Feist. (1980). Characteristics of free radicals in wood. *Wood and Fiber* 12(2):121-130.
Jamison, G. M. (1937). Loss of weight of wood due to weathering. *Journal of Forestry* 35:460-462.
Kalnins, M. A. (1976). Characterization of the attack on wood by marine borer *Limnoria tripunctata* (Menzies) *Proceedings of the American Wood Preservers' Association* 72:1-13.
Koenigs, J. W. (1974). Hydrogen peroxide and iron: A proposed system for decomposition of wood by brown rot Basidiomycetes. *Wood and Fiber* 6(1):66-80.
Kirchman, D., and R. Mitchell. (1983). Biochemical interactions between microorganisms and marine fouling invertebrates. In "Biodeterioration 5" (T. A. Oxley and S. Barry, eds.) 281-290 Wiley, New York.
Kohlmeyer, J., and E. Kohlmeyer. (1979). "Marine Mycology: The Higher Fungi." Academic Press, New York.
Krishna, K., and F. M. Weesner (eds.). (1969). "Biology of Termites." Academic Press, New York.
Kühne, H., U. Leukens, J. Sell, and O. Walchi. (1972). Outdoor weathering tests on wood and exterior wood finishes. *Eidgenossische Materialprufungs-und Versuchsanstalt (EMPA) Dubendorf,* Bericht 198:1-51.
Lane, C. E. (1959). The general histology and nutrition of *Limnoria*. In "Marine Boring and Fouling Organisms" (D. L. Ray, ed.), pp. 34-45 University of Washington Press, Seattle, Washington.
Lyons, J. W. (1970). "The Chemistry and Uses of Fire Retardants." John Wiley, New York.
MacLean, J. D. (1954). Effect of heating in water on the strength properties of wood. *Proceedings of the American Wood Preservers' Association* 50:253-280.
MacLean, J. D. (1951). Rate of disintegration of wood under different heating conditions. *Proceedings of the American Wood Preservers' Association* 47:155-168.

Marion, J. E., and A. Wissing. (1960). The chemical and mechanical deterioration of wood in contact with iron. Part IV. Prevention of deterioration. *Svensk. Papperstid* **63**:174–183.

McLean, J. A. (1985). Ambrosia beetles: A multi-million dollar degrade problem of sawlogs in coastal British Columbia. *Forestry Chronicle* (August):295–298.

Menzies, R. J. (1959). The identification and distribution of the species of *Limnoria*. In "Marine Boring and Fouling Organisms" (D. L. Ray, ed.), pp. 10–33 University of Washington Press, Seattle, Washington.

Meyers, S. P., and E. S. Reynolds. (1957). Incidence of marine fungi in relation to wood borers attack. *Science* **126**:969.

Minuitti, V. P. (1967). Microscopic observations of ultraviolet-irradiated and weathered softwood surfaces and clear coatings. U.S.D.A. Forest Service Research Paper FPL 74. Forest Products Laboratory, Madison, Wisconsin.

Moore, H. B. (1979). "Wood-Inhabiting Insects in Houses: Their Identification, Biology, Prevention, and Control." U.S. Department of Housing and Urban Development, Washington, D.C.

Moore, W. S., and C. S. Koehler. (1980). "Powder-Post Beetles and Their Control." Leaflet 21017. Division of Agricultural Sciences, University of California, Berkeley, California.

Morgan, F. D. (1968). Bionomics of Siricidae. *Annual Review of Entomology* **13**:239–256.

Morrell, J. J., G. G. Helsing, and R. D. Graham. (1984). "Marine Wood Maintenance Manual: A Guide for Proper Use of Douglas-Fir in Marine Exposures." Research Bull. 48. Forest Research Laboratory, Oregon State University, Corvallis, Oregon.

Quayle, D. B. (1959). The early development of *Bankia setacea* Tryon. In "Marine Boring and Fouling Organisms" (D. L. Ray, ed.), 157–174 University of Washington Press, Seattle, Washington.

Ray, D. L. (1959). Nutritional physiology of *Limnoria*. In "Marine Boring and Fouling Organisms" (D. L. Ray, ed.)., University of Washington Press, Seattle, Washington.

Ross, J. D. (1956). Chemical resistance of western woods. *Forest Products Journal* **6**(6):34–37.

Shafizadeh, F., and P. P. S. Chin. (1977). Thermal deterioration of wood. In "Wood Technology: Chemical Aspects, 1977" (I. S. Goldstein, ed.), pp. 57–81. American Chemical Society, Washington, D.C.

Simeone, J. B. (1954). "Carpenter Ants and Their Control." Bulletin 39. State University of New York College of Forestry. Syracuse, New York

Smith, C. A. (1980). Wood for corrosive environments. *Timber Trade Journal 1980* **312**(5404):23, 31.

Snyder, T. E. (1969). "Control of Nonsubterranean Termites." U.S.D.A. Farmers Bulletin 2018, Washington, D.C.

Steiger, F., and G. Horeczko. (1982). The protection of timber piling from marine borer attack by application of plastic barriers. *International Journal of Wood Preservation* **2**(3):127–129.

Thompson, W. C. (1982). Adverse environments and related design considerations—chemical effects. In "Structural Uses of Wood in Adverse Environments" (R. W. Meyer and R. M. Kellogg, eds.), pp. 117–129. Society of Wood Science and Technology. Van Nostrand Reinhold, New York.

Turner, R. D. (1959). The status of systematic work in the Teredinidae. In "Marine

Boring and fouling Organisms" (D. L. Ray, ed.), pp. 124–136 University of Washington Press, Seattle, Washington.

United States Navy. (1965). "Marine Biology Operational Handbook: Inspection, Repair, and Preservation of Waterfront Structures." NAVDOCKS MO-311. Bureau of Yards and Docks. Washington, D.C.

Wangaard, F. F. (1966). Resistance of wood to chemical degradation. *Forest Products Journal* **16**(2):53–64.

Weesner, F. M. (1965). "The Termites of the United States, A Handbook." National Pest Control Association, Dunn Loring, Virginia.

Williams, L. H. (1985). "Integrated Protection against Lyctid Beetle Infestations. Part I. The Basis for Developing Beetle Preventative Measures for Use by Hardwood Industries." U.S.D.A. Southern Forest Experiment Station, New Orleans, Louisiana.

CHAPTER

3

Characteristics and Classification of Fungi and Bacteria

Fungi and bacteria are the principal types of microorganisms that invade wood during its growth, storage, or use, causing decay or other property changes.

An understanding of the decay process and its prevention or control depends, in part, on understanding the features and capabilities of these decay agents. This chapter emphasizes the uniqueness of fungi and their relationship to the other major life forms. It reviews fungal structures, growth patterns, life cycles, reproductive modes, and variable features of fungi, placing emphasis on the wood-inhabitors. Classification systems are presented to facilitate the taxonomic placement and recognition of some of the major wood-inhabiting microorganisms. Since bacteria cause only minor damage to wood, they are only briefly covered.

Fungi in Relation to Other Life Forms

Before the invention of the microscope, all life forms were grouped into the plant and animal kingdoms. The existence of small life forms was unknown. Visible fungal structures were considered to be excrescences of dying or dead plants. Late in the seventeenth century, the development of the microscope and startling reports of Anton van Leeuwenhoek on the *wee animalcules* opened a window to the hidden world of small life forms (the resolving limit of the unaided eye is approximately 0.1 mm). The study of natural materials revealed a microcosmos, teeming with prodigious numbers and diverse kinds of unicellular and other small life forms. The term *microorganism* was introduced, and microbiology began. At first it was believed that the fungus *threads* observed in decayed wood and other small organisms seen in organic materials originated spontaneously or were again excrescences of dying or dead materials. Nearly a century later, the causative role of microorganisms in disease and decay was established by Louis Pasteur, Robert Koch, and others. It was soon recognized that many of the microorganisms involved in fermentation and decay were neither plants nor animals and required a new category. The term *Protista* was proposed for the unicellular or small life forms with unspecialized tissues. In this three-kingdom classification, until a few decades ago, fungi (despite their lack of chlorophyll) were considered to be simple plants because of the similarity of some groups to the algae. Fungi and algae were grouped in the Thallophyta (thallus plants—tissues not specialized into roots, stems, or leaves) and separated by the heterotrophic nature of fungi and photosynthetic nature of algae.

The development of the electron microscope in the 1950s expanded the dimensions of the microscopic realm a thousandfold and led to the startling discovery of two very different cell types—*prokaryotic* cells and *eukaryotic* cells. The prokaryotic cells are small, bacteria. Prokaryotic cells exhibit no mitosis or cytoplasmic streaming, and have no membrane-bound organelles or organized nuclei. Eukaryotic cells are larger and are present in the higher life forms. These cells divide by mitosis, exhibit cytoplasmic streaming, and have organized nuclei with double membranes, mitochondria, and plastids.

More recently, additional evidence clearly established the distinctiveness of fungi from plants, based on differences in cell-wall composition, heterotrophy, the external mode of digestion, and the cytochrome C system (Lindenmayer, 1965).

Based on these characters and others, Whitaker (1969) proposed a new classification that groups living organisms into five kingdoms and places fungi in a separate kingdom.

These five kingdoms are characterized as follows:

Monera   prokaryotic cells (e.g., bacteria, Actinomycetes, and cyanobacteria)
Protista unicellular and closely related organisms with eukaryotic cells (e.g., protozoa, and single-celled and colonial algae)
Fungi   filamentous eukaryotic cells, generally multicellular, heterotrophic, and with external digestion (e.g., Oomycetes, Zygomycotina, Ascomycotina, Basidiomycotina, and Fungi Imperfecti)
Plantae walled eukaryotic cells, multicellular and highly differentiated, and autotrophic (photosynthesis) (e.g., higher algae, liverworts, mosses, ferns, and seed plants)
Animalia wall-less eukaryotic cells, multicellular and highly differentiated, heterotrophic, with ingestion and internal digestion (e.g., invertebrates, vertebrates)

In an evolutionary sense, it is generally accepted that primitive bacteria were among the first life forms. It has been proposed that eukaryotic cells originated from prokaryotic cells by an intriguing series of symbiotic events (Margulis, 1981). The fungi, as a higher life form, are speculated to have evolved from the Protista along with the separate animal and plant lines. Currently there is a diverse range of ideas on the groupings of organisms in the Protista and their evolutionary relationships to the other kingdoms.

Bacteria

Bacteria are unicellular prokaryotes, although in some forms such as the Actinomycetes, chains of cells have a filamentous form. These organisms probably represent the oldest, simplest life forms, and consist of extremely small cells that average only a few μm in length. Common cell shapes are round (cocci), cylindrical (rods), club-shaped (indeterminant rods), and helical (spirilla). The cell wall consists of *peptidoglycan* (a polymer of N-acetylglucosamine, N-acetylmuramic acid, and several amino acids). Bacteria reproduce by transverse binary fission, although budding occurs in one bacterial group. Under ideal growth conditions, cells may divide within 20 min and accumulate in large numbers on suitable substrates or surfaces such as the rhizosphere.

Along with their tremendous reproductive potential, some bacteria are very resistant to environmental extremes. For example, temperatures of 121°C for 15 min are required to sterilize media in the laboratory from some bacterial contaminants. Masses of bacteria appear as small viscous colonies in media cultures. Many bacteria appear to be adapted for growth on surfaces and are able rapidly to exploit a wide range of energy sources. Some are motile, but most are not, and require air, water, surface contacts, or animal vectors for transport. The wood-inhabiting bacteria are heterotrophs. Some, such as the gliding bacteria (*Cytophaga* spp.) are important cellulose decomposers, whereas others, including the cylindrical bacteria (rods), are associated with discolorations in wood (*Clostridium* spp.) or invade and damge parenchyma cells and pit membranes during water storage (*Bacillus polymyxa*). Many bacteria are associated with wood-degrading fungi, but their exact roles are uncertain. Some bacteria are *lithotrophs* obtaining energy by oxidation of various inorganic compounds such as iron or sulphur (see reducing bacteria in Chapter 5). A few bacteria are photosynthetic and utilize H_2S as a hydrogen source. More than 1000 species of bacteria are known, but the bacterial inhabitants of many substrates remain poorly defined. Many bacteria are human, animal, or plant pathogens and are important in medicine, agriculture, or the chemical industry. Very little is known about the identities or roles of bacteria in wood decay. It is generally believed that many new species of bacteria await discovery.

Fungi

Fungi are multicellular eukaryotes. All are heterotrophs and utilize carbon compounds as an energy source. The fungal body (thallus) consists of a series of small interconnected tubelike cells called *hyphae*. The key to understanding the unique fungal *hyphal system* is to remember that it has adapted to penetrate, externally digest, absorb, and metabolize a wide range of organic materials (e.g., plant materials, wood). Masses of hyphae are termed collectively *mycelium*. Fungi reproduce primarily by *spores* formed by fragmentation or abstriction of sections of specialized hyphae.

Fungi play three major roles in the ecosystem. Some fungi are *pathogens* and attack living plants or animals, causing diseases. Other fungi are mutualistic *symbionts* and have developed beneficial associations with other organisms (e.g., mycorrhizae, lichens). Most fungi are *saprobes* and are the principal agents in the ecosystem that decay plant debris, release carbon dioxide, and sustain photosynthesis in green plants. It is in this latter role that fungi are the major decay agents of wood.

Macroscopic Appearances of Fungi

Most of the hyphal systems of fungi that cause disease or decay are internal to the host or substrate, and not readily visible to the unaided eye. At times, however, mycelial masses or structures are visible externally to the unaided eye and may be useful for decay or disease detection (Fig. 3-1). Weblike masses of hyphae known as *mycelial fans* often grow on the surface of decaying wood in the ground or in high humidities and are a useful indicator of internal decay. *Rhizomorphs* are thick, cablelike masses of hyphae that are able to transport water, moisten wood, and facilitate its invasion. They are formed by several of the principal fungi that decay buildings. *Mold* is a general term that describes the visual appearances of colored spore masses or pigmented hyphae that may grow on the surface of wet wood. The term *mildew* describes the appearance of black-pigmented fungi that develop on surfaces such as painted wood in humid environments (Fig. 17-2).

Some fungi form hardened cushions of mycelium known as *stroma* on plant or wood surfaces. Asexual or sexual fruiting structures may develop on or within the stroma. The most common visible form of many of the higher fungi is the *fruiting body* or *sporophore* bearing the sexual spores (e.g., mushrooms, bracket fungi, puffballs). The form of these structures and the manner in which the spores are formed are the principal criteria used in classifying fungi and will be discussed later in this chapter.

Estimates of the number of fungal species range from 45,000 to 100,000. Fungi range in size from a few unicellular groups such as the yeasts to some wood-decay fungi whose individual basidiomata may weigh 45 kg.

Figure 3-1 Some common structures of wood-decay fungi that are visible to the unaided eye. (A) Mycelial fan of *Meruliporia incrassata* on a board surface. Note the advanced brown cubical rot. (B) A rhizomorph of *Meruliporia incrassata* conducting water from a damp basement corner to the floor joists above. (C) The basidioma of *Gloeophyllum sepiarium*, a major decay agent in coniferous wood products, worldwide. (A) and (B) courtesy of Forest Products Laboratory, Madison, Wisconsin; (C) courtesy of Dr. Paul Manion.

Microscopic Features of Fungi

Hyphae are the basic cellular unit of fungal structures. Individual hyphae are small and, with few exceptions, can be seen only after considerable magnification. Individual hyphae range from as small as 0.5 to 20 μm or more in diameter, with most ranging from 2 to 10 μm in diameter. Typical hyphal features seen with the ordinary light microscope include cell walls, cross walls or septa, vacuoles, various inclusions such as fat globules and crystals, and occasionally, nuclei. Most fungal nuclei are very small, and special staining techniques are often required for observation. Hyphal cells may be uninucleate or multinucleate, but many decay fungi have binucleate cells. When the nuclei are different genetically, arise from fusion of two hyphae, and do not fuse, the cells are in the *dikaryon* stage. This nuclear condition is unique for fungi.

Septa or cross walls are present in the higher fungi (Ascomycotina, Basidiomycotina, and Deuteromycotina) but infrequent in the Zygomycotina and Oomycetes of the Mastigomycotina, except where reproductive cells are delimited or the hyphae are injured. Septa or cross walls in the higher forms of fungi probably function as cell-wall-strengthening devices. Hyphal systems that are multinucleate are termed *coenocytic*.

Many septa are perforated, permitting organelles and some nuclei to migrate from cell to cell. Three types of septa are found in fungi. In some cases, the septa are intact (without perforations), such as those found in the Oomycetes and Zygomycotina. Many Ascomycotina have simple one or several perforate septa. Septa with large swollen pore margins called *dolipore septa* characterize many members of the Basidiomycotina (Fig. 3-2). These septa often have an associated amorphous material that appears to plug the pore. The dolipore septum may prevent nuclear migration. Dolipore septa can sometimes be seen at higher magnifications ($\times 600$), and their presence can help to confirm that an isolate is a basidiomycete.

Hyphal growth occurs primarily by apical extension at the tips. Vacuoles usually develop a few cells behind the hyphal tip, and the related turgor pressure within the cell protoplasts is the presumed driving force extending the plastic tip. Hyphal branching is common and usually begins early behind the developing hyphal tips. The fusion of adjacent hyphae of the same genotype by *anastomosing* is also common, resulting in a complex network of interconnected cells. Cytoplasmic streaming appears to move material steadily toward the tips, leaving empty or vacuolated hyphae behind.

In some Basidiomycotina, special hyphal structures called *clamp connections* develop in the dikaryotic hyphae during septa formation. In these cases, a branch develops near the forming septum in the hyphal apex. This branch curves backward and anastomoses or fuses with the cell wall just behind the developing septum. Simultaneously the two nuclei in the cell divide mitotically and migrate separately so that the apical cell and the cell penultimate to it each contain two nuclei genetically identical to the original pair (Fig. 3-3).

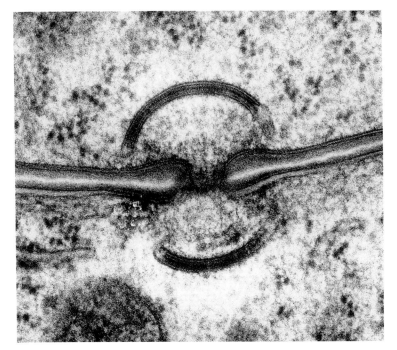

Figure 3-2 The TEM micrograph of a dolipore septum, which characterizes many of the wood-inhabiting Basidiomycotina. The septum is from the mycelium of *Auricularia auricula-judae* prepared by freeze substitution, ×97500. Courtesy of Haisheng Lü and Dr. D. J. McLaughlin (1991) and reprinted by permission of The New York Botanical Garden.

Clamp connections may be single or multiple and may occur at each septum, consistently, randomly, or infrequently, depending on the fungal species. Clamp presence, type, shape, and size are important characters in cultural keys for the identifying of decay fungi (Fig. 3-3).

Hyphal Wall Structure

Fungi, like plants, possess firm cell walls that provide the rigidity needed for the formation of large fungal structures, such as shelving basidiomata or rhizomorphs, and functions, such as the forcible discharge of spores or penetration of plant cell walls. Hyphal walls also serve as important storage reserves for some fungi. Structurally, fungal cell walls consist of an inner network of microfibrils embedded in an amorphus matrix that also forms the outer layers or often lamellae of the wall.

Chemically the walls consist of 80 to 90% polysaccharides, with the remainder composed of proteins and lipids. *Chitin, cellulose,* and in a few cases, *chitosan* form the microfibrils that serve as the skeletal framework of the walls. Chitin is the principal skeletal material and is present in the inner wall of most septate fungi. The chitinous nature of the cell wall is one of the fundamental differences separating most fungi from plants, although some

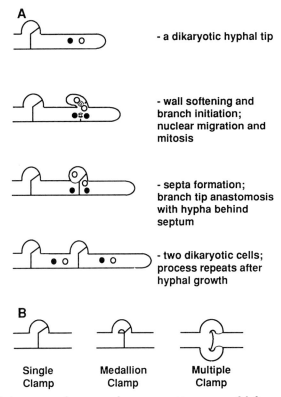

Figure 3-3 Hyphal structures known as clamp connections are useful diagnostic features for many wood-decay fungi in the Basidiomycotina. (A) The developmental stages of a clamp connection. (B) Examples of several types of clamp connections. Drawing courtesy of Dr. J. Worrall.

fungi, such as the Oomycetes, have cellulose in their cell walls, and some yeasts have mannans in the cell wall. Mannans are complex glycoproteins consisting of a 1-6-α-D-mannose backbone and a protein component. In addition, N-acetylglucosamine, the building block of chitin, serves as a bridge between the protein and the mannose. Other fungi, such as the Mucorales, have in their cell wall chitosans, consisting of unacetylated glucosamine monomers. Other sugar monomers and linkages occur in some fungal groups. A basidiomycete, *Schizophyllum commune*, has been reported to have an outer-wall layer consisting of a glucan with an 1-6 linkage (Wessels and Sietsma, 1973). The roles of the proteins and lipids in the walls are uncertain. It has been proposed that some of the protein present may be cell wall-bound enzymes. Other glycoproteins present in walls may function as elicitors in recognition phenomena, whereas others may contribute to the structural integrity of the cell wall.

Fungal cell walls are surprisingly complex, both chemically and structurally. Photomicrographs of the various cell-wall layers, after successive enzymatic treatments, have been assembled by Becket *et al.* (1974) for the representatives of the major fungal groups. Bartnicki-Garcia (1968) suggests that differences in chemical composition of the cell wall may be closely related to taxonomic position. Structural formulas for some of common cell-wall constituents are presented in Fig. 3-4.

Fungal cell walls are readily modified by self-digestion (autolysis) and translocation of wall metabolites to growing hyphal apices. This activity may be an important means of nitrogen conservation. In addition, hyphal fusions, hyphal branching, clamp formation, some spore formations, and spore germination reflect situations in which localized zones of the wall are enzymatically softened and disassembled.

Since the hyphae of most disease- and decay-causing fungi are internal to the host or substrate, it is often difficult to detect or determine the degree of invasion. Procedures have been developed to estimate the fungal biomass in decayed wood by acid hydrolysis of the decayed material and glucosamine detection and quantification (Swift, 1973). Chitin content can be determined directly from the glucosamine residues; however, chitin assays are difficult to perform and not entirely appropriate for quantifying fungal biomass for some species (Wu and Stahmann, 1975). When chitin assays are used, standards must be developed to relate the chitin content to the biomass for the fungi tested. Chitin content of hyphae can also vary substantially with the nutritional status of the fungus.

Ultrastructure of Fungi

The electron microscope has revealed a wealth of information on the fine structure of cells. Discovery of the close relationship between structure and function has done much to clarify the major metabolic events occurring within cells. As eukaryotic cells, fungal cells are similar in the types and roles of the organelles to most plant and animal cells. A review of the major organelles and their function in fungal cells is needed as a background for understanding the metabolic events and mode of toxicant actions discussed in later chapters. A double-layered *cytoplasmic membrane* consisting of lipoproteins lies within the cell wall and regulates the entrance and exit of materials from the cells. In some fungi, the membrane invaginates into aggregations of membranes within the cell wall termed *lomasomes*. Their function is unknown, but they may function in absorption, secretion, or cell-wall modifications. An *endoplasmic reticulum* (ER) is present throughout the cytoplasm and generally concentrated in regions of metabolic activity and cell-wall extension. The ER may have netlike connections with the other cell membranes (nuclear membrane, tonoplast, and cytoplasmic membrane) and plays a key transport role among the organelles. *Ribosomes* are the centers of protein synthesis and appear to be free in the cytoplasm, particularly in regions of hyphal growth. Associations between ribosomes and the ER, termed *rough ER*, which are common in many eukaryotes, are infrequent in

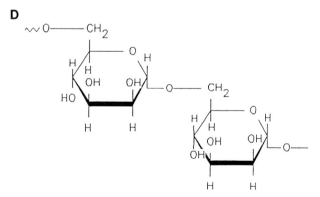

Figure 3-4 Chemical structures of portions or repeating units of several common polysaccharides found in fungal cell walls. (A) Cellulose. (B) Chitin (deacetylation yields chitosan). (C) Glucan. (D) Mannan.

fungi. *Microtubules* are present in the cytoplasm, where they may be associated with protoplasmic streaming, and play a role in nuclear division. Typical *Golgi* apparatus involved in secretory-vesicle assembly in plant and animal cells have not been observed in fungi. *Dictyosomes,* consisting of stacks of platelike membranes, are present and may play this role in the fungi. *Mitochondria* are visible under the light microscope using oil-immersion objectives and are associated with respiration and energy release. Time-lapse microscopy indicates that mitochondria can rapidly change shape and position within the cell. *Vacuoles,* bounded by a membrane called the *tonoplast,* are abundant in older hyphae and generally absent in the tips. They often originate in cells adjacent to the hyphal tips and increase in size in the older hyphal cells.

Vacuolization has been proposed as one method for isolating toxic compounds that have been liberated during the digestion of wood. *Nuclei* are often very small (2–3 µm) and difficult to detect using light optics without special stains. Typical three-layered membranes with pores surround the nuclei. Nuclei also appear to change both size and shape rapidly. At times, and particularly during spore formation, the nuclear membrane appears to be continuous with the ER. In contrast with plant and animal cells, during mitosis in fungal cells, the nuclear membrane does not disassociate but appears only to constrict near the midpoint of the two daughter cells.

A dark region can be seen in the hyphal apex of some septate fungi under the light microscope. In electron micrographs, this dark region is an electron-dense zone closely associated with the hyphal tip and is present only during periods of active growth. This region is believed to contain clusters of transport vesicles involved in cell-apex synthesis. *Crystals* and *globules* containing lipids, glycogen, or other reserve foods are also observed commonly in older hyphae. The ultrastructural features of the hyphal-tip region of a wood-decay fungus are shown in Fig. 3-5.

Hyphal sheaths, consisting of an extracellular matrix with a fine fibrillar or membranous structure have been detected recently in both white- and brown-rot fungi. Their function is uncertain but an interesting transport role in the decay process has been proposed. (See Chapter 9 for additional information.)

Specialized Hyphae

Specialized hyphae have developed for a variety of purposes in some fungal groups. *Appressoria* are flattened, enlarged hyphal tips that adhere to surfaces and facilitate the penetration of fine hyphal pegs through cell walls. They are formed by many pathogens and some wood-staining fungi. *Haustoria* are enlarged, convoluted hyphal cells adapted for absorption, which contact the plasmalemma of host cells after wall penetration. They are present in the rusts, powdery mildews, and some mycorrhizal fungi. Some biotrophic fungi that parasitize other fungi invade their hosts with haustorialike hyphae. Unique cases of specialization are the hyphae that trap nematodes by

Figure 3-5 An electron micrograph of a hyphal tip of the fungus, *Aspergillus niger*, showing the vesicles (V) clustered in the apical zone, ribosomes (R), mitochondria (M), smooth-surfaced Golgi cisternae (G), and the wall zone (W). ×25,000. Courtesy of Drs. S. N. Grove and C. E. Bracker (1970) and permission of the *Journal of Bacteriology*.

the formation of sticky adhesion zones or capture them in rapidly constricting rings. *Chlamydospores* are cells that develop thick walls and are able to withstand long periods of adverse environmental conditions. Some hyphae contain melanin, a dark pigment that protects hyphae growing on surfaces (paint mildew) from ultraviolet light. Melanized hyphae are also present on the surface of some rhizomorphs and fungal propagules, such as microsclerotia, and play a protective function. *Spores* are the reproductive units of fungi. They are specialized hyphae and, because of their importance, are discussed later in a special section.

The large reproductive structures of some Ascomycotina and Basidiomycotina also contain specialized structural hyphae. Aggregations of tightly packed thin-walled hyphae resembling the parenchyma cells of higher plants, called *pseudoparenchyma hyphae,* occur in the soft tissues of some ascomata or fleshy basidiomata. When present in the gill or trama tissue of basidiomata and circular in shape, these hyphae are termed *sphaerocysts* and are useful for identification. Three types of hyphae may form the larger basidiomata of the decay fungi. The thin-walled, septate *generative hyphae* form small, soft basidiomata and are present in all basidiomata, eventually forming the hymenial layers. Leathery basidiomata also contain thick-walled, many branched cells termed *binding hyphae,* which intermesh with the generative hyphae. Large woody basidiomata such as *Ganoderma* (Fomes) *applanatum* also contain *skeletal* hyphae, which are elongate, thick-walled, and rarely branched or septate. Skeletal hyphae resemble wood fibers and provide the hard, woody characteristics of some of the large, shelving basidiomata. The three types of structural hyphae in basidiomata have been key critieria used in recent taxonomic revisions of the wood-decaying Basidiomycotina. Generative hyphae are present in all basidiomata. Others may contain binding or skeletal hyphae or both. This topic will be discussed further in the section on classification.

Cultural Characteristics

A unique and artificial fungal form is the *mycelial mat* formed by a pure culture on a culture medium. Most fungi, other than the obligate parasites, can be grown in the laboratory under axenic culture conditions on various natural or synthetic media. The mycelium of the fungus that develops on the medium is a mixture of the vegetative and reproductive hyphae. Some fungi, such as *Schizophyllum commune,* often develop typical basidiomata after a few weeks in culture. The growth characteristics (macroscopic and microscopic) of the mycelial mats are often distinctive and useful in the identification of the fungus.

The fruiting bodies of decay fungi are frequently absent, ephemeral, or difficult to detect, making it impossible to identify the causal agent. In these cases, the cultural characteristics of fungi isolated from the decayed material may be used for identification. Some useful characteristics employed in cultural identification manuals include mat color, texture, rate of growth, and microscopic features such as the presence of clamp connections, the types and features of specialized hyphae, and the types of asexual spores. Examples of some unique microscopic features of several important wood-decaying fungi are shown in Fig. 3-6. Oxidase reactions (the oxidation and discoloration of gallic and tannic acid media) and growth patterns of fungi on these two media are simple, important tests for separating most brown-rot fungi (no oxidase reaction) from the white-rot fungi (oxidase-positive discoloration of the media). Several useful identification manuals have been developed for various hosts, wood products, or fungal groups; they provide keys and cultural descriptions for many of the major wood-decay fungi (Davidson *et al.,* 1942; Nobles, 1965; Stalpers, 1978; Wang, 1965; Wang and Zabel, 1990).

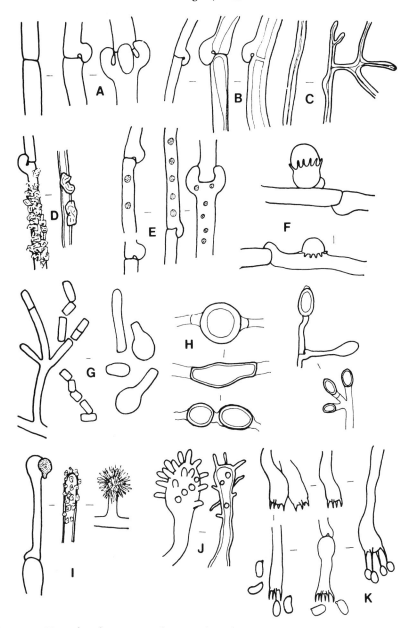

Figure 3-6 Examples of microscopic features of Basidiomycotina hyphae useful in cultural identifications. (A) Variation in hyphal septation. (B) Variation in wall thickness. (C) Fiber hyphae. (D) Encrusted hyphae. (E) Variation in nuclear condition. (F) Stephanocysts (*Hyphoderma* spp.). (G) Arthroconidia [*Bjerkandera adusta* (left); *Gloeophyllum sepiarium* (right)]. (H) Chlamydospores [intercalary (left); terminal (right)]. (I) Cystidia. (J) Acanthophyses (acanthohyphidia). (K) Variations in basidia: *Hyphodontia setulosa, Phlebia brevispora, Antrodia vaillantii* [top] and *Phlebia rufa, Sistotrema brinkmannii*-complex [bottom]. All ×1000. Reproduced by permission of American Type Culture Collection (ATCC) and courtesy of Francis Lombard and Dr. George Chamuris.

Reproduction

Fungi reproduce by unicellular or multicellular spores formed from hyphae. They can be of asexual or sexual origin. Most spores provide a means of dissemination, but others are designed for surviving unfavorable conditions, and a few may function sexually as gametes. Spores are generally very small and readily transported long distances by air currents. The spores of some fungal groups are forcibly ejected into the air. Other spores are sticky and appear to be adapted for dissemination by insect vectors. Asexual spores are formed directly from a hypha without a meiotic division. Sexual spores are formed from hyphal cells where nuclear fusion (karyogamy) and meiosis have preceded the separation into spores. Sexual spores are often termed perfect spores.

Asexual spores Many types of asexual spores are formed by fungi. This review will consider primarily the common types found in wood-inhabiting fungi. *Conidium* is the general term used for most asexual spores formed by members of the Ascomycotina, Basidiomycotina, and Deuteromycotina. Conidia have several very different modes of origin, which are important taxonomically in the Deuteromycotina.

Arthrospores or *oidia* are conidia formed by the fragmentation of an existing hypha by separation of the septal walls. This type of conidial origin is termed *thallic*. Some wood-decay fungi (*Gloeophyllum sepiarium, Phanerochaete gigantea,* and *Bjerkandera adusta*) form abundant arthrospores in culture, and this characteristic facilitates their identification. In some fungi, such as the yeasts and the wood stainer, *Aureobasidium pullulans,* the cell wall softens in a localized zone, balloons out, and forms *blastospores* by a budding process. This type of conidial origin is termed *blastic,* and the conidiogenous cell enlarges generally before septal formation and conidium release. There are several types of blastic conidial development. When the outer and inner walls of the conidiogenous cell are involved in conidia formation, it is called *holoblastic.* When only the inner wall is involved or a new wall is formed, the process is termed *enteroblastic.* Conidia formed by some wood-staining fungi (*Ceratocystis coerulescens*) represent enteroblastic development where spores are formed inside an open-ended hypha and when mature, are ejected into the air. Conidial ontogeny and its importance in taxonomy of the Fungi Imperfecti has been discussed by Hughes (1953) and Kendrick (1971).

Sexual spores Sexual spores result from the fusion of two haploid nuclei in a hyphal cell and subsequent meiosis. Sexual reproduction provides the fungal progeny the great biological advantage of genotypic diversity as a result of crossing over or mixing of chromosome pair portions during meiosis. Genotypic diversity permits high variability for adaptive selections that can aid survival during changing conditions. In many groups of fungi, sexual spores are borne in highly specialized hyphal aggregates called fruiting bodies. Some typical examples of fruiting bodies are puff balls, mushrooms, perithecia, and the conks of wood-destroying fungi. Perfect spores that form inside a specialized terminal hyphal cell, termed the *ascus,* are called *ascospores.* This type of perfect spore formation is endogenous and characterizes the subdivision of fungi known as *Ascomycotina* (Fig. 3-7). This subdivision of fungi contains many of the important wood stainers. A fertile layer of asci on or in an ascoma (ascocarp) is termed the *hymenium.* There are several types of ascomata (Fig. 3-7C). A closed ascoma is called a *cleistothecium.* A similar ascoma with an opening (ostiole) is termed a *perithecium.* The perithecium may be a free structure or embedded in a stroma. When the ascoma is an open or

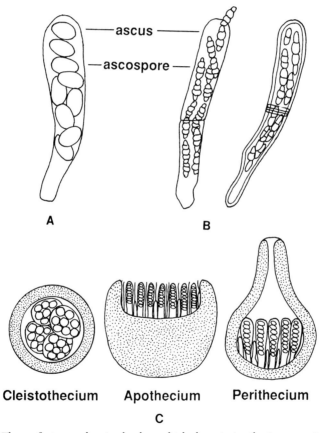

Figure 3-7 The perfect spore-bearing hyphae, which characterize the Ascomycotina. (A) A unitunicate ascus. (B) Several bitunicate asci; the left one is beginning to discharge ascospores. (C) Several typical types of ascomata. Sources are (A) Engler, A. and K. Prantl (1897) *Die natürlichen Pflanzenfamilien,* Engelmann, Leipzig; (B) from Luttrell, E. S. (1960) *Mycologia* **52**:64–69 with permission of the New York Botanical Garden; and (C) adapted from Fig. 4-8, E. Moore-Landecker (1990), *Fundamentals of the Fungi,* with permission of Prentice-Hall, Inc. Drawings courtesy of Dr. George Chamuris.

cuplike structure bearing the hymenium, it is called an *apothecium*. Recently it has been determined that asci are of two types, the *bitunicate asci* with a rigid outer and a flexible inner wall and the *unitunicate* asci with a single wall (Fig. 3-7A,B). This feature and the nature of the ascomata are the principal distinguishing characteristics of the major wood-inhabiting ascomycetes.

Perfect spores formed externally on the tip of a swollen club-shaped hypha, termed a *basidium,* are called *basidiospores* and characterize the subdivision of fungi known as the Basidiomycotina (Fig. 3-8). This subdivision of fungi contains most of the important wood-destroying fungi, and also contains the rusts and smuts, which are major plant pathogens worldwide.

The two classes of Basidiomycotina that contain the wood decayers are the Hymenomycetes and the Gasteromycetes. They are separated from the other two classes, the Urediniomycetes (rusts) and Ustilaginomycetes (smuts), by their well-developed basidiocarps and general saprophytic nature. In the Gasteromycetes, the hymenium is enclosed within the *basidioma*, and the

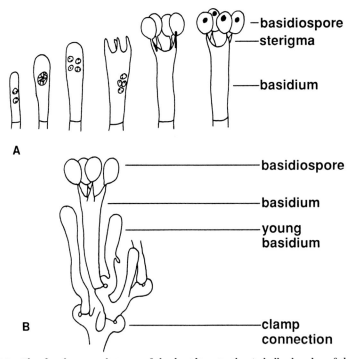

Figure 3-8 The developmental stages of the basidium in the Aphyllophorales of the Basidiomycotina. (A) The developmental stages of a typical basidium (adapted from A. H. Smith (1934) *Mycologia* **26**:305–331) with permission of the New York Botanical Garden. (B) The development of basidia in a representative zone of the hymenium (from E. J. H. Corner (1950) *Ann. Botan. Mem.* **1**:1–74) with permission of Oxford University Press. Drawings courtesy of Dr. George Chamuris.

basidiospores are not discharged forcibly from the basidia. In the Hymenomycetes, the hymenium is exposed and borne on *basidiomata* surfaces, which may have gills, pores, spines, or be warty, or smooth. The form and hyphal structure of the basidioma are major criteria for classifying these fungi. These characteristics will be presented in more detail in the section on classification, since many of the important wood-decay fungi are in the Hymenomycetes.

Life Cycles

Information on the life cycles of the major classes of fungi that attack wood is necessary to understand the nature and great benefits of genetic diversity to fungi. Genetic diversity permits rapid adjustments to new conditions through selection. There are enormous differences in the patterns of sexuality, sexual mechanisms, and the life cycles of the fungi (Raper, 1966). More detailed information is available in the various textbooks on mycology (e.g., Moore-Landecker, 1990; Webster, 1980; Alexopoulos and Mims, 1979). Only a few highlights are presented here. *Sexual* reproduction is an effective means of developing genetic diversity, and it occurs in all fungal classes except the Fungi Imperfecti. Sexuality basically involves the union of gamete protoplasts (plasmogamy) and fusion (karyogamy) of two nuclei (n) to form a diploid nucleus (2n). The diploid nucleus divides by meiosis, and the crossing over or the exchange of genetic material among the paired or homologous chromosomes can occur. Subsequent mitotic divisions then maintain and multiply the new genotypic arrangement. In contrast, only mitotic divisions occur in asexual reproduction, and there is less chance for genetic diversity. In some cases, high reproductive potential coupled with a high frequency of errors or mutations during mitosis can overcome the limitations presented by the absence of a sexual stage.

The life cycle of a typical wood-decaying fungus (Hymenomycete) can be described in steps as follows (Fig. 3-9):

1. A *basidiospore* (haploid nuclear condition) germinates and forms hyphae. Since all the nuclei are the same genetically, it is a *homokaryon* and also a *monokaryon* (n) since each cell contains an identical nucleus.
2. The monokaryotic hypha may form branches that later separate at the septa, forming spores (oidia or arthrospores) that germinate, forming new monokaryons. This is the asexual life cycle.
3. Two compatible monokaryotic hyphae fuse (anastomose) and their protoplasts, with two genetically different nuclei, become associated in a hyphal cell (plasmogamy). This is the beginning of the *dikaryotic* condition (n + n), which is unique to the Basidiomycotina.
4. The dikaryotic condition in the hyphae maintains itself by simultaneous nuclear division with septa formation in new cells. This is facilitated by *clamp connections* in many decay fungi. The major

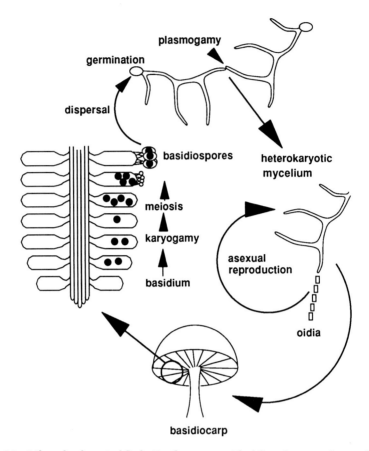

Figure 3-9 Life cycle of a typical fleshy Basidiomycotina. The life cycle is monokaryon-dikaryon and briefly diploid. The vegetative mycelium is dikaryon and may be perennial. Basidiomata (basidiocarp) develop sporadically. Hyphal fusions occur only among compatible mating types. Courtesy of Dr. J. J. Worrall.

vegetative existence of the fungus is in the dikaryotic stage, although in some cases, the monokaryotic stage also causes substantial decay.
5. When the energy reserves and environmental conditions are favorable, the mycelium aggregates and forms a basidioma.
6. Generative hyphae mass on the surface of basidiomata tissues, and the terminal cells in the many hyphal chains become basidia. This zone of parallel-aligned basidia is called the hymenium.
7. Nuclear fusion of the two nuclei occurs, and the basidium briefly enter the *diploid* stage (2n).

8. The diploid (2n) nucleus undergoes meiosis, and four haploid nuclei are formed at the end of the two-stage process. The four nuclei move to the top of the basidium and migrate through sterigmata and into four exogenously produced basidiospores.
9. The basidiospores, when mature, are forcibly ejected. If one lands on a digestible substrate under favorable conditions, the life cycle begins again.

There are many modifications of this cycle. In some Basidiomycotina, the hyphae are heterokaryotic and become dikaryotic only in the cells basal to the basidium. In some cases the basidiospores formed are binucleate and germinate directly to form the dikaryotic stage (Raper, 1978). Occasionally the homokaryon is self-fertile, and clamp connections develop. Bipolar and tetrapolar compatibility patterns occur among the basidiospore isolates. Such patterns, when known, can be very useful in judgments on the origins of decay columns in trees or products.

The heterokaryon is a common nuclear state of the hyphae in the Deuteromycotina and Ascomycotina. Openings in the septal pores appear to facilitate the migration and accumulation of many nuclei in the cells of some species. In many Ascomycotina, two compatible nuclei pair in an *ascogonium*, and by a series of complex nuclear division and wall formations, successive ascogenous hyphae form the ascus, which usually contains eight haploid ascospores. In some species of the Deuteromycotina, a parasexual cycle results in some genetic diversity. The process is complex and beyond our purposes. In essence it starts with heterokaryon formation, the fusion of some nuclei to form diploids, some crossing over of genetic material during their mitotic divisions, subsequent reversion to haploids, and their eventual isolation in conidia or hyphal tips.

Reproductive Capacity

The numbers of spores produced by the various sexual and asexual fruiting structures of many fungi is huge (Webster, 1980). In addition, some fungi limit spore release to periods favorable for germination and subsequent host or substrate invasions. Others produce substantial numbers of resting spores so germination can be delayed until the proper combination of substrate and environmental conditions occurs. It has been estimated that a single basidioma of the wood-decay fungus *Ganoderma applanatum* produces billions of spores daily for extensive periods during the growing season. The large number of spores produced, coupled with their small size and ease of dissemination, means practically that suitable substrates will be showered with spores within a short period and accounts for the high competitive effectiveness of fungi and bacteria as carbon-compound decomposers. Microbial invasion of a substrate is certain to occur wherever and whenever suitable substrates and growth conditions occur.

Variability

Fungi often exhibit wide ranges of variability in physiologic characteristics, appearances, and capacities to act as saprobes, pathogens, or mutualistic symbionts. This variability reflects, in part, the huge numbers of spores produced and the brief reproductive cycles of most fungi. A wide range of tolerances permits fungi and bacteria to respond quickly to changing conditions. Examples include the rapid appearance of fungal strains resistant to new fungicides, or bacterial strains tolerant of new antibiotics.

Variation represents, in part, a sorting out of the many strains or different genotypes composing a species. Selection pressure then favors the steady increase of the strains whose genotype can cope with the new condition. Genotypic variability in fungi arises initially from *mutations* in the genome and *crossing over* or rearrangement of genes on the chromosome during meiosis or parasexuality. It is estimated that some fungi have 20 or more alleles on a single gene pair, providing ample opportunity for developing new allele combinations. *Aneuploidy,* the addition or deletion of parts of a chromosome to the normal chromosome number, is reported to be common in fungi and may represent an important source of variation. The *heterokaryotic* condition is another source of variation, since hyphae within a single mycelial mat may have many different combinations of nuclei with differing physiologic capabilities that affect appearance, growth, and survival. The sectoring of cultures and variations in the appearance of the fungal mats of a species often reflect this condition. For example, the monokaryon and dikaryon mycelial mats of some wood-decay fungi differ in appearance, growth rates, and decay capabilities (Sexton, 1988).

The presence of plasmids or virus particles in a bacterium or fungus may provide an additional source of variation. Some of the most productive strains of penicillin-producing fungi are now known to be virus infected. Some viral-infected bacteria strains are also more virulent. Conversely, plasmids can also render fungi less pathogenic. For example, plamid-infected strains of the chestnut blight fungus, *Cryptonectria (Endothia) parasitica,* are less virulent.

Other factors may be responsible for the variations observed among the isolates of a species. Some fungi are *pleomorphic* and produce different types of asexual spores depending on environmental conditions or age. Some fungi are *dimorphic* and, depending on nutritional sources or environmental conditions, may grow in yeast or mycelial stages.

The ability of fungi to rapidly adapt to new conditions substantially complicates disease- and decay-control programs. at the same time, however, this variation presents tremendous opportunities for industrial uses of fungi and future biotechnology developments.

Growth Requirements

Wood-decay fungi are able to grow on wood and damage or modify it under a wide range of environmental conditions. Four critical requirements for fungal growth in wood are

- Supply of free or unbound *water;*
- Favorable *temperatures* (0–42°C);
- Atmospheric *oxygen;* and
- Digestable *carbon compounds.*

These requisites and their relationship to decay-prevention and -control approaches are discussed in detail in Chapter 4.

A Classification of Fungi

Fungal classification is based primarily on differences in the types of the reproductive structures. The taxonomic study of fungi has two general purposes. One is identification and involves determining distinguishing features and characteristics that will lead to naming the fungal species. *A critical first step in the control of most disease and biodeterioration problems is to determine the identities of the associated microorganisms.* Fungal identities can also clarify ecologic roles, help determine structural and physiological relationships, and assist in the development of rational prevention or control programs. The need to identify fungal associates is even more understandable when we realize the wide array of fungal species and the small size as well as the similarity in general appearance.

A second important purpose of classification is the arrangement of organisms in the order of their phylogenetic origin. Such systems of classification are spoken of as *natural* systems. Related organisms in a natural system genrally have many similarities, so that detailed knowledge of a few often yields general information about many.

Early classifications of wood-decay fungi were based on convenient macroscopic characters such as the size and shape of the fruiting structures and the nature of the surfaces bearing the hymenium (tubes, gills, spines, smooth, etc.). There was only a limited fossil record to suggest possible phylogenetic relationships. Another great difficulty was the evolutionary phenomenon of *convergence,* in which macroscopically similar fungi sometimes have very different origins.

As new information on the microscopic, ultrastructural, physiologic, genetic and biochemical features of fungi has accumulated, major revisions in the classifications of many groups have occurred and are still continuing. Similar changes are occurring in the classifications of the bacteria.

The general classification scheme we shall use for fungi is in the seventh edition of the *Ainsworth & Bisby's Dictionary of Fungi* (Hawksworth *et al.,* 1983). Emphasis in this section will be placed on the groups of fungi that inhabit, modify, or destroy wood. Drastic taxonomic revisions have occurred in the Ascomycotina and several groups of the Basidiomycotina. In some cases in which the revisions are still incomplete, or to provide continuity with groups in the older literature, we use the classification scheme in an earlier *Dic-*

tionary of Fungi (1971). When this is done, the name of the group is placed in brackets to avoid confusion with the 1983 classification.

In the Ascomycotina, whether the ascus was unitunicate or bitunicate became important, and prior groupings based on types of ascomata, less so. Currently all are placed in 37 orders awaiting further study.

Eumycota

The Eumycota include primarily the filamentous fungi. Currently there are five subdivisions, which are listed and characterized in Fig. 3-10. In the

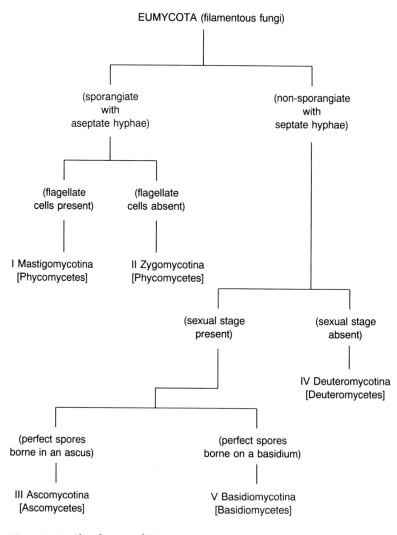

Figure 3-10 Classification of Eumycota.

Basidiomycotina, major revisions have taken place among those with basidiomata. Important differences occur in the hyphal types (generative, skeletal, and binding) and arrangements in the basidiomata that are judged to be of taxonomic significance. In the Deuteromycotina, new information on conidiogenesis has been used to revise older classifications.

Detailed consideration of these changes is beyond the scope of this book. A series of dichotomous keys will be used to characterize the major groups of fungi, with emphasis on those that damage wood. Fungi that are important pathogens or mutualistic symbionts are also included as representatives of the various taxonomic groups. The representative species are preceded by a number to avoid confusing them with key characteristics.

Mastigomycotina and Zygomycotina

The subdivisions Mastigomycotina and Zygomycotina are unimportant in wood products (Fig. 3-11). They are discussed here since some of the Oomycetes are major plant pathogens, and several of the Zygomycotina may cause severe surface molding of lumber. Some taxonomists have regrouped the Oomycetes among the algae.

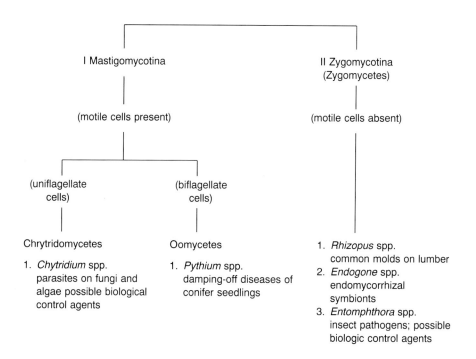

Figure 3-11 Classification of Mastigomycotina and Zygomycotina.

Ascomycotina (Ascomycetes)

The subdivision Ascomycotina contains many of the important wood-stain fungi and soft-rotters. There are many important tree pathogens in the group. Some pathogens causing stem cankers also cause sap rots. There are 36 orders in this subdivision (Fig. 3-12).

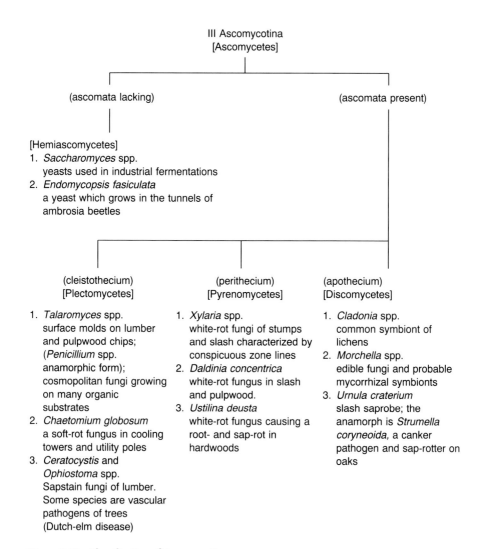

Figure 3-12 Classification of Ascomycotina.

Deuteromycotina (Deuteromycetes or Fungi Imperfecti)

This subdivision contains about 17,000 species. It is characterized by the absence of a sexual stage. Most deuteromycetes are believed to be anamorphs of ascomycetes. The subdivision includes many important plant pathogens, sap-stain fungi, and soft-rotters.

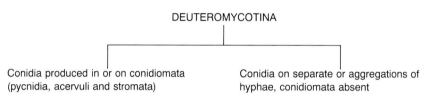

DEUTEROMYCOTINA

Conidia produced in or on conidiomata (pycnidia, acervuli and stromata)

COELOMYCETES

1. *Cytospora* spp.
   stem cankers, sap-rots on stems and brown stain in conifer logs
2. *Mycosphaerella dearnessii*
   brown-rot needle disease of pines
3. *Sphaeropsis (Diplodia) sapinea*
   common stain fungus on lumber
4. *Phoma* spp.
   common plant pathogens

Conidia on separate or aggregations of hyphae, conidiomata absent

HYPHOMYCETES

1. *Hormoconis resinae*
   a stain fungus recently isolated from resinous and creosote-treated wood
2. *Trichoderma* spp.
   common molds on wet lumber; some species are antagonistic to other fungi
3. *Scytalidium lignicola*
   a soft-rot fungus on utility poles; some isolates are antagonistic to other fungi
4. *Phialophora* spp.
   common soft-rot fungi; tolerant of some wood preservatives
5. *Aureobasidium pullulans*
   a principal cause of mildew on painted surfaces and a major blue stain agent

Figure 3-13 Classification of Deuteromycotina.

Basidiomycotina

Subdivision Basidiomycotina is characterized by the basidium and formation of exogenous, perfect spores termed basidiospores. The class Hymenomycetes contains many of the important wood decay fungi. The other classes are the Gasteromycetes, which include some lignicolous fungi and the Teliomycetes, which include the rusts and smuts. The rust and smut fungi are all obligate plant pathogens and cause some major agricultural and forest tree (rusts only) diseases.

Only the Hymenomycetes will be discussed further because of their importance as wood-decay agents.

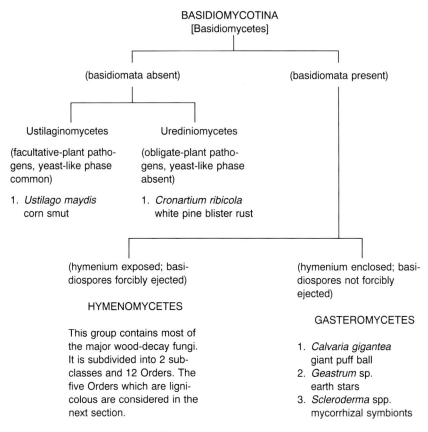

Figure 3-14 Classification of Basidiomycotina.

Hymenomycetes

The Hymenomycetes are characterized by a hymenium borne on the surface of a basidioma. This class is subdivided into two classes based on the presence or absence of primary septa in the basidia.

In the Phragmobasidiomycetidae, the basidium is partially or completely divided by septa. Also, the basidiospores may germinate repetitively as in the rusts and smuts. The basidiomata are often waxy or gelatinous. This subclass contains three orders. Two of them, the Tremellales and Auriculariales, are lignicolous. The third order, Septobasidales, contains fungi that are in symbiotic–parasitic associations with scale insects on living plants.

The Holobasidiomycetidae, the second subclass, is characterized by a nonseptate basidium. It has nine orders, and three of them contain important lignicolous fungi and will be discussed further.

The Aphyllophorales are characterized by a hymenium that adheres firmly to the basidioma surface, which may be smooth, dentate, or poroid. This order contains most of the important wood-decay fungi.

The Agaricales is characterized by a hymenium that detaches readily from the basidioma surface, which may be gills, lamellae, or tubes. Many of the fungi in this order are saprobes of forest litter or symbionts in mycorrhizal associations with higher plants. Some of the species are lignicolous.

The Dacrymycetales is characterized by unique basidia with stout, elongated sterigma that resemble a tuning fork. The basidiomata are waxy or gelatinous. Many of the species are lignicolous.

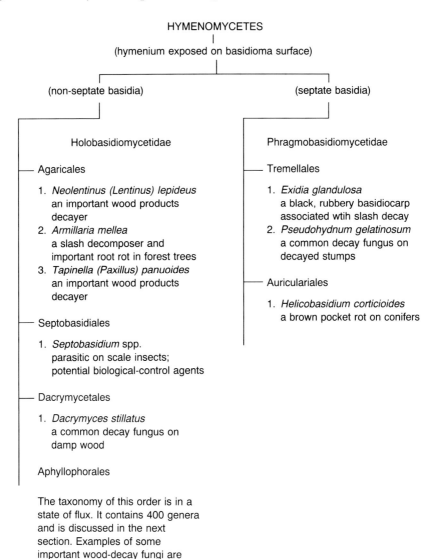

Figure 3-15 Classification of Hymenomycetes.

Aphyllophorales

This order represents a major revision of the Hymenomycetes as originally proposed by Fries (1874). The early groups were based primarily on the macroscopic features of the basidioma and shape of the hymenophore. The original important families of wood-inhabiting fungi were the Thelephoraceae, (leather fungi), hymenophore flat; Polyporaceae (poroid fungi), hymenophore tubes; Agaricaceae (mushrooms), hymenophore gills; the Hydnaceae (dentate fungi), hymenophore downward-directed teeth or projections; and the Clavariaceae (coral fungi), hymenophore upward-directed stalks or colloroid structures.

Corner (1932) studied basidiocarp structure and proposed a classification based on the presence of generative, skeletal, and binding hyphae in the basidioma. The basidiomata of the decay fungi could be grouped into the monomitic, dimitic, or trimitic structural categories discussed earlier in this chapter. Combinations of microscopic, enzymatic, and cytological studies on many of these decay groups indicated that the macroscopic basidiocarp features used in the Friesian system were not a natural grouping for many fungi. The order is in a state of taxonomic flux.

In the revisions, the five original families, listed above, have been redefined and narrowed, based on combinations of characters. The group is still under study, and additional revisions and grouping can be expected. Currently, the 400 genera are placed in 22 families, which are grouped in 6 hymenophore forms.

Some of the important families and their form groupings in the Aphyllophorales are listed below with their affinities to the older Friesian family categories in brackets.

Current Categories	Fries Categories
Thelephoroid forms	[Thelephoraceae]
Coniophoraceae	
Corticaceae	
Punctulariaceae	
Stereaceae	
Thelephoraceae	
Cupuloid forms	
Schizophyllaceae	
Clavaroid forms	[Clavariaceae]
Clavariaceae	
Sparassidaceae	
Hydnoid	[Hydnaceae]
Echinodontiaceae	
Hericiaceae	
Hydnaceae	

Poroid forms   [Polyporaceae]
  Bondarzewiaceae
  Fistulinaceae
  Ganodermataceae
  Hymenochaetaceae
  Polyporaceae

Descriptions and keys for the families in the Aphyllophorales are available in chapters by Talbot, Peterson, Harrison, and Pegler in Ainsworth et al. (1973). A two-volume textbook on the poroid fungi of North America by Gilbertson and Ryvarden (1986, 1987) provides taxonomic keys, macroscopic and microscopic descriptions, and family, genera, and species designations for most of the important members of this group in North America.

There may be some difficulties reconciling older, invalid names of important fungi in the literature with their new taxonomic designations. It is important that we do so, since the evidence is overwhelming that the new categories place many fungi in more natural affinities. A convenient source of current names is Farr et al. (1989). References to the taxonomic literature cited will generally lead to the new accepted name of a wood-decay fungus. Selected examples of the families, genera, and species of the major wood-decay fungi included in this book are listed to facilitate the transition. The older generic designations are placed in brackets.

Polyporaceae (100 or more genera)
  Bjerkandera adusta   [Polyporus]
  Antrodia serialis   [Trametes]
  Antrodia carbonica   [Poria]
  Postia placenta   [Poria]
  Trametes versicolor   [Polyporus]
  Fomitopsis cajanderi   [Fomes]
  Gloeophyllum sepiarium   [Lenzites]
  Irpex lacteus   [Polyporus]
  Hirschioporus abietinus   [Polyporus]
  Heterobasidion annosum   [Fomes]
  Laetiporus sulphureus   [Polyporus]
  Perenniporia subacida   [Poria]
  Cerrena unicolor   [Daedalea]
Hymenochaetaceae (25 genera)
  Inonotus obliquus   [Poria]
  Phellinus pini   [Fomes]
  Phellinus igniarius   [Fomes]
  Phaeolus schweinitzii   [Polyporus]
Ganodermataceae (3 genera)
  Ganoderma applanatum   [Fomes]
  Ganoderma tsugae   [Polyporus]

Corticiaceae (80 genera)
   Piloderma bicolor   [Corticium]
   Sistotrema brinkmannii   [Trechispora]
   Scytinostroma galactinum   [Corticium]
Stereaceae (18 genera)
   Xylobolus frustulatus   [Stereum]
   Amylostereum chailletii   [Stereum]
Coniophoraceae (8 genera)
   Coniophora puteana
   Serpula lacrimans   [Merulius]

A Classification of Bacteria

According to current information, bacteria play only minor roles in the decomposition of wood. This section will be limited primarily to descriptions of the several groups of bacteria in which representatives are known to modify wood or attack plant cell-wall constituents. Detailed information on the kinds of bacteria, their structures and functions, and roles in disease, industry, and the biosphere are available in textbooks on bacteriology or microbiology (e.g., Brock *et al.*, 1984; Nester *et al.* 1983; Cano and Columé, 1986).

    The bacteria are characterized primarily by a prokaryotic cell type and are placed in a separate kingdom, the Monera. Bacteria include an amazing diversity of life forms that have adapted to a wide range of habitats and conditions. There are three major groups of bacteria, based on their energy sources. The *photosynthetic bacteria* utilize light as an energy source for life processes. The *lithotrophic bacteria* obtain their energy from the oxidation of inorganic compounds such as sulphur, ammonia, and hydrogen. The *heterotrophic bacteria* are the third group and obtain their energy from the oxidation of organic compounds. This group plays a complementary role with the fungi as the major decomposers of carbon compounds in the biosphere.

    Bacteria are grouped into genera and species based on morphological, physiological, and biochemical characteristics. *Bergey's Manual of Determinative Bacteriology* (Buchanan and Gibbons, 1974) was the classic publication used for classifyiung and identifying bacteria. A four-volume revision titled *Bergey's Manual of Systematic Bacteriology* (Holt, 1989) takes into account new information on phylogeny among bacteria provided by new molecular and genetic approaches to bacterial taxonomy. These new approaches will eventually provide insights into the phylogeny of bacteria and even hints about the origin of life. However, major reliance for groupings and identification is still placed on traditional morphological, physiological, and biochemical features. Some important morphological bacterial features are the shapes and arrangements of the cells (round, rods, indeterminant, spiral), cell wall-staining properties (gram-positive and gram-negative), the absence or presence and positions of flagella, and the presence or absence of endospores. An important physiological feature is whether the bacterium is aerobic, anaerobic, or microaerobic. The many biochemical features reflect the nutritional requirements, production of specific enzymes, or reactions to various dyes, chemicals, and antibiotics. Hosts, substrates, and locales are also important for identifying some species.

Many bacteria can decompose cellulose in diverse environmental settings such as sea water, soils, and alimentary tracts (Schmidt et al., 1987; Greaves, 1968, 1970; Daniel and Nilsson, 1986; Drysdale et al., 1986; Holt et al., 1979; Rossell et al., 1973). However, the bacterial species that invade and decompose various wood constitutents in living trees and wood products are limited. This scarcity of species reflects our state of knowledge, since traditional microbiology has focused primarily on bacteria important in medicine, agriculture, and the fermentation industry.

Some groups of bacteria of special interest to wood microbiologists are listed and described. The group designations are used as in *Bergey's Manual of Determinative Bacteriology* (Buchanan and Gibbons, 1974).

Part 2: Cytophagales
Small, rod-shaped, chemo-organotrophs, without flagella, yet possess the unusual property of gliding motion. They are morphologically similar to many gram-negative bacteria. *Cytophaga* is an important genus in the group and contains cellulosic decomposers. An interesting feature of various *Cytophaga* spp. is the regular alignment of the bacterial cells on the fiber surface oriented parallel to microfibrillar alignments of the cell wall. *Sporocytophaga myxoccoides* is reported to cause severe soft-rot damage to wood exposed to ocean waters.

Part 8: Gram-negative facultatively anerobic rods
Erwinia nimipressuralis is reported to be associated with wetwood formation in elms, maples, willows, and poplars. In its metabolism, methane is formed, which generates considerable pressure in stems and is partially responsible for copious exudate flows.

Part 7: Gram-negative, aerobic rods, and cocci
Pseudomonas spp. are reported as the initial invaders in stem wounds in forest trees. It has been proposed that they may play a role in the successional aspects of decay development in stems.

Part 15: Endospore-forming rods and cocci. Family I: Bacillaceae
Bacillus polymyxa is reported to decompose the margo fibrils and torus in the bordered pits of submerged pine logs. The damage to the pits is substantial and results in increased penetrability of the wood to liquids (Knuth and McCoy, 1962).

Clostridium quericcolum is reported as the cause of discoloration in living oaks. Many of the cellulolytic anaerobic bacteria in wet conditions are in the genus *Clostridium.*

Part 17: Actinomycetes and related organisms

Coryneform humiferan is associated with wet-wood formation in poplars.

A *Micromonospora* sp. has been isolated from the alimentary tract of termites and is assumed to be involved in cellulose decomposition.

Nocardia spp. and *Streptomyces* spp. are associated with soft-rot development in pines submerged in marine environment.

Roles of Fungi and Bacteria in Ecosystems and Human Affairs

In a book focusing on wood decay and property changes caused by fungi and bacteria, it is natural to equate them primarily with their well-known negative roles of disease, death, and deterioration. It seems appropriate, at this point, to briefly mention some of the broad, positive roles that microorganisms play in ecosystem stability and human affairs for a more balanced perspective. The industrial aspects of these roles are discussed in detail in Chapter 20, "Wood and Biotechnology."

- *Carbon cycle* Fungi and bacteria are the principal decomposers of organic materials, and release to the atmosphere the carbon dioxide that is indispensible for the continued photosynthetic activity of green plants. It is sobering to recognize that this fragile, balanced carbon cycle controls the major energy input to most living systems. The removal of accumulating organic debris is another part of this process. Soil fertility is improved, and humus, an important soil constituent, is formed.
- *Nitrogen cycle* Bacteria play a major role in providing available nitrogen for green plants through nitrogen fixation and cycling.
- *Mutualistic symbionts* Fungi and bacteria are beneficial symbionts with a wide range of other life forms. Symbiotic relationships include the mycorrhizal fungi with tree roots and agricultural plants, fungi with algae in lichens, and nitrogen-fixing bacteria in the root nodules of legumes. The symbiotic activity of bacteria and fungi in the digestive tract of ruminant animals facilitates cellulose decomposition.
- *Food sources* Fungi and bacteria are the principal food sources for a wide range of microfauna including protozoa, ambrosia beetles, and composting ants. Humans consume an estimated 60 to 70 millions pounds of mushrooms per year.
- *Industrial chemicals and medicinals* Fungi and bacteria in controlled

fermentation or transformation processes produce a wide range of useful chemicals and foods, including yeasts, vitamins, alcohols, glycerols, citric acid, enzymes, cortisone, antibiotics, and dairy products. The unique metabolic capabilities of some bacteria and fungi are the basis of a multibillion dollar chemical industry. This industry will grow at a tremendous rate as we learn more precisely to exploit the genetic potential of microorganisms.
- *Biological control of diseases and pests (potential)* Fungi and bacteria are pathogens for some insect pests and are antagonistic to other pathogens or harmful saprobes. Increasing use is being made of them as biological control agents.
- *Biodegradation agents* Fungi and bacteria play major roles in detoxification of industrial poisons and in the breakdown of sewage sludges, garbage, and other wastes into industrially useful or harmless products. They also have great promise for cheaply and effectively modifying materials for subsequent industrial use, e.g., increasing wood permeability for preservative treatment or biological delignification of pulpwood.

Summary

- Fungi are characterized by filamentous eukaryotic cells, a heterotrophic nature, an external mode of digestion, and reproduction by spores. Many fungi have chitin as the skeletal framework of the cell walls. These features now place fungi in a separate kingdom, the Fungi, and change prior views of them as primitive plants.
- The tubular cells of fungi, termed hyphae, are the basic cellular unit of all fungal structures. The *hyphal system* of many fungi appears to be uniquely adapted to penetrate, externally digest, absorb, and metabolize a wide range of plant materials, including wood.
- Fungi reproduce primarily by spores, which are one- to several-celled structures formed from hyphae. Many of the lignicolous fungi produce huge numbers of spores that, coupled with effective dissemination agents, such as wind, water, and insect vectors, assure that most wood substrates are continuously exposed to wood-inhabiting fungi.
- The high reproductive potential of fungi, their short life cycles and great variability provide many fungal species the ability to adapt readily to changing conditions, such as new fungicides or resistant varieties of plants. This large reproductive potential and high variability also offers great potential for improving the capabilities of fungi in biotechnology developments and biological control programs.

Summary continues

- The negative impacts of fungi as disease (pathogen) and biodeterioration agents (saprobes) are well known. Some of the advantageous roles of fungi are carbon and nitrogen cycling in the ecosystem, as symbionts, as agents in production of industrial chemicals and pharmaceuticals, as biological control agents, and as potentially important biodegradation or biotransformation agents.
- An important first step in the understanding and prevention or control of many biodeterioration problems is the identification of the fungi involved.
- The classification of fungi is based primarily on the type and various features of the spore-bearing structures. Early classifications of lignicolous fungi placed major emphasis on the macroscopic features of spore-bearing structures. Recent classification schemes take into account new information on the microstructural, genetic, and biochemical features of fungi. The three major classes of wood-inhabiting fungi are the Ascomycotina, the Deuteromycotina, and the Basidiomycotina. The mode of conidium formation and spore-bearing has become important in the groupings of the Deuteromycotina. The hyphal structure of basidiocarps is an important character in the new groupings of the wood-inhabiting Basidiomycotina.
- In the Basidiomycotina, the order Aphyllophorales contains many of the important wood-decaying fungi. Important families in the order are the Thelephoraceae, Stereaceae, Corticiaceae, Hydnaceae, Polyporaceae, and Hymenochaetaceae. The Polyporaceae contain more than a hundred genera, including many of the fungi causing decay in wood products.
- Key literature sources and references are provided in the text to facilitate the identification and correct naming of the important lignicolous fungi.
- Information on the types and importance of bacteria that invade and damage wood and its products is very limited. We speculate this reflects a failure to thoroughly examine the actual roles of bacteria and their portable interactions with fungi and other organisms in the biodeterioration of wood.

References
Characteristics of Fungi

Aronson, J. M. (1965). The cell wall. In "The Fungi I" (G. C. Ainsworth and A. S. Sussman, eds.), pp. 49–76. Academic Press, New York.

Ballou, C. E. (1976). Structure and biosynthesis of yeast mannan. *Advances in Microbiological Physiology* 14:93–158.

Bartnicki-Garcia, S. (1968). Cell wall chemistry, morphogenesis, and taxonomy of fungi. *Annual Review of Microbiology* **22**:87–105.
Beckett, A., I. B. Heath, and D. J. McLaughlin. (1974). "An Atlas of Fungal Ultrastructure." Longman, London.
Bracker, C. E. (1967). Ultrastructure of fungi. *Annual Review of Phytopathology* **5**: 343–374.
Davidson, R. W., W. A. Campbell, and D. B. Vaughn. (1942). "Fungi Causing Decay of Living Oaks in the Eastern United States and Their Cultural Identification." U.S.D.A. Bulletin No. 785. Washington, D.C.
Griffin, D. H. (1981). "Fungal Physiology." John Wiley, New York.
Grove, S. N., and C. E. Bracker. (1970). Protoplasmic organization of hyphal tips among fungi. *Journal of Bacteriology* **104**(2):989–1009.
Hunsley, D., and Burnett, J. H. (1979). The ultrastructural architecture of the walls of some hyphal fungi. *Journal of General Microbiology* **62**:203–218.
Lindenmayer. A. (1965). Carbohydrate metabolism. 3. Terminal oxidation and electron transport. *In* "The Fungi" (G. C. Ainsworth and A. S. Sussman, eds.), Vol. 1, pp. 301–348. Academic Press, New York.
Lü, H. S. and D. J. McLaughlin. (1991). Ultrastructure of the septal pore apparatus and early septum initiation in *Auricularia auricula-judae*. *Mycologia* **83**(3):322–334.
Margulis, L. (1981). "Symbiosis in Cell Evolution." W. H. Freeman, New York.
Moore, R. T. (1965). The ultrastructure of fungal cells. *In* "The Fungi" (G. C. Ainsworth and A. S. Sussman, eds.), Vol. I, pp. 95–118. Academic Press, New York.
Nobles, M. K. (1965). Identification of cultures of wood-inhabiting Hymenomycetes. *Canadian Journal of Botany* **43**:1097–1139.
Raper, C. A. (1978). Control of development by the incompatibility system in Basidiomycetes. *In* "Genetics and Morphogenesis in the Basidiomycetes" (M. N. Schwalb and P. G. Miles, eds.), pp. 168. Academic Press, New York.
Raper, J. R. (1966). Life cycles, basic patterns of sexuality, and sexual mechanisms. *In* "The Fungi: An Advanced Treatise" (G. C. Ainsworth and A. S. Sussman, eds.), Vol. II, pp. 473–511. Academic Press, New York and London.
Raven, P. H., R. F. Evert, and S. E. Eichhorn. (1986). "Biology of Plants," 4th Ed. Worth Publishers, New York.
Sexton, C. M. (1988). The development and use of breaking radius and impact bending tests for measuring wood strength losses caused by Basidiomycetes isolated from air-seasoning Douglas fir. M. S. thesis. Oregon State University, Corvallis, Oregon.
Smith, A. H. (1934). Investigations of two-spored forms of the genus *Mycena*. *Mycologia* **26**:305–331.
Stalpers, J. A. (1978). "Identification of Wood-Inhabiting Aphyllophorales in Pure Culture. Studies in Mycology No. 16, Centraalbureau voor Schimmelcultures, Baarn, The Netherlands.
Swift, M. J. (1973). The estimation of mycelial biomass by determination of the hexosamine content of wood tissue decayed by fungi. *Soil Biology and Biochemistry* **5**:321–332.
Wessels, J. G. H., and J. H. Sietsma. (1973). Wall structure and growth in *Schizophyllum commune*. *In* "Fungal Walls and Hyphal Growth" (J. H. Burnett and A. P. J. Trinci, eds.), pp. 417. Cambridge University Press, London.
Whittaker, R. H. (1969). New concepts of kingdoms of organisms. *Science* **163**: 150–160.

Wu, L. C., and M. A. Stahmann. (1975). Chromatographic estimation of fungal mass in plant materials. *Phytopathology* **65**:1032-1034.

Classification of Fungi

Ainsworth, G. C., P. W. James, and D. L. Hawksworth. (1971). Ainsworth & Bisby's Dictionary of the Fungi. 6th ed., Commonwealth Mycological Institute, Kew, Surrey, England.

Ainsworth, G. C., F. K. Sparrow, and A. S. Sussman. (1973). "The Fungi, An Advanced Treatise. Volume IVB. A Taxonomic Review with Keys: Basidiomycetes and Lower Fungi." Academic Press, New York.

Alexopoulos, C. J., and C. W. Mims. (1979). "Introductory Mycology," 3rd Ed. John Wiley, New York.

Corner, E. J. H. (1932). A *Fomes* with two systems of hyphae. *Transactions of the British Mycological Society* **14**:263-291.

Corner, E. J. H. (1950). A Monograph of *Clavaria* and Allied genera. Oxford University Press.

Farr, D. F., G. F. Bills, G. P. Chamuris and A. Y. Rossman. (1989). "Fungi on Plants and Plant Products in the United States." American Phytopathological Press, St. Paul, Minnesota.

Fries, E. M. (1874). "Hymenomycetes Europaei." Uppsala, Sweden.

Gilbertson, R. L., and L. Ryvarden. (1986 and 1987). "North American Polypores." Vol. I, 1986, and II. 1987 Fungiflora, Oslo, Norway.

Hawksworth, D. L., B. C. Sutton, and G. C. Ainsworth. (1983). "Ainsworth & Bisby's Dictionary of the Fungi," 7th Ed. Commonwealth Mycological Institute, Kew, Surrey, England.

Hughes, S. J. (1953). Conidiophores, conidia, and classification. *Canadian Journal of Botany* **31**:577-659.

Hunt, J. (1956). Taxonomy of the Genus *Ceratocystis*. *Lloydia* **19**(1):1-58.

Kendrick, B. Ed. (1971). Taxonomy of Fungi Imperfecti, University of Toronto Press, Toronto, Ontario, Canada.

Lentz, P. L. (1955). "*Stereum* and Allied Genera of Fungi in the Upper Mississippi Valley." Agricultural Monograph No. 24, U.S.D.A., Washington, D.C.

Lowe, J. L. (1942). "The Polyporaceae of New York State (except *Poria*)." New York State College of Forestry Technical Publication 60, Syracuse, New York.

Lowe, J. L. (1957). "Polyporaceae of North America. The genus *Fomes*." State University of College of Forestry at Syracuse University Technical Publication 80, Syracuse, New York.

Lowe, J. L. (1966). "Polyporaceae of North America. The genus *Poria*." State University College of Forestry at Syracuse University Technical Publication 90, Syracuse, New York.

Moore-Landecker, E. (1990). "Fundamentals of the Fungi." 3rd Ed. Prentice Hall, Englewood Cliffs, New Jersey.

Slysh, A. (1960). "The Genus *Peniophora* in New York State and Adjacent Regions." State University College of Forestry at Syracuse University Technical Publication 83, Syracuse, New York.

Wang, C. J. K. (1965). "Fungi of Pulp and Paper in New York." State University College of Forestry at Syracuse University Technical Publication 87, Syracuse, New York.

Wang, C. J. K., and R. A. Zabel (ed.) (1990). "Identification Manual for Fungi from Utility Poles in the Eastern United States." American Type Culture Collection, Rockville, Maryland.
Webster, J. (1980). "Introduction to Fungi," 2nd Ed. Cambridge University Press, Cambridge.

Characteristics and Classification of Bacteria

Brock, T. D., D. W. Smith, and M. T. Madigan. (1984). "Biology of Microorganisms," 4th Ed. Prentice-Hall, New York.
Buchanan, R. E., and N. E. Gibbons. (ed.) (1974). "Bergey's Manual of Determinative Bacteriology," 8th Ed. Williams and Wilkins, Baltimore, Maryland.
Cano, R. J., and J. S. Colomé. (1986). "Microbiology." West Publishing Co., Los Angeles, California.
Daniel, G., and T. Nilsson. (1986). "Ultrastructural Observations on Wood-Degrading Erosion Bacteria." International, Research Group on Wood Preservation IRG/WP/1283. Stockholm, Sweden
Drysdale, J. A., P. J. Rutland, and J. A. Butcher (1986). "Isolation and Identification of Bacteria from Degraded Wood—a Progress Report. International Research Group on Wood Preservation. IRG/WP/192. Stockholm, Sweden
Greaves, H. (1970). The effect of selected bacteria and actinomycetes on the decay capacity of some wood rotting fungi. *Material und Organismen* **5**:265-279.
Greaves, H. (1968). Occurrence of bacterial decay in copper-chrome-arsenic treated wood. *Applied Microbiology* **16**:1599-1601.
Holt, J. G. (ed.). (1977). "The Shorter Bergey's Manual of Determinative Bacteriology, 8th Ed. Williams & Wilkins, Baltimore, Maryland
Holt, G. (editor-in-chief). (1989). "Bergey's Manual of Systematic Bacteriology. Vol. 1 (N. R. Krieg, ed.); Vol. 2 (P. H. A. Sneath, ed.); Vol. 3 (J. T. Staley, ed.); and Vol. 4 (S. T. Williams, ed.). Williams & Wilkins, Baltimore, Maryland
Holt, D. M., E. B. Gareth Jones, and S. E. J. Furtado. (1979). Bacterial breakdown of wood in aquatic habitats. *Record of the British Wood Preservers' Association Annual Convention,* 13-22.
Knuth, D. T., E. McCoy. (1962). Bacterial deterioration of pine logs in pond storage. *Forest Products Journal* **12**(9):437-442.
Nester, E. W., E. E. Roberts, M. E. Lidstrom, N. N. Pearsall, and M. T. Nester. (1983). "Microbiology," 3rd Ed. Dryden Press, Hinsdale, Illinois.
Rossell, S. E., E. G. M. Abbot, and J. F. Levy. (1973). A review of the literature relating to the presence, action, and interaction of bacteria in wood. *Journal of the Institute of Wood Science* **32**:28-35.
Schmidt, O., Y. Nagashima., W. Liese, and U. Schmidt. (1987). Bacterial wood degradation studies under laboratory conditions and in lakes. *Holzforschung* **41**(3): 137-140.

CHAPTER

4

Factors Affecting the Growth and Survival of Fungi in Wood (Fungal Ecology)

In a practical sense, the factors that affect the growth and development of fungi in wood substrates can generally be equated with the development of decay or discolorations. When the critical growth requisites of a wood-damaging fungus are known, it is sometimes feasible to modify wood handling or use practices to adversely affect fungus growth, thereby achieving prevention or economic control of the problem. Information on the growth needs of fungi is also useful to optimize cultural conditions for industrial fermentations or other emerging biotechnology processes.

The many chemical and physical factors that affect fungal and bacterial growth are covered in more detail in textbooks on fungal physiology (Lilly and Barnett, 1951; Cochran, 1958; and Griffin, 1981) and microbiology (Stanier et al., 1986; Brock et al., 1984). This chapter emphasizes the growth factors affecting decay fungi, placing emphasis on their limits and citing examples where changes in wood handling or use practices have achieved or may achieve useful control of biodeterioration problems. The following chapter on metabolic activities of fungi demonstrates that most environmental factors affecting fungal growth are ultimately traceable to enzyme characteristics and reactions.

Major Growth Needs of Wood-Inhabiting Fungi

Like all living organisms, fungi have certain requirements for growth and survival. The major growth needs of wood-inhabiting fungi are

1. Water—free water on the surfaces of cell lumina;
2. Oxygen—atmospheric oxygen at relatively low levels for most fungi

and very low levels or chemical oxygen only for some microaerobic and facultative anaerobic fungi;
3. A favorable temperature range—optima for most wood-inhabiting fungi range from 15 to 45°C;
4. A digestible substrate (wood, etc.)—provides energy and metabolites for synthesis via metabolism;
5. A favorable pH range—optima for most wood-inhabiting fungi range from pH 3 to 6; and
6. Chemical growth factors—nitrogen compounds, vitamins, and essential elements.

The last two factors are often included with the substrate. The absence of toxic extractives, although not a requirement, is necessary for growth by most fungi on wood. Visible light is needed by some fungi for the development of spore-bearing structures and may play a role in other physiologic functions. High levels of ultraviolet light are lethal to most fungi.

Water

Water serves four general purposes for fungal growth in wood. Removing moisture or preventing wetting are simple decay-prevention or -control practices in many wood uses.

1. *Reactant in hydrolysis.* Decay involves the exocellular digestion of plant cell-wall carbohydrates by hydrolytic enzymes released by fungi. Water is one of the reactants in the chemical process. If R is used to designate simple sugar monomers connected by glycosidic bonds, forming carbohydrate polymers, the reactions involved in hydrolysis can be simplified as follows:

   Endwise attack (one or two sugar units)

$$\underset{\text{polymer fragment}}{\overset{\downarrow}{R-O-R-O-R-O-R\ldots}} + H_2O \xrightarrow[\text{enzyme}]{\text{hydrolytic}} \underset{\substack{\text{simple}\\\text{sugar}}}{R-OH} + \underset{\text{polymer fragment}}{HO-R-O-R-O-R\ldots}$$

   Random attack

$$\underset{\text{Polymer fragment}}{\overset{\downarrow}{R-O-R-O-R-O-R\ldots}} + H_2O \xrightarrow[\text{enzyme}]{\text{hydrolytic}} \underset{\text{polymer fragments}}{R-O-R-O-R-OH + HO-R\ldots}$$

   The reactions can release smaller fragments of the original polymer or glucose units. The digestion of a cellobiose (2 glucose units) in this hydrolytic reaction by a hydrolytic enzyme (β-glycosidase), as illustrated below, occurs by migration of proton (H^+) and hydroxyl (OH^-) ions from water to the broken glycosidic bond, forming two glucose molecules.

SITE OF ENZYME ACTION

GLUCOSE — GLUCOSE

2. *Diffusion medium for enzymes and solubilized substrate molecules.* Decay is an exocellular process and occurs outside the hyphal cell wall. Free (liquid) water is needed by most wood-inhabiting fungi both as a solvent and diffusion medium for the digestive enzymes (hydrolytic) leaving the hyphae and the solubilized substrate decomposition products subsequently absorbed by the hyphae.
3. *Solvent or medium for life systems.* Water is the indispensable solvent for the many enzymes within the fungal hyphae that are involved in cell metabolic processes such as respiration, synthesis, and growth. It is also the medium supporting colloidal suspensions, which are critical to the structure and functioning of protoplasm necessary for all life forms.
4. *Wood-capillary swelling agent.* In dry wood, water molecules are attracted by hydrogen bonding to the hydroxyl groups on the carbohydrate macromolecules making up the cell walls. As the wood adsorbs water, it swells primarily in the radial and tangential planes. During swelling at the ultrastructural level, small capillaries in the cell wall enlarge and attain a size range permitting the penetration of both free water and fungal enzymes deep into the so-called *transient capillary zone* of the cell wall. This process greatly facilitates the penetration of digestive enzymes (large protein macromolecules) into the cell wall at the ultrastructural level.

A long experience with wood in its many uses indicates that dry wood in protected environments or water-saturated wood seldom decays. So the important questions to many users of wood have been to know the critical wood-moisture limits when decay begins or stops, and how varying amounts of water in wood affect the rate of decay development. These are difficult questions since moisture gradients exist in wood from outer to inner zones; it is difficult to maintain precise moisture contents in wood test samples during the lengthy decay process. Many fungi release metabolic water during the decay process, and there is considerable fungal variability in responses to moisture limits. Also, oxygen and carbon dioxide levels often become confounding factors at the upper moisture-content ranges.

The moisture content of wood is generally expressed as a percentage of the oven-dry weight. This value is used in a physical property sense to indicate the total amount of water present in the wood. Fungal physiologists and wood microbiologists are also interested in the availability of the water in the wood to fungi. In wood, water is tightly adsorbed to cell-wall polymers and unavailable to most fungi at levels below the fiber saturation point. Units used to express the availability of water in wood to a fungus are osmotic pressure, the bar, or Pascal. A bar is defined as 0.987 atmospheres or 0.1 MPa (mega Pascal). Another useful unit is *water activity*, which is defined simply as the vapor pressure of water over the substrate divided by the vapor pressure of pure water (Griffin, 1981). By this measure, wood reaches essentially the fiber saturation point when stored in a confined space in a water-vapor-saturated atmosphere. A more detailed discussion of various wood–water relationships is presented in Chapter 6.

Minimal Moisture Concentrations There are many older reports on the minimal moisture levels in wood necessary to sustain fungus growth or decay development. These are briefly listed in Table 4-1 as to fungi, moisture levels, and sources.

Recently, Griffin (1977) analyzed the water needs of wood-inhabiting fungi for decay initiation, in terms of *water potential*. Water potential is a measurement of available water in soil–water–plant root studies that may be analogous to fungal hyphae–wood–water relationships. Water potential is defined as the sum of the matrix and osmotic potential of the wood–water system expressed in bar units. In a simplified sense, water potential measures the relative availability of the water in the wood to the passive osmotic capabilities

Table 4-1
Minimal moisture levels for decay development

Fungi	Moisture level	Source
Wood-decay fungi	25–32%	Snell, et al. (1925)
Xylobolus (Stereum) frustulatus	16–17%	Bavendamm and Reichelt (1938)
Schizophyllum commune	16–17%	Bavendamm and Reichelt (1938)
Antrodia sinuosa	26%	Freyfeld (1939)
Common wood-decay fungi	22–24% (20% as safe control level)	Cartwright and Findlay (1958)
Ophiostoma (Ceratocystis) piliferum	23%	Lindgren (1942)
Wood-inhabiting fungi	Slightly above the fiber saturation point	Etheridge (1957)

of the fungal hypha in contact with it. Matrix potential depends on the dimensions in the transient capillary system and reflects the strong adhesion forces that retain free water in small capillaries. The osmotic potential depends on the amounts and types of solutes (sugars, extractives, etc.) in the free water in the wood. Griffin defined the fiber saturation point of wood as the condition when all voids of a radius greater than 1.5 µm are devoid of water. He stated that the limiting growth level for fungi is -40 bar, which is consistent with the 30% value often given as the approximate lower limit for decay. He also attests to the extreme difficulty of measuring these values experimentally.

From these data, it is clear that to grow effectively in wood, fungi must have some free water present in the cell lumen. This means that fungi are unable to grow effectively in wood below the fiber-saturation point (f.s.p.). The f.s.p. varies with wood species, but an average figure of 28 to 30% is generally assumed. Since dry (seasoned) wood in structures protected from external water sources can not exceed the f.s.p. (see Chapter 6), a cardinal principle for decay control in wood uses in protected environments appears. As an added safety factor, the prudent decay-control rule prescribed 30 years ago by Cartwright and Findlay (1958), which is still followed by many wood users, is that *to prevent the growth of microorganisms, wood moisture content may not exceed 20% based on oven-dry weight.*

Maximal Moisture Concentrations Most wood-inhabiting fungi are unable to grow effectively in water-saturated wood. A principal reason is that these fungi are obligate aerobic organisms and require moderate amounts of oxygen for respiration. As moisture content increases above the f.s.p., water steadily replaces the air in the cell lumina (atmosphere—21% oxygen), and eventually oxygen becomes limiting. Since the void volume of wood varies inversely with specific gravity, the upper moisture levels limiting fungal growth will be much lower in the high-density woods (Fig. 4-1). Fungi vary in sensitivity to oxygen depletion. For many practical purposes, reducing the void volume of the wood to about 20% of the original void volume causes a rapid reduction in decay rate. Water-saturated wood is resistant to fungal and insect attack and may be stored safely for a number of years; however, more tolerant organisms, including bacteria, may begin to degrade the wood.

Free water contains only a few parts per million of oxygen, and the rate of oxygen diffusion into large wood units such as logs is very slow and insufficient to support active fungal growth. The slow rate of oxygen movement into large wood members may help explain why ponding or continuous water spraying are effective storage methods for veneer bolts and high-quality saw logs. It should be stressed that under these conditions, most fungi will survive but not grow actively. Some soft-rot fungi decay wood submerged in the ocean or wood exposed continuously to near-saturated conditions, such as the baffles in cooling towers or subterranean surfaces of utility poles on wet sites. Some bacteria invade submerged wood and may cause extensive damage to

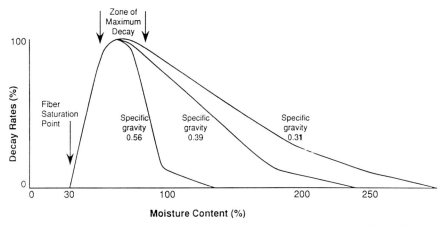

Figure 4-1 An approximation of the cardinal moisture contents for the rates of decay development of several common mesophilic basidiomycete decay fungi as judged from reports in the literature and experience. Maximal moisture contents are calculated based on data from Higgins (1957) and Skaar (1972).

parenchymatous tissue and cell-wall pit membranes. Anaerobic bacteria are also associated with wet-wood formation in living trees.

Optimal Moisture Concentrations Optimal moisture levels for the growth of many wood-inhabiting fungi are not known, but experience with laboratory decay tests suggest that the optimal wood-moisture levels for most decay fungi lie between 40 and 80% (Scheffer, 1973). The optimal moisture values probably vary considerably, and, when known, may explain some of the substrate or condition specificities of fungal types and species. For example, laboratory decay tests with white-rot fungi require more moisture than those with brown-rot fungi to achieve optimal wood-weight losses (Highley and Scheffer, 1970). The differences may partially explain the prevalence of white-rot fungi in wet coniferous chip piles in contrast to the high levels of brown-rot fungi found in drier coniferous slash in the forest or aboveground uses in structures.

Survival at Various Moisture Levels The effect of wood moisture content on the survival of the major types and species of wood-inhabiting fungi remains poorly defined. The survival of heart rot, early decay, or stain fungi during storage or seasoning in the wood product in service may be important in wood uses such as house siding or utility poles where intermittent wetting occurs. In studies on the survival of many wood-decay fungi in ponded pulpwood bolts (*Pinus resinosa*), Schmitz and Kaufert (1938) reported no visible growth, although fungi survived for up to 38 weeks.

Scheffer and Chidester (1948) evaluated the survival of both decay and stain fungi in air-seasoned wood. Boards with viable heart-rot fungi or invaded by the decay or stain test fungi were stored at 27°C and 65% relative humidity. These conditions would bring the boards to an equilibrium moisture content of approximately 12%. The heart-rot fungi died within a year. *Phellinus (Trametes) pini* and *Phaeolus schweinitzii*, two common heart-rot fungi in conifers, survived 6 months or less. Some of the important wood-products decayers, such as *Postia placenta, Gloeophyllum trabeum, Fomitopsis (Fomes) rosea,* and *G. sepiarium,* were still alive after 3 years. *Meruliporia (Poria) incrassata,* an important building rot fungus, died within 25 days. Seven of the eight sapstain fungi died within 7 months. The single exception was *Aureobasidium pullulans,* which survived 1 year. This fungus is a common sapstainer in window sashes and a major cause of the *mildew* (pigmented mycelium) that often disfigures paint surfaces. Survival at low wood-moisture levels may explain its common presence in these two wood uses.

Many fungi appear to invade timber during the period between cutting and final use of the product. Moisture content will play a critical role in the species that colonize the wood and the extent to which these fungi utilize the substrate.

Oxygen

Most fungi are *obligate aerobes* that require free oxygen for several metabolic reactions involving energy release or synthesis. The process of *aerobic respiration* uses atmospheric oxygen as a reactant. The classic summation equation for respiration indicating direct oxygen use is as follows:

$$C_6H_{12}O_6 + 6\ O_2 \rightarrow 6\ CO_2 + 6\ H_2O + \text{energy}\ (\Delta G°—686\ \text{kcal/mol})$$

In the terminal respiration reactions, atmospheric oxygen serves as an electron and proton acceptor. It is important to note that the sugar (glucose) is completely oxidized and a maximum amount of chemical energy is released. Oxygen is also a direct reactant where oxidase enzymes are involved.

Some fungi, such as the yeasts, are able to obtain energy directly from chemical compounds by a series of internal redox reactions in a process termed *fermentation,* when oxygen levels are low or oxygen is absent. Organisms capable of fermentation are called *facultative anaerobes.* The classic simplified fermentation equation is as follows:

$$C_6H_{12}O_6 \rightarrow 2\ C_2H_5OH + 2\ CO_2 + \text{energy}\ (\Delta G°—57\ \text{kcal/mol})$$

In this case, the sugar (glucose) is only partially metabolized, and the process is energy inefficient.

Some bacteria and fungi are *obligate anaerobes* and require no oxygen. *Anaerobic respiration* uses chemical radicals such as NO_3, SO_4, or CO_3 to serve as the electron acceptors in the energy-release process. According to evolutionists, primitive organisms developed anaerobic respiration and fermentation as energy-release processes long before the accumulation of oxygen in the atmosphere as a by-product of photosynthesis by bacteria and plants.

Whereas most of the cellulose and related carbohydrates in the biosphere are oxidized by aerobic microorganisms (principally decay fungi) and are emphasized in this section, it should be remembered that an estimated 5 to 10% of the cellulose produced annually in the biosphere is converted to methane in the absence of oxygen in anaerobic environments such as animal rumens, sediments, and bogs (Vogels, 1979; Ljungdahl and Eriksson, 1986). These anaerobic reactions probably involve several species of bacteria in fermentation sequences, as summarized by the formula:

$$C_6H_{12}O_6 \rightarrow 3\ CH_4 + 3\ CO_2$$

Despite the current abundance of oxygen in the atmosphere (21%), these anaerobic processes are still of major importance in many specialized settings such as waterlogged soil, sediments, bogs, or specialized niches such as ruminant and termite digestion systems.

The effects of atmosphere (with emphasis on O_2 and CO_2) on the growth of fungi in culture has been studied intensively as an important variable in the industrial production of fungal products such as antibiotics, citric acid, and yeasts. A detailed review and extensive bibliography of this literature has been assembled by Tabak and Cooke (1968).

The amount of oxygen consumed by aerobic fungi is directly related to the amount of carbon dioxide produced. The ratio of the volume of CO_2 used to the oxygen consumed is termed the *respiration quotient* and is a useful unit in metabolic studies of fungi. For example, a respiration quotient of 1 would imply that the nutrient source was a simple sugar. Respirometry has been applied in several other interesting ways in wood microbiology. The amounts of carbon dioxide produced or oxygen consumed have been used to monitor decay development, to compare the decay rates of fungi on various substrates (Good and Darrah, 1967; Toole, 1972), and to evaluate wood preservatives (Halabisky and Ifju, 1968; Behr, 1972). Smith (1975) compared an automated analytical procedure to measure respiration and the soil-block test procedure to determine the wood-preservative possibilities of several agricultural fungicides with penta, chromated copper arsenate (CCA), and creosote. He concluded that the respiration method was quicker and provided reliable information in 4 weeks compared to the 12 weeks required by the standard soil-block preservative-evaluation procedure.

The amount of oxygen in the wood or a cultural flask is often simply expressed as the atmospheric pressure in millimeters (mm) of mercury. Conversion of pressure in atmospheres to the recently introduced bar or Pa (Pascal) can be made readily, since one bar equals 0.987 atmospheres or 0.1 MPa (mega Pascal). For example, wood in equilibrium with the atmosphere at sea level pressure would be assumed to contain oxygen at about 160 mm of pressure (21/100 × 760 mm Hg). Another useful measure is the ratio of the oxygen in the substrate to the amount or partial pressure of O_2 in the atmosphere, usually expressed as a percentage. In respirometry studies, the amounts of oxygen consumed or CO_2 formed are measured usually manometrically and expressed as volumes of gas per unit time (e.g., μl/hr).

There are many reports on the effects of atmospheric gases (primarily O_2 and CO_2) on the growth of decay fungi or decay rates (Scheffer and Livingston, 1937; Thacker and Good, 1952; Jensen, 1967, 1969a,b; Toole, 1972; Van der Kamp et al., 1979; Highley et al., 1983; and Scheffer, 1986). These reports indicate that some gaseous oxygen is required for the growth of fungi in virtually all cases of wood biodeterioration; however, the amounts are very low (probably in the range of 1%). Many fungi have optimal growth rates at oxygen levels above 20% of ambient air (Fig. 4-2).

Several papers of historical or special interest are briefly reviewed, with emphasis placed on the amounts of oxygen needed or inhibitory carbon dioxide levels.

Minimal Oxygen Concentrations   In 1929 Snell reported on a series of experiments on several commercial woods in a range of specific gravities to determine the optimal and inhibitory moisture levels for decay development. He concluded that when the accumulating free water in the wood had reduced the air content or void volume to about 20% of the original air volume, that decay development essentially ceased. Oxygen was assumed to be the limiting factor. This choice of 20% of the residual air volume as a decay-limiting level became a frequently cited statistic (Boyce, 1961).

Scheffer and Livingston (1937) determined the effects of mixtures of ni-

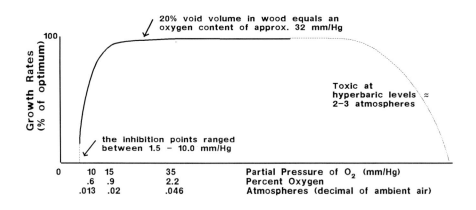

Oxygen Concentration in various units

Figure 4-2 The growth responses of *Trametes versicolor* to various concentrations of oxygen. The lower ranges are taken from the data of Scheffer (1986).

trogen and oxygen on the growth of *Trametes versicolor* in malt agar cultures. Growth first began to decline at an oxygen partial pressure of 37 mm Hg, decreased rapidly below 15 mm Hg, and ceased between 10 and 1.5 mm Hg (1.3 to 0.2% oxygen).

In studies on the effects of various concentrations of oxygen, nitrogen, and carbon dioxide on the biomass of several oak heart-rot fungi and *Trametes versicolor* grown in liquid cultures, Jensen (1967) showed that fungal biomass production decreased rapidly at an oxygen concentration of 15% and ceased in the absence of oxygen. Increasing levels of carbon dioxide acted as a growth inhibitor in oxygen and nitrogen mixtures (Fig. 4-3). Differences were also noted in the sensitivities of some of the fungi to varying concentrations of oxygen and carbon dioxide. Several heart-rot fungi were less sensitive to oxygen deficits and tolerant of the higher levels of carbon dioxide. Tolerance to low oxygen and high carbon dioxide levels in the heartwood zones of living trees may help explain the apparent selectivity of some fungi for this environment.

In subsequent studies, Highley *et al.* (1983) found that low O_2 levels (7.6 mm) and high CO_2 levels (76 mm) substantially reduced decay compared to ambient levels; however, there were no clear-cut differences between the sap-rot and heart-rot fungi. Oxygen levels also strongly influenced decay rates. Below 0.01 atmospheres of oxygen, only slight decay occurred, and it increased progressively as oxygen pressures rose to 0.1 and 0.2 atmospheres.

In a recent study on the minimal oxygen requirements for growth by 48 decay fungi and 6 sapwood-stain fungi, Scheffer (1986) reported that no fungi grew at oxygen levels of 1.5 mm Hg (approximately 0.2% volume). Fungal growth was moderately and then severely retarded at O_2 pressures of 11 mm and 2 mm Hg (1.3–0.3% volume), respectively.

Low oxygen levels in tightly stoppered decay chambers may be responsible for some of the puzzling variations observed in decay tests using similar procedures. In laboratory studies of wood decay, Eades and Roff (1953) found substantial decay reductions when tight rubber stoppers were used to seal the decay chambers. These rate variations were reduced significantly when uniform aeration or venting devices were used.

Optimal Oxygen Concentrations Oxygen-response curves for decay fungi indicate a sharp decrease in growth response between 1 and 2% oxygen (Fig. 4-2). Respiratory oxygen needs appear to be met at relatively low levels. A close inspection of the curves at higher oxygen levels indicates a slow increase in growth rates, suggesting that optimal growth may occur above atmospheric oxygen levels.

Maximal Oxygen Concentrations Since the growth of fungi or decay development at oxygen levels above the 21% of the atmosphere is unlikely to occur in nature, this aspect of fungal growth has been neglected. Highley *et al.* (1983) found that decay rates increased for several of the sap-rot fungi and

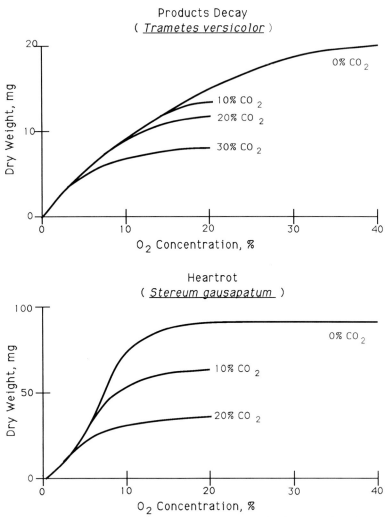

Figure 4-3 The effects of various concentrations of oxygen and carbon dioxide on the growth of a products-decay (*Trametes versicolor*) and a heart-rot fungus (*Stereum gausapatum*). Data from Jensen (1967).

decreased for two heart-rot fungi at oxygen levels of 0.4 atmosphere. Reid and Seifert (1983) also showed that some fungi decayed wood more rapidly in pure oxygen at atmospheric pressure (100% O_2) than air (21% O_2). It is of special interest to note that decay rates increased in the pure oxygen atmosphere, although there was no difference in fungal growth rate. These differences raise questions concerning the reliability of using growth responses of fungi to oxygen or other abiotic growth requisites as measures of the decay process. Hyperbaric levels of pure oxygen exceeding one atmosphere are toxic to several fungi, including *Serpula lacrimans;* however, some *Penicillium* species grow slowly in 3 atmospheres of pure oxygen (Caldwell, 1963). Large accumulations of pyruvic acid were associated with the toxic action of high oxygen concentrations.

Survival at Low Oxygen Concentrations Effects of low oxygen levels are relevant to the major decay and stain fungi, which are all aerobes. In decay tests of several heart-rot fungi, at near anaerobic levels, Van der Kamp et al. (1979) found survival, but *no* wood weight loss after a 10-week incubation period, suggesting a method by which heart-rot fungi survive the occasional anaerobic conditions that occur in living trees. In studies on the survival of fungi in sealed-tube cultures, Scheffer (1986) showed that heart-rot fungi survived longer than sap-and products-decay fungi. Two important building-decay fungi, *Meruliporia (Serpula) incrassata* and *Coniophora puteana*, died within a week, whereas several heart-rot fungi survived 2 years or longer, and the six sap-staining fungi in the test survived from 1 to 6 months. The depletion of oxygen by respiration is a major factor in a sealed-tube test, but other factors such as the accumulation of carbon dioxide or toxic metabolic products and nutrient exhaustion may also affect survival.

Oxygen and CO_2 Levels in Tree Stems Because of the prevalence of heart-rot, considerable attention has been focused on stem conditions that might explain the apparent selectivity of this specialized group of fungi for this environment. Many studies (Chase, 1934; Thacker and Good, 1952; Hartley et al., 1961; Jensen, 1969; Van der Kamp et al., 1979) have established that gases withdrawn from central stem tissues have lower levels of O_2 and higher CO_2 levels than those found in the atmosphere. The living cambium serves as an effective barrier to the diffusion of gases, effectively isolating the interior of the tree. Concentrations of gases in stems varies with position and season of the year.

Carbon dioxide accumulation and oxygen reduction presumably reflect respiration activities of bacteria, fungi, and wood parenchyma cells. Oxygen depletion does not normally reach levels limiting to fungal growth; however, in wet-wood formation, decreases in wood permeability further isolate tissues, and completely anaerobic conditions sometimes develop. Wet wood in *Abies concolor* and other tree species is colonized by bacteria that may inhibit decay fungi by depleting oxygen and producing low-molecular-weight organic acids (Worrall and Parmeter, 1983). Wet wood in black cottonwood stems contained less than 0.1% oxygen for several weeks in the summer season (Van der Kamp et al., 1979). In addition, large volumes of methane may be formed in the wet-wood zone by methanogenic bacteria in the genus *Methanobacterium* (Zeikus, 1974).

Soft-Rot Fungi Soft-rot fungi are often associated with water-soaked wood in baffles in cooling towers or the below-ground sections of posts and poles. In most cases, the soft-rot damage is shallow and limited to surfaces, possibly reflecting anaerobic conditions deeper in the wood. The higher tolerance of soft-rot fungi to low-oxygen concentrations, compared to the white-and brown-rot groups, may explain their prevalence as decay agents in saturated woods (Duncan, 1961).

Carbon Dioxide Relationships Carbon dioxide is toxic to fungi at higher concentrations, but small amounts of CO_2 (0.05%) are essential for fatty acid synthesis in fungi (Griffin, 1981). This requirement suggests a critical growth need that may have been overlooked in previous studies, where pure sources of oxygen and nitrogen were used to prepare gas mixtures; however, growth responses have been noted in the absence of carbon dioxide (Morrell, 1981). In natural aerobic decay, CO_2 always increases in confined zones as O_2 is consumed, and the two factors are confounded and interactions are possible. Increases in CO_2 can also elevate media acidity to undesirable ranges, and this factor has often been overlooked in CO_2 toxicity studies. In addition to its combined toxicity and acidity effects, carbon dioxide also inhibits fruiting in basidiomycetes (Niederpruem, 1963; Taber, 1966).

Ponding and Water Sprays as Storage Practices Long-term protection from insect and fungal attack can be achieved when wood is saturated with water by immersion or spraying. These practices are extensively used to control decay in logs and pulpwood during storage (see Chapter 13). Bridge or building foundation pilings also produce long service lives when the wood remains saturated. In such wood uses, the respiration of living cells in recently felled logs or bolts and/or the resident microbiota may rapidly deplete the oxygen in the stem gases and accumulate toxic levels of CO_2. The low solubility of O_2 in water and slow diffusion rates into the wood minimize replacement, near-anaerobic conditions develop, and aerobic fungal growth ceases. However, facultative anaerobic or microaerophilic bacteria can grow in some wood under these conditions. For example, a bacterium *Bacillus polymyxa* destroys pit membranes in ponded sugar pine logs and drastically increases wood permeability (Ellwood and Ecklund, 1959). Subsequent work has shown that other species and bacteria can damage ponded wood (Fogarty and Ward, 1973).

Some soft-rot fungi are able to attack limited portions of the cell wall on the surfaces of saturated wood when the water is constantly aerated. Soft-rot damage may become important when the wood surface:volume ratio is high, such as in slats in water-cooling towers (Levy, 1965).

Temperature

Temperature directly affects the many integrated metabolic activities of fungi, such as digestion, assimilation, respiration, translocation, and synthesis, that are mediated by enzymes. Within limits, metabolic reaction rates increase with increasing temperature until some reaction in the sequence becomes rate limiting, or the heat denatures the enzymes. In the early logarithmic growth phases of many fungi, the approximate doubling of specific growth rate per 10°C essentially follows the doubling of chemical reaction rates in a similar temperature range.

Because of their many direct applications in the control of disease and biodeterioration and the industrial production of fungal metabolites, the effects of temperature on fungal growth and other metabolic activities have been studied extensively.

Temperature effects on fungi have generally been determined by growth as measured by hyphal extension rates on media surfaces or biomass accumulations. These measures, however, may not accurately reflect the role of temperature in wood degradation.

Cardinal Temperature Levels Each fungus possesses 3 cardinal growth temperatures, a *minimal* level (growth begins), an *optimal* level (best growth), and a *maximal* level (growth ceases). Often the optimal temperature is skewed toward the maximal temperature, particularly for those fungi with the higher optimal growth levels (Fig. 4-4).

Growth curves for fungi should be viewed, in many cases, as approximations only, since many factors, including media, aeration, and method of growth measurement can alter fungal response.

In general, the temperature limits for the growth of most fungi lie between 0 and 45°C (Fig. 4-5). Within these limits, fungi have adapted to utilize various substrates under different temperature conditions. Psychrophiles are considered to have minima below 0°C and maxima of 20°C, with an optimal range between 0° and 17°C. A few *psychrophilic* (cold-loving) fungi in the genera *Cladosporium*, *Sporotrichum*, or *Thamnidium* can cause serious problems on some refrigerated foods even a few degrees below 0°C (Cochran, 1958). Snow-mold fungi, *Typhula* and *Fusarium* species, can grow on cereals

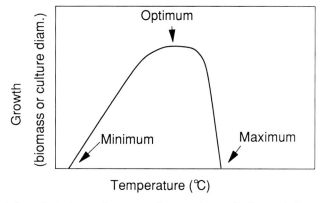

Figure 4-4 A hypothetical growth curve indicating the cardinal growth limits of a typical mesophilic wood-inhabiting fungus.

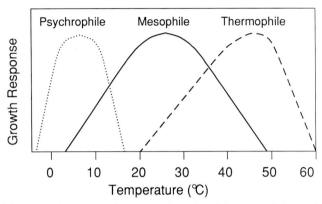

Figure 4-5 The approximate growth ranges of psychrophilic, mesophilic, and thermophilic wood-inhabiting fungi.

and turf grasses under snow cover and cause serious plant diseases (Agrios, 1988). At the other extreme, *thermophilic* fungi tolerate or grow at temperatures above 50°C.

Thermophiles have been defined as having minima at or above 20°C and maxima above 50°C (Cooney and Emerson, 1964). *Chaetomium thermophila* and *Penicillium duponti* are found in decomposing compost or chip piles, where heat builds up as the biomass deteriorates. Several Actinomycetes (bacteria) are true thermophiles, which tolerate high temperatures and cause wood damage in wood-chip piles. Most fungi, including many wood-inhabiting fungi, are *mesophilic* and grow best within a 15 to 40°C temperature range. Their minima are considered to be above 0° and the maxima below 50°C.

The comparative growth rates of decay fungi at 28°C on malt-extract agar are widely used characteristics in cultural identification. Wood microbiologists are particularly interested in the temperature requirements of wood-decay fungi that may explain their distributions and wood-product selectivity, or predict the development and location of decay hazards. Optimal temperature selection is also important for comparative studies of decay fungi, such as toxicity bioassays. Information on temperatures lethal to fungi in wood during processing is very important to the wood processor, since survival of these fungi in the product can lead to subsequent decay problems.

Temperature Grouping of Wood-Inhabiting Fungi Information on the growth rates of 56 species (and several strains) of wood-decay fungi on malt agar at temperatures ranging from 0 to 40°C were assembled by Humphrey and Siggers (1933). Growth measurements at 1- and 2-week intervals were used to determine the minimal, optimal, and maximal temperatures. Of the fungi tested, 12 had optima below 24°C; 42, optima between 24 and 32°C;

and 10, optima above 32°C. At 12°C, growth was entirely inhibited or sharply retarded for all the species tested. No fungus grew above 46°C, and most ceased growth at or above 40°C. A few fungi, such as *Gloeophyllum trabeum* and *G. sepiarium*, had high optimums ranging from 34 to 36°C. Several fungi exhibited bimodal optima. Some examples of common wood-decay fungi are presented with their optimal temperatures in three broad temperature groups in Table 4-2.

Esyln (1986) reported on the temperature relations and decay capabilities of 11 decay fungi (5 to 10 isolates per species) commonly associated with decay development in utility poles and found that growth rates varied widely within a species as well as among species. These differences suggest that caution be used when evaluating temperature optima for testing purposes when only a few isolates of a species are tested. An important conclusion from this study was that optimal decay capabilities of the fungi tested were related generally to their optimal growth rates.

Eleven isolates of blue-stain fungi including seven geographic strains of *Ceratocystis coerulea* were found to have temperature and maxima ranging from 29 to 39°C (Lindgren, 1942). Growth curves were similar in pattern to those for wood-decay fungi.

Temperature and Fungal Wood-Products Selectivity Some fungi appear to be consistently associated with certain wood uses, possibly related in part to temperature optima. The predominance of *G. trabeum* and *G. sepiarium* in the exterior woodwork of buildings reflects one possible association (Eslyn, 1986). Many soft-rot fungi have higher optima and temperature tolerances than Basidiomycetes, which may help explain their selectivity for wooden slats in cooling towers where high temperatures are common (Duncan, 1961). Soft-rot fungi in chip piles demonstrate considerable tolerance to heat (Hulme and Stranks, 1976); however, some soft-rot fungi commonly associated with below-ground zones of utility poles were more damaging at low than high temperatures (Morrell, 1981). It appears that generalizations con-

Table 4-2
Optimal temperatures for common wood decay fungi

Low temperature (24°C and below)	Intermediate temperature (between 24 and 32°C)	High temperature (optimum above 32°C)
Serpula lacrimans 20	Meruliporia incrassata....28	Gloeophyllum trabeum ...34
Phellinus pini 20	Neolentinus lepideus 28	Gloeophyllum
Heterobasidion	Schizophyllum	sepiarium.............36
annosum 24	commune 28	Lentinus strigosus........36
Hirschioporus	Trametes versicolor 28	Trametes hirsuta.........34
abientinus 24	Phellinus igniarius 30	Gloeophyllum striatum ...36
Phaeolus schweinitzii 24		

cerning fungal groups are unwarranted, since even closely related taxa may vary greatly in decay optima.

Time and Temperatures Lethal to Fungi in Wood In general, wood-decay fungi are resistant to prolonged exposure to low temperatures and, conversely, readily killed by short exposure to high temperatures.

Staining fungi are particularly susceptible to high temperatures, and some species are killed by prolonged storage at 35°C. *Schizophyllum commune*, an important product-rot fungus, exhibits unusual resistance to low temperatures. Fruiting bodies of this fungus, frozen at temperatures of −180°C, will produce basidiospores within a few hours after warming.

During the storage of logs and poles, decay fungi may become established and later damage the wood products in service. Information on the time and elevated temperatures necessary to kill fungi in wood during processing such as kiln-drying or preservative treatments is particularly important in large-sized logs, poles, piling, or beams, where storage periods are often prolonged and heat diffusion, slow. Chidester performed comprehensive tests to determine the time and temperature combinations necessary to kill fungi in wood samples previously decayed by several major wood-products fungi (Chidester, 1937, 1939). Green southern pine colonized by *Meruliporia (Poria) incrassata, Neolentinus (Lentinus) lepideus, G. sepiarium, G. trabeum, Fomitopsis (Fomes) rosea,* or *Phlebia subserialis* was exposed to temperatures and times ranging from 40 to 100°C and 6 min to 24 hr, respectively. Since wood rapidly loses strength at elevated temperatures, a principal purpose of this research was to determine the minimal temperature and time combination lethal to decay fungi. The minimal times and internal wood temperatures that safely killed all fungi in greenwood during processing were as follows:

Temperature (°C)	Time (min)
65.6	75
76.7	30
82.2	20
93.3	10
100	5

Temperatures below 65.6°C were judged to be impractical because of the long exposure periods necessary to kill some heat-resistant fungi. *Gloeophyllum trabeum, G. sepiarium, Neolentinus lepideus* were the most heat-resistant fungi, surviving 12 hr at 60°C, 20 hr at 50.5°C, or 24 hr at 50°C. As a result of these tests, 66°C for 75 min was selected as the minimal time–temperature combination for wood sterilization.

Since the recommended temperature is based on the *internal* temperature of the wood, the wood treatment (kilning or preservative treatment) must provide an adequate surface temperature and heating time to permit heat transfer to the center of the wood products. This can be done by determining the rates of heat flow or transfer in woods of various species, sizes, and moisture conditions (MacLean, 1930, 1932, 1934, 1935).

In more recent tests, three of 23 Ascomycotina and microfungi isolated from chip piles, including two soft-rot fungi, survived exposure to temperatures of 65°C or greater for times ranging from 8 to 72 hr (Hulmes and Stranks, 1976). In agreement with the previous studies, nine Basidiomycotina (single exception, *Phanerochaete chrysosporium*) were killed by exposure to temperatures equal to or greater than 50°C. These results clearly illustrate the limits of the blanket sterilization rules.

Substrate (Food Sources)

As heterotrophs, fungi and most bacteria require a food source or substrate that provides three major needs:

1. Energy from the oxidation of carbon compounds;
2. A pool of metabolites for the synthesis of the wide range of compounds needed for growth and development (chitin, glucans, nucleotides, enzymes, proteins, lipids, etc.); and
3. Required vitamins, minor elements, CO_2, and nitrogen.

An indirect substrate requirement is the absence of various growth inhibitors and the physical access of microbial enzymes to the required substrate constituents.

Microbial carbon nutrition has become a large and complicated subject. Essentially all carbon-based compounds are subject to microbial degradation under some conditons. This subject will be discussed in greater detail in Chapter 5 on metabolism.

As a generalization, fungi are eukaryotes that appear to have evolved as scavengers of plant remains (selective for carbohydrates and low pH conditions). Bacteria, as prokaryotes, are the major consumers of animal bodies (selective for proteins and neutral pH conditions). The same generalization holds for the diseases caused by bacteria and fungi. There are, however, many exceptions or crossovers in which bacteria attack living plants or their remains, or fungi attack animals.

Many fungi can degrade and utilize carbohydrates, including cellulose, but only the wood-inhabiting decay fungi—a few thousand species at best— are able to degrade and utilize carbohydrates in the cellulose–hemicellulose–lignin complex composing the wood cell wall.

The *monosaccharide* D-glucose is utilized by essentially all fungi and is a common carbon source in many cultural media. Galactose, mannose, and fructose also are used by many fungi but appear to be initially converted to glucose 6-phosphate and then follow the same metabolic pathways as glucose in the respiration or fermentation processes.

The *oligosaccharides* maltose, cellobiose, and sucrose are also good carbon sources for many fungi. Malt extract is a preferred medium for many wood-decay fungi, providing both glucose and vitamins.

Many fungi are able to utilize polysaccharides, such as cellulose, starches, and various hemicelluloses. The presence of small amounts of lignin as a barrier or shield around clusters of the carbohydrate components apparently drastically limits enzyme access and microbial attack to the small group of wood-inhabiting microorganisms. Some bacteria also degrade wood, but at a very slow rate.

Optimal nutrient sources vary widely for both fungi and bacteria. This variation is exploited in bacterial identification keys and holds high promise in the cultural identification of fungi. Determining optimal nutrient sources and growth conditions for wood-inhabiting microorganisms will help develop a better understanding of probable organismal successions (discussed in Chapter 11) in various stem invasions, heart-rot developments, and preferential attack of various wood products.

Hydrogen Ion Concentration (pH)

Fungi usually have a pH for optimal growth and a minimum and maximum at which no growth occurs. In general, the optimum is skewed toward the maximal value in a manner similar to cardinal temperature requirements. In contrast to vegetative growth, sporulation and spore germination have more restrictive pH tolerances. As a substrate factor, external pH primarily affects substrate availability, rate of exoenzymatic reactions, exoenzyme stability, cell permeability, extracellular components, and solubility of minerals and vitamins. It has little influence on the pH of cytoplasm. Hydrogen ion concentration does not always affect a single characteristic, and low levels may alter exoenzyme activity, whereas high levels might inhibit minor metal solubilities. These effects sometimes produce bimodal pH growth curves.

In general, fungi grow best within a pH range of 3 to 6, whereas many bacteria and actinomycetes grow best at a pH of 7. Some optimal pH values for wood-decay fungi are *Heterobasidion annosum,* 4.6–4.9; *Cerocorticium (Merulius) confluens,* 4.0; and *G. sepiarium, Fomitopsis rosea, Serpula lacrimans,* and *Coniophora (Cerebella) puteana,* 3.0.

Many plant pathogens have optimal growth within a pH range of 5 to 6.5. Wood-decaying Basidiomycotina have pH optima ranging from 3 to 6. Brown-rotters have the lowest optima (around pH 3). Wood-stain fungi are highly pH sensitive, and their growth often diminishes or (ceases) as pH exceeds 5.

Wood-decay fungi decrease the pH of wood during the decay process; this characteristic forms the basis of several chemical indicator tests proposed for detecting incipient decays in pulpwood and utility poles (Eslyn, 1979).

Chemical Growth Factors

Nitrogen Fungi, like other organisms, require substantial amounts of nitrogen for the synthesis of proteins and other cell constituents or products such as nucleoproteins, lipoproteins, enzymes, and the chitin in hyphal cell walls. Many fungi are able to use ammonia, nitrates, nitrites, and urea as sole sources of nitrogen. Ammonia is often the best nitrogen source, but the form in which it is supplied may affect media pH and, hence, growth responses. Certain L-amino acids are utilized by fungi as both carbon and nitrogen sources, whereas others are poor nutrient sources. L-Glutamine, L-asparagine, L-arginine, and L-proline are all good sole sources of nitrogen, whereas D-amino acids are poor nitrogen sources.

Wood-decay fungi can utilize many types of nitrogen but appear to utilize most effectively the amino forms found in wood (Huntgate, 1940). Wood-inhabiting fungi are unique in their ability to obtain their nitrogen needs from the very small amounts generally available in wood. The nitrogen content of wood ranges from .03 to .1%, whereas in herbaceous and other plant forms, it ranges from 1 to 5% (Cowling and Merrill, 1966).

The capacity of decay fungi to meet nitrogen needs wholly from the low amounts available in wood is even more surprising because of the prodigious number of spores released (nitrogen contents of about 3%) and the massive basidiocarps formed by some species, which require substantial amounts of nitrogen-containing chitin in the hyphal walls.

Early researchers suggested that some wood-inhabiting fungi were capable of fixing atmospheric nitrogen. Critical measurements of nitrogen availability in wood and fungi at various stages of decay in pure culture studies have ruled out direct nitrogen fixation (Klingstrom and Oksbjerg, 1963), which appears to be a metabolic capability limited to prokaryotes.

A series of studies on the availability and roles of nitrogen in wood deterioration (Cowling and Merrill, 1966; Merrill and Cowling, 1966a,b,c) summarized by Cowling (1970) showed that decay fungi probably conserve nitrogen by hyphal autolysis and recycling nitrogen toward hyphal tips. Nitrogen contents of the hyphal wall under low nitrogen levels are reduced before any reduction in exoenzyme formation. The close regulation of the cellulase enzyme system in some wood-decay fungi may also serve as a nitrogen-conservation method. Studies have shown that the degree of decay increases with increasing nitrogen content of the wood. Distribution of nitrogen in wood indicates reductions in amounts of nitrogen from outer to inner sapwood, with the lowest amounts in the heartwood. The pith section often contains large amounts of nitrogen. The changes in nitrogen content of a growing tree stem at various developmental stages (Figs. 4–6) may account for some of the puzzling variability in results experienced in standardized wood-decay tests. Nitrogen-content variations may also help to explain the variability and growth patterns of decay fungi in living stems.

Whereas axenic tests of fungi decaying wood have ruled out nitrogen fixation, bacteria are often associated with fungi in the decay process and probably play an important interactive role in some natural wood-decay processes in both nitrogen cycling and fixation. Nitrogen fixation by bacteria associated with fungi in the decayed wood is apparently an important indirect nitrogen source in some ecosystems (Aho et al., 1974; Larsen et al., 1978).

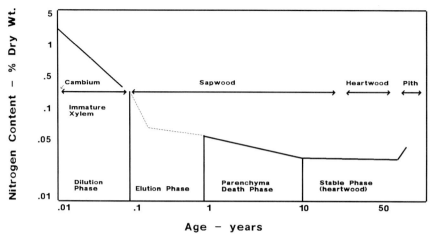

Figure 4-6 The changes in nitrogen content with radial positions and during various development stages of a tree stem. From Cowling (1970).

Vitamins and Minor Metals

Vitamins and minor metals are often necessary components of enzyme systems. The vitamin and minor metal requirements for many wood-decay fungi grown in liquid media have been determined by Jennison et al. (1955). Some fungi are able to synthesize their needed vitamins, but other fungi such as G. trabeum, H. annosum, and N. lepideus are apparently thiamine deficient and require external sources. Many fungi appear to require thiamine (B_1) and biotin, whereas pyridoxine (B_6) is less frequently required. The destruction of thiamine in wood by heat and alkaline treatments has been reported as a potential decay-control measure (Baechler, 1959), but subsequent work showed that the effects were owing to factors other than thiamine reduction (Highley, 1970).

Major mineral elements required by fungi are phosphorus, potassium, magnesium, and sulphur. Trace amounts of minor elements of iron, zinc, copper, manganese, and molybdenum are also required. Many minor metals play essential roles in enzymatic reactions; for example, iron (ferric ion) and manganese have been hypothesized to play key roles in the decay by brown- and white-rot fungi. The provision of minor metals and vitamins is often necessary in critical microbial studies using synthetic media. The reduction or elimination of these minor elements by certain sequestering chemicals or wood treatments may be a promising approach for developing new wood preservatives or protective treatments.

Light

Generally, light is assumed to be harmful to vegetative growth of wood-decay fungi and causes some growth reduction, probably owing to the lethal effects of the ultraviolet portion of light at high intensities. However, some studies suggest that periodic exposure to light may increase decay rates

(Duncan, 1967). In these studies, wood blocks close to a periodic light source developed nearly twice the weight loss of blocks farthest from the light source. No explanation for this carefully verified phenomenon is known, but a temperature increase due to a subtle greenhouse effect is a possibility.

Miscellaneous Factors

Many other environmental factors may affect the growth and reproduction of wood-decay fungi and their capacity to degrade wood, such as osmotic concentrations, atmospheric pressure, sound vibrations, gravitational forces, and radioactivity. Little is known about the effects of these agents on fungal growth and reproduction or the decay process, and they appear to offer interesting research avenues. It is, however, extremely difficult to perform experiments that isolate any of these factors.

Summary

- A review of the factors affecting fungal growth indicates that moisture level, oxygen content, and temperature are the growth requisites that can be most easily altered to adversely affect fungal growth and delay stain and decay development.
- Keeping wood dry or below the f.s.p. (< 28%) eliminates the possibility of effective microbial growth.
- Immersion of wood or the constant spraying of greenwood reduces the oxygen supply necessary for microbial growth and is another effective method for controlling development of stain and decay in stored logs.
- Initiation of storage piles of logs and pulpwood in the cold seasons is also an economical way to reduce biodeterioration damage.
- The use of lethal temperatures during wood processing and treatments is desirable to eliminate possible pretreatment invasions of wood-damaging fungi in standing trees, stored, or seasoning material.
- The addition of poisons or the use of woods containing natural toxicants is the principal method widely used to minimize growth of fungi in wood.
- Further studies on the growth requisites of fungi and interactions with other wood-inhabiting organisms are certain to provide better answers to fungal product selectivities and help to identify more effective means for delaying the inevitable biodeterioration processes.

References

Agrios, G. N. (1988). "Plant Pathology," 3rd Ed. Academic Press, New York. p. 803.
Aho, P. E., R. J. Seidler, H. J. Evans, and P. N. Raju. (1974). Distribution, enumeration,

and identification of nitrogen-fixing bacteria associated with decay in living white fir trees. *Phytopathology* **64**:1413-1420.

Baechler, R. H. (1959). Improving wood's durabiltiy through chemical modification. *Forest Product Journal* **9**:166-171.

Bavendamm, W., and Reichelt, H. (1938). Die Abhängigkeit des Wachstums holzzersetzenden Pilze vom Wassergehalt des Nährsubstrates. *Arch. Mikrobiol.* **9**:486-544.

Behr, E. A. (1972). Development of respirometry as a method for evaluating wood preservatives. *Forest Products Journal* **22**(4):26-31.

Boyce, J. S. (1961). "Forest Pathology," 3rd Ed. McGraw-Hill, New York.

Brock, T. D., D. W. Smith, and M. T. Madigan. (1984). "Biology of Microorganisms," 4th Ed. Prentice-Hall, Englewood Cliffs, New Jersey.

Caldwell, J. (1963). Effects of high partial pressures of oxygen on fungi. *Nature* **197**(4869):772-774.

Cartwright, K. St. G., and W. P. K. Findlay, (1958). "Decay of Timber and Its Prevention." Her Majesty's Stationery Office, 2nd ed. London.

Chase, W. W. (1934). The composition, quantity, and physiological significance of gases in tree stems. *Minnesota Agricultural Experimental Station Technical Bulletin* **99**:1-51.

Chidester, M. S. (1937). Temperatures necessary to kill fungi in wood. *Proceedings of the American Wood Preservers' Association* **33**:316-324.

Chidester, M. S. (1939). Further studies on temperatures necessary to kill fungi in wood. *Proceedings of the American Wood Preservers' Association* **35**:319-324.

Cochran, V. W. (1958). "Physiology of Fungi." John Wiley New York.

Cooney, D. G., and R. Emerson. (1964). "Thermophilic Fungi." W. H. Freeman, San Francisco.

Cowling, E. B., and W. Merrill. (1966). Nitrogen in wood and its role in wood deterioration. *Canadian Journal of Botany* **44**:1533-1544.

Cowling, E. B. (1970). Nitrogen in forest trees and its role in wood deterioration. *Abstracts of Uppsala Dissertations in Science*, 164.

Duncan, C. G. (1961). "Relative Aeration Requirements by Soft-Rot and Basidiomycete Wood-Destroying Fungi." U.S. Forest Service Forest Products Laboratory Report No. 2218, Madison, Wisconsin.

Duncan, C. G. (1967). Effect of light on the rate of decay of three wood-destroying fungi. *Phytopathology* **57**(10):1121-1125.

Eades, H. W., and Roff, J. W. (1953). The regulation of aeration in wood soil contact culture technique. *Forest Products Journal* **3**(3):94-95.

Ellwood, E. L., and Eckland, B. A. (1959). Bacterial attack on pine logs in pond storage. *Forest Products Journal* **9**(9):283-292.

Eslyn, W. E. (1979). Utility pole decay. Part 3. Detection in pine by color indicators. *Wood Science and Technology* **3**:117-126.

Eslyn, W. E. (1986). Utility pole decay Part 4: Growth-temperature relations and decay capability of eleven major utility pole decay fungi. *Holzforschung* **40**: 69-77.

Etheridge, D. E. (1957). A method for the study of decay resistance in wood under controlled moisture conditions. *Canadian Journal of Botany* **35**:615-618.

Fogarty, W. M., and O. P. Ward. (1973). Growth and enzyme production by *Bacillus subtilis* and *Flavobacterium pectinovorum* in *Picea sitchensis*. *Wood Science and Technology* **7**:261-270.

Freyfeld, E. E. (1939). The effect of humidity on the growth of wood-destroying fungi in timber (In Russian). *Review of Applied Mycology* **18**:644 (Abstr.).
Good, H. M., and J. A. Darrah. (1967). Rates of decay in wood measured by carbon dioxide production. *Annals of Applied Biology* **59**:463-472.
Griffin, D. H. (1981). "Fungal Physiology." John Wiley, New York.
Griffin, D. M. (1977). Water potential and wood decay fungi. *Annual Review of Phytopathology* **15**:319-329.
Halabisky, D. D., and G. Ifju. (1968). Use of respirometry for fast and accurate evaluation of wood preservatives. *Proceedings of the American Wood Preservers' Association* **64**:215-223.
Hartley, C., R. W. Davidson, and B. S. Crandall. (1961). "Wetwood, Bacteria, and increased pH in Trees." U.S. Forest Service, Forest Products Laboratory, Rep. No. 2215.
Higgins, N. C. (1957). The equilibrium moisture content–relative humidity relationships of selected native and foreign woods. *Forest Products Journal* **7**:371-377.
Highley, T. (1970). Decay resistance of four wood species treated to destroy thiamine. *Phytopathology* **60**(11):1660-1661.
Highley, T. L., and T. C. Scheffer. (1970). A need for modifying the soil-block test for testing natural resistance to white rot. *Mater. Org. Beih.* **5**:281-292.
Highley, T. L., S. S. Bar-Lev, T. K. Kirk, and M. J. Larsen. (1983). Influence of O_2 and CO_2 on wood decay by heart-rot and sap-rot fungi. *Phytopathology* **73**:630-633.
Hulme, M. A., and D. W. Stranks. (1976). Heat tolerances of fungi inhabiting chip piles. *Wood Science* **8**:237-241.
Humphrey, C. J., and P. V. Siggers. (1933). Temperature relations of wood-destroying fungi. *Journal of Agricultural Research* **47**:997-1008.
Huntgate, R. E. (1940). Nitrogen content of sound and decayed coniferous woods and its relation to loss in weight during decay. *Botanical Gazette* **102**:382-392.
Jennison, M. W., M. D. Newcomb, and R. Henderson. (1955). Physiology of wood-rotting Basidiomycetes. 1. Growth and nutrition in submerged culture in synthetic media. *Mycologia* **47**(3):275-304.
Jensen, K. F. (1969a). Effect of constant and fluctuating temperature on growth of four wood-decaying fungi. *Phytopathology* **59**:645-647.
Jensen, K. F. (1969b). Oxygen and carbon dioxide concentrations in sound and decaying red oak trees. *Forest Science* **15**:246-251.
Jensen, J. F. (1967). Oxygen and carbon dioxide affect the growth of wood-decaying fungi. *Forest Science* **13**(4):384-389.
Klingstrom, A., and E. Oksbjerg. (1963). Nitrogen source in fungal decomposition of wood. *Nature* **197**:97.
Larsen, M. J., M. F. Jurgensen, and A. E. Harvey. (1978). N_2 fixation associated with wood decayed by some common fungi in western Montana. *Canadian Journal of Forest Research* **8**:341-345.
Levy, J. F. (1965). The soft rot fungi—their mode of action and significance in the degradation of wood. *Advances in Botanical Research* **2**:323-357.
Lilly, V. G., and H. L. Barnett. (1951). "Physiology of the Fungi." McGraw-Hill New York.
Lindgren, R. M. (1942). Temperature, Moisture, and Penetration Studies of Wood-Staining *Ceratostomellae* in Relation to Their Control. U.S.D.A. Technical Bulletin 807. Washington, D.C.
Ljungdahl, L. G., and K.-E. Eriksson. (1986). Ecology of microbial cellulose degrada-

tion. In "Advances in Microbial Ecology" (K. C. Marshall, ed.) Vol. 8, pp. 237–299, Plenum Press, New York.

MacLean, J. D. (1930). Studies of heat conduction in wood. Results of steaming green round southern-pine timbers. *Proceedings of the American Wood Preservers' Association* **26**:197–219.

MacLean, J. D. (1932). Studies of heat conduction in wood. Part II. Results of steaming green round southern-pine timbers. *Proceedings of the American Wood Preservers' Association* **28**:303–330.

MacLean, J. D. (1934). Temperatures in green southern pine timbers after various steaming periods. *Proceedings of the American Wood Preservers' Association* **30**:355–373.

MacLean, J. D. (1935). Temperature and moisture changes in coast Douglas fir. *Proceedings of the American Wood Preservers' Association* **31**:77–103.

Merrill, W. and E. B. Cowling. (1966a). Role of nitrogen in wood deterioration. Amounts and distribution of nitrogen in tree stems. *Canadian Journal of Botany* **44**:1555–1580.

Merrill, W., and E. B. Cowling. (1966b). Role of nitrogen in wood deterioration. IV. Relationship of natural variation in nitrogen content of wood to its susceptibility to decay. *Phytopathology* **56**:1324–1325.

Merrill, W., and E. B. Cowling. (1966c.). Role of nitrogen in wood deterioration: amount and distribution of nitrogen in fungi. *Phytopathology* **56**:1083–1090.

Morrell, J. J. (1981). "Soft-rot fungi: Their Growth Requisites and Effects on Wood." Ph.D. thesis. State University of New York, College of Environmental Science and Forestry, Syracuse, New York.

Niederpruem, D. J. (1963). Role of carbon dioxide in the control of fruiting of *Schizophyllum commune*. *Journal of Bacteriology* **85**:1300–1308.

Reid, I. D., and Seifert, K. A. (1983). Effect of one atmosphere of O_2 on growth, respiration, and lignin degradation by white-rot fungi. *Canadian Journal of Botany* **60**:252–260.

Scheffer, T. C. (1973). Microbiological degradation and the causal organisms. In "Wood Deterioration and Its Prevention by Preservative Treatment, Volume 1: Degradation and Protection of Wood," (D. D. Nicholas, ed.) Syracuse University Press, Syracuse, New York.

Scheffer, T. C. (1986). O_2 requirements for growth and survival of wood-decaying and sapwood-staining fungi. *Canadian Journal of Botany* **64**:1957–1963.

Scheffer, T. C., and Chidester, M. S. (1948). Survival of decay blue stain fungi in air-dry wood. *Southern Lumberman* **177**:110–112.

Scheffer, T. C., and B. E. Livingston. (1937). Relation of oxygen pressure and temperature to growth and carbon dioxide production in the fungus *Polystictus versicolor*. *American Journal of Botany* **24**:109–119.

Schmitz, H., and Kaufert, F. (1938). Studies in wood decay. VII. How long can wood-destroying fungi endure immersion in water? *Proceedings of the American Wood Preservers' Association* **34**:83–87

Skaar, C. (1972). "Water in Wood." Syracuse University Press, Syracuse, New York.

Smith, R. S. (1975). Automatic respiration analysis of the fungitoxic potential of wood preservatives including an oxathiin. *Forest Products Journal* **25**(1):48–53.

Snell, W. H. (1929). The relation of the moisture contents of wood to its decay. III. *American Journal Botany* **16**:543–546.

References / 115

Snell, W. H., N. O. Howard, and M. V. Lamb. (1925). The relation of moisture contents of wood to its decay. *Science* **62**:377-379.

Stanier, R. Y., J. L. Ingraham, M. L. Wheelis, and P. R. Painter. (1986). "The Microbial World," 5th ed. Prentice-Hall, New York.

Tabak, H. H., and W. B. Cooke. (1968). The effects of gaseous environments on the growth and metabolism of fungi. *Botanical Review* **34**:126-252.

Taber, W. A. (1966). Morphogenesis in basidiomycetes. *In* "Fungi" Vol. 2, (G. C. Ainsworth and A. S. Sussman, eds.), pp. 387-412. Academic Press, New York.

Thacker, D. G., and H. M. Good. (1952). The composition of air in trunks of sugar maple in relation to decay. *Canadian Journal of Botany* **30**:475-485.

Toole, E. R. (1972). Oxygen utilization and weight loss associated with decay by wood-decaying fungi. *Wood Science* **6**(1):55-60.

Van der Kamp, B. J., A. A. Gokhale, and R. S. Smith. (1979). Decay resistance owing to near-anaerobic conditions in black cottonwood wetwood. *Canadian Journal of Forest Research* **9**(1):39-44.

Vogels, G. D. (1979). The global cycle of methane. *Antonie Leeuwenhoek J. Microbiol. Seral.* **45**:347-352.

Worrall, J. J., and J. R. Parmeter, Jr. (1983). Inhibition of wood decay fungi by wetwood of white fir. *Phytopathology* **73**:1140-1145.

Zeikus, U. G. (1974). Methane formation in living trees: A microbial origin. *Science* **184**:1181-1183.

CHAPTER
5

Fungal Metabolism in Relation to Wood Decay

A brief review of selected aspects of fungal metabolism is necessary to set the stage for an understanding of the enzymatic nature of wood decay by microorganisms and why preservatives are toxic to fungi. Emphasis is placed on fungal metabolism since it is similar to that of the principal wood-damaging bacteria.

Metabolism is a complex topic and broadly includes all the chemical reactions occurring in living systems. General reviews of the energy release and synthesis aspects of cell and organism functioning are available in many modern textbooks on biology and microbiology (Raven *et al.*, 1986; Cano and Colomé, 1986; Brock *et al.*, 1988). More comprehensive coverage is available in textbooks on biochemistry (Stryer, 1981) or fungus physiology (Griffin, 1981; Garraway and Evans, 1984). These sources provide references to the many specialized topics composing the vast literature that has accumulated on microbial metabolism. In this chapter, emphasis will be placed on digestion and energy release from organic carbon compounds, since these two processes play key roles in the decay process.

Metabolism Defined

Biochemistry is the principal discipline dealing with the study of the chemical processes occurring in living systems (metabolism).

In the 1800s, many scientists studying the chemical functioning of cells believed in a mysterious and unique property of living materials that they termed *vitalism*. As the knowledge of cell systems improved, the concept of vitalism declined, and scientists accepted that all metabolic reactions are explainable as chemical reactions that follow chemical laws.

Metabolic reactions can be classified as *catabolic* reactions, which release energy, or *anabolic* reactions, which require an energy source. Respiration represents a series of catabolic reactions. Photosynthesis is a classic anabolic reaction whereby light energy (photons) is transformed and stored in carbon compounds as chemical energy and becomes the major energy source for most living organisms.

Some examples of major metabolic activities occurring, often simultaneously, in microorganisms are digestion (an external process in fungi leading to substrate degradation), absorption, respiration, synthesis of cell components, growth, storage of food reserves, and reproduction. In terms of the specific growth requisites discussed in Chapter 4, metabolism includes the synthesis of digestive enzymes; their transport and release through cell membranes and walls; the absorption of H_2O, O_2, and the solubilized wood components; the release of CO_2; the synthesis of needed vitamins; and the absorption of minor metals. All of these reactions occur within a relatively narrow range of temperature and pH conditions.

Energy Sources, Transfer, and Storage

The fungi and most bacteria are heterotrophs, which require external organic energy sources. They obtain this needed energy from the respiration of organic compounds as chemical bonds are broken and reformed in chemical reactions. In this process, electrons drop to a lower energy level and generally are transferred to another element or compound. In many metabolic reactions, the chemical energy associated with these electron changes is temporarily transferred to special compounds termed *electron carriers*. Electron carriers play a key role in the timely transfer, storage, and use of chemical energy in many metabolic reactions. The free energy change (ΔG) involved in a chemical reaction between reactants and their products determines whether energy will be released (exergonic) or required (endergonic). In respiration the products of the reaction are at a lower energy state and energy is released ($-\Delta G$).

Oxidation–Reduction Reactions

In chemical reactions, the loss of electrons is termed *oxidation* and is associated with energy release. In organic materials (carbon compounds), oxidation often also involves loss of protons (hydrogen ions). The gain of electrons is termed *reduction* and requires energy. In organic compounds, reduction also often involves the simultaneous gain of protons. In reduction, the electrons involved in the chemical bonds of the products are raised to higher energy levels. The transfer of electrons between atoms, elements, or compounds is always coupled. One reactant in the reaction gains electrons, while the other loses them in oxidation–reduction reactions (the term is usually abbreviated to *redox* reactions). Redox reactions are emphasized in this section because of their importance in respiration. Carbohydrates are good electron donors that are easily oxidized in cell respiration and serve as a prime energy source for many fungi. Atmospheric oxygen (O_2) is an excellent electron acceptor, relative to most other elements or compounds, explaining its universal role in aerobic respiration and the term oxidation.

Electron Carriers and High-Energy Compounds

Several complex molecules function as electron carriers between donor and acceptor compounds in metabolic reactions. Some carriers appear to facilitate enzyme function and are also termed *coenzymes*. Coenzymes may serve as carriers of small molecules from one enzyme to another. Most are synthesized from vitamins. Electron carriers and high-energy compounds temporarily store and transport chemical energy released during respiration so it can be available when needed for critical cell functions requiring energy, such as enzyme synthesis, active solute transport, or growth.

Adenosine triphosphate (ATP) serves as the major energy provider for most living organisms. The structure of ATP consists of ribose, adenine, and ester linkages connecting three phosphates (Fig. 5-1). Energy is stored in the last two phospho-diester bonds. Successive loss of these groups in phosphorylation reactions, with electron transfer to other compounds, forms first the related compounds ADP (adenosine diphosphate) and then AMP (adenosine monophosphate). Phosphorylation reactions, whereby a phosphate group is added enzymatically to another compound, are key energy transfer mechanisms in respiration.

An important electron carrier in respiration is nicotinamide adenine dinucleotide (NAD) and a phosphorylated form NADP (Fig. 5-1). These two compounds also transport protons; NAD is a major transporter of protons during respiration, whereas NADP plays a similar role in many synthetic reactions.

Enzymes

Catalysts in chemical reactions are substances that accelerate the rates of reactions without being permanently altered themselves. Biocatalysts or enzymes are involved in most chemical reactions in living systems.

Structure and Mode of Action

Enzymes are proteins consisting of one or more polypeptide chains folded in a complex tertiary structure connected by disulfide linkages. Binding sites control enzyme specificity, and a unique topography corresponding to the reactive zones of the substrate determines the *active site*. Recent studies have substantially improved our understanding of the structure and function of enzymes, but the *lock-and-key* model proposed by Emil Fisher half a century ago still provides a useful way to explain the close structural relationship between enzyme and substrate (Fig. 5-2A). Often critical regu-

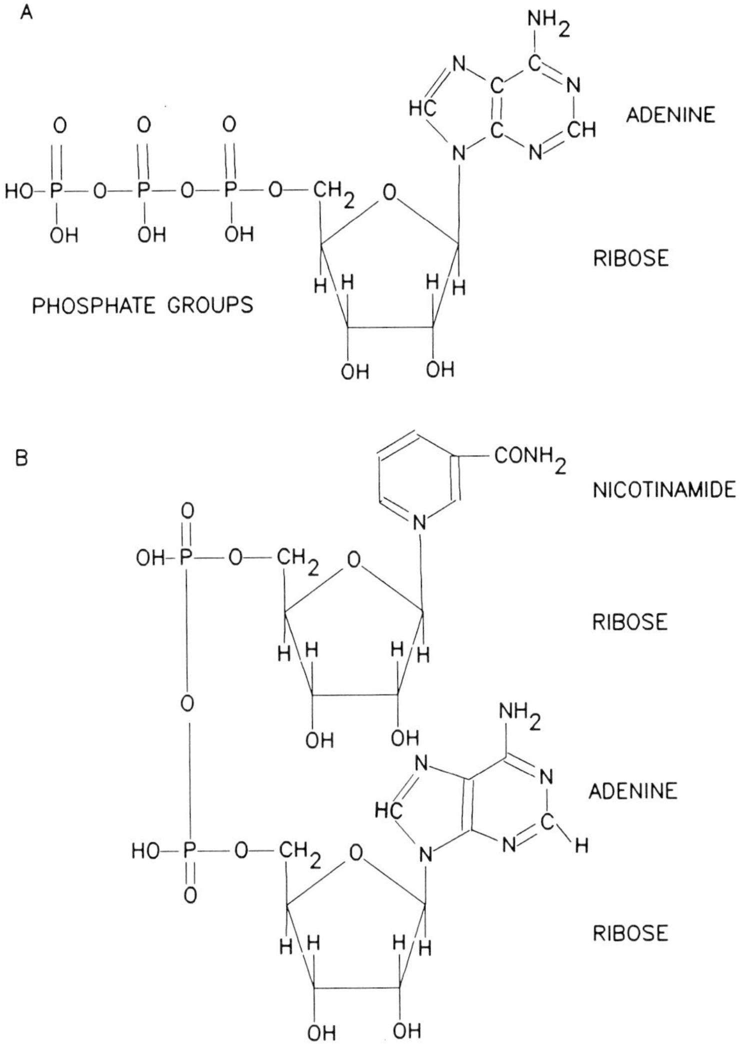

Figure 5-1 The structural formulas of two major compounds involved in energy transfers in respiration, showing their constituents and close relationship. (A) Adenosine triphosphate (ATP) has two high-energy phosphate bonds, which are major energy sources for many metabolic reactions. (B) Nicotinamide adenine dinucleotide (NAD) is an important electron carrier.

lators or cofactors that alter the enzyme structure so the active site meshes more precisely with a portion of the substrate molecule are associated with an enzyme. Copper or iron are two common enzyme cofactors.

The energy required to break a chemical bond so a reaction can proceed is the *energy of activation*. A loose analogy may be the push necessary to start a large boulder rolling down a steep hill, causing a large amount of potential en-

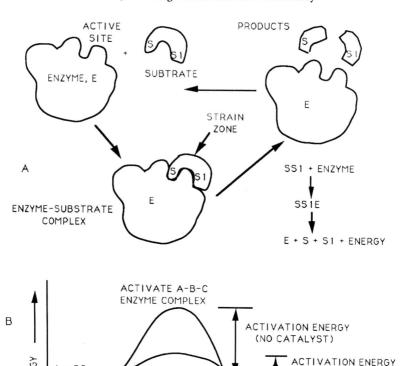

Figure 5-2 (A) Traditional *lock and key* model, suggesting the mechanism of enzyme action. (B) Representation of the energy required to start the reaction with and without a catalyst.

ergy to be released as kinetic energy at the bottom. Enzymes reduce the activation energy, thus increasing the rate of the reaction (Fig. 5-2B). The number of substrate molecules catalyzed per second per enzyme molecule has been reported to be as high as 10^5 (also known as turnover) for some enzymes. Enzymes also mediate many reactions at room temperatures and near pH neutrality that otherwise could occur only under extreme temperature or pH conditions. Enzymes are highly specific to the substrate used and the reaction catalyzed. They are easily activated, modified, or inactivated by a variety of cell controls. The range of reactions carried out by different types of enzymes is broad, and collectively they are very versatile.

Enzymes also enable organisms to transform energy from one form to another. Autotrophic plants can convert light energy (photons) into chemical bond energy, providing a basic energy pool that sustains heterotrophic life forms.

Whereas enzymes are highly versatile, they cannot alter the direction of a reaction, which is determined by ΔG, and only greatly accelerate the reaction rate. Many enzymes are fragile and inactivated by temperatures above 50°C, changes in pH, or the presence of heavy metal cations of copper (Cu), mercury (Hg), and lead (Pb).

The mechanisms by which enzymes function are still not fully known. The enzyme binds to the substrate forming an intimate union with it called the *enzyme-substrate complex*. Presumably, certain chemical groups on the substrate are stressed by the electrostatic forces involved, and bonds are strained and broken. The active sites on the enzyme are probably where this molecular distortion occurs (Fig. 5-2A).

Prosthetic groups attached to the enzyme are necessary in some cases for activation. Vitamins such as thiamine may play this role. Other compounds associated, but not attached, that are necessary for the reaction to proceed are termed *cofactors*. Magnesium cations, for example, are necessary cofactors for phosphorylation reactions catalyzed by kinases. Important factors affecting the rates of enzyme reactions are enzyme and substrate concentrations, pH, and temperature.

Types and Classifications of Enzymes

The first few enzymes discovered were unique and assigned trivial names such as diastasis, pepsin, or ferment. In 1878, Kuhn proposed that the suffix *ase* be added to the substrate modified. This practice was quickly adopted and is still in use for the general or trivial name of many enzymes or enzyme complexes, e.g., cellulases and hemicellulases. By the 1950s, many enzymes had been discovered, and considerable confusion developed, with similar names used for different enzymes and the use of different nomenclature schemes. In 1955, the International Union of Biochemistry established a Commission on Enzymes to develop a uniform classification system and standardized nomenclature, which is still in use today. The principal grouping was based on the type of chemical reaction involved. The six main groups are as follows.

1. Oxido-reductases — oxidation–reduction reactions involving basically the loss and gain of electrons
2. Transferases — transfer of radicals such as amino, methyl, or acetyl from one compound to another
3. Hydrolases — separation of compounds at the oxide bridge, often into monomers or dimers, by the addition of water
4. Lyases — removal of groups leaving a double bond or the reverse: addition of groups to a double bond
5. Isomerases — catalysis of an internal rearrangement in a molecule to form an isomer

6. Ligases (synthetases)   catalysis of the joining of molecules to form a new compound, usually requiring chemical energy from ATP

Hydrolases and oxido-reductases are the major enzyme types involved in decay.

Some other common descriptive groupings of enzymes in general use are the following:

1. site of action—*endocellular* and *exocellular* enzymes act inside or outside the cell, respectively (hydrolases are exocellular, whereas most oxido-reductases are endocellular)
2. constitutive or induced—whether the enzymes are always present in the cell (constitutive) or appear *only* in the presence of a specific substrate (induced)
3. terms combining the name of the substrate altered and the chemical reaction involved—glucose oxidase, pyruvate decarboxylase, and glyceraldehyde 3-phosphate dehydrogenase

Knowledge about enzymes and their function has increased enormously in the past few decades. Whereas a single volume summarized enzymology in the early 1960s (Dixon and Webb, 1964), nearly 200 volumes in a series currently present the methods of enzymology, and a score of volumes now compose a definitive treatment titled *The Enzymes* (Sigman and Boyer, 1990).

Digestion and Hydrolases

The major cell-wall constituents of wood are all large macropolymers that are insoluble in water. The principal source of nutrition for most wood-inhabiting microorganisms is the carbohydrates present in cell walls and storage tissues. Cellulose, hemicelluloses, and pectins are the major polymeric carbohydrates, composing about 75% of the cell wall. Lignin, an aromatic heteropolymer, makes up most of the remaining cell-wall substance. Small amounts of lipids and proteins are also used by some bacteria and fungi. Starches and lipids are the principal nutrition sources in the storage tissues. A wide range of compounds is present also in the wood extractives in rather small amounts. Lignin forms a shield or barrier around the cellulose and hemicellulose microfibrils in the cell wall that must be altered or removed before the microfibrils can be enzymatically digested. For many years, lignin was seen as a barrier that the fungus had to remove or modify to gain access to the cellulose; however, the fact that many fungi selectively remove lignin has altered this assumption.

Cellulose, hemicelluloses, and lignin are complex polymers that must be reduced to small diffusible units before they can be absorbed and utilized.

This process of external *digestion* is carried out largely by *hydrolases* and *oxidases*. It is an exocellular process suggesting that the fungal hypha or bacterial cell be in a film of water for the diffusion and transport of the breakdown products. The enzymes attacking the various substrates also require either a water film to diffuse to the active sites on the substrate molecules or direct contact between the hyphal surface or bacterial walls and the wood cell wall. Membranous structures associated with hyphal sheaths have been proposed recently as another means of degrading agent delivery into the cell wall (Highley, 1989). Hydrolases are enzymes that, by the addition of H_2O (proton and hydroxyl), break the polymer into smaller units. Simplified hydrolytic (digestive) reactions are shown for a sugar, polypeptide, and lipid in Fig. 5-3.

In general, large polymers are attacked by several specific hydrolytic enzymes. First the polymer complex in the cell wall is exposed from the others and simplified by chain separations and removal of side chains. The exact nature of these initial steps is poorly defined. Once the polymer is exposed, enzymes known as *exohydrolases* attack the ends of the chains, usually releasing monomers or dimers. Other enzymes known as *endohydrolases* randomly attack the chains and in effect form many more ends that are accessible to exohydrolases, greatly accelerating the decomposition process. This brief review of digestion is limited to cellulases, hemicellulases, and ligninases

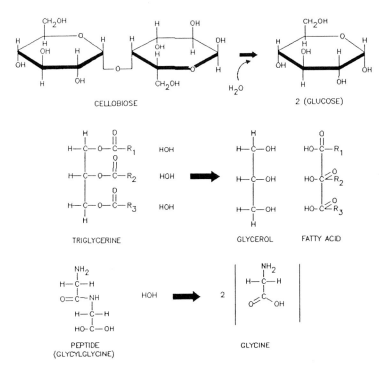

Figure 5-3 Hydrolytic reactions involving an oligosaccharide, a lipid, and a peptide.

because of their central role in wood decay. This topic is considered in more detail in a later chapter on wood decay.

Cellulose

This polymer is decomposed by a multienzyme complex known generally as *cellulase*, which consists of at least three enzymes that may vary with fungi (decay type) and type of substrate. Endo-1,4-β-D-glucanases act randomly on exposed molecule surfaces, randomly cleaving the cellulose polymer. The exposed ends are attacked by exo-1,4-β-glucanases (cellobiohydrolase) forming glucose or cellobiose. A 1,4-β-glucosidase then converts the cellobiose into glucose, which can be absorbed by the microorganism. In some fungi, oxidative enzymes may also be involved.

Hemicelluloses

These polymers are also decomposed by a multienzyme complex. Hemicelluloses are complex heteropolymers containing xylose, galactose, glucose, and mannose with acetyl, methyl, and short oligosaccharide units as side chains. Many of the hemicellulase enzymes are poorly understood. Those that attack the polymer backbones are presumed to be β-D-galactanases, β-D-mannanases, and β-D-xylanases. Since the hemicelluloses are noncrystalline and more exposed in the transient capillary zone, only 1,4-β-D endoenzymes are involved and attack the polymers randomly. Separate enzymes may be necessary to remove the side chains, but the coordination between polymer decomposition and side-chain attack is uncertain. Exoglycosidases then attack the oligosaccharide residues, and the xylose, mannose, glucose, and galactose units can be absorbed and utilized.

Lignin

Lignin biodegradation is the unique capability that distinguishes wood-decay fungi from most other microorganisms. At present, the process is only partially understood for several white-rot fungi and is still under intensive study. Enzymatic breakdown of lignin is an oxidative process in sharp contrast to that of the hydrolytic enzymes that predominate in the degradation of cellulose and hemicelluloses. Lignin is included in this section primarily because its enzymatic removal or alteration is necessary for carbohydrate digestion (degradation) to proceed. A key part of the breakdown is the oxidative separation of carbon to carbon bonds and ether linkages among certain phenyl propane units by peroxidases. An interesting feature of this enzymatic reaction is the requirement for an extracellular source of H_2O_2. The enzymatic breakdown of lignin will be covered in some detail in Chapter 8 on the chemical aspects of wood decay.

Absorption of Digestion Products

Digestion begins and ends at the plasmalemma (cytoplasmic membrane) just within the cell wall. It begins with the release of exoenzymes by exocytosis and diffusion into the surrounding liquid medium to the substrate. Digestion ends when the soluble products of digestion diffuse back to the hyphal wall. These polymer residues of the decay process are primarily simple sugars (glucose, xylose, galactose, etc.), which are small enough to readily pass through the cell wall but not the plasmalemma, which is semipermeable and highly selective. There is some uncertainty as to the membrane transport mechanism used to move sugars into the hyphae. In yeasts and some molds, both *facilitated diffusion* and *active transport* are reported, depending on the type of simple sugar and fungal species. In the decay of wood, solute concentrations are very low, and *active transport* is the probable method. In this process, the sugar is phosphorylated and transported by a protein carrier (permease) across the plasmalemma into the cytoplasm, where respiration can begin. This process requires energy, which is obtained generally from the high-energy compound, ATP.

Aerobic Respiration

Most fungi undergo *aerobic respiration* in the presence of atmospheric oxygen (O_2) by synthesizing ATP (storage of chemical energy) from the oxidation of glucose. The ATP is the principal energy source for a wide range of energy-dependent metabolic activities. Aerobic respiration has three major phases, glycolysis, the citric acid cycle, and the electron-transport chain (Fig. 5-4). The process begins with glucose and ends with CO_2, H_2O, and chemical energy stored in ATP. The process also provides a rich array of metabolites for synthesis.

Glycolysis

Glycolysis takes place in the cytoplasm. This phase of respiration begins by the enzymatic phosphorylation of glucose at the 1 and 6 carbons, successively, using ATP. These reactions are carried out by kinase enzymes with Mg^{++} as a critical cofactor. Thus, magnesium is a critical growth requisite for some fungi. The phosphorylated glucose is then changed into fructose, which is further phosphorylated and broken into two triose sugars. Then, a series of enzymatic oxidations and molecular rearrangements converts the triose sugar glyceraldehyde-3-phosphate to pyruvic acid. In this process, 2 moles of ATP are gained per mole of glucose, and one mole of NAD is reduced (protons acquired). Pyruvic acid is a key intermediate in several biochemical pathways and has been termed a *biochemical turntable* in cell metabolism. It is interesting to note that the oxidation–reduction reactions releasing chemical energy for transfer to ATP in glycolysis occur on the glucose molecule. Some portions

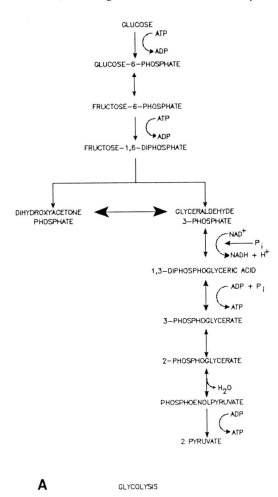

Figure 5-4 The three major biochemical pathways in aerobic respiration. (A) In glycolysis, glucose is oxidized to pyruvate. (B) In the citric acid cycle, the electron carriers NAD and FAD are reduced and donate electrons to the electron-transport system. (C) In the electron-transport system, the electrons passed through the various cytochromes provide the energy to synthesize ATP. These cycles produce 38 molecules of ATP per molecule of glucose oxidized to CO_2 and H_2O. *(Figure continues)*

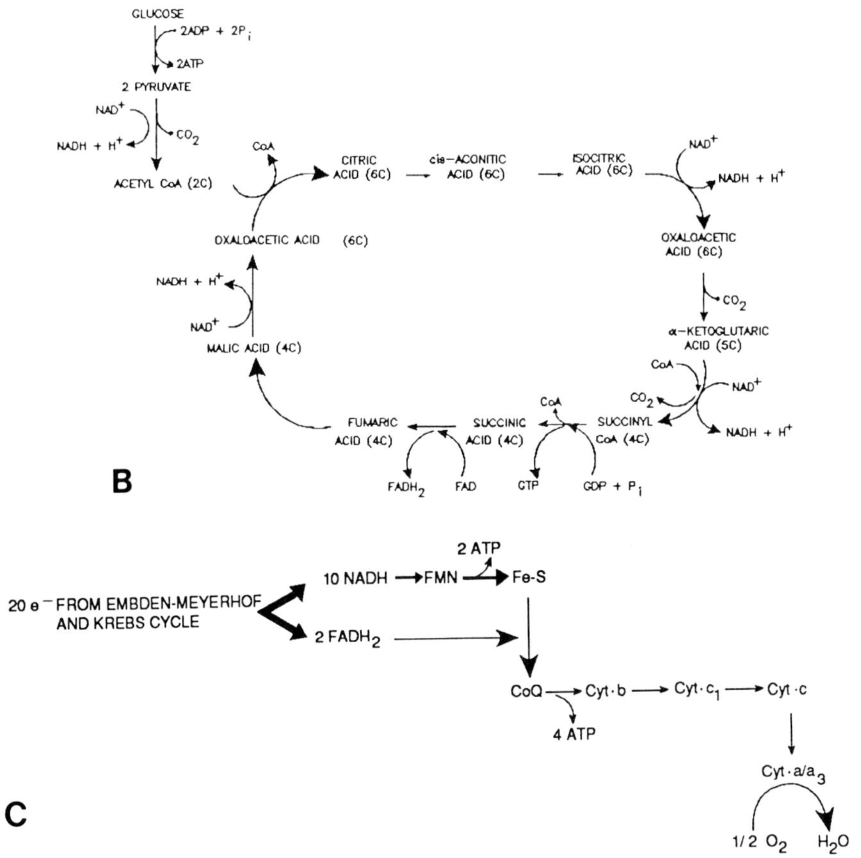

Figure 5-4 (continued)

become highly oxidized, forming CO_2, while other portions become further reduced to methyl groups in the acetyl radical. In aerobic respiration, the pyruvate is further oxidized and then decarboxylated, forming CO_2 and an acetyl radical (CH_3CO^-). In this process, two molecules of reduced NAD ($NADH^+$) are formed per mole of glucose.

Citric Acid Cycle

The citric acid or Krebs cycle takes place in the inner matrix of the cristae in the mitochondria. In this phase of aerobic respiration, coenzyme A (CoA) combines with an acetyl group, temporarily forming an acetyl–coenzyme A complex. The acetyl group in this complex then combines with oxaloacetate (a 4-carbon acid) to form citrate (a 6-carbon acid). In a further series of oxidations and molecular rearrangements, 2 moles of ATP are formed, and 6 moles of NAD are reduced per mole of glucose. Also, 2 moles of another electron carrier, flavine adenine dinucleotide (FAD), are reduced per mole of glucose. The oxaloacetate is regenerated and available for further reactions with the acetyl radical.

Electron Transport Chain

At this stage of respiration, the electrons and protons released during the oxidation of glucose are transported by the electron carriers NAD and FAD to a series of electron carriers embedded in walls of the inner mitochondrial membranes. These electron carriers differ in the energy levels at which the electrons are held. The electrons move down the sequence of carriers, and at the end, are accepted by oxygen. Oxygen then combines with the available protons to form water. The major electron carriers are the cytochromes, iron–sulphur proteins, and quinones. The quinones are able to transport protons across the mitochondrial membranes. The proton (H^+) gradient or proton-motive force established across these membranes is another energy source for some metabolic reactions and is believed to play a role in energy transfer during oxidative phosphorylation.

The energy released at various levels in this sequence of electron energy levels in the presence of the enzyme ATP synthetase regenerates ADP to ATP in a process termed *oxidative phosphorylation*. Most of the energy obtained from glucose respiration is derived from the electron and proton transfers that occur in this final stage of aerobic respiration. This phase of aerobic respiration occurs in an array of enzymes and enzyme carriers on the surface of the cristae in the mitochondria.

Pentose Shunt

Another important aerobic respiratory pathway for the utilization of glucose is the hexose-monophosphate pathway or pentose shunt. The pentose shunt provides the 5-carbon sugars, ribose, and deoxyribose needed for the synthesis of DNA, RNA, ATP, CoA, and NAD^+. It also is an energy-yielding process and reduces NADP to NADPH, which serves as the major proton and electron donor in many synthetic reactions. In this pathway, glucose 6-phosphate is oxidatively decarboxylated in several steps to form ribose 5-phosphate and CO_2. In these reactions, NADP is reduced to $NADPH^+$ and utilized as a readily availabe source of reducing power for many biosynthetic reactions, or oxidized in the mitochondria and its chemical energy transferred to ATP. Later stages of this pathway form intermediates that are precursors for the synthesis of the aromatic amino acids or glyceraldehyde 3-phosphate, which connects the pathway with glycolysis. The pentose shunt is present in most organisms and is an interesting example of a pathway's playing an important role in both catabolism and anabolism. This is also an excellent example of how many of the major metabolic pathways are connected by common intermediate compounds, creating alternate routes and regulation possibilities.

Fermentation

Simple sugars such as glucose and some other carbon compounds can be utilized by some bacteria and fungi such as the yeasts in the absence of oxygen in the process termed *fermentation*. In this type of metabolism there is no external electron acceptor, and the oxidation–reduction reaction that releases energy takes place between parts of the same substrate molecule. As such, fermentation is an inefficient process that yields only a portion of the potential energy available in a compound. Fermentation is believed to be a primitive biochemical pathway that originated in microorganisms before the accumulation of oxygen in the atmosphere from photosynthesis. The glycolysis phase of fermentation, however, is still retained as a critical first step in aerobic respiration.

The biochemical pathway of the fermentation of glucose is the same as that in aerobic respiration to the formation of pyruvic acid, but there are differences in subsequent steps depending on the fermenting organisms.

For some yeasts, pyruvate is decarboxylated enzymatically, releasing CO_2, and acetaldehyde is subsequently reduced by NADH to form ethanol. In the fermentation process, a net of only 2 moles of ATP is generated per mole of glucose. Other compounds that may be produced fermentatively are alcohols, such as butanol; organic acids, such as acetic acid; and ketones, such as acetone. For some bacteria and animals, the final product of glucose fermentation under anaerobic conditions is lactic acid.

Anaerobic Respiration

Anaerobic respiration is important in the decomposition of cellulose and other carbon compounds that accumulate in oxygen-deficient environments, e.g., animal rumens, wetwood in trees, and intestinal tracts (herbivores and termites). *Anaerobic respiration* can be defined as a nutritional biochemical pathway in which compounds other than oxygen are used as electron acceptors.

The denitrifying bacteria that reduce NO_3 to NO_2 or nitrogen in their decomposition of organic materials are an interesting example of anaerobic respiration and an important source of fertility loss in some soils. The methanogenic bacteria are of special importance in decomposing cellulose in anaerobic environments such as animal rumens, bogs, and water-logged sediments. In these fermentations, methane (CH_4) gas is most commonly formed from the reduction of either CO_2 or acetic acid (CH_3COOH). Cellulose breakdown is actually carried out by a consortium of anaerobic bacteria, and their relative roles in the decomposition are currently under investigation.

Sulphate (SO_4) reduction is also carried out by a variety of anaerobic bacteria that utilize organic acids and alcohols as electron donors. Some bacteria also can use hydrogen as the donor. Sulphate-reducing bacteria often accumulate in environments that become anaerobic as a result of active microbial decomposition. Two major groups are recognized that reduce sulphate (SO_4) to a gas (hydrogen sulphide) or sulphides.

Anaerobic respiration also occurs in the inner stem zones of trees when a condition known as *wetwood* occurs. In this condition, considerable amounts of methane are formed, and an exudate known as *slime flux* develops on the trunk surface from cracks in the stem.

Enzyme Inhibitors

The extreme complexity of metabolism and the many processes involved provide numerous potential points that can be altered or blocked chemically. Many chemical compounds are known to inactivate enzyme activity. If other properties, such as cost, ease of handling and treatment, and safety to other life forms, are acceptable, selective enzyme inhibitors or poisons represent potential fungicides or preservatives. Many of the heavy metals such as Hg^{2+}, Pb^{2+}, Cr^{3+}, and Ag^+ are broad-spectrum toxicants that disrupt many enzymes and damage the physical structure of proteins. In more specific toxic actions, tributyltin oxide blocks the exchange of electrons and ions across mitochondrial membranes in the electron-transport process; pentachlorophenol is an uncoupling agent and disrupts the formation of ATP from ADP; and arsenic blocks the pyruvate dehydrogenase enzyme system in glycolysis and also serves as a competitive inhibitor of phosphorus in both substrate and oxidative phosphorylation reactions. Cytochalasins inhibit the transport of cellulase across cell membranes, and antibiotics, such as penicillin, disrupt cell-wall synthesis in some types of bacteria. This topic will be discussed in more detail in the chapter on wood preservatives.

Nutrition in Relation to Fungal Growth Requisites and Decay Control

A general diagram of the nature and site of these many degradative and respiratory activities is summarized in Fig. 5-5. In this model, the related roles of the ecologic factors affecting the growth of fungi and decay rates are re-emphasized. It is interesting to note that the effect of each growth factor can be explained at the molecular level by an enzymatic reaction or its requirements. Important points to stress are

1. Water is the diffusion medium for enzymes and O_2, a reactant in hydrolysis, and the solution medium for all cell chemistry.
2. Oxygen (free) is the ultimate electron and hydrogen acceptor in the energy-yielding aerobic oxidation–reduction reactions, forming H_2O.

3. Temperature controls the rate of reactions and, at higher levels, disrupts the stability of enzyme structures.
4. The substrate provides the basic energy, the pool of metabolites for synthesis, and in many cases, the vitamin and nitrogen sources for fungi.
5. Minor metals and vitamins play critical roles as cofactors or coenzymes in the many enzymatic reactions.
6. Hydrogen ion concentration (pH) defines the optimal level for many enzyme reactions and protein stability.

Summary

- *Metabolism* broadly includes all the chemical reactions occurring in living systems. Nutrition, involving the digestion, absorption, and respiration of energy-rich organic compounds, is particularly important to the process of wood decay and its control (Fig. 5-5).
- Fungi and most bacteria are heterotrophs, which require energy from organic sources. This energy is released in the process of *respiration*. The basic source is electron donors (organic compounds) and involves the release and transfer of energy when chemical bonds are broken and reformed during the exchange of electrons in *redox* reactions.
- Adenosine triphosphate (ATP) is a major energy-rich compound, which stores and provides the energy released by respiration for many cellular activities. Nicotinamide adenine dinucleotide (NAD) is one of the important electron carriers in respiration reactions and a phosphorylated form (NADP) plays a similar major role in many synthetic reactions.
- Many chemical reactions in metabolism are facilitated by *enzymes*. Enzymes are biocatalysts that accelerate the rates of reactions without being permanently altered themselves. Enzymes are complex proteins that are readily destroyed by adverse temperature, pH, and many toxic chemicals.
- *Hydrolases* and *oxido-reductases* are the major types of enzymes involved in decay and cell respiration.
- Wood decay can be considered to be the external digestion of the large macropolymers (cellulose, hemicelluloses, and lignin) of the cell wall that are insoluble in water. The digestion is carried out primarily by hydrolases and oxidases, which reduce the complex polymers to diffusible units that can be absorbed and respired for energy and metabolites for synthesis.
- Most fungi utilize the *aerobic respiration* pathway, which requires atmospheric oxygen (O_2) as the final electron acceptor in the oxidation of glucose and the synthesis of ATP (storage of chemical energy). The

Summary continues

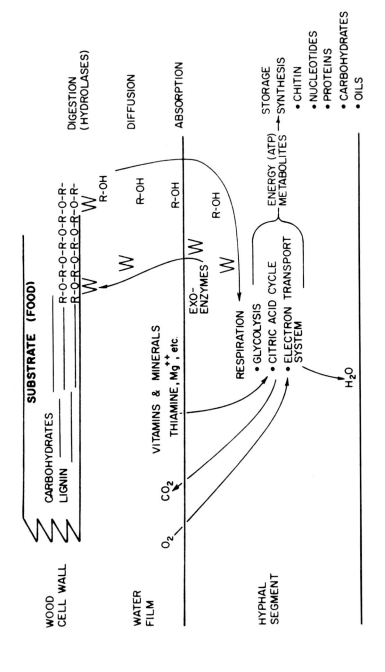

Fungus Growth Requisites
1. Accessible food (substrate)
2. Water (above the f.s.p.)
3. Oxygen (more than 20% of the wood-void volume)
4. Temperature (15–35°C)
5. pH (3–6 units)
6. Minor metals and vitamins
7. Absence of natural substrate toxicants

Decay Control (Reciprocals)
1. Infusion with toxicants or chemical modifications
2. **Keep wood dry** (below f.s.p.)
3. Submerge or spray wood in water
4. Concentrate roundwood storage in a cold season; heat to sterilize
5. Alkaline solution dip treatments for stain control
6. Sequestering preservatives; heat-treat wood to destroy vitamins
7. Use of durable woods

Figure 5-5 A simplified model of the major nutritional activities of a decay fungus, including digestion, absorption, and respiration and the relationships between growth requisites and decay-control practices.

three major phases of respiration are glycolysis, the citric acid cycle, and the electron-transport chain.
- Some fungi and bacteria utilize nutrients by *fermentation*. In this biochemical pathway, there is no external electron acceptor, and the electron transfers that release energy take place between portions of the substrate molecules.
- Some bacteria utilize an *anaerobic respiration* pathway, in which compounds other than oxygen are the electron acceptors. Some bacteria in this group are able to decompose cellulose in anaerobic environments and animal rumens.
- The requisites for fungal growth can often be explained at the molecular level by the related enzymatic reactions involved or its requirements. Practical prevention or control of decay can often be achieved by adversely affecting a critical growth factor (Fig. 5-5).

References

Abelson, J. N. and M. I. Simon. (1990). "Methods in Enzymology." Volume 175. Academic Press, San Diego.

Ainsworth, G. C., and A. S. Sussman. (1965). "The Fungi." Vol. 1. Academic Press, New York.

Blumenthal, H. J. (1965). Carbohydrate metabolism. *In* "The Fungi, An Advanced Treatise." (G. C. Ainsworth, and A. S. Sussman, eds.), pp. 229–348. Academic Press, New York.

Brock, T. D., D. W. Smith, and M. T. Madigan. (1988). "Biology of Microorganisms," 5th Ed. Prentice-Hall, Englewood Cliffs, New Jersey.

Cano, R. J., and J. S. Colomé. (1986). "Microbiology." West Publishing, New York.

Colowick, S. P., and N. O. Kaplan. (1988). "Methods in Enzymology, Volume 160 Biomass. Part A, Cellulose and Hemicellulose." (W. A. Wood and S. T. Kellogg, eds.), p. 774. Academic Press, New York.

Corbett, J. J. (1974). "The Biochemical Mode of Action of Pesticides." Academic Press, New York.

Dixon, M., and E. C. Webb. (1964). "Enzymes." 2nd Ed. Academic Press, New York.

Fersht, A. (1977). "Enzyme Structure and Mechanism." W. H. Freeman, San Francisco.

Garraway, M. O., and R. C. Evans. (1984). "Fungal Nutrition and Physiology." John Wiley, New York.

Griffin, D. H. (1981). "Fungal Physiology." John Wiley, New York.

Highley, T. L. (1989). Wood decay: new concepts and opportunities for control. *Proc. Amer. Wood Pres. Assoc.* **85**:71–77.

International Union of Biochemistry. (1961). "Report of the Commission on Enzymes." Pergamon Press, New York.

Raven, P. H., R. F. Evert, and S. E. E. Eichhorn. (1986). "Biology of Plants," 4th Ed. Worth Publishers, New York.

Sigman, D. S., and P. D. Boyer (Eds.). (1990). "The Enzymes. Vol. XIX. Mechanisms of Catalysis" 3rd Ed. Academic Press, San Diego.

Smith, J. E., and D. R. Berry. (1976). "The Filamentous Fungi. Volume II. Biosynthesis and Metabolism." John Wiley, New York.

Stryer, L. (1981). "Biochemistry," 2nd Ed. W. H. Freeman, San Francisco.

CHAPTER
6

The Decay Setting: Some Structural, Chemical, and Moisture Features of Wood in Relation to Decay Development

In order to understand the anatomical and chemical aspects of the decay process, a brief review of wood structure is needed, emphasizing the types and locations of the major chemical constituents of cells walls and the dimensions of the various openings and capillaries of the wood system that control enzyme access and water accumulations.

These topics are presented in detail in the various books and review articles cited on wood anatomy, wood ultrastructure, wood technology, wood–liquid relations, reaction woods, and wood chemistry (Côté, 1965, 1976, 1977, 1981; Core et al., 1976; Fengel and Wegener, 1984; Higuchi, 1990; Higuchi, 1985; Panshin and de Zeeuw, 1980; Skaar, 1972, 1988; Siau, 1984; Timell, 1964, 1965a, 1982; and Zimmerman, 1964).

For our purposes wood will be defined broadly as the accumulated xylem of the aerial stems of perennial plants. Commercial woods come primarily from the Gymnosperms (conifers) and dicotyledons (hardwoods) of the Angiosperms. In the tropical zones, a few monocotyledons (palms and bamboo) are also used for structural purposes. As a plant material originating from the stems of many different species, wood is a complex and highly variable material both *structurally* and *chemically*.

Wood Functions

The functions of wood in the living plant provide useful insights into its structure and help to explain how saprobes discolor and decay wood products. The wood-tissue system consists of clusters of specialized xylem cells carrying out four major functions.

1. *Conduction* of water and various solutes through tracheids or vessels from the roots to the leaves.
2. *Support* to hold erect a large assemblage of branches and leaves in a competitive position for light with those of other trees in drastically varying weather conditions. This function requires a supporting material with high tensile, compressive, and bending strength. The stem of a tree closely resembles a loaded vertical cantilever beam in carrying out this function.
3. *Storage* of water and various translocates (e.g., reserve foods, hormones) in both radial and longitudinal parenchyma tissues.
4. *Protection* (durability) of the energy-rich accumulating stem tissues from pathogen invasions and disease or destruction by decay fungi.

Tracheids carry out both the conduction and strength functions in conifers. Conduction is carried out primarily by vessels, whereas the fibers provide support in hardwoods. The parenchyma cells in both groups serve a storage function and provide protection against biological attack by deposition of protective toxicants when the stem is injured or during heartwood formation.

Structural Features of Wood

Growth Patterns and Microscopic Features

Wood represents the annual accumulations of cone-shaped increments of xylem cells originating from a lateral meristem or *cambium.*

The cambium is a continuous ring of meristematic cells that forms around the outer circumference of the developing stem. Cambium originates from the interfascicular cambium of the procambial strands and the ground meristem. The procambial strands and ground meristem originate from the apical meristem in the growing tip of the stem and are primary tissues (Esau, 1965). The cambial cells divide periclinally and form xylem cells (wood) to the inside and phloem cells (bark) to the outside. One cell of the dividing pair remains meristematic. Occasional anticlinal divisions of the cambial cells account for the increases in stem girth. There are two general types of cambial cells based on shape and the tissues formed. Vertically elongated cambial cells, termed *fusiform initials*, form the longitudinal parenchyma, the tracheids in conifers, and the vessels and fibers in hardwoods. Horizontally elongated or cuboid cells, termed *ray initials*, form the radially aligned wood rays, consisting of ray parenchyma and, in conifers, also ray tracheids.

The vascular cambium divides only during the growing season. In temperate zones, the annual accumulations of xylem often result in abrupt *annual rings*, usually consisting of a zone of rapid growth (*earlywood*) in the spring and early summer and slower growth (*latewood*) later in the growing season. In tropical zones, annual rings are not so apparent in many species, and the differences in growth accumulations probably reflect regular seasonal patterns in rainfall. Annual-ring elements in conifers consist primarily of tracheids in uniform radial rows. In angiosperms annual rings consist of vessels, tracheids, and fibers arranged in ring-porous, semiporous, or diffuse porous patterns, depending on the size and distribution of the vessels. Parenchyma cells are intermixed with these conductive and strength-providing tissues in longitudinal and ribbonlike radial distributional patterns. Adjacent wood cells are interconnected by thin zones in the wall and cell ends called *pits*, which are discussed in a later section on the cell wall.

The outer zone of the stem, which contains many living parenchyma cells, is the *sapwood*. Sapwood is white in color in most species and functions for conduction, food storage, and stem protection. The *pith* is a small zone at the center of the stem, consisting of parenchyma cells and originating as a primary tissue from the ground meristem. As the girth of the tree expands and the inner sapwood tissues age and recede from the phloem, increasing numbers of parenchyma cells slowly die, and the tissue develops into heartwood. The transition zone between the sapwood and heartwood is sometimes termed *intermediate wood*. *Heartwood* is the internal core of dead tissue in the stem, which may or may not be colored. It slowly expands outward as the tree ages and growth diminishes. The transition zone may be abrupt or gradual and does not necessarily form uniformly in the same annual ring. The heartwood is deeply colored in some species of *Quercus, Juglans,* and *Fraxinus*, and similar to sapwood in color in species of *Abies, Populus, Picea,* and *Tsuga*. The sapwood is wide in some species such as ponderosa pine (*Pinus ponderosa*) and loblolly pine (*Pinus taeda*) and narrow in Douglas fir (*Pseudotsuga menziesii*) and longleaf pine (*Pinus palustris*). Current theory on heartwood origins suggests successive formation of small air bubbles (embolisms) in the older vessels and tracheids, which isolate the adjacent parenchyma cells from food sources, ultimately resulting in their death. Whereas most cells are dead in the older heartwood, a few isolated parenchyma cells are reported to remain alive for years in some species. Significant chemical and structural changes during the transformation of sapwood into heartwood include the loss of starch, the deposition of extractives, and the aspiration of pits in conifers or the formation of tyloses in hardwoods. These changes may render the wood more resistant to biological attack or decrease the permeability, making preservative treatment more difficult. The common macroscopic features of wood are illustrated in Fig. 6-1.

Coniferous wood consists of tracheids and parenchyma cells. Tracheids function in both conduction and structural support. Hardwoods are more complex and are considered to be taxonomically more advanced. Short barrel-shaped cells called vessels form the conductive tissue for hardwoods and long, thick-walled cells called fibers provide strength. Parenchyma cells

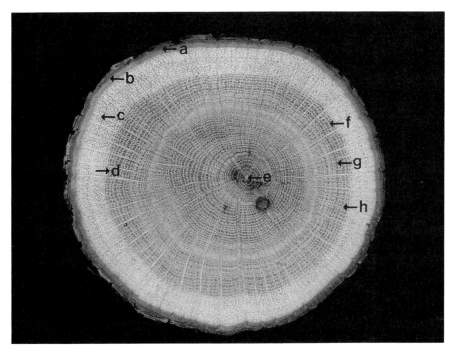

Figure 6-1 Cross section (transverse) of an oak stem (*Quercus alba*) showing the bark (a), cambial zone (b), sapwood (c), heartwood (d), pith zone (e), earlywood (f) and latewood zones of annual rings (g), and a large compound wood ray (h).

carry out food storage, translocation, and protective functions in both wood types. The major cell types in wood are illustrated in Fig. 6-2. Parenchyma cells are arranged radially in wood rays, but are also axially oriented in woods that form longitudinal parenchyma, which often surrounds the vessels. Longitudinal parenchyma in some conifers, such as the pines, larches, spruces, and Douglas fir, form large resin canals that are visible to the unaided eye. Characteristic features of conifers are the arrangement of tracheids in radial rows of uniform width and inconspicuous wood rays. In contrast, the *wood rays* are a prominent feature of many hardwoods, approaching 25% of the wood volume in some species such as the oaks and beeches.

It is convenient to describe wood anatomically as the tissues appear in the three planes of a precisely cut wood cube. The *cross* or *transverse section* represents the plane oriented at a right angle to the longitudinal axis of the stem. The transverse face exposes the open ends of the longitudinally aligned or axial cells. The *radial section* is the surface exposed by a plane running from the outer stem and bisecting the stem center or pith zone. In this plane, the radial surface of the wood rays is exposed. The *tangential* plane is oriented at a tangent to the outer stem surface, at a right angle to the radial section. These

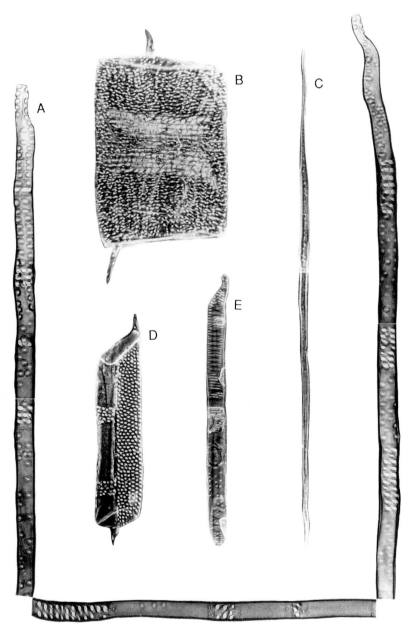

Figure 6-2 A radial view of the principal cell types of a conifer and a hardwood indicating their relative sizes and the types and frequency of pitting: (A) conifer tracheid, (B) earlywood vessel, (C) libriform fiber and (D) and (E) latewood vessels. Note the greater length of the conifer tracheid and the abundance of pitting in the vessels. The zones of abundant pitting in the tracheid are where ray parenchyma were in contact. Courtesy of Dr. W. A. Côté, from Kollman and Côté (1968), with permission of Springer-Verlag.

wood planes are illustrated by photomicrographs of a typical conifer and hardwood (Fig. 6-3).

Cell-Wall Ultrastructure

Wood cell walls consist primarily of cellulose, hemicellulose, and lignin, in the form of large biopolymers. The chemical structures and interrelationships of these components are discussed later in this chapter.

Cellulose molecules in the cell wall form microfibrils, which are surrounded by hemicellulose. Microfibrils are laid down successively and form layers or laminae in the cell wall as growth occurs. After cell maturation, lignin deposition occurs. Wardrop (1964) has proposed the useful terms of *framework* for the cellulose microfibrils, *matrix* for the hemicellulose, and *encrustant* for the lignin, to explain both their functions and interrelationships in cell-wall structure.

Tracheids in conifers and fibers in hardwoods are the predominant cell types affecting strength and the many use properties of wood, so this section will be limited to their cell-wall structure. A cross section of a typical fiber or tracheid cell wall reveals a common organizational pattern. The *middle lamella* (ML) is a narrow zone between contiguous cells and consists primarily of pectins and lignin. The middle lamella is derived from the new cell plate that forms during the mitotic division of a cambial cell into another cambial cell and a cambial initial (which develops into either a xylem cell on the inside or phloem cell on the outside). In carefully stained and prepared sections, the middle lamella may be seen as a fine dark line between contiguous cells. A thin primary wall (PW) is initially laid down during enlargement and maturation of the cambial initial into a tracheid or fiber. The wall consists of a loose network of mostly axially oriented cellulose microfibrils. The secondary wall (SW) develops next and consists of three successively formed layers termed the S1, S2, and S3. The S1 and S3 are narrow zones in which the cellulose microfibrils are arranged in a flat helix. The S2, which composes the bulk of the cell wall, consists of microfibrils arranged in a steep helix, oriented nearly parallel to the longitudinal axis of the cell. The S2 layer is the most important zone of the cell wall and is responsible for a majority of wood-strength properties, particularly its remarkable tensile strength. The S2 cell wall composes most of the cell wall seen in cross sections under the light microscope. In some woods, a warty layer develops on the S3 surface. This layer represents either additional depositions of the S3 wall material or accumulations of protoplasmic debris after cell death. Discerning individual layers (PW and the S1, S2, and S3 of the SW) requires the higher magnification of the electron microscope. The various parts and layers of the cell wall are illustrated in Fig. 6-4. The inner cell cavity is termed the cell *lumen*. In the living tree or the wood product, the lumen is inert space occupied by air and/or water. The lumen volume, collectively, in most woods is large, and will be seen later as the critical cell-wall zone where most decay fungi initiate the decay process.

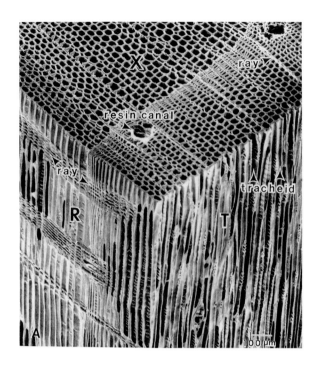
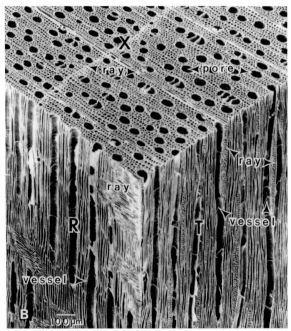

Figure 6-3 The minute features of wood are illustrated generally by surfaces through the cross or transverse (X) radial (R), and tangential (T) planes. Cubes of (A) eastern white pine (*Pinus strobus*) and (B) red maple (*Acer rubrum*) indicate the planes and several important microscopic features. Courtesy of Dr. W. A. Côté (1981), with permission of Springer-Verlag for (B).

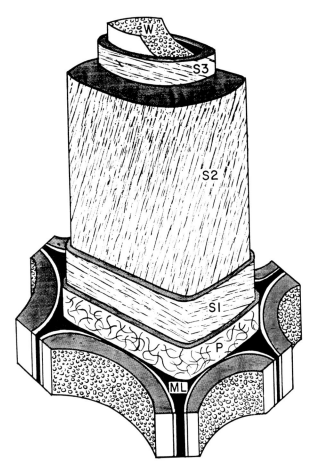

Figure 6-4 A model of several contiguous wood cells showing the organization and microfibrillar orientation of the major cell-wall layers. The layers are identified from the middle lamella (ML) and outward as primary wall (P), the S1, S2, and S3 composing the secondary wall, and the warty (W) lining of the lumen surface. Courtesy of Dr. W. A. Côté (1976), and with permission of the University of Washington Press.

Cell-Wall Pittings

Conduction of water and various solutes between adjacent cells occurs across contiguous thin zones in the wall termed *pits*. Pits are the principal cell-wall zone initially penetrated by hyphae during wood colonization and degradation by stain and decay fungi. Pits are gaps in the secondary wall containing a modified portion of the adjacent primary walls called the *pit membrane*. The types of pits vary with cell types and plant species.

The tracheids of many commercial conifers (Pinaceae) and fibers possess *bordered pits*. In the bordered pit, the membrane consists of cellulosic strands (margo) and a thickened central portion termed a *torus* (Fig. 6-5). The spaces between the strands of the membrane are large enough to permit liquid flow and small particle passage (up to 1 µm) between adjacent cells. During heartwood formation, these openings may become occluded with extractives. The secondary wall of each adjacent cell in the pit pair forms a partial arch around the connecting opening or *aperture* of each cell member. The torus may block the aperture after cell death, changes in moisture content, or

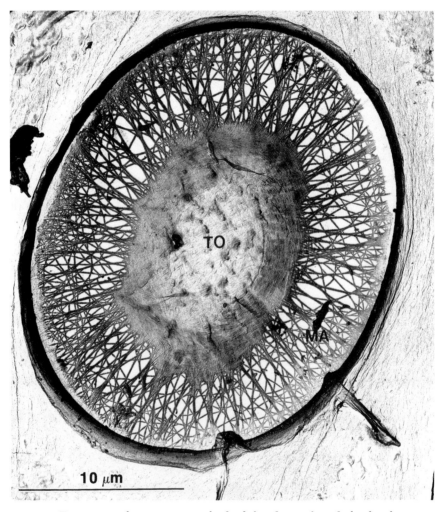

Figure 6-5 Transmission electron micrograph of radial surface replica of a bordered pit membrane from an earlywood tracheid of eastern white pine (*Pinus strobus*). Note the nature and dimensions of the openings among the margo (MA) and the thickened center or torus (TO). Courtesy of Dr. H. Core, *et al.* (1976), with permission of Syracuse University Press.

alterations in air pressure. Wood with many blocked (aspirated) pits is difficult to season or treat with preservatives.

The pits between ray parenchyma cells and adjacent tracheids are termed *semibordered pits*. The membranes in these pits lack a torus and are continuous. The pits between parenchyma cells are termed *simple pits* and contain numerous small plasmodesmatal pores (Harada and Côté, 1985). The pits between vessels and adjacent parenchyma cells in some hardwoods form tyloses. These *tyloses* are formed when the pit membrane bulges or balloons into the vessel lumen, occluding fluid flow. In some species such as the white oaks, tyloses may completely block the large earlywood vessels (Fig. 6-6). Tylosis formation followed by wilting is a common symptom of many severe diseases of forest trees, such as oak wilt or Dutch elm disease.

Major Chemical Constituents of Wood

Cellulose, the hemicelluloses, and lignin are the three major types of chemical constituents of wood cell walls. They are large macromolecules or bio-

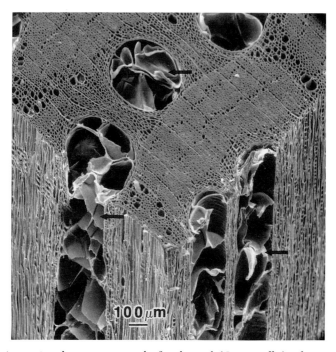

Figure 6-6 A scanning electron micrograph of a white oak (*Quercus alba*) indicating numerous tyloses (identified by arrows), which are occluding the large vessels in the earlywood zones. Courtesy of the Nelson C. Brown Center for Ultrastructure Studies, SUNY College of Environmental Science and Forestry, Syracuse, New York. ×13,140.

polymers that are closely associated physically or covalently bonded in the case of lignin and the hemicelluloses. Collectively, these cell-wall polymers represent the major organic compounds in the biosphere and are a principal carbon sink in terrestrial ecosystems. Cellulose and the hemicelluloses are carbohydrates that are readily digested by many organisms. Lignin is an aromatic heteropolymer consisting of condensed phenylpropane units and is a recalcitrant compound that can be degraded by only a few groups of specialized fungi (the wood decayers) or bacteria and often over long periods.

Cellulose

Cellulose is a long, linear homopolymer consisting of β-D-glucose residues connected by (1→4) glycosidic linkages (Fig. 6-7A). The surfaces of the cellulose molecules contain three exposed hydroxyl groups per anhydroglucose unit, which control the structural properties in the cell wall as well as many physical and chemical properties of wood (Fig. 6-7B). An average degree of polymerization (DP) of around 10,000 has been determined for bark and wood cellulose (Goring and Timell, 1962), but the degree of polymerization of plant celluloses can vary between 3000 and 26,000. The cellulose chains are aligned in parallel and form the microfibrils discussed above. Microfibrils can be observed using the transmission electron microscope (TEM), and these structures seem to be the common form of all natural or native celluloses.

Microfibrils consist of highly organized or crystalline zones interspersed with noncrystalline or unorganized zones. The degree of crystallinity for cellulose in wood ranges from 60 to 70%.

A single cellulose molecule in a microfibril may continue through several crystalline and noncrystalline zones. X-ray diffraction studies of the crystalline zones have defined the precise dimension of a unit cell in the crystal (Fig. 6-7C). The surfaces of the microfibrils are surrounded by hemicelluloses (Fig. 6-8A). The polysaccharides are laid down successively and form layers or lamina in the cell wall as growth and cell enlargement occurs. As the cell matures, lignin deposition occurs. The hemicellulose and lignin are associated primarily with the noncrystalline zones that occur within and between the microfibrils (Fig. 6-8B).

Hemicellulose and Other Carbohydrates

Hemicelluloses are polymers of various pentose and hexose sugar units. The major sugar residues in the polymer backbones are glucose, xylose, mannose, galactose, arabinose, rhamnose, and uronic acids. These polymers differ from cellulose in having short chain lengths, side chains that are sometimes branches, and sugar monomers other than glucose. The types and amounts of the hemicelluloses present in the cell walls of hardwoods and conifers differ. Softwoods contain fewer hemicelluloses than do hardwoods, with mannose being the most common constituent of conifer hemicellulose. Xylose is a

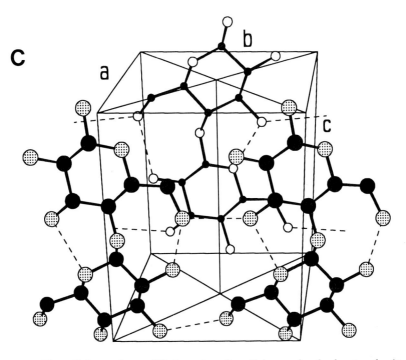

Figure 6-7 The cellulose polymer. (A) A portion of a cellulose molecule showing the (1→4) glycosidic linkage between two glucose units. (B) The same portion drawn in a chair configuration showing the equatorial position of the hydroxyls on the number 2, 3, and 6 carbons. (C) A unit cell of parallel-chain ramie cellulose I showing the dimensions and relative positions of the glucose units in the crystalline structure. The unit cell parameters are a = 7.78 Å, b = 8.20 Å, c (fiber axis) = 10.34 Å, = 96.5. The chains in the 100 and 200 planes (corner and center chains, respectively) are hydrogen-bonded into sheets, whereas there are no hydrogen bonds between the sheets. (Hydrogen atoms are not shown; hydrogen bonds are indicated with dashed lines). (A) and (B), courtesy of Dr. T. Timell and (C), Dr. A. Sarko (Woodcock and Sarko, 1980).

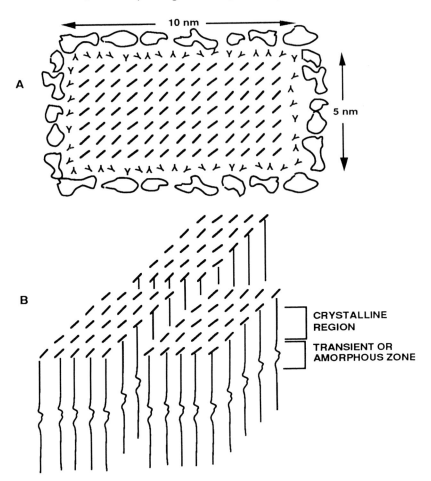

Figure 6-8 A schematic illustration of the postulated arrangement of the long, linear cellulose molecules in a microfibril. (A) A cross section through the crystalline zone of a microfibril. Representations are lines, cellulose; Ys, hemicelluloses; and amorphous units, lignin. (B) A longisection of several microfibrils showing the associations and relative sizes of the crystalline and amorphous zones. The hemicelluloses and lignin are omitted in this sketch.

major constituent of hardwood hemicellulose, which also contain more acetyl groups than that of conifers (Table 6-1).

Xylan, the predominant hemicellulose in hardwoods, is a homopolymer of β-D-xylose monomers connected by (1→4) glycosidic bonds. Side branches of 4-*O*-methyl-α-D-glucuronic acid are attached by (1→2) linkages to some xylose units, and *O*-acetyl groups substitute some hydroxyls (Timell, 1964) (Fig. 6-9A). *Glucomannan*, the major hemicellulose in conifers (up to 20% of cell wall) is a heteropolymer with a backbone containing β-D-glucose and β-D-mannose units connected by (1→4) glycosidic bonds (Fig. 6-9B). Acetyl

The Major Chemical Constituents of Wood / 149

Table 6-1
Major differences in the sugar residues from hemicellulose of three conifers and three hardwoods[a]

Species	Content in extract-free stemwood		
	Xylose (%)	Mannose (%)	Acetyl group (%)
Abies balsamea	5.2	10.0	1.4
Pinus sylvestris	7.6	12.4	1.6
Picea glauca	7.0	12.0	1.2
Fagus grandifolia	21.7	1.8	4.3
Populus tremuloides	12.2	3.5	3.9
Betula papyrifera	23.9	2.0	3.9

[a]Summarized from the date of Côté et al. (1966) and Timell (1969).

groups and galactose residues are attached to some monomers in the backbone (Timell, 1964, 1965a, b). The xylan in conifers has arabinose side chains instead of acetyl groups. Larch trees often contain large amounts of an arabinogalactan in their heartwood.

Many other types of hemicellulose occur in smaller amounts in wood materials. Pectin is found in the middle lamella zone and in the torus of bordered pits. A *glucan*, starch, is an important food reserve in plants and is often found in abundance in wood parenchyma cells. One constituent of starch,

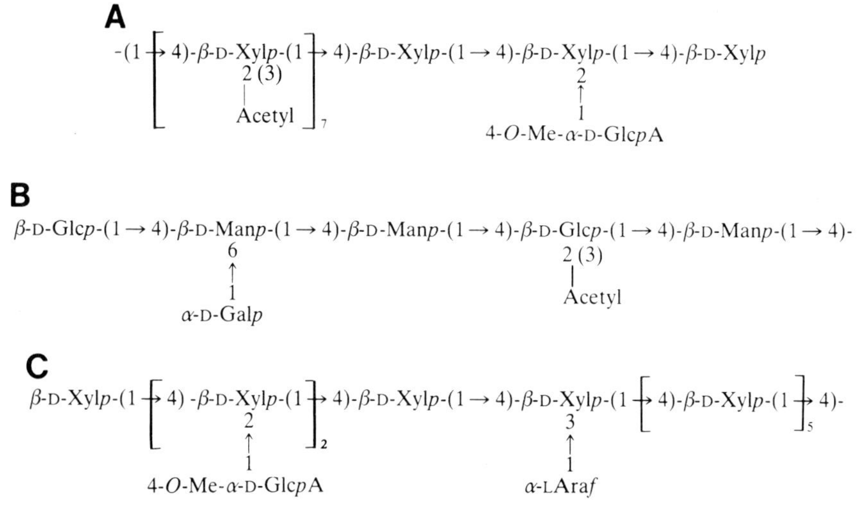

A
$$-(1 \rightarrow 4)\text{-}\beta\text{-}D\text{-}Xylp\text{-}(1 \begin{bmatrix} \rightarrow 4)\text{-}\beta\text{-}D\text{-}Xylp\text{-}(1 \\ 2\,(3) \\ | \\ \text{Acetyl} \end{bmatrix}_7 \rightarrow 4)\text{-}\beta\text{-}D\text{-}Xylp\text{-}(1 \rightarrow 4)\text{-}\beta\text{-}D\text{-}Xylp\text{-}(1 \rightarrow 4)\text{-}\beta\text{-}D\text{-}Xylp$$
$$\begin{array}{c} 2 \\ \uparrow \\ 1 \\ 4\text{-}O\text{-}Me\text{-}\alpha\text{-}D\text{-}GlcpA \end{array}$$

B
$$\beta\text{-}D\text{-}Glcp\text{-}(1 \rightarrow 4)\text{-}\beta\text{-}D\text{-}Manp\text{-}(1 \rightarrow 4)\text{-}\beta\text{-}D\text{-}Manp\text{-}(1 \rightarrow 4)\text{-}\beta\text{-}D\text{-}Glcp\text{-}(1 \rightarrow 4)\text{-}\beta\text{-}D\text{-}Manp\text{-}(1 \rightarrow 4)\text{-}$$
$$\begin{array}{cc} 6 & 2\,(3) \\ \uparrow & | \\ 1 & \text{Acetyl} \\ \alpha\text{-}D\text{-}Galp & \end{array}$$

C
$$\beta\text{-}D\text{-}Xylp\text{-}(1 \begin{bmatrix} \rightarrow 4)\text{-}\beta\text{-}D\text{-}Xylp\text{-}(1 \\ 2 \\ \uparrow \\ 1 \\ 4\text{-}O\text{-}Me\text{-}\alpha\text{-}D\text{-}GlcpA \end{bmatrix}_2 \rightarrow 4)\text{-}\beta\text{-}D\text{-}Xylp\text{-}(1 \rightarrow 4)\text{-}\beta\text{-}D\text{-}Xylp\text{-}(1 \begin{bmatrix} \rightarrow 4)\text{-}\beta\text{-}D\text{-}Xylp\text{-}(1 \\ \end{bmatrix}_5 \rightarrow 4)\text{-}$$
$$\begin{array}{c} 3 \\ \uparrow \\ 1 \\ \alpha\text{-}L\text{-}Araf \end{array}$$

Figure 6-9 The partial chemical structures of the predominant hemicellulose polymers commonly found in wood. (A) A xylan in hardwoods (O-acetyl-4-O-methylglucuronoxylan); (B) A galactoglucomannan in softwoods (O-acetyl-galactoglucomannan); and (C) A xylan in softwoods (arabino-4-O-methylglucuronoxylan). Courtesy of Dr. T. Timell (1986b).

amylose, consists of β-D-glucose units linked by (1→4) glycosidic bonds and is linear, whereas *amylopectin*, the major starch component is branched at position C-6.

Hemicelluloses in the wood cells probably serve a structural function by coating and binding the cellulose microfibrils into a common matrix, and may also serve to prevent the cellulose from becoming too crystalline. The short chain lengths (DP 200 in most cases), which increase hemicellulose solubility and the outer exposed position on the surface of the microfibrils, may explain why these polymers are among the first cell-wall components attacked by decay fungi.

Lignin

Lignin, the most complex of the cell-wall constituents, is a polyphenolic polymer formed from three types of phenyl propane units (Fig. 6-10). These monomeric units condense by free radical polymerization to form the huge, heterogeneous aromatic biopolymer. Lignin composes about 20 to 30% (in a few species the lignin content approaches 40%) of the wood cell wall and is a constant feature of all vascular plants (ferns, fern allies, and seed plants). Lignin provides both mechanical strength and protection (durability) to stem tissues. Lignification may be viewed as a remarkable evolutionary event that permitted the development of aerial plants from which the major timber species originated with their large, vertical, perennial stems.

The structure of lignin varies between conifers and hardwoods. Guaiacylpropane units are the principal repeating monomer in conifers, whereas

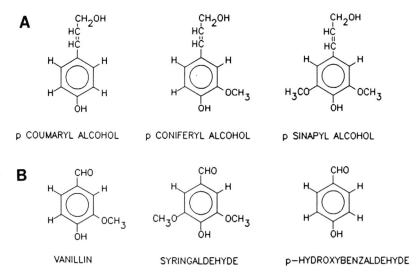

Figure 6-10 (A) The principal precursors of lignin, coniferyl, sinapyl, and *p*-coumaryl alcohols. (B) Several typical lignin degradation products.

both guaiacyl and syringylpropane units are present in hardwoods. Lignins also contain small amounts of *p*-hydroxyphenylpropane units, a lignin monomer common in monocotyledonous plants. The principal precursors of lignin formation in the cell wall are three *p*-hydroxycinnamyl alcohols; coniferyl, sinapyl, and *p*-coumaryl alcohols (Fig. 6-10). Dehydrogenation of these alcohols forms phenoxyradicals that, by subsequent dehydrogenation, polymerize to form lignin. The polymerization process of these phenoxy radicals is random, and the lignin macromolecule formed has none of the predictable fixed or repeating structures that are found in the other cell-wall constituents, cellulose and hemicelluloses.

The principal linkages among the various phenylpropane units are carbon to carbon (C—C) and ether (C—O—C) bonds. Of the many interunit types of linkages present in lignin, the β-aryl ether linkage (β-O-4) occurs more than 50% of the time (Fig. 6-11). Lignin deposition begins after the new xylem cell has enlarged and the polysaccharides have been laid down in the outer part of the secondary wall. The process proceeds slowly, following successive formation of the layers of the secondary wall (S1, S2, and S3), and then progresses more rapidly after these layers are complete. Lignin is probably interspersed in the spaces within and between the hemicellulose-coated cellulose microfibrils, forming an interpenetrating polymer system with the hemicelluloses. Lignin links covalently with some hemicellulose units. Lignification provides cohesion and strength to the cell wall and serves as an effective barrier against microbial access and digestion of the carbohydrates. Lignin is an amorphous polymer whose chemical structure can be modeled only on the basis of the kinds of units, frequency of linkages, and the nature of some of its degradation products. Expanded models of lignin, based on a computerized analysis program involving 94 phenylpropane units, have been developed (Glasser *et al.*, 1981).

Miscellaneous Cell-Wall Chemicals

Extractives A number of other compounds are present in the cell wall, including extractives. As the term implies, extractives are mostly low-molecular-weight compounds that are readily extracted from the wood by solvents such as water, alcohol, benzene, or ether. These compounds are found in parenchyma cells and in the lumen of other cells in amounts ranging from 1 to 5% of the total wood weight. In exceptional cases, extractives may represent 10–40% of the wood weight. Extractives represent many classes of compounds including a large number that are species specific. A few examples are carbohydrates, such as starch, glucose, fructose, and sucrose; phenolic compounds, such as stilbenes, tannins, phlobaphenes, flavanoids, and lignans; oils and waxes; esters of organic acids; alkaloids; and tropolones.

Extractives affect many wood properties, and two of special concern in this textbook are susceptibility to sapstains (both biotic and abiotic stains) and the natural durability of the heartwood. These effects will be considered in

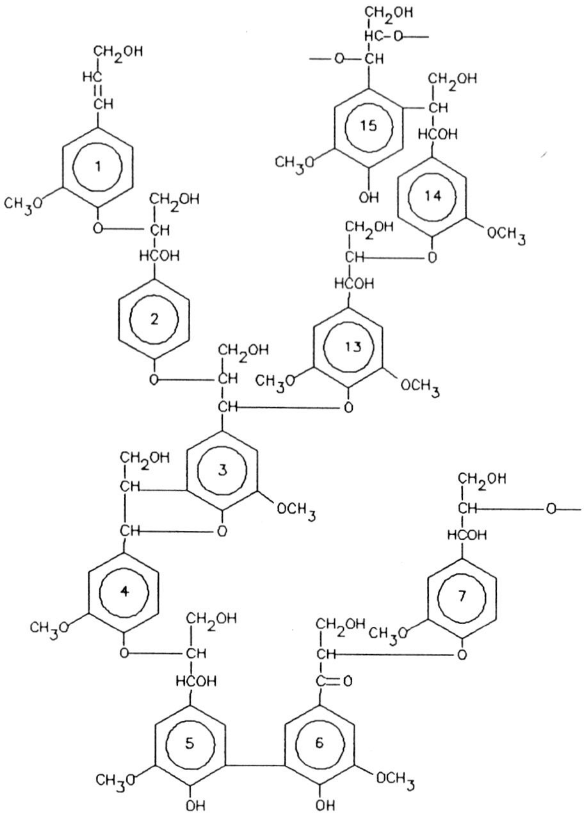

Figure 6-11 A structural model of spruce lignin. Examples of the common guaiacylglycerol-β-aryl ether linkage (β-O-4) can be seen between units 1 and 2. A syringyl phenyl propane unit with two methoxyl groups (OCH_3) in the benzene ring can be seen in unit 13, and biphenyl linkages (C-C) can be seen between units 5 and 6 (Adler, 1977).

more detail in the later chapters, "Wood Sapstains" (Chapter 14) and "Wood Durability" (Chapter 18).

Ash Content The inorganic or mineral content of wood is low and rarely exceeds 1% in temperate zone species. Some tropical hardwoods contain high levels of silica, which improves the resistance to marine borers. Mineral content is determined by incinerating the wood under controlled temperature conditions to reduce the losses of volatile ash components. The principal elements present are calcium, potassium, and magnesium. Other common elements include manganese, sodium, phosphorus, and chlorine, as well as low concentrations of trace elements (Young and Quinn, 1966). Several minerals play significant direct or indirect roles in the development of decay and wood defects. *Mineral stain*, a serious discoloration in some hardwoods, is characterized by an abnormally high calcium carbonate content and excessive warping (Fig. 14-2). It is believed to originate as an injury response to outer sapwood in the living tree. Calcium is reported to increase the disease resistance of tissues in the living stem to pathogen attack in herbaceous plants (Agrios, 1988). Both iron and manganese have been proposed to play roles as oxidants in some decays and are discussed further in Chapter 9. It is theorized that the release of cell-wall cations during decay is one of the factors responsible for increases in electrical conductivity (Shortle, 1982), a characteristic that has been proposed as a decay-detection device (Shigo and Shigo, 1974). Historically, it is interesting to note that the alkaline nature of leached wood ash (Ca, K) was used in colonial times to saponify fats for the preparation of soap, and large acreages of the original eastern forests were burned for commercial potash production.

Proteins Small amounts of protein are present in wood cells, with the largest amounts in the cambium, early xylem derivatives, and parenchyma cells. Only trace amounts of protein are present in the dead xylem cells. The protein is the principal source of the nitrogen in wood, which is very low compared with that in other plant forms, ranging from only 0.03% in heartwood to as high as 0.1% in the young sapwood. Nitrogen is discussed in more detail in Chapter 4 as a necessary growth factor for fungi in wood.

Amounts and Distributions of Cell-Wall Components

Wood is not only highly heterogeneous in the various arrangements of the cell types, but also in the amounts and distributions of the major cell-wall components. Information on the distribution of cell-wall components helps explain the differing ways various microorganisms attack and differentially utilize cell-wall parts. For example, some bacteria selectively attack the cellulosic strands in the margo in the bordered pits of some pines; some soft-rot fungi selectively attack the S2 of the secondary walls in conifers, and individual wood cells are detached in the intermediate stages of white rot caused by *Phellinus (Fomes) pini.*

Although the distribution of chemicals in cell walls differs by species, stem position, rate of growth, or the presence of heartwood/sapwood, some generalizations concerning wood chemistry can still be made (Table 6-2). Cellulose content is relatively constant across all species and represents 40–50% of cell-wall substance. In temperate zone woods, lignin is present in higher levels in conifers than in hardwoods; the difference is made up by the higher hemicellulose content in hardwoods. The chemical components of five hardwoods and five conifers from the temperate zone indicate the relative uniformity of cellulose levels in all the species studied as well as the higher lignin and lower hemicellulose contents in the conifers as contrasted with lower lignin and higher hemicellulose contents in hardwoods (Table 6-3). Tropical hardwoods have higher lignin and ash contents than most temperate zone woods.

Table 6-2
Chemical composition of wood from five hardwoods and five conifers[a]

Component	Hardwoods				
	Acer rubrum L.	Betula papyrifera Marsh.	Fagus grandifolia Ehrh.	Populus tremuloides Michx.	Ulmus americana L.
Cellulose	45	42	45	48	51
Lignin	24	19	22	21	24
O-Acetyl-4-O-methyl-glucurono-xylan	25	35	26	24	19
Glucomannan	4	3	3	3	4
Pectin, starch, ash, etc.	2	1	4	4	2

Component	Conifers				
	Abies balsamea (L.) Mill	Picea glauca (Moench) Voss	Pinus strobus (L.) Carr.	Tsuga canadensis L.	Thuja occidentalis L.
Cellulose	42	41	41	41	41
Lignin	29	27	29	33	31
Arabino-4-O-methyl-glucurono-xylan	9	13	9	7	14
O-Acetyl-galacto-glucomannan	18	18	18	16	12
Pectin, starch, ash, etc.	2	1	3	3	2

[a]All values given in percentage of extractive-free wood. Data from Kollmann and Côté (1968). Results provided by T. E. Timell.

Distributions of the Major Chemicals in the Wood Cell Wall

Individual chemical components vary widely across cell walls, but some generalizations can be made regarding the distribution of the principal chemical constituents among cell types and within cell-wall layers. Cellulose is present at the highest levels in the secondary wall and is least abundant in the compound middle lamella (middle lamella and the adjoining primary walls). Hemicellulose levels are highest in the S1 and lowest in the S2 of the secondary wall tracheids and fibers. The hemicellulose content is higher in parenchyma cells, but hemicellulose types and distributions vary greatly among species. The ray parenchyma cells have a higher xylan content than do tracheids and fibers.

The removal of cellulose by acid hydrolysis and careful sectioning for electron microscopic study of the residual lignin clearly demonstrate the high levels of lignin present in the middle lamella and primary wall as well as the lower and relatively uniform distribution in the secondary wall (Fig. 6-12).

Table 6-3
Major chemical components of a representative group of temperate zone hardwoods and conifers contrasted with several tropical hardwoods[a]

Name	Amount of chemical component (%)							
	Holocellulose	Cellulose	Hemicellulose	Pentosans	Lignin	Ethanol-benzene extracts	Hot-water extracts	Ash content
Temperate zone—conifers								
Abies balsamea	70.0	49.4	15.4	7.0	27.7	4.3	3.6	0.4
Picea abies	80.9	46.0	15.3	8.3	27.3	2.0	2.0	—
Pinus sylvestris	14.3	52.2	13.5	8.2	26.3	—	—	—
Pseudotsuga menziesii	67.0	50.4	—	6.8	27.2	4.4	5.6	0.2
Sequoia sempervirens	71.8	49.9	16.7	—	37.0	13.5	8.7	0.2
Temperate zone—hardwoods								
Populus tremuloides	80.3	49.4	21.2	17.2	18.1	3.8	2.8	0.4
Fagus sylvatica	85.8	44.5	30.2	20.6	22.2	—	—	—
Quercus sp.	73.2	40.5	23.3	17.5	22.2	—	—	—
Acer rubrum	71.0	44.5	—	17.1	22.8	2.5	4.4	0.7
Robinia pseudoacacia	81.7	50.1	—	23.7	20.6	2.8	4.6	0.3
Tropical hardwoods[b]								
Obeche	77.2	47.8	20.1	16.8	21.3	12.6	4.2	1.8
Kefe, awari	78.1	44.9	25.1	15.8	22.7	2.6	2.6	1.3
Teak	—	39.1	—	13.0	29.3	13.0	1.8	0.7
Mahogany	—	43.9	—	16.0	28.2	3.5	3.3	1.1
Balsa	—	52.0	—	19.0	24.5	2.6	2.8	1.6

[a]Data selected from Fengel and Wegener (1984).
[b]Scientific names for the tropical hardwoods selected are obeche (*Triplochiton sceroxylan* K. Schum.); Kefe, awari (*Pterogota macrocarpa* K. Shum.); teak (*Tectona grandis* L); African mahogany (*Khaya anthotheca* C.D.C); and balsa (*Ochroma lagopus* SW).

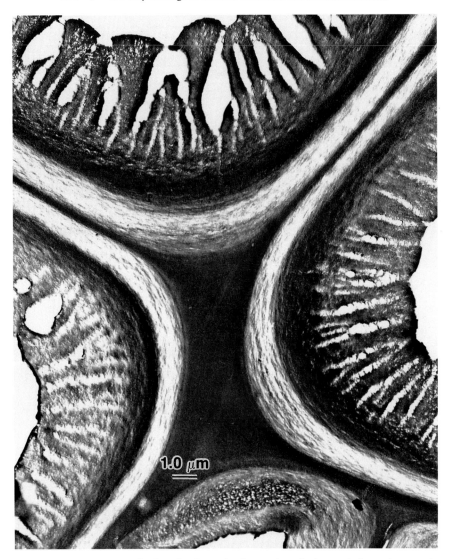

Figure 6-12 Cross-section of a Douglas fir tracheid, revealing the lignin skeleton that remains after removal of the carbohydrates by successive exposures to HF. The concentration of the lignin in the middle lamella and network distribution in the secondary wall are indicated. Courtesy of Dr. W. A. Côté (1976), with permission of the University of Washington Press.

The selective removal of carbohydrates by brown-rot fungi has also been used to prepare residual lignin for study (Côté et al., 1966). New detection techniques for lignin involving either ultraviolet microscopy or bromination followed by energy dispersive x-ray analysis (TEM-EDXA) (Saka and Goring, 1985) now permit the detailed mapping of lignin composition in the cell-wall layers of conifers and hardwoods. Substantial differences occur in various cell types. Ray parenchyma cells and the secondary walls of hardwood fibers primarily contain a syringyl-type lignin, whereas vessel walls contain mostly a guaicyl-type lignin.

In addition to the major components, pectins are present in the middle lamella zone. Starch and extractives are found predominantly in the parenchyma cells of the living sapwood, although the extractives generally occur in the greatest amounts in the outer zones of the heartwood.

Reaction wood in conifers, commonly known as *compression wood*, develops on the lower or compression side of leaning stems or branches. In hardwoods, reaction wood is termed *tension wood* and develops on the upper or tension side of leaning stems or branches. Compression wood is characterized macroscopically by a dark color, wide growth rings, and microscopically, by intercellular spaces between the rounded tracheids. Cells in these zones contain excessive amounts of lignin in the secondary wall, and the S3 layer is lacking. The cellulose content is low. Finished lumber containing tension wood has a silvery sheen and a fuzzy surface due to pulled fibers. At the anatomical level, tension wood contains so-called gelatinous fibers. These cells have a cell wall layer (G layer) that is unlignified and consists of cellulose. The G layer may develop after deposition of any one of the three wall layers, S1, S2, and S3 (usually S2) (Fig. 6-13). Tension wood is more susceptible to some decay fungi, whereas compression wood is more resistant than normal wood, probably because of its high lignin content (35–40%).

The literature concerning the origins, anatomical features, and chemical constituents of compression wood, and the effects of these characteristics on its utilization, has been summarized by Timell (1982, 1986a).

Organizational Levels in the Cell Wall

There are three general levels of organization in the cell wall, and each profoundly affects access of microbial enzymes and moisture to the chemical constituents. The *gross capillary zone* consists of the cell lumen, pit chambers, pit apertures and pit pores and has general size dimensions ranging from about 10 µm to 2000 Å for the smallest pit pores. It has been estimated that a gram of wood possesses a square meter of gross capillary surface. The *amorphous* or *transient* capillary zone between the crystalline portions of the microfibrils has very small capillaries (*up to 200 Å in size*). This zone is accessible to water in vapor form and is the chemically reactive portion of cell-wall material. The transient zone is accessible to some enzymes and is the area where initial decay often develops. A gram of wood at this level is estimated to possess 300 M^2 of surface. The *crystalline* zone within the highly ordered cellulose crystals is initially inaccessible to water or enzymes. The dimensions of this zone are the distance between cellulose chains in unit cells and are in the order of 8 to 10 Å (Fig. 6-7C). The mode of access of the enzyme molecules to the inner structure of the cell wall is an interesting question and can be introduced briefly at this point. These potential modes include

158 / 6. Decay Setting: Structural, Chemical, and Moisture Features

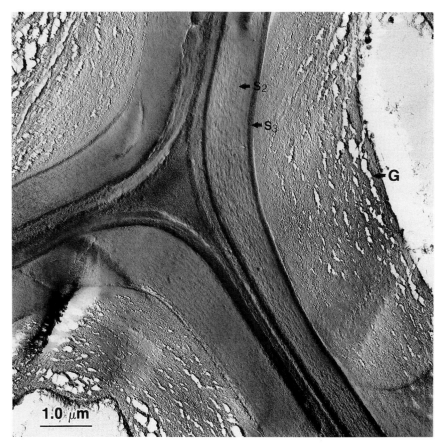

Figure 6-13 A cross section of several gelatinous fibers of *Celtis occidentalis* showing the deposition of a G layer after formation of the S2 and S3. TEM, × 14,400. Courtesy of Dr. W. A. Côté and Arnold Day (1962), with permission of Forest Products Research Society.

delivery by hyphal structures and progressive digestion inward through the wall from the lumen surface, or rapid diffusion of enzymes in a water film into the openings and cavities in the cell wall. The relative sizes of the available openings and the digestion enzymes are discussed further in Chapters 8 and 9 on wood decay.

Wood–Water Relationships

As previously discussed (Chapter 4), certain levels of moisture in wood are required for decay development. Moisture levels also determine the degree of wood swelling and shrinking. These changes can lead to deep checks, particularly in roundwood products or poorly fitting joints in structures, which, as we shall see in the later chapters on decay, are zones where decay hazards may develop. Understanding the wood–moisture relationship is the key for decay control in many wood uses.

The moisture properties of wood result from both its chemical constituents and the capillary nature of the amorphous zones of the microfibrils that make up the cell walls. Water may occur in wood in *liquid* or *vapor* phases in the lumina of the cells and the pit cavities (*gross capillary system*). Water also may occur as *bound* (hydrogen bonding) water on the surfaces, within and between the microfibrils in the amorphous zones. The zone between the microfibrils is termed the *transient capillary system* since it contracts when wood is dried and expands when it is moistened. Changing moisture levels in this zone affect many wood properties such as strength, swelling and shrinking, and electrical conductivity. This zone is also the point where fungal enzymes gain access inside the cell wall and initiate decay.

The principal water sources in wood are residual, from the living tree; liquid water, from the atmosphere; and contact with wet porous materials, such as soil, condensation, and water vapor from the atmosphere.

To understand the various forms of water in wood in relation to various use and treatments, it is useful to start with a wet board with a moisture content of 100%. Liquid water is removed from the surface under low-humidity and/or high-temperature conditions via capillary flow mechanisms and by evaporation from the capillary menisci as water vapor. When water is depleted from the cell lumina (gross capillary system), but the cell walls remain saturated and swollen, the moisture level is termed the *fiber saturation point* (f.s.p.). The f.s.p. ranges from 25 to 35%, depending on wood species and extractive content. As we shall see later, the f.s.p. is a transition zone and might better be defined simply as the level at which many wood properties change. Wood begins to shrink below the f.s.p., and most strength properties increase. Next the bound water in the transient capillary system evaporates from the wood via the now open gross capillary system as water vapor. This process requires that humidity near the surface be lower than that of the wood, and is termed *desorption*. Initially, the water molecules removed are attached to other water molecules by hydrogen bonding, but the residual water molecules are held tenaciously to the surface of the microfibrillar capillary zone; their removal requires lower humidities or higher temperatures. About one third of the moisture below the f.s.p. is held tenaciously as a monomolecular layer by hydrogen bonding to the exposed hydroxyls on the cellulose and other biopolymers in the cell wall. The final bound water can be removed only by prolonged heating or after long periods over desiccants and under vacuum conditions. Elevated temperatures increase the kinetic energy of the bound water molecules, enhancing their release from the wood. Wood in which all the bound water is removed is called oven dry, and the term *oven-dry weight* (o.d.w.) designates this condition.

The reverse process of exposing oven-dry wood to water vapor is termed *adsorption*. A typical adsorption–desorption curve for wood at a range of relative humidities from 0 to 100% is shown in Fig. 6-14A. Wood below the f.s.p. establishes a series of *equilibrium moisture contents* (e.m.c.) at various tem-

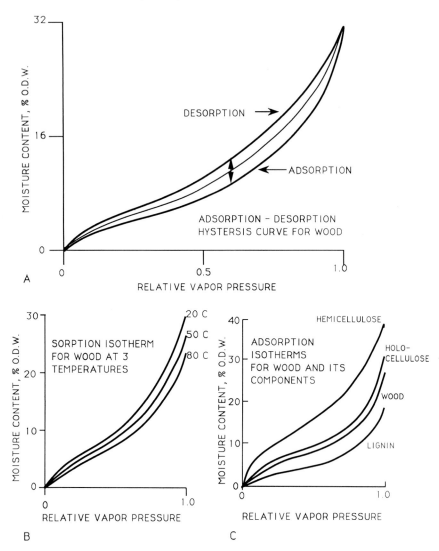

Figure 6-14 Wood–water sorption curves showing (A) equilibrium moisture content curves for wood, illustrating desorption and adsorption curves at various water vapor pressures at a given temperature; (B) the effects of temperature on the sorption characteristics of wood, and (C) a series of curves indicating the different adsorption isotherms for the major wood-cell components. Courtesy of Dr. C. Skaar (1972), with permission of Syracuse University Press.

peratures and relative humidities, termed sorption isotherms (Fig. 6-14B).

The e.m.c. attained are 1–2% higher when wood is equilibrated by desorption (from f.s.p. to e.m.c.) than adsorption (o.d.w. to e.m.c.). This is termed *hysteresis* and is a characteristic of other colloidal materials, such as agar (Fig. 6-14A). The hysteresis effect is a common source of error in wood-decay experiments, when differences in e.m.c. weights before and after decay are used to determine decay capabilities without repeating the original drying direction.

The various chemical constituents in the wood cell have substantial e.m.c. differences. Hemicelluloses equilibrate at the highest, and lignin, the lowest e.m.c. values (Fig. 6-15C). Some other interesting features of wood sorption isotherms include the following.

1. Wood at low moisture contents is very hygroscopic and gains weight rapidly in the presence of water vapor.
2. Wood that has been heated or kiln-dried subsequently attains lower e.m.c. values than does unheated wood.
3. At the same relative humidity conditions, wood attains higher e.m.c. values at low temperatures than at high temperatures.
4. Wood with high extractive contents generally attains lower e.m.c. values.
5. Hardwoods attain higher e.m.c. values than conifers, probably reflecting the higher hemicellulose and lower lignin contents of hardwoods.
6. The slope of the sorption isotherm curves is sigmoidal, which may reflect the difficulty of breaking hydrogen bonding of water molecules to cell-wall-constituent hydroxyls at the low end of the curve, and the ease of water removal from the gross capillary system at the high end of the curve.

The sorption characteristics of wood reflect the strong attraction of water vapor molecules for the exposed hydroxyl (OH) groups that stud the surfaces of the cellulose and hemicellulose molecules in the noncrystalline (amorphous) zones of the microfibrils. Examples of hydrogen bonding between two hydroxyl groups and between an OH and H_2O molecules are diagrammed in Fig. 6-15. Intermolecular hydrogen bonding between hydroxyls on adjacent cellulose molecules plays a role in hysteresis and in the aggregations of cellulose molecules into the supramolecular structures termed *microfibrils*. The amorphous zone retains about one water molecule per available hydroxyl, accounting for approximately 25% of the f.s.p.

Wood swells or shrinks as it gains or loses moisture. The swelling begins just above o.d.w. condition and ends when the f.s.p. condition is attained. Shrinkage begins at the f.s.p. and ends at the o.d.w. condition. The dimensional changes associated with variations in moisture content are essentially limited to the tangential and radial planes of the wood and are caused by the swelling by the transient capillary system as it adsorbs or desorbs water molecules. No significant dimensional changes occur in the longitudinal wood plane, since the cellulose, a long, linear polymer, contains thousands of covalently linked monomer units per molecule.

The differential dimensional changes in wood tangentially and radially lead to the formation of radial splits or checks during the seasoning or use of

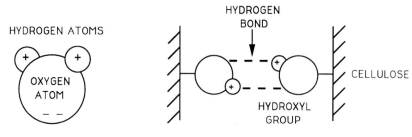

A. Water – a Bipolar Molecule

B. Hydrogen Bond Between Adjacent Cellulose Molecules

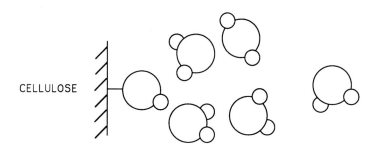

C. Water Molecules Clustering Around a Hydroxyl Group on a Cellulose Molecule

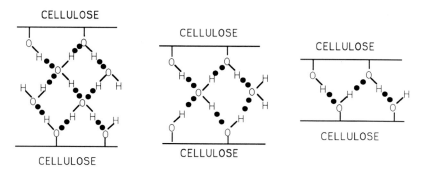

D. Change in Position of Adjacent Cellulose Polymers with Water Loss.

Figure 6-15 Diagrams illustrating the causes and nature of hydrogen bonding. (A) The bipolar nature of the water molecule; (B) hydrogen bonding between hydroxyls on adjacent cellulose molecules; (C) water molecules clustering by hydrogen bonding around a hydroxyl on a cellulose molecule; and (D) reduction of hydrogen bonding as water molecules evaporate during wood drying. (A), (B), and (C) courtesy of Dr. C. Skaar (1972), with permission of Syracuse University Press, and (D) from Fengel and Wegener (1984), with permission of Walter de Gruyter & Company.

some wood products. Checks may become severe in roundwood materials such as poles or piling, particularly when they develop after preservative treatment. Checks in relation to decay development are discussed in Chapters 13 and 15.

A major decay-control principle is evident from the water-sorption characteristics. *Dry wood protected from free water sources never exceeds the f.s.p.*, and *wood kept below the f.s.p. cannot decay*. Some estimated moisture levels of wood used for various interior purposes in several sections of the country are shown in Table 6-4.

Wood Variability

Wood is an extremely variable material that, when used in structures, often requires the application of substantial safety factors in critical uses. As reviewed in this chapter, wood variability is both *structural* and *chemical*. Structural variability reflects the many cell types and organizational patterns, whereas chemical variation represents differences in the types and amounts of cell-wall constituents. The selection of certain species from the hundreds of tree species for special uses, such as lumber, pulpwood, veneer, piling, and poles, reflects properties such as strength, grain pattern, density, hardness, treatability, and form that are specific for that species.

Many wood properties are based on genetic traits and may vary considerably within a species. These traits form the basis for tree-breeding or tree-improvement programs and the selection of seed stock with rapid growth rates, high density, disease resistance, durability, and improved stem form.

Variations in wood properties within a *single stem* are of special interest because they may affect patterns of decay development or decay susceptibility in various wood products. These variations also make it important to carefully select representative wood samples for various testing purposes. Some of the many variations in wood properties include the following.

Table 6-4
Moisture levels of wood used for interior purposes at two seasons in several geographic regions[a]

Region	July	January
Atlanta, Georgia	11.5	8.5
Albuquerque, New Mexico	6.0	7.0
Boston, Massachusetts	13.0	7.0
Madison, Wisconsin	10.0	6.0
New Orleans, Louisiana	13.5	12.5
New York, New York	12.5	7.0
San Francisco, California	10.5	10.5
Washington, D.C.	11.0	8.0

[a]Summarized from the data of Peck (1950).

1. Stemwood is very different structurally and chemically from rootwood and upper branch wood.
2. Nitrogen distribution varies with radial position and is highest in the outer sapwood and pith zones.
3. Extractive contents show extreme variability within the heartwood, both radially and longitudinally.
4. Many species contain a juvenile wood in the stem center with lower fiber lengths and densities.
5. Reaction wood, depending on stem position and wood origin, can severely alter wood properties.
6. Wood densities vary with growth rates and, within certain ranges, rapid growth is associated with high-density woods in hardwoods and low-density, in conifers.
7. The handling and processing of wood may alter properties, e.g., high temperature decreases hygroscopicity, and pond storage may increase porosity.
8. The prior invasion by nonpigmented molds or the early stages of invasion by decay fungi may significantly affect properties such as porosity and strength.

These examples highlight wood variability and stress the need for having a thorough knowledge of wood properties whenever designs incorporating this material are considered.

Summary

- Wood consists primarily of a series of thick-walled, elongated cells, with the dual functions of fluid conduction and structural support for the tree stem. These functions are carried out by tracheids in conifers and by vessels (conduction) and fibers (strength) in hardwoods. About 90% or more of the biomass of most commercial woods is composed of tracheids or fibers.
- Parenchyma cells perform storage (starch and oils) or secretory functions (resins and gums) and form the wood rays, which are radially aligned tissues interspersed between the axial elements.
- Wood cells are interconnected by a series of pits in the secondary walls that permit passage of water and solutes among contiguous cells. The pits are the primary invasion routes for many wood-inhabiting fungi. The parenchyma cells are the major colonization site for the fungi that discolor the sapwood.
- Wood cell walls consist primarily of the large macropolymers, cellulose, hemicelluloses, and lignin. Cellulose is a long, linear polymer consisting of β-D-glucose units with $(1 \rightarrow 4)$ glycosidic linkages, and

Summary continues

forms the cell-wall *framework*. Cellulose occurs in plant-cell walls as bundles of parallel-aligned molecules termed microfibrils. The microfibrils contain alternating crystalline and noncrystalline or amorphous zones. The hemicelluloses are shorter, linear molecules containing hexose or pentose sugars as the monomer units. The monomer units in the main chain are, as in cellulose, connected by (1→4) glycosidic linkages. Some hemicelluloses are branched, and most contain side chains. The hemicelluloses are deposited around the microfibrils and form the cell-wall *matrix*. Lignin is an aromatic polymer formed by free radical polymerization of three types of cinnamyl alcohols. This constituent is a huge amorphous polymer without a regular structure and forms an interpenetrating polymer system around and between the hemicellulose-coated microfibrils of cellulose. As a *cell-wall encrustant*, lignin provides stiffness and strength. Lignin also is a very durable material and acts as a barrier against microbial attack of the more vulnerable carbohydrates in the cell wall.
- The cell walls of tracheids or fibers consist of a primary wall (PW) laid down initially during cell maturation, followed successively by the three layers of the secondary wall (S1, S2, S3). The S2 is the major cell-wall zone, and the microfibrils in this zone are oriented nearly parallel with the stem axis. The S2 layer of the wall has the most significant impact on wood strength, density, and moisture properties.
- The types and distribution of the major chemical constituents in the wood cell wall vary among and within species, as well as within a single cell. Cellulose content is relatively constant across all wood species and makes up about 40 to 50% of wood biomass. Hemicelluloses are more abundant in hardwoods than in conifers, whereas lignin is present at higher levels in conifers than in hardwoods. Exceptions to these trends occur in some tropical woods. In addition to varying levels of constituents, the types of hemicelluloses and lignin differ between conifers and hardwoods. In the cell wall, cellulose levels are highest in the secondary wall, hemicelluloses levels are highest in the S1, and lignin is present at highest levels in the middle lamella and primary wall (compound middle lamella).
- The three organizational levels in the cell wall are the *gross capillary*, *transient capillary*, and *crystalline zones*. Liquid water or water vapor can occur in the gross capillary zone, which consists of the lumina of cells and pit cavities. Water occurs in the transient capillary zone as bound water (hydrogen bonding). This zone includes the amorphous or noncrystalline zones of the microfibrils. The transient capillary zone contracts as wood dries, and expands as it is moistened between the oven-dry and fiber-saturation levels.
- Wood attains equilibrium-moisture contents at various temperature–vapor pressure conditions within the oven-dry and f.s.p. range.

Summary continues

- Many special properties of wood can be explained in part by the structure of cellulose and its primary orientation in the cell wall. The anisotrophy or swelling and shrinking of wood in only two planes (radial and tangential) is the result of the unique exposed position of the hydroxyls on the cellulose molecules and their hydrogen bonding with hydroxyls on adjacent cellulose molecules or available water molecules. The high tensile strength of wood is the result of the longitudinal orientation of the cellulose molecules, their high degree of polymerization, and the covalent bonds between the glucose units.
- The sorptive properties of wood clearly indicate that wood protected from free water sources can *not* attain sufficient moisture to sustain decay or stain development.

References

Adler, E. (1977). Lignin chemistry—past, present, and future. *Wood Science Technology* 11(3):169–218.

Agrios, G. N. (1988). "Plant Pathology." 3rd Ed. Academic Press, New York.

Core, H. A., W. A. Côté, and A. C. Day. (1976). "Wood Structure and Identification." Syracuse University Press, Syracuse, New York.

Côté, W. A., Jr. (ed.). (1965). "Cellular Ultrastructure of Woody Plants." Syracuse University Press, Syracuse, New York.

Côté, W. A. (1976). "Wood Ultrastructure—An Atlas of Electron Micrographs." University of Washington Press, Seattle, Washington.

Côté, W. A., Jr. (1977). Wood ultrastructure. *In* "Recent Advances in Phytochemistry Vol. II. The Structure, Biosynthesis, and Degradation of Wood." (F. A. Loewus and V. C. Runeckles, eds.), pp. 1–44. Plenum, New York.

Côté, W. A. (1981). Ultrastructure—critical domain for wood behavior. *Wood Science Technology* 15:1–29.

Côté, W. A., and A. C. Day. (1962). The G layer in gelatinous fibers—Electron microscope studies. *Forest Products Journal* 13(7):333–338.

Côté, W. A., A. C. Day, and T. E. Timell. (1968). Distribution of lignin in normal and compression wood of tamarack. *Wood Science Technology* 2:13–37.

Côté, W. A., T. E. Timell, and R. A. Zabel. (1966). Studies on compression wood. Part 1. Distribution of lignin in compression wood of red spruce, *Picea rubens* Sarg. *Holz als Roh-Werkstoff* 24:432–438.

Crawford, R. L. (1981). "Lignin Biodegradation and Transformation." John Wiley, New York.

Esau, K. (1965). On the anatomy of the woody plant. *In* "Cellular Ultrastructure of Woody Plants" (W. A. Côté, Jr., ed.), pp. 35–50. Syracuse University Press, Syracuse, New York.

Fengel, D., and G. Wegener. (1984). "Wood Chemistry, Ultrastructure, Reactions." Walter de Gruyter, New York.

Glasser, W. G., H. R. Glasser, and N. Morohoshi. (1981). Simulation of reactions with lignin by computer 1981 (SIMREL). *Macromolecules* 14:253–262.

Goring, D. A. I., and T. E. Timell. (1962). Molecular weight of native celluloses. *TAPPI* 45:454–460.

Harada, H., and W. A. Côté, Jr. (1985). Structure of wood. In "The Biosynthesis and Biodegradation of Wood Components" (T. Higuchi, ed.), pp. 1–42. Academic Press, New York.

Higuchi, T. (ed.) (1985). "Biosynthesis and Biodegradation of Wood Components." Academic Press, New York.

Higuchi, T. (1990). Lignin biochemistry: Biosynthesis and degradation. *Wood Sci. Technol.* **24**:23–63.

Kollman, F. F. P., and W. A. Côté, Jr. (1968). "Principles of Wood Science and Technology, Vol. I. Solid Wood." Springer-Verlag, Berlin.

Panshin, A. J., and C. deZeeuw. (1980). "Textbook of Wood Technology," 4th ed. McGraw-C. Hill, New York.

Peck, E. C. (1950). Moisture content of wood in use. U.S.D.A., Forest Service, Forest Products Laboratory Report 768. Madison, Wisconsin.

Saka, S., and D. A. I. Goring. (1985). Localization of lignins in wood cell walls. In "Biosynthesis and Biodegradation of Wood Components." (T. Higuchi, ed.), pp. 51–62. Academic Press, New York.

Shigo, A. L., and A. Shigo. (1974). Detection of discoloration and decay in living trees and utility poles. U.S.D.A., Forest Service Research Paper, NE-294, Northeastern Forest Experiment Station, Bromall, Pennsylvania.

Shortle, W. C. (1982). Decaying Douglas fir wood: Ionization associated with resistance to a pulsed current. *Wood Science* **15**(1):29–32.

Siau, J. F. (1984). "Transport Processes in Wood." Springer-Verlag, New York.

Skaar, C. (1988). "Wood-Water Relations." Springer-Verlag, New York.

Skaar, C. (1972). "Water in Wood." Syracuse University Press, Syracuse, New York.

Timell, T. E. (1964). Wood hemicelluloses. Part I. *Advances in Carbohydrate Chemistry* **19**:247–302.

Timell, T. E. (1965a). Wood hemicelluloses. Part II. *Advances in Carbohydrate Chemistry* **20**:409–483.

Timell, T. E. (1969). The chemical composition of tension wood. *Svensk Papperstidning.* **72**:173–181.

Timell, T. E. (1965b). Wood and bark polysaccharides. In "Cellular Ultrastructure of Woody Plants" (W. A. Côté, ed.), pp. 127–156. Syracuse University Press, Syracuse, New York.

Timell, T. E. (1982). Recent progress in the chemistry and topochemistry of compression wood. *Wood Science Technology* **16**:83–122.

Timell, T. E. (1986a). "Compression Wood in Gymnosperms." Springer-Verlag, Heidelberg.

Timell, T. E. (1986b). Wood: Chemical composition. In "Encyclopedia of Materials Science and Engineering." (M. B. Bever, ed.), pp. 5402–5408. Pergamon Press, New York.

Wardrop, A. B. (1964). The structure and formation of the cell wall in xylem. In "The Formation of Wood in Forest Trees" (M. H. Zimmerman, ed.), pp. 87–134. Academic Press, New York.

Woodcock, C., and A. Sarko. (1980). Packing analysis of carbohydrates and polysaccharides. II. Molecular and crystal structure of native ramie cellulose. *Macromolecules* **13**:1183–1187.

Young, H. E., and V. P. Quinn. (1966). Chemical elements in complete mature trees of seven species in Maine. *TAPPI* **49**:190–197.

Zimmerman, M. H. (ed.) (1964). "The Formation of Wood in Forest Trees." Academic Press, New York.

CHAPTER

7

General Features, Recognition, and Anatomical Aspects of Wood Decay

The characteristic softening, discoloration, and eventual disintegration of wood in some uses is well known as *decay* or *rot* to most wood processors and users. Colloquial terms used to describe decay in some regions include *doze*, *dote*, and *punk*. A broad dictionary definition of decay implies slow changes in a material that reduce its useful properties, resulting from organism actions, e.g., insects, borers, and fungi, or abiotic factors such as weathering.

Decay has a more restrictive and specific meaning in wood microbiology. In this book we shall define decay as *significant changes* in the *physical* and *chemical properties of wood* that are *caused* by the *chemical* (primarily enzymatic) *activities* of *microorganisms* (primarily higher fungi). Since *rot* is commonly used by laymen to describe decay, we shall continue to use the two terms interchangeably.

In this chapter the general features of wood decay will be reviewed. Emphasis will be placed on decay types, decay detection from various macroscopic and microscopic evidences, and the anatomical features of decay development.

The Dual Nature of Decay

It is important to note initially that the term *decay* is used two ways. One way describes the *physical condition* of wood. For example, an inspector judges a board to be sound or decayed. In this sense, wood decay is viewed as a property of a material related to its usefulness. The second, broader meaning is the *process* of wood digestion by microorganisms including the enzymes, the oxidants, and the changes in the wood constituents involved. In this sense, decay is a verb and can be viewed as the external digestion of a complex organic material by microorganisms. Decay generally is a slow process requiring weeks and sometimes years to disintegrate wood, but small wood blocks, under ideal decay conditions, may be consumed totally within a month. At the other extreme, decays developing in the heartwood (heart rots) of durable trees may develop for hundreds of years in some cases. Also, by definition, decay must cause a significant change in wood properties. We generally consider these changes to include a measurable loss in cell-wall biomass.

General Features of Wood Decay

Stages

The decay process, under ideal conditions, is a linear *continuum* that begins with a few innocuous spores and ends when the wood is destroyed or mineralized. Points along the continuum have been selected arbitrarily to designate various stages in decay development, which, as we shall see later, can be related to various use properties (Fig. 7-1). Decay begins when the hyphae of decay fungi penetrate wood, initiate colonization, and release enzymes. At this early colonization phase, damage is limited, and since there are no visible evidences, it is termed the *incipient* or *hidden stage* of decay. As decay develops, slight changes in color, wood texture, and fiber brashness may appear, and these changes constitute the *early stage* when decay is detectable, but not obvious. As decay continues to the *intermediate stage*, obvious changes in wood color and texture are evident, but the gross structure still remains intact. The *late stage* is when the wood structure is totally disrupted and the residual wood has become a brownish amorphous, whitish punky,

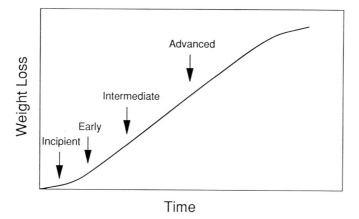

Figure 7-1 A diagram illustrating the continuum of the decay process under ideal decay conditions and its subdivision into several decay stages.

or fibrous material. Some fungi (white rots) may completely degrade the wood, producing weight losses approaching 96–97%. Other decay fungi degrade only the carbohydrate portions of the wood cell walls and cause maximum weight losses of 60 to 70%.

Macroscopic Features

The appearances of the intermediate and late stages of decay are extremely variable. The residual wood may be a white, gray, or brown, and the color, consistent or mottled. The wood texture may be soft, spongy, feathery, or fibrous for the white and gray decays or consist of loosely adhering soft brown cubes for the brown decays. Narrow black to brownish *zone lines* or the *black lines* of *decay* may be associated with the white decays. They may represent either boundaries of undecayed wood between genetically different fungi, or mycelial responses to surface drying (Campbell, 1933; Hopp, 1938). The wood may be uniformly decayed or concentrated in localized pockets. In some cases, the decay may be concentrated along the earlywood bands of the annual ring, forming a condition called *ring rot*. When the decay leads to separations along the annual rings, the term *laminated rot* is often used. In other decays, wood rays or fibers may be selectively removed, forming shotholes or producing a stringy texture in the residual wood. Many macroscopic features of decay appear to be characteristic of the causal fungi [e.g., *Phellinus (Fomes) pini* causes a white-pocket decay and often a ring shake in conifers; *Postia amara (Polyporus amarus)* causes a brown-cubical pocket decay in incense cedar; *Echinodontium tinctorium* forms a brown-stringy decay in western hemlock; and *Ganoderma (Fomes) applanatum* causes a white-mottled decay and abundant zone lines in hardwoods]. Some of these macroscopic features of decay and their causal fungi are shown in Fig. 7-2.

Decay Patterns

The location and stages of decay in various wood types and uses may vary greatly with degree of decay hazard and the initial invasion point. Stem decays in living trees are often concentrated in the more susceptible heartwood. A cross section through a stem decay often shows all decay stages ranging from none in the outer sapwood to complete wood removal in the heartwood (Fig. 7-3). In treated woods such as railroad ties or poles, deep-drying checks often penetrate through the outer treated shell, and a characteristic decay pattern forms between the treated and untreated zones, punctuated by advanced decay where the check enters the untreated wood. Decay patterns will be discussed in more detail in the subsequent chapters on decays associated with various wood uses. Understanding decay patterns provides useful clues to the wood user about when and where a decay fungus entered or why a protective treatment failed.

General Features of Wood Decay / 171

Figure 7-2 Some typical features of several major types of decay based on color and texture. (A) A brown cubical rot. (B) A white pocket rot, probably an advanced stage of *Phellinus pini* decay. (C) A white spongy rot caused by *Trametes versicolor* (arrows indicate associated basidiomata and zone lines). (D) A brown cubical pocket rot in incense cedar caused by *Postia amara*.

Figure 7-3 Cross section of a beech *Fagus grandifolia* stem showing the various decay stages and macroscopic features of *Phellinus igniarius,* a major heart-rot fungus in hardwoods. Letters indicate (A) advanced decay, (B) early decay, (C) zone lines, and (D) sound sapwood. × .33

Major Disadvantages of Decay

Strength and volume (biomass) reductions are the principal losses associated with decay. Drastic strength losses occur in the incipient (hidden) decay stage with some decays. In some high-stress uses of wood, such as ladder rails or poles, failure to detect early decay can lead to serious losses. In the intermediate and late stages of decay, drastic alterations in the dimensions and chemical and physical properties of wood adversely affect many use properties, including decreased volume and mass resulting in dimensional collapse of supporting timbers or loss of raw materials for chemical processes; reductions in many strength properties (tension, bending, and compression); increased hemicellulose solubility, leading to substantial losses during pulping and related water pollution problems; increased permeability to liquids, leading to excessive absorption of preservatives or coatings; changes in electrical properties (increased conductivity); and aesthetic losses due to abnormal colors, rough textures, or pulled fibers. These property changes are discussed in more detail in Chapters 8 through 10.

Decay Losses

Collectively, decay losses are very large and include those incurred during timber growth, wood storage, processing, and in many product uses. Loss estimates are available only for some parts of the total. Losses in standing timber due to stem decays are estimated to be 15–25%. These losses will probably decrease as over mature stands are utilized and the average timber age decreases. Losses during the storage and conversion of logs, pulpwood, and chipwood piles are estimated to be 15%. The major decay losses, however, occur in wood products and have been estimated to represent a replacement cost of 10% of the annual cut. This cost does not include the labor of replacement, service interruptions, inconvenience, or the serious injuries and losses that can occur when a structure fails. Decay in structures often occurs in locations where access and repair is difficult (e.g., foundation plates, floor joists, support beams, etc.), and replacement costs can sometimes greatly exceed the original wood cost. The specific decay losses associated with various product uses are discussed in subsequent chapters. Accurate decay-loss estimates are necessary to sustain decay-reduction programs in the major wood-use industries and to justify research for more effective decay controls. Substantial reductions in decay losses could be effected if known information and recommendations on decay control were applied by wood users and the designers of wooden structures.

Whereas this book stresses the economic losses associated with some wood uses, in the aggregate, it should be remembered that decay is beneficial in forest ecosystems. Slash removal, soil enrichment, and returning carbon dioxide to the atmosphere for green plant use in the carbon cycle are all dependent on wood decay.

Recognition of Decay (Visual Evidences)

The rapid detection of decay in the field is important during the purchase of roughwood (e.g., logs, poles, pulpwood, posts), and the selection of wood for some products (e.g., veneer, structural lumber). Decay detection is particularly important during the grading and selection of wood destined for high-stress uses such as ladder rails, laminated beams, and poles. Early decay detection is also important during the periodic inspections of wood used under high-decay-hazard conditions such as utility poles, pilings, and cooling towers. Failure to recognize decay in the wood product when the visual evidences are evident often becomes a key issue in litigation, and is used to determine responsibility for a wood failure and related damages.

Decay detection is intrinsically difficult since it occurs primarily in the interior of an opaque material, often in inaccessible zones (below ground). Decay evidences in the incipient and early stages may be subtle and difficult to distinguish from natural variations present in sound wood.

In those cases, where the interior wood is exposed on cut surfaces such as log ends, boards, or increment cores, the later stages of decay in the interior zones can easily be determined visually. However, visual detection of decay in the early stages is difficult and requires both a familiarity with the macroscopic features of early decay as well as an awareness of the natural variations in wood.

Macroscopic Decay Evidences

Some of the useful *early* visual decay evidences on wood surfaces include the following.

1. Changes from normal wood color patterns. Common colors of early decays are red, pink, purple, yellow, gray, brown, and white. Generally the colors do not follow annual rings, often distinguishing them from sapwood–heartwood color differences. A bleached or mottled appearance is a common feature of the early stages of some white rots (Fig. 7-4A) (e.g., *Trametes versicolor, Ganoderma applanatum*).
2. Black lines of decay (zone lines) between the sound and decayed wood or between different decays (Fig. 7-4B). Zone lines are a positive evidence of early decay and occur most commonly with white rots on hardwoods.
3. Abnormal shrinkage patterns, as evidenced by wood collapse and development of numerous, small surface checks. The checks generally form at right angles to the longitudinal plane of the wood and are most evident after the wood surface has dried. Cross checks are a feature of brown rots (Fig. 7-4C).
4. Brashness or brittleness of the wood as evidenced by a roughened saw-cut on exposed log ends, pulled fibers during lumber surfacing, or extreme brittleness of long slivers lifted from the wood surface with a probe or pick (Fig. 7-4D, Fig. 16-1).
5. The physical presence of the fungi on the wood, such as mycelial fans on board surfaces, punk knots in boards, or basidiocarps on logs. The mycelial fans that form on wood surfaces under moist conditions or between boards often assume intricate fanlike patterns and are known among wood tradesmen as *fungal flowers* (Fig. 3-1A). A characteristic fungus odor (similar to commercial mushrooms) is often associated with mycelial fans. Decayed knots in boards associated with abnormal discolorations often signal extensive zones of incipient decay.

Color changes and brashness are only *presumptive* evidences of early decay. Zone lines and the physical presence of the fungus can generally be safely assumed to be *definitive* indicators of early decay. Combinations of decay indicators often occur, increasing the reliability of a positive decay diagnosis.

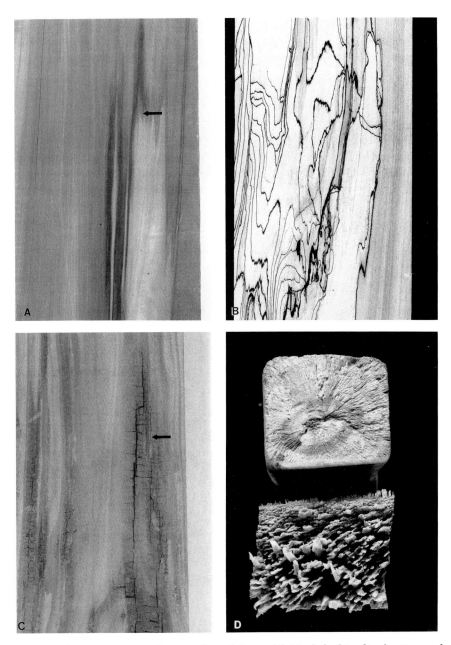

Figure 7-4 Some macroscopic features of wood decay. (A) Mottled white discolorations and zone lines associated with stem decay in sweet gum. (B) Black zone lines associated with the white rot of *Xylaria polymorpha*. (C) Abnormal shrinkage cracks on the surface of a Douglas fir board that had been solid piled for several months after inadequate seasoning. (D) A brash break in decayed beech (top) contrasted with a fibrous or splinter break (bottom) in sound wood.

Microscopic Decay Evidences

Microscopic features of early decay, such as cell-wall erosion, bore holes, or hyphae with clamp connections, are *definitive* evidences of early decay (Fig. 7-5). Other useful decay criteria are the localized destruction of ray-parenchyma cells or the formation of diamond-shaped or long, linear cavities, axially aligned in the secondary wall. Microscopic features of decay are discussed in more detail in the later section of this chapter on the anatomical aspects of decay. Microscopic features can reliably confirm decay and are useful to confirm presumptive macroscopic features. It should be remembered that decay does not occur uniformly throughout the wood. Reliable decay detection depends on adequate sampling to minimize the chances of a negative diagnosis when early decay is present.

Other Decay-Detection Procedures

Detection of early decay can also be accomplished by several nonvisual methods. The isolation of the associated fungi and their identification from cultural features is a useful method (Wang and Zabel, 1990; Eslyn, 1979; Graham and Corden, 1980). Chemical-spot tests with various dyes have been proposed for the detection of brown rots in some wood products (Cowling and Sachs, 1960; Eslyn, 1979). Various physical tests based on changes in the acoustic, X-ray, electrical, and strength properties of decayed wood have also been proposed (Miller *et al.*, 1965; Shigo and Shigo, 1974; Stoker, 1948). The theory and methodologies of decay detection are presented in detail in Chapter 16 ("Detection of Decay").

Types and Classifications of Wood Decay

Decays have been grouped in a variety of often confusing and overlapping ways in the literature. The major subdivision into *white* and *brown* rots was made in 1874 by Hartig, based on the colors of the residual woods and the assumptions that the whitish material was cellulose (white rot) and the brownish material, lignin (brown rot). Subsequent research established that white-rot fungi actually utilized all cell-wall constituents (cellulose, hemicelluloses, and lignin) (Campbell, 1932; Scheffer, 1936), but the early dogma persisted for many years. The grouping of decays into white- and brown-rot categories remains a major subdivision, reflecting very different chemical processes in the decay of wood. Another important early contribution of Hartig, which had a great impact on decay classification, was the demonstration that a *specific fungus* caused a *specific type of decay* and that the decay was *similar* in other woods. This observation opened the door to decay detection in standing timber, since identification of basidiocarps on the stem permitted

Types and Classifications of Wood Decay / 177

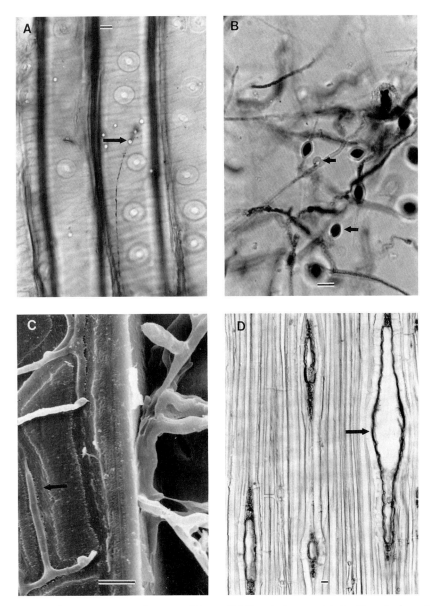

Figure 7-5 Some microscopic features of decay. (A) Bore holes associated with early decay in Douglas fir caused by *Postia placenta*. (B) The hyphae of *Antrodia carbonica* showing a typical clamp connection and numerous ovoid-shaped chlamydospores. (C) Cell-wall erosion associated with early decay in southern yellow pine caused by *Trametes versicolor* (TEM). (D) The selective destruction of ray parenchyma cells caused by *Postia placenta* in the early decay of southern yellow pine. Bars equal 10 μm. (C) reprinted with permission of Electric Power Research Institute, Palo Alto, California.

estimates of the associated internal-cull columns. The early interest of forest pathologists and foresters was primarily the heart rots in standing timber. Decay classifications proposed by Hubert (1924) were based on decay color (white or brown), textural features (e.g., stringy, spongy, pocket, cubical, mottled), and stem location.

The early textbooks in forest pathology essentially followed these decay groupings, and emphases were placed on the macroscopic features of advanced decay, basidioma appearances, and associated cull (Hubert, 1931; Boyce, 1961; Baxter, 1952). In 1938, the classic publication by Davidson *et al.* on the differentiation of wood-decay fungi into two groups by their staining reactions when grown on gallic- and tannic-acid media (originally determined by Bavendamm in 1928) defined the white- and brown-rot groups and further substantiated the basic chemical differences between them.

In 1958, Cartwright and Findlay published the first comprehensive textbook on timber decay. The decays were grouped into various timber type and wood use categories as follows:

Decays of standing timber (conifers and hardwoods);
Decays of felled timber and timber in service in open;
Decays of timber in buildings and structures;
Decays in timber during storage, conversion, and shipment; and
Decays of timber in various uses.

Their treatment again emphasized identification of the causal organisms and decay descriptions; however, it was the first comprehensive grouping of decays by major wood-processing and use categories.

Soft Rots—A New Decay Type

In 1954, Savory described an unusual type of wood decay caused by *Chaetomium globosum*, an Ascomycotina. The decay is characterized macroscopically by surface softness and microscopically by selective attack of the interior zone of the secondary wall. Savory coined the term *soft rot* to describe this new decay type. Before this discovery, it had been assumed that only Basidiomycotina were capable of decaying wood. Soft rots were initially judged to be an interesting oddity, primarily because of the unique shapes and patterns of the longitudinal bore holes.

The discovery of soft rots was soon followed by a series of studies, indicating many Ascomycotina and Deuteromycotina were capable of decaying wood (Liese, 1955; Duncan, 1960, Merrill and French, 1965; Lundström, 1973; Nilsson, 1974).

Initially, soft rots were defined by the formation of unique longitudinal bore holes in the secondary cell wall; however, it was determined by Corbett (1965) that some non-Basidiomycotina eroded the secondary wall in a manner similar to that of some white-rotters. Corbett proposed that the cavity-formers be called Type 1 soft-rotters, and the cell-wall eroders, the Type 2 soft-rotters. A further complication to this classification was the ultrastructural studies of Liese and Schmid (1962) that demonstrated that small cavities were formed in the secondary walls at later decay stages by several Basidiomycotina decayers. Some groupings have limited the term decay to wood destruction by *Basidiomycotina* and proposed soft rot as a category for wood damaged by other fungi and bacteria. This is unsatisfactory since it provides different terms for cases in which wood is severely and similarly damaged by microorganisms in different taxonomic groups.

In a review article on woody-tissue disintegration by microorganisms, Liese (1970) presented a classification of decay types. We shall follow that system with some modifications in this book. In this separation, major reliance is placed on the order or sequence of cell-wall constituents utilized or altered. Secondary emphasis is placed on the mode of hyphal penetration of prosenchyma cells (fibers and/or tracheids) and the types of tissue damaged. The major types of woody-tissue destruction by fungi and bacteria are listed in decreasing severity of cell-wall damage, and examples are given of typical causal organisms. The major non-decay categories of wood-inhabiting microorganisms are also included to emphasize their distinctive features. The wood-decay group are characterized by significant weight and/or strength losses resulting from destruction of prosenchyma tissue, and hyphal bore holes as large as, or larger than hyphal width. The nondecay group (wood modifiers) are characterized by tissue destruction limited primarily to parenchyma cells, and hyphal bore holes absent or, when present, narrower than hyphal width.

A Classification of Wood Modifications by Microorganisms

Wood Decayers

1. *Simultaneous white-rotters* attack all cell-wall constituents, essentially uniformly during all decay stages (e.g., *Trametes versicolor, Irpex lacteus*).
2. *Sequential white-rotters* attack all cell-wall constituents; however, the initial attack is selective for hemicelluloses and lignin (e.g., *Phellinus pini, Heterobasidion annosum*). Other sequences of attack of cell-wall constituents may occur.
3. *Brown-rotters* primarily attack the cell-wall carbohydrates, leaving a modified lignin at the end of the decay process (e.g., *Gloeophyllum sepiarium, Serpula lacrimans*).
4. *Soft-rotters* preferentially attack cell-wall carbohydrates in the S2 layer of the secondary cell wall, forming longitudinal cavities (Type 1) or eroding the wood cell wall from the lumen surface in hardwoods (Type 2) or the S2 in conifers (e.g., *Chaetomium globosum, Alternaria alternata*). Some bacteria are known to cause typical soft-rot cavities, and related tunnel- or cavitation-type cavities in cell walls.

Nondecaying Wood Inhabitors

1. *Sap stainers* primarily attack the parenchyma cells and discolor the sapwood by pigmented hyphae. Hyphal penetration of the wood occurs primarily through pits. Weight losses are minimal (e.g., *Ophiostoma piliferum, Aureobasidium pullulans*). Mold is a related term used to describe the surface discolorations of wood caused by colored spores or mycelial masses (e.g., *Trichoderma* spp.).
2. *Scavengers* primarily utilize the simple carbon compounds that are available in the wood rays, longitudinal parenchyma, and lumen surfaces or are released during decay by other organisms. Hyphal penetration of the wood also occurs primarily through pits. No appreciable weight losses occur; however, pit penetration may increase wood permeability. A wide range of wood-inhabiting fungi and bacteria fall into this group, and their roles and interactions are discussed in Chapter 11 (e.g., *Penicillium* spp., *Gliocladium virens, Rhinocladiella atrovirens*). Some bacteria cause erosion channels on the lumen surface and over long periods may cause considerable wall damage, but these organisms are included with the scavengers since this appears to be their major role in wood. The characteristics and features of the major types of wood-inhabiting microorganisms are summarized in Table 7-1.

There are some inconsistencies and neglected areas in this grouping of the four major decay types. For example, some basidiomycetes form small cavities in the cell wall. These resemble Type 1 soft-rot attack, and the cell-wall erosion of the Type 2 soft-rotters in hardwoods resembles an early stage of white rot. Recent studies of decays caused by several of the Xylariaceae (Ascomycotina) have shown that wood damage by these fungi is similar to that produced by the simultaneous white rots, but also contains typical soft-rot cavities (Kistler and Merrill, 1968; Nilsson *et al.*, 1989). Other groups of fungi associated primarily with slash decomposition in the forest floor such as the Tremellales, Dacrymycetales, Auriculariales, and the Gasteromycetes have been neglected as probable agents of decay and may provide new insights into the chemical aspects of decay. In a sense, the similarities in decay patterns produced by divergent microorganisms probably reflect a convergence in attack strategies, determined by wood structure. As more becomes known about the chemical processes involved in decay, the decay groupings will certainly change to more closely follow the phylogenetic groupings of fungi and provide insights into the origin and development of the decay process itself.

Other Common Wood-Decay Groups

Several other common decay groupings based wholly on the status of the wood in the forest or during its processing and use are largely self-explanatory but warrant a listing and brief description as follows:

Table 7-1
A summary of the anatomical and chemical features of the major types
of wood-inhabiting microorganisms

Wood-inhabiting microorganisms	Cell-wall constituents used	Anatomical features	Causal agents
Decayers (Cell-wall erosion and/or large bore holes formed > 2 μm)			
Simultaneous white rots	All	Cell walls attacked progressively from lumen surface	Basidiomycotina Some Ascomycotina
Sequential white rots	All, but hemicelluloses and lignin used selectively initially	Cell walls attacked progressively from lumen surface	Basidiomycotina Some Ascomycotina
Brown rots	Carbohydrates, but lignin modified	Entire wall zone attacked rapidly	Basidiomycotina
Type 1 soft rot	Carbohydrates	Longitudinal bore holes develop in secondary wall	Ascomycotina Deuteromycotina Some bacteria
Type 2 soft rot	Carbohydrates	Secondary wall erosion from lumen surface (in conifers mainly the S2)	Ascomycotina Deuteromycotina Some Bacteria
Nondecayers (No cell-wall erosion and occasional bore holes are minute < 1 μm)			
Sapstainers	Wood extractives	Invade parenchyma cells in sapwood primarily	Ascomycotina Deuteromycotina
Molds	Wood extractives	Surface growth on wet wood	Zycomycotina Ascomycotina Deuteromycotina
Scavengers	Wood extractives and decay residues	Penetrate wood cells primarily through pits	Bacteria Zygomycotina Ascomycotina Basidiomycotina Deuteromycotina

1. Stem decays in standing timber are grouped into *heart rots* and *sapwood decays* of the living trees; *slash rots* are those that develop in down timber, branches, and logging slash;
2. *Storage decays* occur during the storage and processing of pulpwood, pulp chips, logs, and piling;
3. Special commodity and use categories are the *building-rot decays*, decays of *utility poles, mine timbers*, and piling; and
4. *Conifer decays* are often contrasted with *hardwood decays*.

Manion (1991) has separated decays into two groups: the *heart rots* in living stems and *saprobic* decays in wood products.

Some Anatomical Features of Wood Decay

Anatomical study of wood decay is one of the oldest aspects of wood microbiology. Curious observers probably were looking at decayed wood shortly after the discovery of the microscope (*circa* 1650). Early observers assumed the hyphal filaments were a stage of the decayed wood rather than the cause. As indicated in Chapter 1, Hartig first clearly established the causal role of fungi (hyphae) in decay in 1874, and in 1878, described and illustrated in detail the hyphal distribution, bore holes, and cell-wall erosion associated with a major building rot fungus, *Serpula lacrimans*. A large literature has accumulated on the anatomical features of decay. The older literature has been reviewed by Hubert (1924) and more recently by Wilcox (1968; 1970) and Erikson *et al.* (1990). We shall limit this section to a review of the significant contributions that lead to a clearer understanding of microscopic features of decay, a summary of the anatomical features of the three major decay types (white, brown, and soft rots), and end with some additional research needs. The related ultrastructural aspects of decay are covered later in Chapter 9.

An Early History and Major Contributions to the Anatomy of Decay

Hubert (1924) assembled a classic publication on the anatomical features and diagnosis of decay in wood, describing many decays (heart rots, slash rots, and products rots) microscopically, and suggested that the important decay fungi could be identified from anatomical features such as bore-hole types, cell-wall erosion patterns, the sequences of tissues attacked, and hyphal distribution patterns in the wood.

In the period from 1915 to 1940, a popular approach to research in forest pathology was preparation of a biology of an important decay fungus. Some of those that contain detailed observations on the microscopic features of several products-rot fungi are listed by author and fungus as follows: Buller (1905), *Neolentinus* (*Lentinus*) *lepideus*; Hirt (1928, 1932), *Phellinus* (*Polyporus*) *gilvus* and *Trametes suaveolens*; Rhoads (1918), *Trichaptum biforme* (*Polyporus pargamenus*); Spaulding (1911), *Gloeophyllum* (*Lenzites*) *sepiarium*; and White (1920), *Ganoderma* (*Fomes*) *applanatum*.

In 1936, Scheffer reported on the progressive effects of *Trametes* (*Polyporus*) *versicolor* decay in red gum, studying, for the first time, the anatomical, physical, and chemical features of a decay. He observed a uniform thinning of cell walls from the lumen toward the middle lamella and found that the majority of cell-wall penetrations took place through pits. Chemical analyses of the wood at various stages of decay clearly showed a uniform consumption of *all* cell-wall constituents by this white-rot fungus, and finally dispelled the stubborn myth that white-rotters decayed only lignin.

In 1937, Bailey and Vestal carefully described the longitudinal bore holes in the secondary walls of several decayed woods and reported that the cavities often contained or were connected by small hyphae. The unique diamond shapes of the cavities and their periodicity were a curiosity that stimulated much research interest. Several years later Barghoorn and Linder (1944) culturally confirmed several fungi as the cause of similar longitudinal cavities in wood submerged in the ocean.

In 1954, Savory demonstrated that several Ascomycotina and Deuteromycotina caused surface decay of cooling tower wood, which was characterized by the unique longitudinal cavities. He coined the term *soft rot* for the new decay type because it was associated with a surface softening of the wood. It was now clear that non-Basidiomycotina could also decay wood; decay was no longer the unique capability of some Hymenomycetes.

In 1961, Cowling completed a comparative study of the anatomical, physical, and chemical properties of sweet gum sapwood colonized by a brown rot *Postia (Poria) placenta* and a white rot *Trametes (Polyporus) versicolor*. Based on anatomical and chemical changes in cell walls, hypotheses were developed on the relative sizes and sequences of the cellulose- and lignin-destroying enzymes involved in the two decay types. This paper stimulated extensive chemical and ultrastructural studies of decay.

In the period from 1959 to 1968, several researchers (Ellwood and Ecklund, 1959; Knuth and McCoy, 1962; Greaves and Levy, 1965; Greaves, 1969; Boutelje and Bravery, 1968) clearly demonstrated that under some conditions of use, bacteria were able to damage pit membranes and cause some surface erosion of wood.

In 1968, Nicholas and Thomas demonstrated that filtrates and enzymes from decay fungi can induce typical decay damage in wood. This evidence opened the door to fundamental studies of the anatomy of decay and future biotechnological applications.

Wilcox (1964, 1970) improved procedures for the microscopic study of decayed wood and compared the major anatomical features of wood damaged by white-, brown-, and soft-rot fungi and bacteria.

Principal Anatomical Features of Decay

There are many conflicts and inconsistencies in the literature on some anatomical features of decay reflecting, in part, the extreme difficulty of the anatomical study of decay with the light microscope. Most fungal hyphae are hyaline and require special stains for detection against the wood background. Some hyphal features are at the resolving limit of the light microscope. In preparing sections for study, it is difficult to maintain the hyphal filaments in their precise original position and, at times, to distinguish decay damage from sectioning or prior wood damage. Also since decay is a process, some microscopic features may vary with stage of decay. Finally, there has been a tendency to study a limited number of fungi representative of a decay type, and to expect all in the type to have similar features.

In this section we shall limit the review to the white rots, brown rots, and soft rots. Anatomical features of the sap stains and molds are covered in Chapter 14.

Entrance and Early Colonization

Wood-inhabiting microorganisms enter wood primarily through the torn cell walls of the wood rays and axial cells (e.g., tracheids, vessels) exposed on the surfaces of various wood products or damaged wood in the living tree. Microfauna and insects are effective vectors for some decayers (e.g., horntail wasps vector *Amylostereum chailletti* into dying and dead balsam fir) and many sapstain fungi. The inoculum sources are airborne spores or spores and mycelial fragments from contacts with soil, water, or wood-processing machinery. Spore germination and/or mycelial growth occurs when suitable conditions for decay are present (Chapter 4). Occasional reports of fungi in the monokaryon condition suggest occasionally that single basidiospores are able to initiate decay (Zabel *et al.*, 1980; Morrell *et al.*, 1987; Przybylowicz *et al.*, 1987). The entrance and early colonization phases are sustained by food reserves in the inoculum and the available simple carbon compounds in the cell lumina. The hyphae of wood-inhabiting fungi in the colonization phase appear to grow longitudinally along the lumina surface and pass from cell to cell through the pits. Growth rates longitudinally through the wood, under ideal conditions such as wood blocks in decay chambers, approach growth rates on malt-agar medium. Rapid-growing fungi, such as *Irpex lacteus*, can penetrate a 1-cm wood block longitudinally in several days. Whether decay begins directly after hyphal entrance or is delayed until the available simple carbon compounds are utilized is unknown and probably varies with the fungal species and its wood-inhabiting role. *Chaetomium globosum* was reported to cause extensive soft-rot in beech sapwood blocks within 7 days in malt-agar decay chambers (Jutte and Zabel, 1974), whereas some decay fungi appear to have a long lag period before causing extensive wood degradation. There is some selection of the cell types initially invaded. In hardwoods, white-rot fungi often develop initially in the vessels and wood rays, whereas the wood rays and longitudinal parenchyma are first colonized in conifers. Brown-rot fungi are less selective, and hyphae are often present in most cells. There are exceptions, such as *Postia placenta*, which initially attacks the wood rays (Fig. 7-5D). Sapstain fungi confine their invasions primarily to parenchyma cells. Soft-rot fungi also initially appear to invade primarily wood rays and axial parenchyma. Hyphae are often more abundant in the cells damaged by white rots than in those damaged by brown rots or soft rots. No correlation is evident between the abundance of hyphae in cells and decay severity.

Cell-Wall Penetration

There has long been an interest in how the fragile-appearing hyphae penetrate woody cell walls. Nutman (1929) described in detail the initial cell-wall penetrations of *Inonotus* (*Polyporus*) *hispidus* in ash. At the point of contact between the hyphal tip and the cell wall, a fine hair-like penetration peg emerged, which enlarged after wall passage. After penetration, cell-wall erosion began on each corresponding wall, and eventually an hourglass-shaped bore hole was formed, and the initial fine penetration peg swelled to regular hyphal size. The penetration process was believed to be enzymatic because of the nature of the hourglasslike bore hole. Cartwright (1930), in studies of decay in spruce caused by *Postia placenta* (*Poria monticola*), reported that the hyphal tips developed minute projections that formed nicks on lumen wall contact and then rapidly penetrated. The bore holes later enlarged uniformly to diameters several times greater than the original hyphal size, suggesting that the penetration process must be primarily enzymatic, owing to the borehole enlargement.

In a major study of cell-wall penetration, Proctor (1941) used six decay fungi, four wood species, and both ultraviolet microscopy and polarized light techniques. He concluded after critical study of the hyphal tip–cell wall contact zones that enzymes were secreted primarily at the hyphal tip zone and that dissolution of the cell wall by enzymes preceded any physical hyphal contact. A transpressorium (a small stalk with a pointed tip) has been observed on the hyphal tips of some stain fungi, suggesting primarily a mechanical mode of the cell-wall penetration (Liese and Schmid, 1964).

The question of the cell-wall penetration method, however, still lingers. It is well known that some stain fungi are able to penetrate thin silver or aluminum foils. This seems to indicate that mechanical pressure also may play a role in those stain and decay fungi that form appressoria at hyphal tips or form penetration pegs. There may be a close analogy between the penetration made through thick cuticle and epidermal cell walls of plants and that made by wood-inhabiting fungi through woody cell walls. In the case of plant pathogens, a fine hypha or penetration peg develops from the contact surface of an appressorium and penetrates the cell wall by mechanical force and enzymatic softening of the cell wall. The penetration peg resumes the normal hyphal size after the wall passage. The cell-wall passages of many soft-rot and sapstain fungi appear to be similar. The penetration modes of wood-inhabiting fungi are probably primarily enzymatic for the decay fungi that often form bore holes in the size range of hyphal diameters and primarily mechanical for the sapstainers, soft-rotters (Type 2), and others that form the narrow penetration pegs (Fig. 7-6A). The type of bore hole formed is a useful way to separate stains, molds, and decay fungi. The enlargement of the bore holes during or after penetration to normal hyphal diameter appears to be a special trait of the decay fungi. This type of bore hole appears to develop primarily enzymatically. Bore holes formed by fine penetration pegs characterize the other wood-inhabiting fungi. These holes do not enlarge after penetration, and their development appears to be primarily mechanical. Pits are also enlarged enzymatically, and it is often difficult to distinguish degraded pits from the larger bore holes. Enlarged pits are commonly associated with some of the white rots.

Soft-rot fungi (Type 1) initially form penetration pegs into the cell wall. When the tip of the peg reaches the S2 layer of the cell wall, it may branch or turn at a right angle and continue to penetrate the S2 longitudinally. Hyphal tips that divide in the S2 form structures termed *T-cells* and initiate two penetrating hyphae (Fig. 7-6C). Periodically, enzymes are secreted from the tips of these hyphae, and their action on the wood produces the typical longitudinal soft-rot cavities described elsewhere (Fig. 7-6B).

It should be stressed that the formation of longitudinal bore holes also occurs in some Basidiomycotina. Minute pockets and cavities with conical ends have been reported in the secondary wall for several white rots (Liese and Schmid, 1962), but these cavities are much smaller and more irregular than typical soft-rot cavities. Duncan (1960) has reported that *Rigidoporus crocatus* (*Poria nigrescens*) forms longitudinal cavities in the secondary wall. White-rot fungi produce a gradual thinning of the cell walls from the lumina toward the middle lamella as decay advances. This attack pattern occurs most commonly in hardwoods and has been related to complete utilization of all cell-wall constituents as the decay develops. Thinning implies that the principal decay action occurs on a surface. In some sequential white rots (e.g., *Phellinus* (*Fomes*) *pini*), cell detachment and shape changes occur in the pockets in the early decay stages (Fig. 7-7). In contrast, brown rots show no change in wall thickness until the late decay stages when the residual lignin develops shrinkage cracks (initially in the vicinity of pits) and collapses. This implies that the carbohydrate portions of the wall are quickly removed, whereas the lignin initially maintains the cell shape and dimension. The decay damage in white rots also appears to be rather uniform in adjacent cells when viewed microscopically in cross section. In contrast, the decay damage is often erratic and localized in both the brown and soft rots. Soft rots in conifers appear to preferentially attack the S2 of the secondary wall. Type 1 soft-rot fungi form the longitudinal bore holes previously described, whereas Type 2 soft-rotters selectively attack the secondary cell wall in a manner very similar to that of brown rots of conifers. In Type 2 attack in conifers, the S3 may become detached and appear as a small ring in the decayed cell-wall material (Fig. 7-6D). Soft rots in hardwoods are generally more severe, and whereas the Type 1 longitudinal bore holes predominate in the S2, in the Type 2s, both the S2 and S3 are eroded by Type 2 soft-rot fungi, and the damage resembles white-rot attack. These various modes of cell-wall attack for white-, brown-, and soft-rot fungi are illustrated in Fig. 7-8.

The cell-wall layers exhibit striking differences in decay susceptibility. Cowling (1965) proposed that the selective action of some fungi against various cell-wall layers could be useful for their selection and structural study. Generally the S2 is most susceptible. In conifers, the compound-middle lamella and S3 are more resistant to both brown and soft rots. In contrast, the compound-middle lamella zone in conifers is quickly decayed by the sequential white rotters, often resulting in cell detachment early in the decay process. In hardwoods, the compound-middle lamella zone is also resistant to brown rotters.

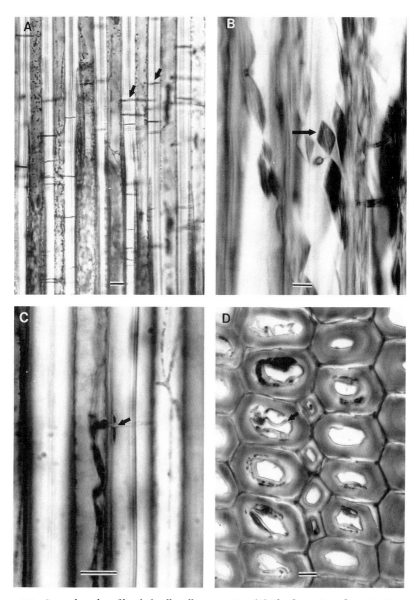

Figure 7-6 Several modes of hyphal cell-wall penetration. (A) The formation of penetration pegs by *Alternaria alternata* in a series of successive cell-wall penetrations. (B) Diamond-shaped cavities formed by a *Phialophora* sp. (C) The formation of a *T-cell* and initiation of a longitudinal bore hole. (D) Type 2 rot caused by *A. Alternata*, indicating the separation of rings of the S3. Bars equal 10µm.

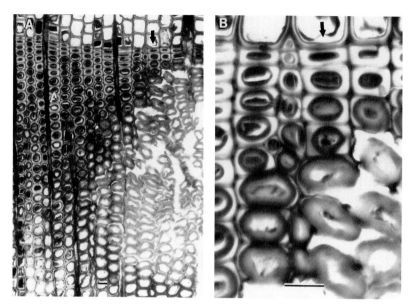

Figure 7-7 The sequential stages of decay development in a white pocket rot of red spruce caused by *Phellinus pini*. Vertical arrows identify the same radial row of tracheids. (A) A cross section through a decay pocket. (B) The successive delignification of the tracheid walls from the lumen surface toward the middle lamella and then the detachment and rounding of cells. Bar equals 20 µm. Courtesy of Dr. Susan Anagnost.

Some Research Needs

A major research need is the development of a more rational system of decay classification that combines chemical changes in cell-wall constituents with taxonomy. The current decay-classification system is based on chemical studies of a rather limited number of fungi and neglects several important taxonomic groups. This system fails clearly to separate some Type 2 soft rots from brown rots, and some white rots (Xylariaceae) from Type 1 soft rots. There is a need to study the major representatives of all major decay groups and search for a more definitive classification system that better reflects current taxonomic classification.

Some additional research needs can be posed through a series of questions. What is the fundamental difference in the nature of the enzymes produced by white rots and brown rots, limiting the former to surface action whereas the latter rapidly and deeply penetrate the wood-cell wall? What limits the effective distance an enzyme can diffuse from a hypha and still be effective? How do brown rot and the other wood-inhabiting fungi penetrate cell walls in the absence of lignin-destroying enzymes? How and where are enzymes released by the hyphae? How can the conical ends and puzzling periodicity of successive longitudinally aligned cavities be explained for the

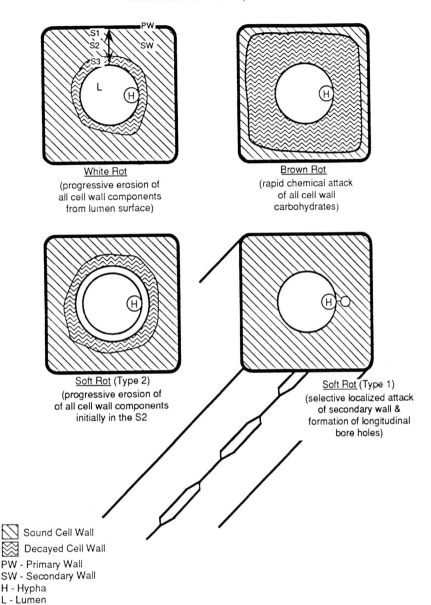

Figure 7-8 Diagrams showing the various modes of cell-wall destruction for white rots, brown rots, and the two types of soft rots.

soft-rotters? Is there an evolutionary sequence hidden in the many types and degrees of cell-wall disintegrations caused by microorganisms? A final research need is to develop methodologies to simplify and increase the accuracy of early decay detection, microscopically.

Summary

- Decay is defined as those significant changes in the physical and chemical properties of wood that are caused by the chemical (enzymatic) activities of microorganisms.
- The term *decay* is used both as a description of the condition of wood and the process of its external digestion by microorganisms.
- The decay process has stages that have been arbitrarily designated as incipient (hidden), early, intermediate, and late.
- The major disadvantages of decay are losses in strength and biomass.
- Decay losses are large in the aggregate and estimated at 15 to 25% of the gross volume of standing timber, 15% of the storage volume annually, and 10% of the annual cut as a replacement for decay of wood in service, plus the replacement costs.
- Decay can be recognized *macroscopically* by color and textural changes and *microscopically* by characteristic hyphal features, bore holes, and cell-wall erosion.
- Common decay groups based on the wood location and use are stem decays, heart rots, slash rots, storage decays, and products rots, including special uses such as building rots. Wood-inhabiting fungi can be grouped into decay and nondecay categories.
- Decay groups, based on the cell-wall constituents utilized and the type of causal agents, are the *white rots* (attack of all cell components by Basidiomycetes) *brown rots* (attack primarily of carbohydrates by Basidiomycetes); *soft rots* (attack primarily of carbohydrates by Ascomycotina and Deuteromycotina).
- The major categories of the nondecay wood-inhabiting fungi are sapstains, molds, scavengers, and bacteria.
- During early colonization, most wood-inhabiting fungi pass from cell to cell through the pits. Decay fungi are characterized by forming large bore holes through cell walls, primarily enzymatically in the later decay stages. Bore holes are a useful microscopic feature of decay.
- Nondecay fungi may form narrow threadlike hyphal filaments called *penetration pegs*, which pass through cells mechanically and enzymatically.

Summary continues

- White rots utilize all cell-wall constituents and are characterized by a uniform thinning of the cell walls from the lumen toward the middle lamella as decay develops.
- Brown rots display no change in wall thickness until the late decay stages when the residual lignin shrinks and collapses.
- Soft rots are subdivided into two types. Type 1 soft rots are characterized by longitudinal bore holes in the S2 of the secondary wall. Type 2 soft rots erode the cell wall similarly to brown rots in hardwoods, but are more confined to the S2 in conifers.

References

Bailey, I. W., and M. R. Vestal. (1937). The significance of certain wood-decaying fungi in the study of the enzymatic hydrolysis of cellulose. *Journal of Arnold Arboretum* 18:196-205.

Barghoorn, E. S., and D. H. Linder. (1944). Marine fungi: Their taxonomy and biology. *Farlowia* 1:395-467.

Bavendamm, W. (1928). Über das Vorkommen und den Nachweis von Oxydasen bei Holz zerstörenden Pilze. Z. *Pflanzenkrankeheiten u. Pflanzenschutz* 38:257-276.

Baxter, D. V. (1952). "Pathology in Forest Practice," 2nd Ed. John Wiley, New York.

Boutelje, J. B., and A. F. Bravery. (1968). Observations on the bacterial attack of piles supporting a Stockholm building. *Journal of the Institute of Wood Science* 20: 47-57.

Boyce, J. S. (1961). "Forest Pathology," 3rd Ed. McGraw-Hill, New York.

Buller, A. H. R. (1905). The destruction of wood paving blocks by the fungus *Lentinus lepideus* Fr. *Journal Economic Biology* 1:2-13.

Campbell, A. H. (1933). Zone lines in plant tissues: I. The black lines formed by *Xylaria polymorpha* (Pers.) Grev. in hardwoods. *Annals of Applied Biology* 20: 123-145.

Campbell, W. G. (1932). The chemistry of the white rots of wood. III. the effect on wood substances of *Ganoderma applanatum* (Pers.) Pat., *Fomes fomentarius* (Linn.) Fr., *Polyporus adustus* (Willd.) Fr., *Pleurotus ostreatus* (Jacq.) Fr., *Armillaria mellea* (Vahl.) Fr., *Trametes pini* (Brot.) Fr., and *Polystictus abietinus* (Dicks.) Fr. *Biochemistry Journal* 26:1829-1838.

Cartwright, K. St. G. (1930). "A Decay of Sitka Spruce Timber, Caused by *Trametes serialis*, Fr. A Cultural Study of the Fungus." Department of Scientific and Industrial Research, Forest Products Research Bulletin No. 4 (London).

Cartwright, J. St. G., and W. P. K. Findlay. (1958). "Decay of Timber and Its Prevention." Her Majesty's Stationery Office, London.

Corbett, N. H. (1965). Micro-morphological studies on the degradation of lignified cell walls by Ascomycetes and Fungi Imperfecti. *Journal of the Institute of Wood Science* 14:18-29.

Cowling, E. B. (1961). "Comparative Biochemistry of the Decay of Sweetgum Sapwood by White-Rot and Brown-Rot Fungi." U.S.D.A. Technical Bulletin 1258.

Cowling, E. B. (1965). Microorganisms and microbial enzyme systems as selective tools in wood anatomy. *In* "Cellular Ultrastructure of Woody Plants" (Côté, W., ed.), pp. 341-368. Syracuse University Press, Syracuse, New York.

Cowling, E. B., and I. B. Sachs. (1960). Detection of brown rot. *Forest Products Journal* **10**(11):594-596.

Davidson, R. W., W. A. Campbell, and D. J. Blaisdell. (1938). Differentiation of wood-decaying fungi by their reaction on gallic or tannic acid medium. *Journal of Agricultural Research* **57**:683-695.

Duncan, C. G. (1960). "Wood-Attacking Capacities and Physiology of Soft-Rot Fungi." Forest Products Laboratory Report 2173. Madison, Wisconsin.

Ellwood, E. L., and B. A. Ecklund. (1959). Bacterial attack of pine logs in pond storage. *Forest Products Journal* **9**:283-292.

Eriksson, K.-E. L., R. A. Blanchette, and P. Ander. (1990). "Microbial and Enzymatic Degradation of Wood and Wood Components." Springer-Verlag, New York.

Eslyn, W. E. (1979). Utility pole decay. Part 3: Detection in pine by color indicators. *Wood Science and Technology* **13**:117-126.

Graham, R. D., and M. E. Corden. (1980). "Controlling Biological Deterioration of Wood with Volatile Chemicals." Electric Power Research Institute Final Report EPRI-EL-1480, Palo Alto, California.

Greaves, H. (1969). Micromorphology of the bacterial attack of wood. *Wood Science and Technology* **3**:150-166.

Greaves, H., and J. F. Levy. (1965). Comparative degradation of the sapwood of Scots pine, beech, and birch by *Lenzites trabea, Polystictus versicolor, Chaetomium globosum,* and *Bacillus polymyxa. Journal of the Institute of Wood Science* **15**: 55-63.

Hartig, R. (1874). "Wichtige Krankheiten der Waldbäume." J. Springer, Berlin.

Hartig, R. (1878). "Die Zerzetzungerscheinungen des Holzes der Nadelholzbäume und der Eiche in forstlicher, botanischer, und chemischer Richtung." J. Springer, Berlin.

Hirt, R. R. (1928). "The Biology of *Polyporus gilvus* (Schw.) Fries." Tech. Pub. 22, pp. 27. New York State College Forestry, Syracuse, New York.

Hirt, R. R. (1932). "On the Biology of *Trametes suaveolens* (L.) Fries. Tech. Pub. 37, pp. 36. New York State College Forestry, Syracuse.

Hopp, H. (1938). The formation of colored zones by wood-destroying fungi in culture. *Phytopathology* **28**:601-620.

Hubert, E. E. (1924). The diagnosis of decay in wood. *Journal of Agricultural Research* **29**:523-67.

Hubert, E. E. (1931). "An Outline of Forest Pathology." John Wiley, New York.

Jutte, S. M., and R. A. Zabel. (1974). Initial wood decay stages as revealed by scanning electron microscopy. *Proceedings Scanning Electron Microscopy* (Part II):445-452.

Kistler, B. R., and W. Merrill. (1968). Effects of *Strumella coryneoidea* on oak sapwood. *Phytopathology* **58**:1429-1430.

Knuth, D. T., and E. McCoy. (1962). Bacterial deterioration of pine logs in pond storage. *Forest Products Journal* **12**:437-442.

Liese, W. K. F. (1955). On the decomposition of the cell wall by microorganisms. Record of the Annual Convention British Wood-Preservation Association, Cambridge, 59-160.

Liese, W. (1970). Ultrastructural aspects of woody tissue disintegration. *Annual Review of Phytopathology* **8**:231-258.

Liese, W., and R. Schmid. (1962). Elektronenmikroskopische Untersuchungen über den Abbau des Holzes durch Pilze. *Angew. Bot.* **36**:291-298.

Liese, W. and R. Schmid. (1964). Über das Wachstum von Bläuepilzen durch verholtze Zellwände. *Phytopathol. Z.* **51**:385–393.

Lundström, H. (1973). "Studies on the Wood-Decaying Capacity of the Soft-Rot Fungi *Allescheria terrestris, Phialophora (Margarinomyces) luteo-viridis,* and *Phialophora richardsiae.*" Research Note R 87. Department of Forest Products, Royal College of Forestry, Stockholm, Sweden.

Manion, P. (1991). "Tree Disease Concepts," (2nd Ed.) Prentice-Hall, New York.

Merrill, W., and French, D. W. (1965). Wood fiberboard studies. 3. Effect of common molds on the cell wall structure of the wood fibers. *Tappi* **48**:653–654.

Miller, B. D., F. L. Taylor, and R. A. Popek. (1965). A sonic method for detecting decay in wood poles. *Proceedings of the American Wood Preservers' Association* **61**: 109–113.

Morrell, J. J., M. E. Corden, R. D. Graham, B. R. Kropp, P. Przybylowicz, C. M. Sexton, and S. M. Smith. (1987). Basidiomycete colonization of Douglas fir poles during air seasoning and its effect on wood strength. *Proceedings of the American Wood Preservers' Association* **83**:284–296.

Nicholas, D. D., and R. J. Thomas. (1968). The influence of enzymes on the structure and permeability of loblolly pine. *Proceedings of the American Wood Preservers' Association* **64**:70–76.

Nilsson, T. (1974). "Studies on wood degradation and cellulolytic activity of microfungi." Stud. For. Suec. 104. Royal College of Forestry, Stockholm, Sweden.

Nilsson, T., G. Daniel, T. K. Kirk, and J. R. Obst. (1989). Chemistry and microscopy of wood decayed by some higher Ascomycetes. *Holzforschung* **43**:11–18.

Nutman, F. J. (1929). Studies of wood-destroying fungi. I. *Polyporus hispidus* (Fries). *Annals of Applied Biology* **16**:40–64.

Procter, P., Jr. (1941). "Penetration of the Walls of Wood Cells by the Hyphae of Wood-Destroying Fungi." Yale University School of Forestry Bulletin No. 47. New Haven, Connecticut.

Przybylowicz, P. R., B. R. Kropp, M. E. Corden, and R. D. Graham. (1987). Colonization of Douglas fir poles by decay fungi during air-seasoning. *Forest Products Journal* **37**(4):17–23.

Rhoads, A. S. (1918). "The Biology of *Polyporus pargamenus* Fries." New York State College, Forestry Tech. Pub. 11, Syracuse, New York.

Savory, J. G. (1954). Breakdown of timber by Ascomycetes and Fungi Imperfecti. *Annals of Applied Biology* **41**:336–347.

Scheffer, T. C. (1936). "Progressive Effects of *Polyporus versicolor* on the Physical and Chemical Properties of Red Gum Sapwood." U.S.D.A. Technical Bulletin No. 527. Washington, D.C.

Shigo, A. L., and A. Shigo. (1974). "Detection of Discoloration and Decay in Living Trees and Utility Poles." U.S.D.A. Forest Service Research Paper, NE-294.

Spaulding, P. (1911). The timber rot caused by *Lenzites sepiaria* U.S.D.A. Bureau Plant Industry Bul. 214. Washington, D.C.

Stoker, R. S. (1948). X-ray pole inspection. *Proceedings of the American Wood Preservers' Association* **44**:298–313.

Wang, C. J. K., and R. A. Zabel. (1990). "Identification Manual for Fungi from Utility Poles in the Eastern United States." American Type Culture Collection, Rockville, Maryland.

Wilcox, W. W. (1964). "Preparation of Decayed Wood for Microscopical Examination." U.S.D.A. Forest Service Forest Products Laboratory Research Note FPL-056. Madison, Wisconsin.

Wilcox, W. W. (1968). "Changes in Wood Microstructure through Progressive Stages of Decay." U.S.D.A. Forest Service Research Paper FPL-70. Madison, Wisconsin.

Wilcox, W. W. (1970). Anatomical changes in wood cell walls attacked by fungi and bacteria. *Botanical Review* **36**(1):1–28.

White, J. H. (1920). On the biology of *Fomes applanatus* (Pers.) Wallr. *Transactions of the Royal Canadian Institute* **12**:133–174.

Zabel, R. A., F. F. Lombard, and A. M. Kenderes. (1980). Fungi associated with decay in treated Douglas fir transmission poles in northeastern United States. *Forest Products Journal* **30**(4):51–56.

CHAPTER
8

Chemical Changes in Wood Caused by Decay Fungi

The chemistry of wood decay is a subject of growing interest and importance. It is of interest because of the natural challenge of any complex problem. It is potentially of great economic importance since increased knowledge of the successive chemical steps involved in the decay process may permit wood to be protected more effectively for high-decay-hazard uses. It may also lead to more economical methods of biopulping, pulp bleaching, and the separation of carbohydrates from the wood complex for chemical purposes, or new uses of *enzyme-modified wood* as cheap animal food or energy sources.

Great progress has been made in the past few decades toward understanding the chemistry of wood decay. These results are summarized in recent textbook chapters, textbooks, and symposia (Kirk, 1971, 1973; Loewus and Runeckles, 1977; Kirk et al., 1980; Crawford, 1981; Sjöström, 1981; Fengel and Wegener, 1984; Higuchi, 1985; Eriksson et al., 1990). Key contributions that either set the stage for a better understanding of the decay process or added significant new insights include the following: Reese (1963, 1977) characterized the enzymes involved in cellulose degradation; Cowling and Brown (1969) examined the dimensional relationships between the molecular architecture of the cell wall and the puzzle of its accessibility to enzyme molecules; Timell (1967) elucidated the complex structures of the hemicelluloses; Côté (1977) clarified the ultrastructure of wood; Adler (1977) studied lignin structure; Tien and Kirk (1983) identified the first enzymes involved in the enzymatic attack of lignin; Eriksson (1978) identified the *cellulase* enzymes associated with some white-rot fungi; Higuchi (1985) studied lignin biodegradation; and Ruel and Barnard (1985) and Blanchette et al. (1991) described the micromorphological sequences of many decays, which suggested that some nonenzymatic events occurred in white rot.

The major wood cell-wall constituents involved in the decay process are cellulose, the hemicelluloses, and lignin. Decay occurs primarily from the enzymatic activities of a few groups of specialized fungi.

Previous chapters have reviewed the general chemical nature and structure of the three major cell components, considered their interrelationships at various levels of organization, and the spatial dimensions of wood at the microscopic and ultrastructural levels, and reviewed enzyme functions and related them to the major metabolic events taking place in and around hyphae. The sequential chemical and enzymatic events involved in the decay process are, of course, the basic cause of the grosser anatomical and wood physical property effects discussed in Chapters 7 and 10.

A useful approach to this complex topic will be to look first at the effects of decay on the major wood components for the three major decay types (white, brown, and soft rots), then to consider the sites and sequences of action of the enzymes involved, and to end with a summary of current understandings of the decay process and its effects on various important wood properties.

Changes in Cell-Wall Components by Decay Type

The cell-wall components of wood are utilized in varying orders and rates by different decay fungi. These differences are the basis for the chemical classification of decay presented in Chapter 7. The general depletion patterns of the cell-wall components at progressive stages of decay provide useful insights into the types and sequences of the enzymes involved. The depletion pattern can be monitored by comparing the weight of the cell-wall constituents in original wood to their weights in the residual decayed wood. Comparative weight-loss curves are developed by examining weight losses at various times. For example, lignin content steadily decreases as decay develops in a white rot, whereas essentially no lignin loss occurs with brown rots (Table 8-1). Another useful method for monitoring cell-wall depletion is to express the level of each component at the various decay stages on the basis of the changes in the original amount present in sound wood. This method permits an easy comparison of the percentage change in the cell-wall constituents with the total weight loss percentage to detect differences in rate of component utilization (Tables 8-2, 8-3, and 8-4). Since the precise gravimetric determination of the major cell-wall constituents is both time consuming and difficult, some quicker techniques involving chromatographic quantification of the major sugars have been developed. For example, the analysis of glucan, xylan, and mannan in decayed coniferous wood provides estimates of cellulose, galactoglucomannan, and arabino-4-O-methylglucuronoxylan content (Tables 8-2 and 8-3). Hydrolysis of these carbohydrates and their groupings into glucan, xylan, and mannan does not segregate the arabinose or galactose and may lead to overestimates of certain fractions, but minor analytical errors introduced are judged to be insignificant (Kirk and Highley,

Table 8-1

Chemical composition of sweet gum sapwood in progressive stages of decay caused by white- and brown-rot fungi[a]

Average weight loss (%)	Total carbohydrate	Composition of residual material (%)					
		Glucan	Galactan	Mannan	Xylan	Araban	Lignin
Trametes (Polyporus) versicolor (white rot)							
0	77.1	52.3	1.1	2.7	20.1	0.9	22.9
25.3	58.0	34.5	.6	2.6	14.9	.4	16.7
55.2	33.9	22.8	.3	1.7	8.9	.3	10.9
79.0	14.9	9.7	.2	.9	3.9	.2	6.1
96.7	2.5	2.0	.03	.1	.3	.03	.8
Postia placenta (Poria monticola) (brown rot)							
0	77.1	52.3	1.1	2.7	20.1	0.9	22.9
20.1	56.8	40.1	.5	1.3	14.5	.4	23.1
44.8	32.6	22.2	.2	.7	9.2	.2	22.6
69.5	9.2	5.4	.1	.2	3.4	.1	21.3

[a]The weight loss percentages are based on the moisture-free weight of the original sound wood. Data selected from Cowling (1961).

Table 8-2
Weight losses of wood biomass and the major cell-wall components of conifers and hardwoods caused by several white-rot fungi

Substrate	Fungus	Weight loss (%)[a]				
		Total (biomass)	Glucan	Mannan	Xylan	Lignin
		Conifers[b]				
Pinus monticola (western white pine)	Trametes versicolor	13	4	13	13	27
		22	17	22	21	33
		43	43	47	47	52
		61	65	68	67	62
		83	85	89	89	86
	Ganoderma applanatum	16	12	16	19	26
		43	42	52	51	57
Picea sitchensis (Sitka spruce)	Trametes versicolor	6	10	13	3	5
		21	27	29	27	20
		40	39	50	45	52
		61	65	72	66	65
	Ganoderma applanatum	11	13	18	17	11
		35	33	47	44	48
		Hardwoods[c]				
Betula alleghaniensis (yellow birch)	Trametes versicolor	21	20	26	26	31
		36	39	54	39	39
	Ganoderma applanatum	17	28	27	14	18
		32	38	50	36	37
	Bondarzewia berkeleyi	8	16	8	4	31
		22	31	33	30	42
		39	44	51	40	63
	Ischnoderma resinosum	11	17	25	10	35
		22	30	27	21	44

[a]The weight-loss percentages are based on the original amount of each component in sound wood.
[b]Data selected from Kirk and Highley (1973).
[c]Data selected from Kirk and Moore (1972) and rounded to the nearest percentage.

Table 8-3
Weight losses of the major cell-wall components in conifers and hardwoods caused by several brown-rot fungi[a]

Substrate	Fungus	Weight loss (%)				
		Total (biomass)	Glucan	Mannan	Xylan	Lignin
Conifers[b]						
Tsuga heterophylla (western hemlock)	Postia placenta	6	4	11	16	8
		26	26	38	39	4
		46	79	88	75	2
Picea engelmannii (Englemann spruce) Postia placenta		12	13	26	22	6
		26	29	41	69	10
		49	81	93	79	−13
Picea engelmannii (Englemann spruce)	Gloeophyllum trabeum	10	12	14	18	9
		19	22	47	37	4
		43	55	80	65	11
Picea sitchensis (Sitka spruce)	Neolentinus lepideus	6	12	6	19	3
		27	37	68	49	−4
		45	57	78	64	6
Pinus taeda (loblolly pine)	Postia placenta	9	13	25	1	−4
		24	29	58	26	2
		45	68	81	69	4
Hardwoods[c]						
Liquidambar styraciflua (sweet gum)	Postia placenta	20	23	52	30	0
		45	58	74	55	0
		70	90	93	85	−9

[a]The weight-loss percentages are based on the original amount of each component in sound wood.
[b]Data selected from Kirk and Highley (1973).
[c]Data selected from Cowling (1961).

1973). Lignin determinations also vary substantially with method of preparation, and these variations are discussed further in the lignin section.

When evaluating changes in cell-wall component ratios, it is prudent to remember that the decayed sample itself contains all the different cell and tissue types that have varying levels of each wood component; that microorganisms may selectively attack specific cell types; that substantial decay gradients can occur even in small samples of decayed wood; and that at some of the later decay stages, the fungus may be simultaneously assembling storage sugars or polyphenols that may be inadvertently detected as cell-wall components.

The general depletion patterns of cell-wall components for white-, brown-, and soft-rot fungi in conifers and hardwoods are presented in Tables 8-1 through 8-4. General information on each decay type is briefly summarized.

White-rot fungi are able to attack and metabolize all major wood constituents. In this regard, the white-rot fungi are unique among most microorganisms in their capacity to depolymerize and metabolize lignin. Numerous studies now indicate, however, that the major components are used in varying orders and rates by different white-rot fungi, suggesting that these fungi probably represent a heterogeneous group with widely varying enzymatic capabilities. These differences were used by Liese (1970) to group white-rot fungi into the *simultaneous* rotters, which utilized the components uniformly, and *white-rotters*, which utilized lignin initially more rapidly than cellulose.

Early, differential lignin attack by some white-rotters has important potential commercial applications, and attempts are now being made to find isolates or develop new strains for possible use in biopulping, the biobleaching of pulps, or for increasing the biodegradability of wood for use in industrial fermentation or ruminant feed. *Trametes (Polyporus) versicolor* is an example of a fungus that uses all wood components essentially uniformly. *Phellinus (Fomes) pini, Heterobasidion (Fomes) annosum,* and *Bondarzewia (Polyporus) berkeleyi* are examples of fungi that remove the lignin at a faster rate than cellulose or the hemicelluloses in early decay stages. *Ganoderma applanatum* is an interesting example of the few white-rot fungi that initially remove cell-wall carbohydrates somewhat more rapidly than lignin.

Some general features of utilization of wood constituents by white-rot fungi are summarized as follows:

1. All cell-wall components are presumably ultimately consumed, with the exception of the minor minerals. There is considerable variation in the sequence and rate of component utilization by both species and fungal strains within a species. In most cases, the hemicelluloses are utilized preferentially in early decay stages. Weight losses may approach 95–97% of the original wood material when prolonged exposures under optimal decay conditions prevail.

Table 8-4
Weight losses of the major cell-wall components of conifers and hardwoods caused by several soft-rot fungi[a]

Substrate	Fungus	Weight loss (%)[b]				
		Total	Glucan	Mannan	Xylan	Lignin
Conifers						
Pinus monticola	*Papulospora* sp.	15	18	13	18	12
(western white pine)	*Paecilomyces* sp.	10	7	6	8	14
	Thielavia terrestrius	7	3	−3	9	14
Hardwoods						
Alnus rubra	*Papulospora* sp.	10	15	14	6	9
(red alder)		17	23	18	8	12
	Paecilomyces sp.	15	21	28	25	11
		25	37	20	23	13
		41	60	22	44	19
	Thielavia terrestrius	7	10	19	1	9
		28	40	21	29	17
Populus balsamifera	*Papulospora* sp.	10	14	14	17	0
(balsam poplar)		21	27	25	29	4
	Paecilomyces sp.	14	15	28	35	10
		28	41	30	37	11
	Thielavia terrestrius	10	11	29	12	6
		23	25	43	27	13

[a] Data selected from Eslyn et al. (1975).
[b] The weight-loss percentages are based on the original amount of each component in sound wood.

2. At all stages of decay, the residual wood has a low solubility in 1% sodium hydroxide (alkali solubility), suggesting that the breakdown products of decay are utilized by the fungi as rapidly as they are released.
3. The cellulose, hemicelluloses, and lignin remaining in the undecayed portions of the wood appear to be essentially unaltered, suggesting that white-rot fungi concentrate their attack on exposed cell-wall surfaces. Thus, the enzymes slowly erode their way into the cell walls from lumina surfaces.

Brown-rot fungi primarily decompose the cell-wall carbohydrates, leaving behind a modified, demethoxylated lignin residuum (Tables 8-1, 8-3). Selective removal of carbohydrates in the later stages of brown-rot attack has been used to study the distribution of lignin in the cell wall (Côté *et al.*, 1966). The hemicelluloses are removed more rapidly than cellulose by brown-rot fungi in the early decay states. Highley (1977) has shown that carbohydrate supplements such as mannan are necessary for depolymerization of pure cellulose by *Postia (Poria) placenta*, a brown-rotter. Brown rots differ from white rots in the extensive depolymerization of the carbohydrates in the secondary wall early in the decay process (Kirk and Highley, 1973). Whereas lysis zones are closely associated with white-rot hyphae, substantial cell-wall damage at distances up to several cell widths has been noted for some brown rots (Eriksson *et al.*, 1990).

Brown-rot fungi modify wood in the following ways during progressive decay development.

1. All carbohydrates are ultimately consumed, leaving a residuum of modified lignin in the cell wall.
2. Large increases in water solubility and solubility in 1% NaOH occur at early decay stages, owing to the rapid carbohydrate depolymerization in the early decay stages and increased lignin solubility in the later decay stages. Brown-rot fungi appear to depolymerize wood in the early decay stages much more rapidly than the decay products can be metabolized. The occurrence of excess wood-decomposition products may help explain the frequent presence of other wood scavengers in brown-rotted wood.
3. The decay process rapidly involves the S1 and S2 layers of the cell walls, but develops irregularly and lacks the hyphal-associated lysis zones typical of white-rot fungi.
4. There appears to be much less variation in the sequential attack of cell-wall constituents by brown-rot fungi as compared to the white-rotters.

Soft-rot fungi display considerable variation in their effects on cell-wall constituents during decay development (Tables 8-4, 8-5). For many species, the principal targets are the carbohydrates, whereas lignin attack is limited to minor demethoxylation. Some soft-rotters, however, selectively removed

Table 8-5
Chemical composition of European beech in progressive stages of decay by *Chaetomium globosum*, a soft-rot fungus[a]

Average weight loss (%)	Composition of residual material (%)			
	Cross and Bevan cellulose[b]	Alpha cellulose[c]	Pentosans	Lignin
0	56.2	42.2	27.8	21.6
29.9	34.3	26.5	18.4	19.4
31.7	34.7	25.7	17.6	19.3
40.6	28.2	19.9	14.6	17.6
49.1	22.4	16.3	12.5	16.7

[a]The weight-loss percentages are based on the dry weight of the original undecayed wood, corrected for mineral content. From Savory and Pinion (1958).
[b]Roughly equivalent to total carbohydrate, as listed in Tables 8-1 to 8-3.
[c]Roughly equivalent to glucan, as listed in Tables 8-1 to 8-3.

more lignin than carbohydrate from coniferous wood, in a manner similar to that of some white-rotters (Eslyn et al., 1975). Most of the soft-rot fungi tested remove glucans at a faster rate than the hemicelluloses, although several exceptions were noted. Type 1 soft-rot fungi are able to degrade crystalline cellulose, as reflected by the formation of characteristic cavities in the S2 zone of the secondary wall. Wood decayed by soft-rot fungi resembles that of white-rot-degraded wood in having low alkali solubility, indicating that the degradation products are used at the same rate as they are released. In conifers, the S3 zone of the secondary wall is resistant to soft-rot attack, but delignification substantially increases decay susceptibility and may shift the fungus from cavity formation (Type 1) to the wall erosion (Type 2) mode (Morrell and Zabel, 1987). The soft-rot fungi appear to be variable in their effects on the cell-wall components and share some features of both white and brown rots. It is probable that this broad group, now defined primarily as non-basidiomycete decayers, includes fungi typified by several different patterns of cell-wall decomposition.

Chemical Mechanisms of Wood Decay

Whereas it is important to understand the relative differences in wood degradation, it is equally important to consider how these drastic changes in the various cell-wall components are accomplished by the three major types of decay fungi. Carbohydrate components are depolymerized primarily by a series of *hydrolytic enzymes*, whereas lignin is degraded primarily by *oxidative enzymes*. There is also evidence that powerful, nonenzymatic oxidizing agents may initiate the first steps in the brown rots by depolymerizing the crystalline zones of the cellulose. As we examine the various fungal enzymes, it is important to note that the enzymes active in decay are secreted and act outside of the hyphae (exoenzymes). These enzymes must physically bind or form a complex with the portion of the polymer or macromolecule they attack. Since the large macromolecules composing the cell wall are nonsoluble, the enzymes involved in decay must either diffuse into the cell wall to contact the susceptible bonds or be delivered there through direct contact with the fungal hyphae.

Cellulose Decomposition

Whereas many fungi are able to decompose modified cellulose products, only a limited number are able to decompose cellulose in its native, highly crystalline form (cotton, ramie, and wood pulp). These species are termed *cellulolytic* fungi.

Initial Concepts on Cellulolytic Enzymes (C_1-C_x Hypothesis)

The fungi capable of cellulolytic attack on fabrics have been studied intensively since the 1950s. Lists of cellulolytic fungi and reviews of the early research on their enzyme systems were assembled by Reese (1963), Norkrans (1963, 1967), and Eriksson *et al.* (1990).

In 1950, Reese *et al.*, working with several *Trichoderma* species, postulated that cellulolytic fungi secreted an enzyme (C_1), which was able to separate the cellulose molecules in the crystalline regions, and enzymes (C_x), which then both randomly and endwise depolymerized the separated chains into glucose, cellobiose, and oligosaccharides. This concept stimulated additional studies to identify and characterize the enzymes in the *cellulase* complex of several species of *Trichoderma* and other cellulolytic fungi (Iwasaki *et al.*, 1964; Halliwell, 1965; Selby and Maitland, 1967; King and Vessal, 1969).

Until recently, cellulase was considered to be a multienzymatic system consisting of at least three enzymatic components that converted cellulose to glucose. C_1 was posed as the component that attacked crystalline cellulose and separated the chains by an unknown mechanism. To function, C_1 required the presence of the C_x enzyme components, consisting of two types of β-1-4 glucanases. *Exo*-β-1,4 glucanase successively removes single cellobiose or glucose units from the nonreducing end of the cellulose chain, whereas *endo*-β-1,4 glucanases cleave cellulose at random sites along the chain (Fig. 8-1A). β-Glucosidases are attached to the hyphal wall and further decompose cellulose fragments by hydrolyzing the various oligomers, including cellobiose, which is converted to two molecules of glucose. β-Glucosidases hydrolyze the smaller cellulose fragments, and exoglucanases, the larger ones. The exoglucanases depolymerize by inversion and release α-glucose units.

It was generally presumed that cellulase enzymes were also involved in cellulose breakdown in wood decay, even though their identities and modes of action were based largely on studies of pure cotton cellulose. However, there was some question about the validity of applying this model completely to cellulose decomposition *in situ* in wood, since cellulose microfibrils in wood are surrounded by hemicelluloses and high concentrations of lignin. Furthermore, the inability of purified C_1 enzyme to depolymerize unmodified cellulose in the absence of C_x enzymes was puzzling and raised questions concerning the role of C_1 in decay initiation. When examining the role of cellulase in wood decomposition, it is helpful to evaluate systems for white-, brown-, and soft-rot fungi separately.

White-Rot Fungi Studies of the white-rot fungus, *Phanerochaete chrysosporium*, first suggested that the C_1 enzyme was actually the exoenzyme that detaches both glucose and cellobiose from the ends of the cellulose molecules (Eriksson, 1969).

In subsequent reports, the enzymes and the major steps in the enzymatic breakdown of cellulose by a white rot fungus were elucidated and involved a series of hydrolytic and oxidative reactions (Eriksson, 1978). These steps can be outlined as follows:

I. Initial cellulose decomposition pathways (hydrolytic)

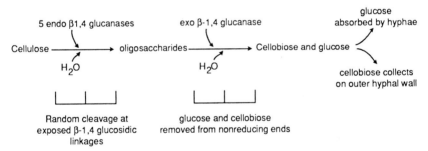

II. Secondary decomposition pathways (hydrolytic and oxidative)

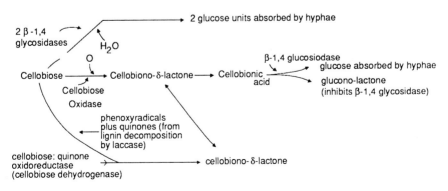

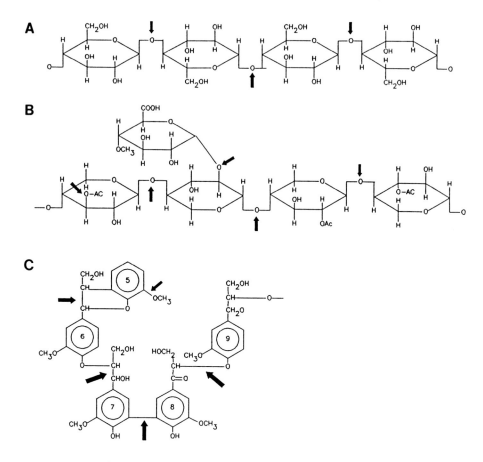

Figure 8-1 General sites for enzymatic cleavage of major wood cell-wall components during decay by a white-rot fungus. (A) hydrolytic cleavages of glucose, cellobiose, and glucose oligomers from cellulose by exo- and endo-glucanases; (B) hydrolytic cleavages of xylose, xylobiose, and xylose oligomers from a xylan O-acetyl-4-O-methylglucuonoxylan) by endoxylanase and xylosidase. The methylglucose side branch and the acetyl substituents are removed by other hydrolytic enzymes; and (C) oxidative random cleavage of a spruce lignin (after Adler, 1977) at C_α–C_β bonds by a ligninase that releases phenolic fragments.

Strong synergism has been reported between the endoglucanases and exoglucanases on some substrates (Streamer and Peterson, 1975). The molecular weights of the exoglucanase are around 50,000, whereas endoglucanases range from 30,000 to 35,000. The two 1,4-β-glucosidases are larger enzymes with molecular weights of 165,000 and 182,000. It is interesting to note that the smaller enzymes are those that penetrate the cell wall. Several proteases formed by this fungus are reported to enhance the activity of the endoglucanase (Eriksson and Peterson, 1982), suggesting that modification of the native enzyme enhances activity.

A series of reactions appears to regulate the enzymatic decomposition of cellulose and control the rate of glucose release to the capacity of the fungus to utilize it. The regulation reactions controlling glucose release from cellulose degraded by white-rot fungi are briefly outlined.

1. An excess of glucose inhibits the release of the 1,4-β-endoglucanases (and possibly the exoglucanases) by catabolite repression.
2. An excess of glucose induces the formation of gluconolactone.

$$\text{glucose (excess)} \xrightarrow{\text{glucose oxidase}} \text{gluconolactone (a β-glycosidase inhibitor)} \rightarrow \text{gluconic acid (may be absorbed by hyphae and utilized)}$$

3. When β-glycosidase is inhibited by gluconolactone, cellobiose accumulates.

$$\text{cellobiose (excess)} \xrightarrow[\text{transglucosylation}]{\text{endoglucanases}} \text{oligosaccharides} + H_2O$$

4. Transglycosylation and the formation of storage products (dextrins, etc.) may occur when an excess of simple sugars is present.
5. An excess of simple sugars may also inhibit the release of hemicellulases.

The regulation processes reduce excesses of simple sugars that may limit competition for the substrate from other wood-inhabiting saprobes. The absence of excess sugars also explains the absence of solubility changes in wood decayed by white-rot fungi.

Brown-Rot Fungi As would be expected from the drastic strength losses and rapid drops in degree of polymerization (DP) exhibited at early decay states, the brown-rot fungi appear to extensively degrade cellulose in a manner that differs from that of the white-rot fungi. Brown-rot fungi utilize endo-1,4-β-glucanases, but lack exo-1,4-β-glucanases (Highley, 1975). The mechanism of attack on crystalline cellulose by brown-rot fungi is unknown, and the production of a nonenzymatic depolymerizing agent has been proposed, since the dimensional constraints of the crystalline zones preclude enzyme access by diffusion (Cowling and Brown, 1969). Koenigs (1974) has suggested that oxidation by an H_2O_2/Fe^{2+} system may provide a nonenzymatic cellulose depolymerization agent. This process may also be related to early hemicellulose attack in brown rots. Lyr (1960) has shown that glucomannan, an important hemicellulose in conifers, is initially removed more rapidly than cellulose, and its decomposition is associated with the release of H_2O_2. It has also been shown that some brown-rot fungi, such as *Postia (Poria) placenta*, can utilize pure cellulose in culture in the presence of hemicelluloses (Highley, 1978). There is uncertainty about the source and role of H_2O_2 since some brown-rot fungi produce insignificant amounts (Highley, 1982). Recently, the hydroxy radical has been detected in brown rotted wood (Illman and Highley, 1989). After the cellulose chains in the crystalline zones are separated, endo-1,4-β-glucanases similar to those in white-rot fungi randomly cleave the cellulose molecule, and 1,4-β-glucosidases convert the cellobiose to glucose. In contrast to the regulated process of cellulose decomposition associated with white-rot fungi, excesses of glucose and cellodextrins are present in brown-rotted wood, but appear to play no role in cellulase regulation. Also, the endoglucanases appear to be less substrate-specific (Keilich *et al.*, 1970) and a multifunctional endoglucanase is reported from *Postia placenta* (Highley and Wolter, 1982).

Soft-Rot Fungi These fungi resemble the brown-rot fungi in that their attack is limited primarily to cell-wall carbohydrates, with demethoxylation of lignin. However, soft-rot fungi also resemble the white rotters in having similar hydrolytic enzyme systems.

Much of the early research on cellulolytic enzymes involved fungi now known to have soft-rot capabilities, e.g., *Trichoderma* spp. Considerable efforts have been devoted to developing mutant strains of *T. reesei* that produce high levels of cellulase for industrial use (Montenecourt and Eveleigh, 1977). Some soft-rot fungi utilize exo-1,4-β-glucanases, endo-1,4-β-glucanases, and 1,4-β-glucanases to degrade cellulose. Other species utilize primarily the endo-1,4-β-glucanases and 1,4-β-glucosidases and appear to limit their attack to the amorphous cellulose zones in the microfibrils. As in the white-rot fungi, synergism has been reported among the enzymes in the cellulase complex of soft-rot fungi (Mandels and Reese, 1964). These differences suggest that the soft-rot fungi, now defined broadly as *non-basidiomycete* decayers, are a cosmopolitan group of fungi with rather diverse mechanisms for degrading cellulose.

Hemicelluloses Decomposition

Many bacteria, yeasts, and fungi are able to degrade and utilize the hemicelluloses in plant tissues. Hemicelluloses are often the first cell-wall components degraded by decay fungi, probably owing to their shorter chain lengths, solubility, and exposed locations around the cellulose microfibrils.

The hemicellulose-degrading enzymes are primarily hydrolytic in nature, and the attack patterns are analogous to cellulase degradation of cellulose. However, exoenzymes are absent, probably reflecting the low degree of polymerization (DP) (<200) of hemicelluloses. Since the major wood hemicelluloses are heteropolymers consisting of several sugars, side branches, and substituent groups, the enzymatic processes involved in degradation are much more complicated and are just beginning to be elucidated. As a result, no comparisons among decay types will be made.

We shall limit our attention to the enzymes involved in the digestion of xylan (*O*-acetyl-4-*O*-methyl-glucurono-β-D-xylan) and mannan (galactoglucomannans). Enzymatic degradation of the other hemicelluloses has been reviewed by Dekker (1985) and Eriksson *et al.* (1990).

Xylans

The breakdown of xylan requires the action of several xylanolytic enzymes (Figure 8-1B). These enzymes and their functions are as follows:

1. Endo-1,4-β-xylanases separate the polymer backbone into xylose and its oligomers. Endoxylanases have been isolated from *Trichoderma* spp., *Coniophora (cerebella) puteana*, *Postia (Poria) placenta*, and *Irpex lacteus*. These enzymes are analogous to the endoglucanases in the cellulase enzyme complex.
2. 1,4-β-Xylosidases hydrolyze the xylo-oligosaccharides or xylan fragments to xylose. These enzymes have been obtained from fungi in a wide range of genera (Reese et al., 1973) and appear to be inducible enzymes. The enzymes are cell-wall bound and analogous to the β-glucosidases in the cellulase enzyme complex.
3. α-Glucuronidase separates 4-O-methylglucuron side chains from the xylan backbones and releases glucuronic acid units.
4. α-Arabinosidase removes L-arabinose side chains, greatly increasing the effectiveness of the endoxylanases.
5. Acetyl esterase removes the acetyl substituent groups from the xylose. Xylan esterases have been obtained from *Trichoderma reesi* and *Schizophyllum commune*.

Mannans

The mannans in wood are heteropolymers, and their degradation requires the combined action of several enzymes to remove the various sugars composing the polymer backbone and the galactose and acetyl side chains. The enzymes required have modes of action similar to those of the xylanases and include endo-1,4-β-mannanase, β-mannosidase, β-glucosidase, α-galactosidase, and acetyl esterase.

Endomannanases have been obtained from a wide range of organisms, including white-, brown-, and soft-rot fungi. The sequences of enzymatic attack on mannans are unknown, but various patterns of degradation have been proposed by Dekker (1985).

The utilization rates of the various wood hemicelluloses by white-rot fungi suggest that there are substantial differences among fungal species in the amounts of hemicellulase produced and the sequences by which these enzymes degrade the substrate.

Hemicellulases have come under more intensive study because of their potential commercial applications for removing residual xylans from pulp and for converting pentose sugars in agricultural wastes to proteins.

Lignin Decomposition

Lignin is resistant to degradation by most microorganisms and, indeed, its primary role in the wood cell wall is to protect the carbohydrates from microbial attack. Lignin is efficiently degraded in nature primarily by white-rot fungi, including many litter-decomposing basidiomycetes. Owing to its complex chemical structure, we have only recently begun to understand fungal mechanisms of lignin decomposition.

As expected, there are major differences between fungal degradation of lignin and carbohydrates. Lignin breakdown is accomplished by a limited group of specialized fungi (white-rotters), whereas many microorganisms successfully degrade and utilize wood carbohydrates. Lignin decomposition proceeds by oxidative reactions that separate carbon-to-carbon bonds or ether linkages and separate various functional groups, side chains, and aromatic rings randomly from the huge, amorphous lignin macromolecule (Fig. 8-1C) rather than the uniform hydrolytic cleavages of β-1,4-glycosidic bonds common to the carbohydrates. White-rot fungi can completely degrade lignin in wood in laboratory decay tests. Yet, the end product of lignin decomposition in nature contains partially decomposed, fragmented lignin (humus), which enters the soil cycle, and the rapid evolution of H_2O and CO_2 as decomposition products is lacking. Soil bacteria, microfauna, and even physical processes may play important roles in the final breakdown of humus. This is a slow process, and some lignin-degradation products have soil-residence times of centuries (Zeikus, 1981).

Great progress has been made in the past decade toward understanding the biodegradation of lignin. Recent review articles or textbook chapters now present detailed accounts of the many chemical changes in the decay of lignin, the enzymes involved, and tentative biodegradation pathways (Kirk and Farrell, 1987; Kirk, 1988; Chen and Chang, 1985; Higuchi, 1985; Gold *et al.*, 1989; Eriksson *et al.*, 1990). For our purposes, coverage of this complex topic will be limited to the major chemical changes, the enzymes involved in lignin degradation, and a comparison of lignin degradation by the three major groups of decay fungi.

Lignin Determinations

Caution should be exercised in judgments on lignin degradation and metabolism since lignin determination can vary considerably with the analytical method employed. Gravimetric determination of the sulphuric acid insolubles in the wood as lignin (Klason lignin) is widely used because of its simplicity, but this method includes some extractives and degraded products in the soluble fraction, artificially inflating lignin losses. This method also dramatically alters the chemical structure of the residual lignin, making it less useful for studying the structure of decomposed lignin. As an alternative, a lignin can be synthesized by oxidation of the lignin precursors with H_2O_2 and peroxidase. This material, called dehydrogenation polymerizate (DHP) lignin, has particular advantages in that carbohydrates, which are difficult to remove from native lignins, are absent. Lignin structures have also been studied using milled wood lignin (MWL) (Björkman lignin), which is extracted from finely ground or milled wood by solvents; MWL is generally accepted as a preferred lignin-preparation method for critical studies. Other lignin-preparation methods and their advantages and disadvantages have been reviewed by Crawford (1981).

Useful clues on the lignin-degradation process have come from the analysis of residual products in partially decayed lignin, the identification of the many low-molecular-weight degradation products, and the use of isotope-labeled dimeric lignin model compounds.

Chemical Modifications in Decayed Lignin

The analysis of residual lignin at various stages of decay by white-rot fungi indicates a steady loss of methoxyl groups and increases in oxygen and hydroxyl content. The major structural changes first suggested by Kirk (1975) include

1. Demethylation

2. Oxidation of α-carbon atoms

3. Cleavage of the side chains between the α and β carbons of the phenyl propane units.

4. Direct cleavage of an arylglycerol-β-aryl ether linkage

5. Hydroxylation and dioxygenase cleavage of aromatic rings

Degradation products in the low-molecular-weight solubles include vanillin, syringaldehyde, coniferyl aldehyde, vanillic acid, syringic acid, and a wide range of aliphatic or aromatic acids and phenols (Figs. 8-2, 8-3).

Lignin-Degrading Enzymes from White-Rot Fungi

In the past few years remarkable progress has been made in determining the enzymatic nature and probable pathways of lignin digestion since the initial identification of a *ligninase* (Tien and Kirk, 1983, 1984; Gold et al., 1984).

The principal enzymes now believed to play a role in this complicated process are lignin peroxidase (LIP), manganese-peroxidase (MnP), H_2O_2 producing enzymes, and other phenol oxidizing enzymes such as laccase (Kirk and Shimada, 1985; Kirk, 1989). The two peroxidase enzymes require H_2O_2 in their catalytic reactions with lignin.

Lignin peroxidase appears to be the key lignin degrading enzyme in *Phanerochaete chrysosporium*, the most studied white-rot fungus to date. It has been detected also in *Trametes versicolor* and several other white-rot fungi. Early research indicated cleavage of the C_α—C_β chemical bond between carbons in phenyl propane units of dimer models of lignin representing common linkages by the enzyme (Fig. 8-3). The enzyme is reported also to oxidize the phenolic units of lignin to aryl cation radicals which then cleave both C—C and C—O bonds in a series of non-enzymatic reactions (Datta et al., 1991). The nonspecific oxidations of the aromatic nuclei produce a wide array of products. The nonspecificity of these reactions and their variety led Kirk and Farrell (1987) to describe the process with the novel term *enzymatic*

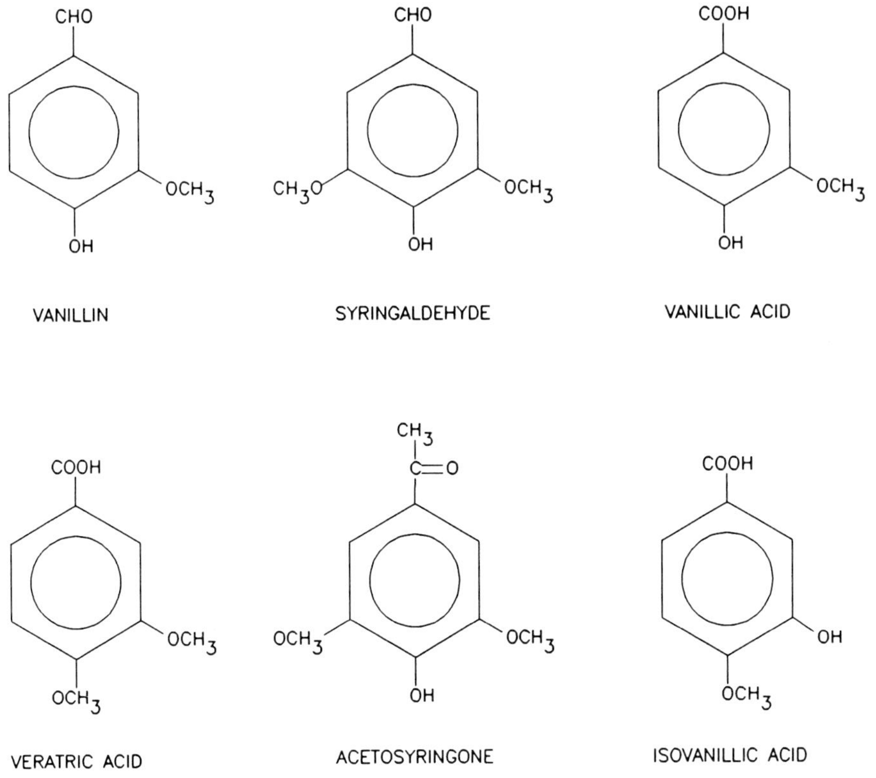

Figure 8-2 Examples of some of the common low-molecular-weight products obtained from partially decayed lignin.

combustion. In cultural ligninolytic studies with *P. chrysosporium*, the ligninase is a secondary metabolite, released in liquid cultures only when nitrogen becomes limiting.

Manganese peroxidase was the major enzyme involved in lignin degradation by *Dichomitus squalens (Polyporus anceps)* and in this case nitrogen levels were not limiting (Périé and Gold, 1991). The presumed role of MnP is to oxidize Mn^{2+} to Mn^{3+} which then oxidizes the various phenolic structures in lignin (Gold, 1987). MnP has been reported to oxidize syringyl lignin units better than the corresponding guaiacyl units (Glenn and Gold, 1985).

Laccase and other oxidases such as aryl alcohol oxidase have been detected in other white-rot fungi (Périé and Gold, 1991). The presence of phenol oxidizing enzymes in white rot fungi and the color changes they induce in gallic and tannic acid is the basis of the well-known Bavendamm reaction for separating the white-rot fungi from the brown-rot fungi (Davidson *et al.*, 1938). Their production by most white-rot fungi has long suggested some role in lignin decomposition (Kirk and Kelman, 1965).

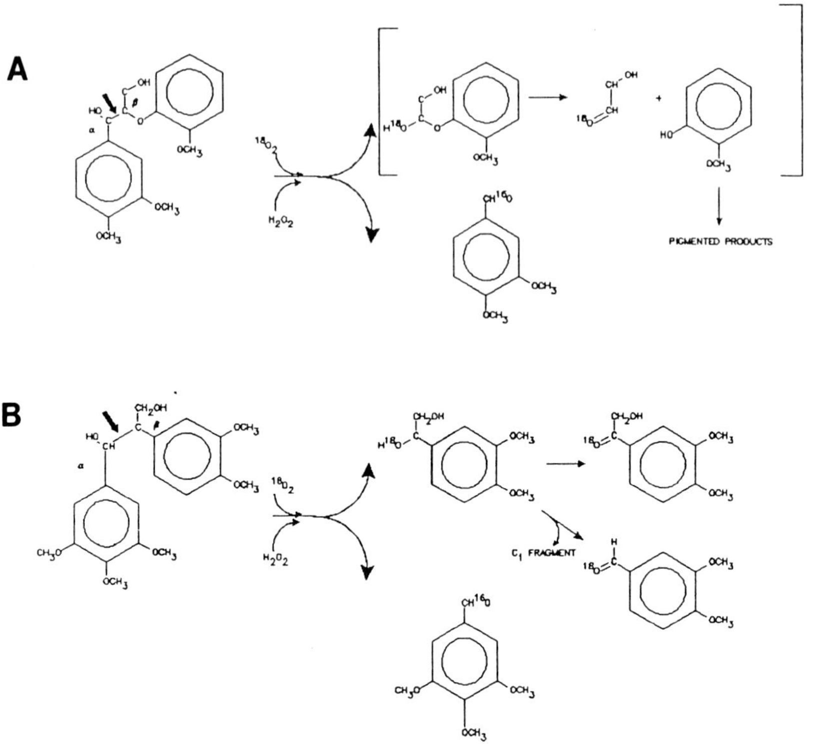

Figure 8-3 Cleavage of the C_α—C_β chemical bond between the carbons in the propane chains of two dimeric models that represent two major linkage patterns between phenyl propane units in lignin and their probable degradation products. (A) Cleavage in a β-O-4 model representing the syringyl glycerol-β-aryl-ether bond. (B) Cleavage in a β-1 model representing the diarylpropane bond (Tien and Kirk, 1984).

Laccase is reported to cause C_α-oxidation, demethoxylation, cleavages in phenyl groups and C_α—C_β cleavage in syringyl structures (Kirk and Farrell, 1987). Laccase also provides the phenoxyradicals and quinones from lignin decomposition which play a key role in the decomposition of cellobiose by cellobiose dehydrogenase. One puzzle in the function of laccase is the mechanism that prevents the rapid repolymerization of the phenoxyradicals back into lignin, since phenoxyradicals are starting materials for lignin biosynthesis. The removal of toxic quinones may prevent repolymerization in the degradative reaction. A number of white-rot fungi lack laccase, but other polyphenol oxidases can apparently carry out the same reactions.

H_2O_2-forming enzymes are required for the ligninase reactions described above. Glucose oxidase and other oxidases have been proposed as the enzymes that form H_2O_2 from various sugars possibly in peroxisome-like organelles in the periplasmic zones adjacent to the hyphal wall. Recently glyoxal oxidase, an extracellular H_2O_2 producing enzyme, has been detected in ligninolytic cultures of *P. chrysosporium* and may be a source of H_2O_2 or play a regulatory role in lignin decomposition (Kersten, 1990).

These data suggest that several systems of peroxidase and oxidase enzymes may be involved in lignin decomposition by the many white-rot fungi. An interesting question will be to see if the various degrading systems can be grouped and related eventually to the phylogeny of the white-rot fungi.

Lignin Degradation by Non-White-Rot Fungi

Brown-Rot Fungi Whereas brown-rot fungi are not considered to be lignin degraders, chemical analyses of residual lignin from brown-rotted wood indicates some degradation and substantial increases in solubility. The principal changes included decreased methoxyl content, oxidations of some alcohol and aldehyde groups to carboxyls, and the introduction of some phenolic hydroxyls (Kirk, 1975). There appears to be no significant separation of the guaiacyl and syringyl units composing the amorphous macromolecule.

Lignin losses during decay have been reported, however, for some brown-rot fungi (Enoki *et al.*, 1988). Whether these losses represent natural variation in lignin degradation within brown-rot fungi or differences in the method of lignin detection is uncertain. Degradation studies of labeled dimeric lignin models also indicate a substantial variation among brown-rot fungi (Tanaka *et al.*, 1986). *Gloeophyllum trabeum*, *Neolentinus (lentinus) lepideus*, and *Pholiota adiposa* were the most active of the 11 brown-rotters tested, suggesting that more detailed studies may reveal some intermediates between the white-rot and brown-rot fungi. The classification of white- and brown-rot fungi is an artificial system, and the occurrence of species with characteristics intermediate between both groups is not surprising.

Soft-Rot Fungi Soft-rot fungi have been reported to cause considerable lignin degradation (Savory and Pinion, 1958; Levi and Preston, 1965; Eslyn *et al.*, 1975). Some species degrade more lignin than do the brown-rot fungi (Seifert, 1966) and preferential attack of the syringyl units has been observed (Nilsson *et al.*, 1989). The enzymes involved in soft-rot attack of lignin are not known, but the yeast phases of many soft-rot fungi may be extremely useful in lignin-degradation schemes, since filamentous fungi often pose problems in fermentation systems. As a group, the soft-rot fungi exhibit considerable variation in their attack on cell-wall components and appear to be intermediate in their lignin-degrading capabilities between the white- and brown-rot groups.

Roles of Bacteria in Wood Decomposition

Bacteria are commonly present in wood during all the stages of its decomposition, and their many roles can be grouped conveniently into four general categories: (1) primary log invaders that damage pits in parenchymatous tissues, increasing wood permeability; (2) heartwood invaders in living stems that cause wetwood and often decrease wood permeability; (3) species associated with other wood-inhabiting flora or fauna, which may antagonistically or synergistically interact with them; and (4) species capable of degradation of wood or its isolated major cell-wall components. This latter role is discussed in this section, whereas the others are reviewed in Chapter 11.

Although some bacteria can degrade wood, the damage is often slight and develops very slowly compared to that of fungal attack. Bacteria are unicellular and are unable to invade wood substrates as rapidly as filamentous fungi. Soil-inhabiting Actinomycetes, however, can assume a mycelial form, and some species cause soft-rot damage to wood in marine use or under extreme nutrient/moisture regimes.

Many bacteria decompose the pectins and hemicelluloses in plant materials. Cellulose is also readily decomposed by some specialized aerobic bacteria and also by consortia of bacteria with other microorganisms, anaerobically during ruminant and termite digestion. Bacteria growing on hemicelluloses and cellulose are a major source of the troublesome slime deposits that develop during the pulp- and paper-making processes. Bacteria are also able to utilize many of the low-molecular-weight lignin-degradation products in partially decayed lignin and pulping-waste liquors. Bacteria can cleave benzene rings, and some species are being proposed as agents for detoxifying chlorinated phenols. Crawford (1981) has reported the ability of some strains of bacteria to degrade lignin, lignin model compounds, and DHP lignin in a manner similar to that of the white-rot fungi by a process involving substituent oxidations and cleavage of both 2-carbon fragments from propyl side chains and aryl ether linkages.

In wood, however, bacterial damage is generally minor compared to fungal damage, and is largely limited to cell-wall surface etching and erosion in sapwood zones. Decay tests on large numbers of wood-inhabiting bacteria have generally been negative (Schmidt and Dietrichs, 1976). Lignification appears to be the major factor limiting significant bacterial attack in most woods, and delignification of wood substantially increases the severity of bacterial degradation.

Several recent studies have shown that *Streptomyces* species (Actinomycetes) cause soft rot (Type 1) in wood in some marine exposures (Cavalcante and Eaton, 1980; Baecker and King, 1982; Holt, 1983). A unique tunneling type of bacterial damage has been noted in wood cell walls and has been attributed to unidentified bacteria, probably in the Myxobacteriales or Cytophagales (Daniel *et al.*, 1987). The damage to the cell wall is severe, and cultural studies demonstrate degradation of radioisotope-labeled lignin.

These reports suggest that bacteria in the Actinomycetes and Eubacteria may play important roles in wood decomposition in certain environments.

The significance of these organisms in large-scale deterioration of wood products deserves further studies to more accurately quantify their roles.

A Decay Model and Related Research Needs

A tentative model of the major steps involved in the decay process for a typical brown and white rot can both emphasize current views and demonstrate how much is still unknown and awaiting additional research.

1. The decay process begins when the hyphal tip (of a fungus with decay capabilities) enters a wood cell and lies along the lumen surface in a moisture film (above the f.s.p.), which bathes the hypha and serves as a transport medium.
2. Under favorable growth conditions, the presence of certain soluble cell-wall compounds elicits synthesis and secretion of enzymes through the hyphal walls by exocytosis. The enzymes or chemical agents either diffuse with moisture to the wood cell wall or are carried there in a hyphal sheath.
3. An unknown enzyme, characteristic of all decay fungi, alters the chemical bonds in the lignin shield or those between lignin and the hemicelluloses to facilitate enzyme access to carbohydrates; the lignin barrier around the hemicelluloses and cellulose microfibrils is not complete, and some enzymes are able to access the unprotected zones; or oxidizing agents or the concerted action of several enzymes operating simultaneously exposes carbohydrates.
4. Hemicelluloses in the amorphous zones of the microfibrils are attacked initially. Whether the removal of substituent groups and polymer debranching occurs before, during, or after depolymerization of the polymer backbone is unknown. The sequences of the removal of various sugars in the several heteropolymers composing the different hemicelluloses in various wood species remain poorly understood.
5. Subsequent events vary with decay type and are discussed in parallel to emphasize both the differences and similarities of the two systems. Some events are still speculative.

Brown-Rot Fungi	*White-Rot Fungi*
a. H_2O_2 is formed from hemicellulose sugars and may react with Fe^{2+} to form an unidentified oxidizing agent which diffuses into microfibrils to separate some cellulose molecules in the crystalline zones and randomly break some glycosidic bonds. b. Glucose induces cellulase formation. c. Cellulase and hemicellulase enzymes	aa. H_2O_2 is formed from oxidases (GLOX) associated with hemicellulose degradation and ligninase (LIP) and manganese peroxidase (MnP) are synthesized and released. In concert with H_2O_2, free radicals are formed and lignin degradation begins, and involves separation of propyl chains between $C_\alpha-C_\beta$ carbons, aryl ether

are released and are primarily endoenzymes. Synergism probably occurs between the enzyme types for the 1,4-β-glycosidic bond. In some fungi, the enzymes may occur as an enzyme complex.
d. Glycosidases on the hyphal wall depolymerize the cellobiose and the dimers of the hemicellulose fragments into glucose or equivalent simple sugars and excesses accumulate.
e. Lignin degradation, via an unknown enzyme or oxidizing agent, occurs by demethoxylation, some propyl side-chain splitting, oxidations involving formation of carbonyl groups, and hydroxylations in the benzene rings. Whereas the lignin monomer groups still remain generally linked, some fragmentation occurs, and their protective role for the carbohydrates is eliminated.
f. A degraded lignin is all that remains at the end of the decay process.

bond separation, and some benzene-ring splitting.
bb. Cellobiose induces cellulase formation. Cellulase and additional hemicellulases are then released as the lignin barrier is broken. Synergism occurs between the 1,4-β-exo and endonucleases. Several control mechanisms appear to limit the accumulation of excess free sugars.
cc. Cellobiose is oxidized to cellobionic acid and other compounds by an oxido-reductase enzyme via the reduction of lignin fragments such as quinones by *laccase*. This may be a detoxification mechanism to remove toxic lignin-breakdown products.
dd. Laccase and other oxidase enzymes also play a role in lignin degradation.
ee. Lignin is degraded and metabolized. Syringyl monomer units appear to be degraded more readily than the guaiacyl units. It is not known whether lignin degradation is an energy-providing or -consuming process.
ff. At the end of the decay process, only a few residual minerals remain.

6. The simple sugars and hydrocarbon fragments from the decay process are absorbed by active hyphal transport processes. The external digestive process (decay) is complete. The simple carbon compounds enter various metabolic pathways and are converted to adenosine triphosphate (ATP) for synthesis and energy purposes or stored as food reserves.
7. Differing sequences and amounts of the various enzymes and chemical agents involved in the decay process probably explain the many types of decay. Since the enzymes are under gene control, the developing methods of biotechnology offer exciting possibilities for developing decay fungi for specific commercial purposes.

Most decay-mechanism studies have been based on single enzyme tests on substrates in fermentation flasks. The natural decay processes probably involve a cascade of simultaneous reactions that are highly interactive. The situation is further complicated by microbial interactions, which are either competitive or synergistic. Thus, whereas progress has been made in the past decade toward understanding decay, it is clear that much remains to be elucidated.

Summary

- White-rot fungi degrade and metabolize all major cell-wall constituents, proceeding from the lumen surface inward, while leaving the residual cell-wall material intact. Wood degraded by white-rot fungi is characterized by low alkaline solubility and only minor reductions in the degree of polymerization (DP) of the residual cellulose. White-rot fungi vary in the degree of early attack on lignin and cellulose. These differences may be exploited by identifying species or developing strains that preferentially remove lignin for biopulping and biobleaching possibilities. Strength losses in white-rotted wood are approximately proportional to specific gravity reductions.
- Brown-rot fungi depolymerize and metabolize primarily the structural carbohydrates in the wood cell wall. Lignin is partially degraded, but remains as an unutilized residuum. Brown-rot fungi develop rapidly in the S2 and S1 zones of the cell wall, causing dramatic reductions in DP and increasing alkali solubility. These changes are also associated with early, drastic reductions in wood-strength properties.
- The soft-rot fungi have capabilities that appear to lie between those of the white- and brown-rot fungi. These fungi concentrate their attack on the cell-wall carbohydrates but cause more lignin degradation than do brown-rot fungi. Soft-rot degradation is primarily localized in the S2 zone of the cell wall. As with the white rots, the soft-rotted wood is characterized by only minor reductions in alkali solubility.
- Structural carbohydrates in all decays are depolymerized primarily by *hydrolytic* enzymes. The major cellulose degrading enzymes are endo-1,4-β-glucanase, exo-1,4-β-glucanase, and 1,4-β-glucosidase. Hemicelluloses are more structurally complex and require a more complex set of enzymes to debranch and remove substituent groups as well as depolymerize the polymer backbone. Some of the enzymes involved are endo-1,4-β-xylanase, 1,4-β-xylosidase, their mannan equivalents, and several acetyl esterases.
- Small, *diffusible* oxidants that separate and depolymerize cellulose molecules in the crystalline zones may be responsible for the early sharp reductions in wood strength and DP are believed to be produced by brown-rot fungi.
- Lignin is degraded by all three decay types, but the degradation is minor in the brown- and soft-rots and involves primarily demethoxylation. Lignin is completely degraded and metabolized by white-rot fungi. *Oxidation* initiated by several peroxidase enzymes is the primary breakdown mode. The enzymatic degradation of lignin is just beginning to be understood and has been studied for only a few white-rot fungi. The major enzymes involved directly or indirectly in

Summary continues

the random oxidation process are ligninase peroxidase, manganese peroxidase, other polyphenol-oxidizing enzymes, and H_2O_2-forming enzymes.

- Whereas great progress has been made in the past decade in determining the degradative sequences and the enzymes involved in the decomposition of cellulose, the hemicelluloses, and lignin, much remains to be clarified. Major research is still needed to better understand the concerted action and coordination of all the constituent enzymes on the wood. There are also a number of questions concerning the nature of and mechanisms by which rapid cellulose depolymerization occurs in early stages of brown-rot development. Finally, the protective role of lignin and the occurrence of uneven deposition zones around the microfibrils that may provide sites for enzyme contact merit further research.

It is prudent to remember that our current ideas about the chemical mechanisms of wood decay are based on detailed studies of only a few fungi in each major decay type. As more fungi are studied, including the neglected higher Ascomycotina, the litter-decomposing Basidiomycotina, and soil fungi, we may find that the fungi have solved the lignin barrier to *energy-rich carbohydrates* in a variety of other interesting and novel ways.

References

Adler, E. (1977). Lignin chemistry—past, present, and future. *Wood Science and Technology* 11:169–218.

Baecker, A. A. W., and B. King. (1982). Soft rot in wood caused by *Streptomyces*. Journal of the Institute of *Wood Science* 10:65–71.

Blanchette, R. A. (1991). Delignification by wood-decay fungi. *Annual Review of Phytopathology*. 29:381–398.

Cavalcante, M. S., and R. A. Eaton. (1980). "The Isolation of Actinomycetes from Wood in Ground Contact and the Sea." International Research Group Wood Preservation, Document IRG/WP/1110. Stockholm, Sweden.

Chen, C.-L., and H.-M. Chang. (1985). Chemistry of lignin biodegradation. *In* "Biosynthesis and Biodegradation of Wood" (T. Higuchi, ed.), pp. 535–556. Academic Press, New York.

Côté, W. A. (1977). Wood ultrastructure in relation to chemical composition. *In* "Recent Advances in Phytochemistry: The Structure, Biosynthesis and Degradation of Wood" (F. A. Loewus and V. C. Runeckles, eds.). Plenum Press, New York.

Cowling, E. B. (1961). Comparative biochemistry of the decay of sweetgum sapwood by white-rot and brown-rot fungi. U.S.D.A. Technical Bulletin No. 1258.

Cowling, E. B., and W. Brown. (1969). Structural features of cellulosic materials in relation to enzymatic hydrolysis. *In* "Cellulases and Their Applications." (G. J.

Cowling, E. B., and W. Brown. (1969). Structural features of cellulosic materials in relation to enzymatic hydrolysis. *In* "Cellulases and Their Applications." (G. J.

Hajny and E. T. Reese, eds.), pp. 152–187. Advanced Chemistry Series 95. American Chemical Society Press, Washington, D.C.

Crawford, R. L. (1981). "Lignin Biodegradation and Transformation." John Wiley, New York.

Daniel, G. F., T. Nilsson, and A. P. Singh. (1987). Degradation of lignocellulosics by unique tunnel-forming bacteria. *Canadian Journal of Microbiology* 33:943–948.

Datta, A., A. Bettermann, and T. K. Kirk. (1991). Identification of a specific manganese peroxidase among ligninolytic enzymes secreted by *Phanerochaete chrysosporium* during wood decay. *Applied Envir. Microbiology* 57(5):1453–1460.

Davidson, R. W., W. A. Campbell, and D. J. Blaisdell. (1938). Differentiation of wood-decaying fungi by their reactions on gallic or tannic acid medium. *Journal of Agricultural Research* 57:683–695.

Dekker, R. F. H. (1985). Biodegradation of hemicelluloses. In "Biosynthesis and Biodegradation of Wood Components" (T. Higuchi, ed.), pp. 505–533. Academic Press, New York.

Enoki, A., H. Tanaka, and G. Fuse. (1988). Degradation of lignin-related compounds, pure cellulose, and wood components by white-rot and brown-rot fungi. *Holzforschung* 42:85–93.

Eriksson, K. E. (1978). Enzyme mechanism involved in cellulose hydrolysis by the white-rot fungus, *Sporotrichum pulverulentum*. *Biotechnology Bioengineering* 20: 317–332.

Eriksson, K.-E. L., and B. Peterson. (1982). Purification and partial purification of two acidic proteases from the white-rot fungus *Sporotrichum pulverlentum*. *European Journal of Biochemistry* 124:635–642.

Eriksson, K.-E. L., R. A. Blanchette, and P. Ander. (1990). "Microbial and Enzymatic Degradation of Wood and Wood Components." Springer-Verlag, New York.

Eslyn, W. E., T. K. Kirk, and M. J. Effland. (1975). Changes in the chemical composition of wood caused by six soft-rot fungi. *Phytopathology* 65:473–476.

Fengel, D., and G. Wegener. (1984). "Wood Chemistry, Ultrastructure, Reactions." Walter de Gruyter, Berlin.

Glenn, J. K., and M. H. Gold. (1985). Purification and characterization of an extracellular Mn(II)-dependent peroxidase from the lignin-degrading basidiomycete, *Phanerochaete chrysosporium*. *Archives of Biochemistry and Biophysics* 242:329–341.

Gold, M. H., M. Kuwahara, A. A. Chin, and J. K. Glenn. (1984). Purification and characterization of an extracellular H_2O_2-requiring diaryl propane oxygenase from the white-rot basidiomycete, *Phanerochaete chrysosporium*. *Archives Biochemistry and Biophysics* 234:353–362.

Gold. M. H., W. Warüshi, and K. Valii. (1989). Extracellular peroxidases involved in lignin degradation by the white rot basidiomycete *Phanerochaete chrysosporium*. *Amer. Chem. Soc. Symp. Ser.* 389:127–140.

Halliwell, G. (1965). Total hydrolysis of celluloses by cell-free preparations from soil microorganisms. *Biochemical Journal* 94:270–289.

Highley, T. L. (1975). Properties of cellulases of two brown-rot fungi and two white-rot fungi. *Wood and Fiber* 6(4):275–281.

Highley, T. L. (1977). Requirements for cellulose degradation by a brown-rot fungus. *Material und Organismen* 12:25–36.

Highley, T. L. (1978). Degradation of cellulose by culture filtrates of *Poria placenta*. *Material und Organismen* 12(3):161–174.

Highley, T. L. (1982). Is extracellular hydrogen peroxide involved in cellulose degradation by brown-rot fungi? *Material und Organismen.* **17**:205-214

Highley, T. L. and K. E. Wolter. (1982). Properties of a carbohydrate-degrading enzyme complex from the brown-rot fungus *Poria placenta*. *Material und Organismen.* **17**:127-134.

Higuchi, T. (ed.). (1985). "Biosynthesis and Biodegradation of Wood Components." Academic Press, New York.

Holt, D. M. (1983). Bacterial degradation of wood cell walls in aerobic aquatic habitats: Decay patterns and mechanisms proposed to account for their formation. *Journal of the Institute of Wood Science* **9**:212-223.

Illman, B. L., D. C. Meinholtz, and T. L. Highley. (1989). Oxygen free radical detection in wood colonized by the brown-rot fungus, *Postia placenta*. *Biodeterioration Research.* **2**:497-509.

Iwasaki, T., K. Hayashi, and M. Funatsu. (1964). Purification and characterization of two types of cellulase from *Trichoderma konigii*. *Journal of Biochemistry* **55**: 209-212.

Keilich, G., P. Bailey, and W. Liese. (1970). Enzymatic degradation of cellulose, cellulose derivatives, and hemicelluloses in relation to fungal decay of wood. *Wood Science and Technology.* **4**:273-283.

Kersten, P. J. (1990). Glyoxal oxidase of *Phanerochaete chrysosporium*: Its characterization and activation by lignin peroxidase. *Proc. Natl. Acad. Sci. U.S.A.* **Vol. 87**: 2936-2940.

King, K. W., and M. I. Vessal. (1969). Enzymes of the cellulase complex. *In* "Cellulases and Their Applications" (G. J. Hajny and E. T. Reese, eds.), pp. 7-25. Advanced Chemistry Series 95, American Chemical Society, Washington, D.C.

Kirk, T. K. (1971). Effects of microorganisms on lignin. *Annual Review of Phytopathology* **9**:185-210.

Kirk, T. K. (1973). Chemistry and biochemistry of decay. *In* "Wood Deterioration and Its Prevention by Preservative Treatments" (D. D. Nicholas, ed.), pp. 149-181. Syracuse University Press, Syracuse, New York.

Kirk, T. K. (1975). Effects of a brown-rot fungus, *Lenzites trabea*, on lignin in spruce wood. *Holzforschung* **29**(3):99-107.

Kirk, T. K. (1988). Lignin degradation by *Phanerochaete chrysosporium ISI Atlas Sci. Biochem.* **1**:71-76.

Kirk, T. K. (1975). Lignin-degrading enzyme system. *In* "Cellulose as a Chemical Energy Resource" C. R. Wilke, ed.), pp. 139-150. Biotechnology and Bioengineering Symposium, No. 5. John Wiley & Sons, New York.

Kirk, T. K., and R. L. Farrell. (1987). Enzymatic "combustion": The microbial degradation of lignin. *Annual Review of Microbiology* **41**:465-505.

Kirk, T. K., and T. L. Highley. (1973). Quantitative changes in structural components of conifer woods during decay by white- and brown-rot fungi. *Phytopathology* **63**(11):1338-1342.

Kirk, T. K., T. Higuchi, and H. M. Chang. (1980). "Lignin Biodegradation: Microbiology Chemistry and Potential Applications," Vols. 1, 2. CRC, Boca Raton, Florida.

Kirk, T. K., and A. Kelman. (1965). Lignin degradation as related to the phenol oxidases of selected wood-decaying basidiomycetes. *Phytopathology* **55**:739-745.

Kirk, T. K., and W. E. Moore. (1972). Removing lignin from wood with white-rot fungi and digestibility of resulting wood. *Wood Fiber* **4**:72-79.

Kirk, T. K., and M. Shimada. (1985). Lignin biodegradation: The microorganisms involved and the physiology and biochemistry of degradation by white-rot fungi. In "Biosynthesis and Biodegradation of Wood Components" (T. Higuchi, ed.), pp. 579-605. Academic Press, Orlando, Florida.

Koenigs, J. W. (1974). Hydrogen peroxide and iron: a proposed system for decomposition of wood by brown-rot basidiomycetes. Wood and Fiber 6:66-80.

Levi, M. P., and R. D. Preston. (1965). A chemical and microscopic examination of the action of the soft-rot fungus (Chaetomium globosum on beechwood (Fagus sylvatica). Holzforschung 19:183-190.

Liese, W. (1970). Ultrastructural aspects of woody tissue disintegration. Annual Review of Phytopathology 8:231-258.

Loewus, F. A., and V. C. Runeckles. (eds.). (1977). "The Structure, Biosynthesis, and Degradation of Wood" Plenum Press, New York.

Lyr, H. (1960). Formation of ecto-enzymes by wood-destroying and wood-inhabiting fungi on various culture media. V. Complex medium as carbon source. Archives of Microbiology 35:258-278.

Mandels, M., and E. T. Reese. (1964). Fungal cellulases and the microbial decomposition of fabrics. Developments Industrial Microbiology 5:5-20.

Montenecourt, B. S., and D. E. Eveleigh. (1977). Preparation of mutants of Trichoderma reesii with enhanced cellulase production. Applied Environmental Microbiology 34:777.

Morrell, J. J., and R. A. Zabel. (1987). Partial delignification of wood: Its effect on the action of soft-rot fungi isolated from preservative-treated southern pines. Material und Organismen 22:215-224.

Nilsson, T., G. Daniel, T. K. Kirk, and J. R. Obst. (1989). Chemistry and microscopy of wood decay by some higher Ascomycetes. Holzforschung 43:11-18.

Norkrans, Birgitta. (1963). Degradation of cellulose. Annual Review of Phytopathology 1:325-350.

Norkrans, Birgitta. (1967). Cellulose and cellulolysis. Advances in Applied Microbiology 9:91-130.

Périé, F. H., and M. H. Gold. (1991). Manganese regulation of manganese peroxidase expression and lignin degradation by the white-rot fungus Dichomitus squalens. Applied Envir. Microbiology 57(8):2240-2245.

Reese, E. T. (1963). "Advances in Enzymatic Hydrolysis of Cellulose and Related Materials." Pergamon, New York.

Reese, E. T. (1977). Degradation of polymeric carbohydrates by microbial enzymes. In "Recent Advances in Phytochemistry: The Structure, Biosynthesis and Degradation of Wood" (F. A. Loewus and V. C. Runeckles, eds.), pp. 311-367. Plenum Press, New York.

Reese, E. T., A. Maguire, and F. W. Parrish. (1973). Production of β-D-xylopyranosidases by fungi. Canadian Journal of Microbiology 19:1065-1074.

Reese, E. T., R. G. H. Siu, and H. S. Levinson. (1950). The biological degradation of soluble cellulose derivatives and its relationship to the mechanism of cellulose hydrolysis. Journal of Bacteriology 59:485-497.

Ruel, K., and F. Barnoud. (1985). Degradation of wood by microorganisms. In "Biosynthesis and Biodegradation of Wood Components" (T. Higuchi, ed.), pp. 441-467. Academic Press, New York.

Savory, J. G., and L. C. Pinion. (1958). Chemical aspects of decay of beech wood by Chaetomium globosum. Holzforschung 12(4):99-103.

Schmidt, O., and H. H. Dietrichs. (1976). Zur Aktivität von Bakterien gegenüber Holzkamponenten. *Material und Organismen (Suppl.)* 3:91-102.

Seifert, K. (1966). Die chemische Veränderung der Bucherholz-Zellwand durch Moderfäule (*Chaetomium globosum* Kunze). *Holz als Roh-und Werkstoff* 24: 185-189.

Selby, K., and C. C. Maitland. (1967). The cellulase of *Trichoderma viride* cellulolytic complex. *Biochemistry Journal* 104:716-724.

Sjöström, E. (1984). "Wood Chemistry: Theory and Applications." Academic Press, New York.

Streamer, M., K.-E. Eriksson, and B. Peterson. (1975). Extracellular enzyme system utilized by the fungus *Sporotrichum pulverlentum* (*Chrysosporium lignorum*) for the breakdown of cellulose. IV. Functional characterization of five-endo-1,4-β-glucanases and one exo-1,4-β-gluconase. *European Journal of Biochemistry* 59:607-613.

Tanaka, H., A. Enoki, and G. Fuse. (1986). Correlation between ethylene production from α-oxo-γ-methylthiobutyric acid and degradation of lignin dimeric model compounds by wood-inhabiting fungi. *Mokuzai Gakkaishi* 32:125-135.

Tien, M., and T. K. Kirk. (1983). Lignin-degrading enzyme from the Hymenomycete *Phanerochaete chrysosporium* Burds. *Science* 221:661-663.

Tien, M., and T. K. Kirk. (1984). Lignin-degrading enzyme from *Phanerochaete chrysosporium*: Purification, characterization, and catalytic properties of a unique H_2O_2-requiring oxygenase. *Proceedings of the National Academy of Science U.S.A.* 81:2280-2284.

Timell, T. (1967). Recent advances in the chemistry of wood hemicelluloses. *Wood Science and Technology* 1:45-70.

Ziekus, J. G. (1981). Lignin metabolism and the carbon cycle: Polymer biosynthesis, biodegradation, and environmental recalcitrance. *Advances in Microbial Ecology* 5:211-243.

CHAPTER

9

Ultrastructural Features of Wood Decay

The general anatomical features of decay as seen with the light microscope were presented in Chapter 7. This chapter reviews the ultrastructural features of wood decay and stresses the new insights they offer into the chemical nature of the decay processes.

The nature of fungal attack through an opaque, solid material such as wood makes it extremely difficult to follow the sequences of cell-wall degradation. Most evidence of early fungal attack is invisible to the naked eye; hence, the term *incipient decay*. Examination of thin sections cut from decaying wood, using the light microscope, increases our ability to study the characteristics of decayed wood and to examine the hyphae associated with this damage; however, hyphae are generally very small, and their thickened cell walls have limited the detailed studies of internal function that have been performed with animals and higher plants. Also, the cell-wall changes associated with the early stages of fungal colonization are generally subtle and not easily detected with the light microscope. Furthermore, determining the spatial relationships between associated fungi in the wood tissues is difficult using thin sections. Whereas modifications such as the ultraviolet (UV) microscopy and phase-contrast optics have enhanced the usefulness of the light microscope, these techniques are still limited to useful maximum magnifications of about ×1500.

Electron Microscopy

Many of the difficulties of studying fungal attack of wood can be overcome by the use of transmission electron microscopy (TEM) or scanning electron microscopy (SEM). In electron microscopy, a beam of electrons is substituted for the wavelengths of light. This provides for better resolution and a larger magnification factor. For TEM, electrons passing through a thin specimen reflect the relative electron density of a given specimen, whereas electron scattering by thicker specimens is used to construct surface images for SEM. The principles of operation and practical protocols for routine laboratory use of electron microscopy are available in several textbooks and laboratory manuals (Dawes, 1979; Postek et al., 1980; Hyat, 1981; and Robards, 1985).

Transmission Electron Microscopy

TEM was first described in 1932 by Knoll and Ruska, but the resolution of early models was no better than that of existing light microscopes. TEM resolution exceeded that of light microscopy within 3 years and, by 1946, resolution to ten Ångstroms (1×10^{-10}m) had been achieved. In TEM operation, a tungsten filament, heated to incandescence, emits electrons accelerated by high voltage. A series of condenser lenses is then used to focus the electron beam to a small zone on the specimen (3 to 5 μm). The lenses used in electron microscopy are electromagnetic, not the glass or quartz lenses used in light or UV microscopy. After penetrating the specimen, the beam passes through intermediate and projective lens, and finally the image is viewed on a fluorescent screen or a photographic plate. TEM now permits examination of samples at magnifications exceeding ×1,000,000. Unlike light microscopy, in which absorption of light decreases image brightness, deflection of electrons from the beam as it passes through the specimen plays a major role in image contrast. Heavier elements are more electron dense and are often used as stains [e.g., potassium permanganate ($KMnO_4$), uranyl acetate, lead citrate] to improve contrast in biological samples.

The successful development of TEM opened a vast new dimension (from micrometers to Ångstroms) to microscopists studying wood and its degradation, permitting observations of sequences and interactions that could previously only be imagined. Initially, techniques for embedding were imprecise, and it was at times difficult to distinguish between real differences in structures and those that occurred from the embedding process or damage from the electron beam (artifacts). As researchers began to understand the system, however, the number of artifacts decreased, and TEM became an extremely powerful tool for studying wood and fungal ultrastructure and interactions between wood and fungi during the decay process.

There are some disadvantages and difficulties with TEM. All materials must be dehydrated and viewed under high vacuum. The process requires extensive, time-consuming, preparatory steps. In some instances, researchers were primarily interested in biological events on surfaces and not internal structure. The replica technique, whereby samples were coated with thin layers of heavy metal and then treated to dissolve the sample, permitted such studies, but again, the technique is time-consuming and limited to rigid materials that could withstand the required coating and stripping steps. Furthermore, these techniques are limited by the shallow depth of field of the TEM.

Scanning Electron Microscopy

In SEM, an electron beam is focused to a small diameter (10–20 nm) on the specimen surface by two condenser lens. Biological specimens are often coated with either carbon or a heavy metal (gold-palladium) to reduce the build-up of electrical charges that obscure the surface image. Two pairs of scanning coils in the second condenser lens deflect the focused beam so it can scan a square region of the sample surface. As the beam penetrates the sample, secondary electrons, emitted from the surface, are collected and counted by an electron detector located above and to one side of the specimen. The detector uses a grid with accelerating potential in front of a scintillation counter to convert electrons to light. The light is conducted through a photomultiplier tube and visualized with a cathode-ray tube. The SEM produces useful magnifications ranging from ×50 to ×20,000 (more than 10 times those of the light microscope). The main advantages of SEM are direct observation of surfaces at great magnifications with excellent depth of field and easier sample preparation.

In addition to its topographic capabilities, the electron beam generates X-rays characteristic of the various elements present in the material being observed. Energy dispersive X-ray analysis (EDXA) is a technique that collects these X-rays and identifies the various elements present in the specimen. EDXA is extremely useful for determining relative levels of various elements in a sample and can be used, for example, to study distribution of some preservatives in the wood.

The combination of light microscopy, TEM, and SEM now permits detailed examination of fungal–wood interactions. A wealth of literature related to the ultrastructure of the decay process has developed.

Some Wood and Fungal Ultrastructural Features

Several of the ultrastructural discoveries in wood and fungi that helped develop a deeper understanding of the decay process are reviewed briefly here (see also Chapters 3 and 6). Electron microscopes were largely unavailable commercially until the early 1950s.

Wood Ultrastructure

The electron microscope rapidly provided new information on wood ultrastructure, including studies of differences between the cell types of each wood species, the presence of crystals in the wood, a warty layer on the lumen surface, pit membrane structure, and, most important, clear delineation of the cell-wall layers (Côté, 1967; Core et al., 1976; Parham and Gray, 1984). The new information on the cell wall set the stage for the studies of enzymatic degradation of wood and illustrated the important effects of variations of the major chemical constituents in the cell-wall layers. Studies examining the cellulose microfibril (Hirada and Côté, 1985) permitted precise delineation of microfibrillar orientation in each cell-wall layer. This orientation explains, for example, the characteristic spiraling of soft-rot cavities as they closely follow the microfibrillar angle of the S2 cell-wall layer.

Subsequent research examined the relative distribution and localization of chemical components in the wood cell wall (Kerr and Goring, 1975; Parameswaran and Liese, 1982). For example, staining with potassium permanganate provided important clues concerning the distribution of lignin in the wood cell wall (Bland et al., 1971). In later studies, bromination of the lignin would be used to study its distribution in various cell-wall layers (Saka and Goring, 1985; Otjen and Blanchette, 1988). The distribution of hemicellulose, which is an important component in the initial phases of some fungal attacks, has been studied by first treating the wood with xylanase, and then staining with thiocarbohydrazide and silver proteinate (Ruel and Joseleau, 1984; Joseleau and Ruel, 1985). The weak reactivity of cellulose to the periodic acid-thiocarbohydrazide silver proteinate (PAT Ag) staining technique for the hemicelluloses also provided a useful contrast for inferring cellulose locations. So the three major cell-wall constituents can be identified and changes detected microscopically as decay develops.

Fungal Ultrastructure

At the same time that wood anatomists were unlocking the secrets of the wood cell wall, mycologists were also using electron microscopy to study the ultrastructure of fungal hyphae (see Chapter 3). Most early studies were taxonomic in purpose and descriptive in nature. Some were of limited value since it was often difficult to separate fungal activity from preparative artifacts; however, these early studies began to provide much useful information on the structure and physiology of fungal hyphae. The ultrastructural aspects of fungi have been the subject of extensive research; for example, even by 1967, Bracker cited nearly 180 papers dealing with this subject.

Mycologists exploring fungal ultrastructure were able to determine the structure of the rigid cell wall, the plasmalemma, the formation of clamp connections, nuclear behavior, and septa (Beckett et al., 1974; McLaughlin, 1982; Thielke, 1982). An atlas of fungal ultrastructure was assembled by Beckett et al. in 1974. Whereas there have been numerous ultrastructural studies, it is prudent to remember that thousands of fungi remain unexplored. The wide array of structural variations noted in those species already examined suggest that many new variations in fungal structure remain to be discovered. The widespread availability of electron microscopes and the general familiarity of many researchers with electron microscopy techniques resulted in a virtual explosion in the use of electron microscopy to explore changes in fungal structure and function over the course of wood degradation. The large and recent literature on this subject makes it difficult to provide a comprehensive review. Our approach will be to briefly review those TEM and SEM studies that have provided new insights into the decay process for the three major decay types.

Wood Ultrastructural Changes during Decay

White-Rot Fungi

Of the major decay types, the white-rot fungi, owing to their potential value in biopulping, bleaching, and other lignin-removal schemes, have been most extensively studied (Fig. 9-1). In early TEM studies, the reaction of phenolic groups in lignin with osmium tetroxide was used to monitor the

Figure 9-1 Scanning electron micrograph of *Acer* sp. colonized by a white rot fungus causing a white stringy rot with nearly complete removal of the fibers and parenchyma ×30. From Blanchette *et al.* (1988).

removal of lignin from the wood cell wall (Liese, 1970). These studies illustrated the gradual loss of lignin from the lumen outward, associated with *Trametes (Polyporus) versicolor* and the rapid, widespread lignin removal noted with *Phanerochaete chrysosporium* (Highley and Murmanis, 1987; Blanchette, 1984; Blanchette et al., 1985, 1987; Ruel and Barnoud, 1985). These studies suggested a selective removal of lignin before cellulose attack. Further decay studies with cellulase-less mutants suggested that some cellulose attack is necessary for uniform lignin removal (Ruel *et al.*, 1984, 1986; Eriksson *et al.*, 1980; Eriksson, 1981). The seemingly widespread lignin degradation in the cell wall, often associated with the presence of electron-dense particles, led some researchers to suggest that the lignin-degrading enzymes are more mobile than previously believed (Ruel *et al.*, 1981). Examination of wood decayed by white-rotters also revealed that hemicellulose is removed before or concurrently with lignin (Hoffman and Parameswaren, 1976; Blanchette and Abad, 1988; Blanchette *et al.*, 1989a) (Fig. 9-2). The utilization of hemicellulose has been proposed as an important first step in the degradation of lignin (Ruel *et al.*, 1981). This suggests that a decay sequence for many white-rot fungi may be hemicellulose, lignin, and then cellulose degradation.

Along with the chemical studies, a large literature has developed recently to describe the varying patterns by which fungi cause white rot (Adaskaveg and Gilbertson, 1986; Blanchette, 1980; Blanchette and Reid, 1986; Blanchette *et al.*, 1988; Dill and Kraepelin, 1986; Murmanis *et al.*, 1984; Otjen and Blanchette, 1986, 1987; Reid, 1985). Some of these studies indicate extensive cell-wall degradation occurs away from the vicinity of fungal growth. In a recent summary of a series of classic studies Blanchette (1991) and co-workers have demonstrated ultrastructurally the initial selective delignifying role of some white-rot fungi. From earlier light microscopy and chemical studies, white-rot fungi were presumed to cause damage nearer to the zone of fungal growth, whereas brown-rot fungi produced enzymes that diffused for greater distances through the wood (Cowling, 1961). Ultrastructural studies now seem to negate these observations for some white-rot fungi and suggest that cell-wall constituents are affected to a significant degree at the early stages of colonization by both brown- and white-rot fungi.

Studies of fungal hyphae associated with white-rot fungi indicate an extensive hyphal sheath around the hyphal tip (Palmer *et al.*, 1983b; Highley and Murmanis, 1984; Foisner *et al.*, 1985a,b) and the presence of osmiophilic particles (Messner and Stachelberger, 1984). The function of these tip structures is unknown and the subject of much debate. These structures may retain fungal enzymes near the hyphae, prevent desiccation, and provide a diffusion channel for movement of wood-decomposition products into the fungal cell. The sheath also could provide a contact point between the wood and the fungi and serve as a concentrating point for enzymatic attack to minimize dilution or an external point of assembly for fungal enzymes that are too large to be excreted directly through the hyphal wall.

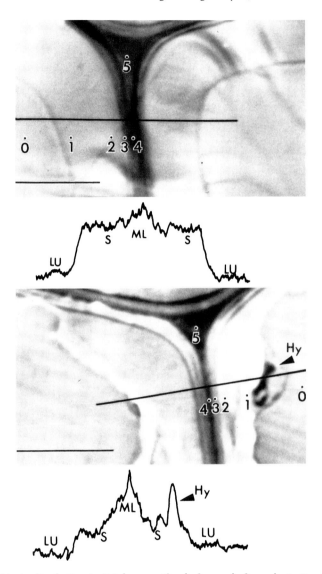

Figure 9-2 Lignin distribution in *Betula papyrifera* before and after colonization by *Trametes versicolor* as shown by EDXA-STEM analysis following bromination of the lignin residues. Lu, lumen; S, secondary cell wall; ML, middle lamella; Hy, hyphae. From Blanchette et al. (1987).

The effects of fungal enzymes on the wood cell wall have also come under increasing scrutiny as new biochemical tools are developed for identifying and quantifying fungal enzymes (Forney et al., 1982). Recent studies suggest that fungal enzymes that attack lignin, cellulose, or hemicellulose may be too large to diffuse directly into sound wood (Srebotnik et al., 1988a). Thus, a small, nonenzymatic entity has been proposed that initiates wood cell-wall decomposition and expands the capillary system. Once the wood structure becomes modified and accessible, the fungal enzymes can then begin to directly interact with the various cell-wall components.

Perhaps the most dramatic influence of ultrastructural studies on wood degradation has been the development of immunocytological techniques. Formerly, TEM or SEM observations were limited to describing changes in the wood structure or fungal hyphae, sometimes aided by staining with specific chemical reagents, but it was generally not possible to determine the biochemical nature of the changes at specific sites in the wood. The development of antibodies with high specificity for fungal enzymes, coupled with the development of specific chemical staining, or colloidal gold coupled antibody techniques, have now permitted detection of enzyme activity within specific areas of the fungus or the wood cell (Daniel *et al.*, 1989a,b; Blanchette *et al.*, 1988, 1989a; Garcia *et al.*, 1987; Srebotnik and Messner, 1988; Srebotnik *et al.*, 1988b; Blanchette *et al.*, 1989a,b). These techniques have identified and located lignin peroxidase in the fungal cell and have shown that this enzyme is capable of moving into decayed wood, but not sound wood, supporting the existence of nonenzymatic decay factors previously discussed (Fig. 9-3). Despite the high sensitivity of immunocytological techniques, caution must still be exercised when preparing materials for these techniques. For example, the fixation process can alter immunoreactivity, dramatically influencing the results (Srebotnik *et al.*, 1988b).

Ultimately, immunocytological studies will help delineate the pathways for enzyme synthesis and secretion, providing important clues for the selection of control chemicals.

Brown Rots

The ultrastructure of brown-rotted wood has received less attention, but similar studies to delineate the nature of brown-rot decay have important applications in the control of many important wood-products fungi. Such information may also prove useful for altering wood or modifying cellulose to make it more digestible for ruminant animals. Studies show some cellulose dissolution adjacent to the hyphal tip (Fig. 9-4) as well as the depletion of cellulose in the secondary cell wall at great distances from the sites of hyphal growth. This latter loss is reflected in large decreases in wood strength at very early stages of fungal colonization. Early TEM studies (Chou and Levi, 1971) suggested that enzymes diffused into the cell wall from the margins of bore holes and not from hyphae in the lumen. They also noted the presence of a gelatinous sheath around the hyphae. The hyphal sheath has been detected in other studies (Palmer *et al.*, 1983a, 1984), but its function also remains unknown (Fig. 9-5). TEM studies showed that the hyphal sheath contains low-molecular-weight β-1,3 glucans (Green *et al.*, 1989). Other studies suggest that the sheath contains contents from autolyzed hyphae or provides a site for translocation of enzymes (Highley *et al.*, 1983a,b; Palmer *et al.*, 1983a). The

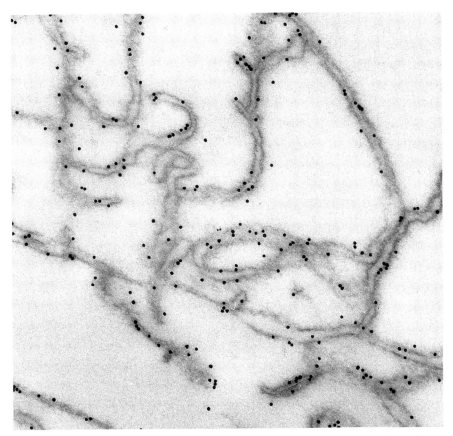

Figure 9-3 TEM of a thin section through pine sapwood colonized by *Phanerochaete chrysosporium* showing localization of colloidal gold-coupled lignin peroxidase within the fungal-associated extracellular slime and membranous structure within the lumina of degraded wood fibers. From Daniel *et al.* (1989a).

sheath may also function to retain and concentrate enzymes released from the hyphae, as described for the white rots.

Early in the study of brown-rotted wood, researchers noted that large quantities of modified lignin with a reduced methoxyl content and increased solubility remained in the wood cell walls, which were largely devoid of cellulose and hemicelluloses. These results suggested that brown-rot fungi lacked significant lignin-degrading capability, while rapidly utilizing cellulose. However, it was puzzling that attempts to grow brown-rot fungi on pure cellulose, either in cotton fibers (Highley *et al.*, 1983b) or in wood delignified by white-rot fungi (Blanchette, 1983), were unsuccessful. These results suggested that some lignin or other growth factor must be present for these fungi

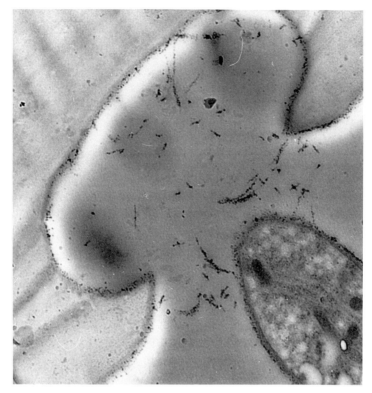

Figure 9-4 Dissolution of a tracheid wall of eastern white pine associated with a hyphal tip of *Postia placenta*. (TEM micrograph (×9225) from unpublished data of S. Jutte and R. Zabel.)

to initiate the cellulose attack. Recent TEM studies have suggested that the cellulase molecules are too large to diffuse into sound wood and support the theory that a nonenzymatic factor also initiates brown-rot decay (Srebotnik and Messner, 1988).

The nature of brown-rot enzymes has also become the subject of extensive research. Several studies have detected electron-dense particles in wood attacked by a variety of brown-rot fungi or by cellulase preparations obtained from *Trichoderma reesi* (Murmanis *et al.*, 1987, Messner and Stachelberger, 1982, 1985). The electron-dense particles were later suggested to be lignin-decomposition products (Srebotnik and Messner, 1988).

The effects of brown-rot fungi on lignin remain largely unknown. The presence of a peroxide-based lignin-degradation system has been proposed for the white-rot fungi (see Chapter 8), but no peroxide was detected in cultures of *Postia placenta* (Highley and Murmanis, 1985, 1986; Murmanis *et al.*, 1988a). However, recently weak reactivity to *Phanerochaete chrysosporium* lignin peroxidase has been reported at the hyphal tips of *Fomitopsis*

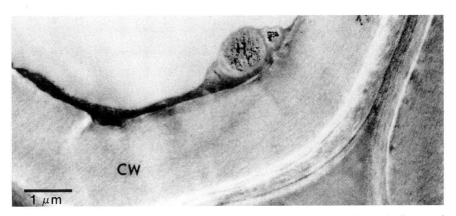

Figure 9-5 TEM of *Ganoderma applanatum* hyphae showing hyphal sheath attached to wood cell wall of *Tsuga heterophylla* (× 19,800). Murmanis *et al.* (1985).

pinicola (Blanchette *et al.*, 1989a) and peroxide has been detected in *Antrodia (Poria) carbonica*. This suggests that some brown-rot fungi may retain a limited ability to partially degrade lignin, a finding clearly supported by chemical analysis of brown-rotted wood.

The attack of hemicellulose by brown-rot fungi remains the most poorly understood process in the degradation of wood by these fungi, despite the premise that some hemicellulose degradation is essential for and may precede the attack of other wood components. Extensive hemicellulose degradation has been noted with *F. pinicola* using the PAT Ag staining technique, with higher levels of attack in the middle lamella (Blanchette and Abad, 1988). However, xylanase and endoglucanase activity were not detected in wood colonized by this fungus (Blanchette *et al.*, 1989b). These studies highlight the difficulty of using electron microscopy for chemical and immunocytological analyses.

It is readily apparent that our knowledge of the enzyme systems responsible for brown-rot degradation has many gaps. Furthermore, even this knowledge is limited to studies on a few species that are easily manipulated under laboratory conditions (e.g., *Postia placenta*). Yet brown-rot fungi, because of their prevalence in many wood products and their drastic effects on wood strength reduction, are an important group that merits intensive study. Elucidating the initial pathways of fungal degradation may provide improved strategies for preventing or controlling colonization. Conversely, an improved knowledge of the enzyme systems could be exploited to produce chemical feedstocks or to improve the digestibility of wood for animal feeds.

Soft Rots

The diamond-shaped cavities formed within the S2 cell wall layer by Type 1 soft-rot fungi have long intrigued wood anatomists who have used light and electron microscopy to explore the unique morphology of these cavities (Figs. 9-6, 9-7). Most studies have been descriptive in nature, owing to the relative scarcity of research on the enzyme systems of the soft-rot fungi (Mouzouras, 1989; Unligil and Chafe, 1974; Daniel and Nilsson, 1988, 1989; Nilsson et al., 1989; Jutte and Zabel, 1974).

For example, Crossley and Levy (1977) noted a mucilaginous sheath that was extruded by the cavity hyphae. Earlier, Unligil and Chafe (1974) had noted that T-branch formation by soft-rot fungi involved enzymatic activity. Cavity initiation was reported to occur by dissolution, not physical pressure, and the wood and fungus in these cavities were connected by fine strands of amorphous material (Hale and Eaton, 1981, 1985a,b). Numerous ribosomes were found in the hyphal tip, suggesting active enzyme synthesis. An electron-opaque halo was detected around the hyphal tips, which was proposed to contain an exocellulase (Fig. 9-8). Also, these cavities appeared to contain lignin residues and extracellular polysaccharides.

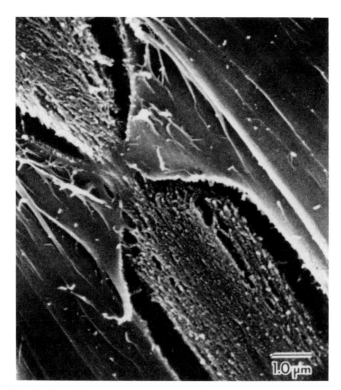

Figure 9-6 Type 1 soft rot damage in a creosote-treated southern pine pole showing an enlarged section of a hyphae in two contiguous cavities and the typical conical zones (SEM). Source Zabel et al. (1985) and permission of the Society of Wood Science and Technology.

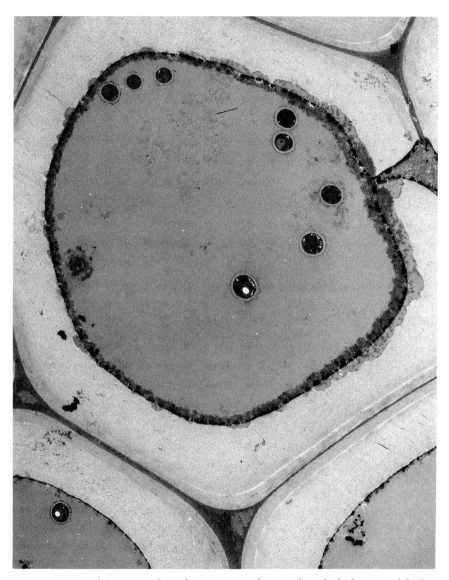

Figure 9-7 TEM of CCA-treated *Betula verrucosa* and exposed to *Phialophora mutabilis* for 1 month showing erosion of the S2 cell-wall layer away from fungal hyphae as well as CCA residue along the S3 cell-wall layer. From Daniel and Nilsson (1989).

Electron microscopy has also been used to understand the selectivity of soft-rot attack on certain treated wood species. Ryan and Drysdale (1988) examined the relationship between chromated copper arsenate (CCA) loading in the S2 cell-wall layer and soft-rot attack and found that this relationship varied with wood species. Daniel and Nilsson (1988, 1989) examined wood

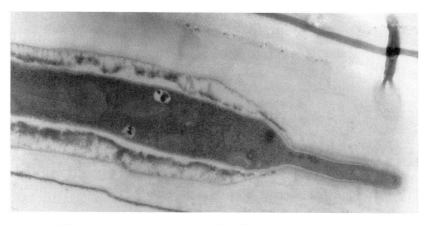

Figure 9-8 TEM of a Type 1 soft-rot cavity produced by *Phialophora hoffmannii* in *Betula pendula* showing a diamond-shaped cavity with fine proboscis hyphae and halo surrounding the hyphal tip and the tip of a cavity prior to the extension of the proboscis hypha. From Hale and Eaton (1981).

attacked by *Phialophora mutabilis* and reported that electron-dense particles in the soft-rot cavities contained high levels of CCA and phenolics.

Bacterial Erosion

Whereas basidiomycetes cause more dramatic wood changes, the nature of bacterial attack has also been studied at the ultrastructural level. Bacteria were not considered to be capable of wood damage until the late 1950s, when studies on sinker logs (from log-storage ponds) showed that bacterial attack of the pit membranes was the cause of excessive water absorption (Ellwood and Ecklund, 1959). Wood infected with bacteria clearly showed (in SEM examination) the severe pit-membrane damage (Levy, 1975; Liese and Greaves, 1975; Greaves, 1968, 1969; Schmidt *et al.*, 1987). Further examination of bacterially infected wood from marine environments has shown that bacteria cause damage similar to that caused by soft-rot fungi, although the rates of attack were extremely slow. Initially, bacteria were shown to form erosion channels on the lumen surface of the cell wall in a manner similar to that found with Type 2 soft-rot fungi on hardwoods (Holt *et al.*, 1979; Greaves, 1969; Holt, 1983). Recently, bacteria have been found to produce tunnels similar to Type 1 soft-rot cavities in the wood, particularly under conditions where excessive moisture and nutrients are present (Nilsson and Singh, 1984). Cavitation and tunneling produce numerous cavities in the S2 cell-wall layer, as well as degradation of the primary cell wall and middle lamella. These studies indicate that some bacteria have the capability to degrade wood components under special conditions.

Although the economic importance of bacteria in wood degradation is somewhat limited, these organisms have rapid life cycles and are easily manipulated in the laboratory. Because some species have the ability to degrade all wood components, they may be useful agents for further elucidation of the mechanisms of decay.

Microbial Interactions in Decay

Transmission electron microscopy and SEM can provide important clues about the decay process and the nature of microbial interactions (see Chapters 8 and 11). Ultrastructure may be particularly useful for exploring the interactions that occur when organisms compete for the same substrate and may provide important clues concerning the outcomes of various microbial combinations. Observations of hyphal coiling, cell lysis, and the production of toxic metabolites may all be visualized using electron microscopy. Electron microscopy may be especially useful for locating potential biological-control agents, particularly when immunocytological techniques are used to localize various enzymatic components of the interacting organisms in natural systems.

Whereas SEM has been used to study spatial interactions among fungi in wood, there are relatively few TEM studies of decay in mixed cultures. Murmanis et al. (1988b) examined *Trichoderma harzianum, Trichoderma polysperum*, and several basidiomycetes in mixed cultures and found evidence of cell disruption in the basidiomycete. More detailed studies, in combination with biochemical tests, may provide important clues concerning the mechanisms by which biological control agents function. In the laboratory, mixed cultures of bacteria, yeasts, and basidiomycetes associated with conifers have been shown both to stimulate mycelial growth and to increase decay weight losses (Blanchette and Shaw, 1978).

Detection and Quantification of Wood Preservatives

In addition to its usefulness for studying the ultrastructure of decay, SEM has also been used to determine distribution of various preservatives in the wood cell wall and the subsequent effects of these preservatives on fungal colonization and wood degradation. Resch and Arganbright (1971) evaluated EDXA for determining the distribution of pentachlorophenol in the wood cell wall and reported some penetration into the secondary cell-wall layers when liquified petroleum gas was used as the carrier, as did Wilcox and Parameswaran (1974) and Wilcox et al. (1974). Similar studies have been performed on fungi in wood treated with CCA (Chou et al., 1973; Daniel and Nilsson, 1989; Ryan and Drysdale, 1988; Ryan, 1986).

Summary

- The resolution of the TEM permitted researchers to view the internal structure of the fungal–wood interactions at the molecular level. However, the technique is not without difficulties. The process of dehydrating, embedding, and ultrathin sectioning requires extensive and time-consuming procedures and results in a static two-dimensional image. If researchers are primarily interested in biological events on surfaces and not internal structures, the replica technique is utilized. This technique produces a 3-D view of the surface. As a result of its great resolving power, the TEM has become an extremely powerful tool for studying wood and fungal ultrastructure and the interactions between wood and fungi during the decay process.
- Electron microscopic studies of wood have provided a wealth of new and confirming information about the layered nature of the cell wall and microfibrillar orientations that, when coupled with new staining techniques, provides the locations of cellulose, hemicelluloses, and lignin in the cell wall. This information has set the stage for a new level of understanding of the decay process.
- Ultrastructural studies of white rots suggest that hemicellulose attack precedes lignin decomposition and that cellulose decomposition follows early lignin attack. Cell-wall decomposition occurs both in the vicinity of some hyphae and at considerable distances away, implying diffusible enzymes or an unknown nonenzymatic factor. Staining tests specific for lignin suggest that the ligninase enzymes do not penetrate sound wood, supporting the hypothesis of a nonenzymatic, presumably oxidative factor that modifies sound wood before enzymatic entry and attack.
- Ultrastructural studies of brown-rotted wood indicate substantial decomposition of cellulose at considerable distances from the zone of hyphal growth. This early loss of cellulose is very important in the brown rots, as it is associated with drastic strength reductions at very early decay stages. Recent TEM studies using special staining techniques suggest that cellulases are too large to enter sound wood until it has been modified by a postulated nonenzymatic decay agent. They also suggest that some brown rots, although they do not metabolize lignin, may substantially modify it and may retain low levels of lignin peroxidase activity. This implies that the decay mechanisms in white- and brown-rotters may not be so different as is generally believed.
- Ultrastructural studies indicate the presence of a *gelatinous* sheath (whose functions are unknown) around the hyphae of brown-rotters and the hyphal tips in white-rotters. They may play some role in enzyme or nonenzymatic factor secretion and delivery to the cell-wall surfaces.

Summary continues

- Ultrastructural studies of soft-rotted wood (Type 1) have been largely descriptive in nature. Recent TEM studies indicate that T-branches and cavities form enzymatically and that the debris within some cavities is residual lignin.
- Ultrastructural studies have confirmed that, under some conditions, bacteria erode cell-wall surfaces in a manner similar to that of Type 2 soft-rotters in hardwoods. Also, under special conditions, some bacteria can tunnel in the cell wall and form cavities.
- Electron microscopy may be a useful way to locate antagonists in wood–microorganism interactions with biological-control potential and also to detect and quantify some of the metallic preservatives in wood.

References

Adaskaveg, J. E., and R. L. Gilbertson, (1986). In *vitro* decay studies of selective delignification and simultaneous decay by the white-rot fungi *Ganoderma lucidum* and *G. tsugae*. *Canadian Journal of Botany* **65**:1611–1619.

Beckett, A., I. B. Heath, and D. J. McLaughlin. (1974). "An Atlas of Fungal Ultrastructure." Longman Group Limited, London.

Blanchette, R. A. (1980). Wood decomposition by *Phellinus (Fomes) pini*: A scanning electron microscopy study. *Canadian Journal of Botany* **58**:1496–1503.

Blanchette, R. A. (1983). An unusual decay pattern in brown-rotted wood. *Mycologia* **75**(3):552–556.

Blanchette, R. A. (1984). Screening wood decayed by white-rot fungi for preferential lignin degradation. *Applied and Environmental Microbiology* **48**(3):647–653.

Blanchette, R. A. (1991). Delignification by wood-decay fungi. *Annual Review of Phytopathology* **29**:381–398.

Blanchette, R. A., and A. R. Abad. (1988). Ultrastructural localization of hemicellulose in birch wood (*Betula papyrifera*) decayed by brown- and white-rot fungi. *Holzforschung* **42**(6):393–398.

Blanchette, R. A., and I. D. Reid. (1986). Ultrastructural aspects of wood delignification by *Phlebia (Merulius) tremellosus*. *Applied and Environmental Microbiology* **52**(2):239–245.

Blanchette, R. A., and C. G. Shaw. (1978). Associations among bacteria, yeasts, and basidiomycetes during wood decay. *Phytopathology* **68**:631–637.

Blanchette, R. A., L. Otjen, M. J. Effland, and W. E. Eslyn. (1985). Changes in structural and chemical components of wood delignified by fungi. *Wood Science Technology* **19**:35–46.

Blanchette, R. A., L. Otjen, and M. C. Carlson. (1987). Lignin distribution in cell walls of birch wood decayed by white-rot basidiomycetes. *Phytopathology* **77**:684–690.

Blanchette, R. A., J. R. Obst, J. I. Hedges, and K. Weliky. (1988). Resistance of hardwood vessels to degradation by white-rot Basidiomycetes. *Canadian Journal of Botany* **66**:1841–1847.

Blanchette, R. A., A. R. Abad, R. L. Farrell, and T. D. Leathers. (1989a). Detection of lignin peroxidase and xylanase by immunocytochemical labeling in wood decayed by basidiomycetes. *Applied and Environmental Microbiology* **55**(6):1457–1465.

Blanchette, R. A., A. R. Abad, K. R. Cease, R. E. Lovrien, and T. D. Leathers. (1989b). Colloidal gold cytochemistry of endo-1,4-β-glucanase, 1,4-β-D-glucan cellobiohydrase, and endo-1,4-β-xylanase: Ultrastructure of sound and decayed birch wood. *Applied and Environmental Microbiology* **55**(9):2293–2301.

Bland, D. E., R. C. Foster, and A. F. Logan. (1971). The mechanism of permanganate and osmium tetroxide fixation and the distribution of lignin in the cell wall of *Pinus radiata*. *Holzforschung* **25**(5):137–143.

Bracker, C. E. (1967). Ultrastructure of fungi. *Annual Review of Phytopathology* **5**: 343–374.

Chou, C. K., and M. P. Levi. (1971). An electron microscopical study of the penetration and decomposition of tracheid walls of *Pinus sylvestris* by *Poria vaillantii*. *Holzforschung* **25**(4):107–112.

Chou, C. K., R. D. Preston, and M. P. Levi. (1973). Fungitoxic action of copper-chromium arsenate wood preservatives. *Phytopathology* **64**:335–341.

Core, H. A., W. A. Côté, Jr., and A. C. Day. (1976). "Wood Structure and Identification." Syracuse University Press, Syracuse, New York.

Côté, W. A., Jr. (1967). "Wood Ultrastructure—an Atlas of Electron Micrographs." University of Washington Press, Seattle, Washington.

Cowling, E. B. (1961). "Comparative Biochemistry of the Decay of Sweetgum Sapwood by White-Rot and Brown-Rot Fungi." U.S.D.A. Technical Bulletin 1258, Washington, D.C.

Crossley, A., and J. F. Levy. (1977). Proboscis hyphae in soft-rot cavity formation. *Journal Institute of Wood Science* **7**:30–33.

Daniel, G., and T. Nilsson. (1985). "Ultrastructural and T.E.M. EDXA Studies on the Degradation of CCA-Treated Radiata Pine by Tunneling Bacteria." International Research Group on Wood Preservation Document No. IRG/WP/1260.

Daniel, G., and T. Nilsson. (1986). "Ultrastructural Observations on Wood-Degrading Erosion Bacteria." International Research Group on Wood Preservation Document No. IRG/WP/1283. Stockholm, Sweden.

Daniel, G. F., and T. Nilsson. (1988). "Interactions between Soft-Rot Fungi and CCA Preservatives in *Betula verrucosa*." The International Research Group on Wood Preservation, Document No. IRG/WP/1367, Stockholm, Sweden.

Daniel, G. F., and T. Nilsson. (1989). Interactions between soft-rot fungi and CCA preservatives in *Betula verrucosa*. *Journal of the Institute for Wood Science* **11**(5): 162–171.

Daniel, G. F., T. Nilsson, and B. Pettersson. (1989a). Intra- and extracellular localization of lignin peroxidase during the degradation of solid wood and wood fragments by *Phanerochaete chrysosporium* by using transmission electron microscopy and immuno-gold labeling. *Applied and Environmental Microbiology* **55**(4): 871–881.

Daniel, G. F., B. Pettersson, T. Nilsson, and J. Volc. (1989b). "Cytoplasmic and Extracellular Localization of Manganese II-Dependent Peroxidase(s) in White-Rot Fungi during Degradation of Wood Materials." The International Research Group on Wood Preservation, Document No. IRG/WP/1416, Stockholm, Sweden.

Dawes, C. J. (1979). "Biological Techniques for Transmission and Scanning Electron Microscopy." Ladd Research Industries, Inc. Burlington, Vermont.

Dill, I., and G. Kraepelin. (1986). Palo Podrido: Model for extensive delignification of wood by *Ganoderma applanatum*. *Applied and Environmental Microbiology* 52(6):1305-1312.

Ellwood, E. L., and E. A. Ecklund. (1959). Bacterial attack of pine logs in pond storage. *Forest Products Journal* 9:283-292.

Eriksson, K.-E., A. Grunewald, T. Nilsson, and L. Vallander. (1980). A scanning electron microscopy study of the growth and attack on wood by three white-rot fungi and their cellulase-less mutants. *Holzforschung* 34(6):207-213.

Eriksson, K.-E. (1981). Fungal degradation of wood components. *Pure and Applied Chemistry* 53:33-43.

Foisner, R., K. Messner, H. Stachelberger, and M. Röhr. (1985a). Wood decay by basidiomycetes: Extracellular tripartite membranous structures. *Transactions of the British Mycological Society* 85(2):257-266.

Foisner, R., K. Messner, H. Stachelberger, and M. Röhr. (1985b). Isolation and characterization of extracellular three-lamellar structures of *Sporotrichum pulverulentum*. *Journal of Ultrastructure Research* 92:32-46.

Forney, L. J., C. A. Reddy, and H. S. Pankratz. (1982). Ultrastructural localization of hydrogen peroxide production in ligninolytic *Phanerochaete chrysosporium* cells. *Applied and Environmental Microbiology* 44(3):732-736.

Garcia, S., J. P. Latge, M. C. Prevost, and M. Leisola. (1987). Wood degradation by white-rot fungi: Cytochemical studies using lignin peroxidase–immunoglobulin–gold complexes. *Applied and Environmental Microbiology* 53(10):2384-2387.

Greaves, H. (1968). Occurrence of bacterial decay in copper-chrome-arsenic treated wood. *Journal of Applied Microbiology* 16:1699-1701.

Greaves, H. (1969). Micromorphology of bacterial attack of wood. *Wood Science and Technology* 3:150-166.

Green, F., M. J. Larsen, L. L. Murmanis, and T. L. Highley. (1989). "Proposed Model for the Penetration and Decay of Wood by the Hyphal Sheath of the Brown-Rot Fungus *Postia placenta*." The International Research Group on Wood Preservation, Document No. IRG/WP/1391, Stockholm, Sweden.

Hale, M. D., and R. A. Eaton. (1981). "Soft-Rot Ultrastructure." The International Research Croup on Wood Preservation, Document No. IRG/WP/1138, Stockholm, Sweden.

Hale, M. D., and R. A. Eaton. (1985a). The ultrastructure of soft rot fungi. I. Fine hyphae in wood cell walls. *Mycologia* 77(3):447-463.

Hale, M. D., and R. A. Eaton. (1985b). The ultrastructure of soft-rot fungi. II. Cavity-forming hyphae in wood cell wall. *Mycologia* 77(4):594-605.

Harada, H., and W. A. Côté, Jr. (1985). Structure of Wood. *In* "Biosynthesis and Biodegradation of Wood Components" (T. Higuchi, ed.), pp. 1-42. Academic Press, New York.

Highley, T. L., and L. L. Murmanis. (1984). Ultrastructural aspects of cellulose decomposition by white-rot fungi. *Holzforschung* 38(2):73-78.

Highley, T. L., and L. L. Murmanis. (1985). "Involvement of Hydrogen Peroxide in Wood Decay by Brown-Rot and White-Rot Fungi." The International Research

Group on Wood Preservation, Document no. IRG/WP/1256, Stockholm, Sweden.
Highley, T. L., and L. L. Murmanis. (1986). Determination of hydrogen peroxide production in *Coriolus versicolor* and *Poria placenta* during wood degradation. *Material und Organismen* 20(4):241-252.
Highley, T. L., and L. L. Murmanis. (1987). Micromorphology of degradation in western hemlock and sweetgum by the white-rot fungus *Coriolus versicolor*. *Holzforschung* 41(2):67-71.
Highley, T. L., L. Murmanis, and J. G. Palmer. (1983a). Electron microscopy of cellulose decomposition by brown-rot fungi. *Holzforschung* 37(6):271-277.
Highley, T. L., J. G. Palmer, and L. L. Murmanis. (1983b). Decomposition of cellulose by *Poria placenta*: Light and electron microscopy study. *Holzforschung* 37(4): 179-184.
Hirada, H., and W. A. Côté, Jr. (1985). Structure of Wood. *In* "Biosynthesis and Biodegradation of Wood Components" (T. Higuchi, ed.), pp. 1-42. Academic Press, New York.
Hoffmann, P., and N. Parameswaran. (1976). On the ultrastructural localization of hemicelluloses within delignified tracheids of spruce. *Holzforschung* 30(2):62-70.
Holt, D. M. (1983). Bacterial degradation of lignified wood cell walls in aerobic habitats, decay patterns and mechanisms proposed to account for their formation. *Journal of the Institute of Wood Science* 9:212-223.
Holt, D. M., E. B. Gareth-Jones, and S. E. J. Furtado. (1979). Bacterial breakdown of wood in aquatic habitats. *Record, British Wood Preservers' Association Annual Convention*, 13-22.
Hyat, M. A. (1981). "Principles and Techniques of Electron Microscopy Biological Applications," Vol. 1, 2nd Ed. University Park Press, Baltimore, Maryland.
Joseleau, J. P., and K. Ruel. (1985). A new cytochemical method for ultrastructural localization of polysaccharides. *Biology of the Cell* 53:61-66.
Jutte, S. M., and R. A. Zabel. (1974). Initial wood decay stages as revealed by scanning electron microscopy. *Scanning Electron Microscopy* 1974 (Part II) 445-452.
Kerr, A. J., and D. A. I. Goring. (1975). The ultrastructural arrangement of the wood cell wall. *Cellulose Chemistry and Technology* 9:563-573.
Levy, J. F. (1975). Bacteria associated with wood in ground contact. *In* "Biological Transformation of Wood by Microorganisms" (W. Liese, ed.), pp. 64-73. Springer-Verlag, New York.
Liese, W. (1970). Ultrastructural aspects of woody-tissue disintegration. *Annual Review of Phytopathology* 8:231-257.
Liese, W., and H. Greaves. (1975). Micromorphology of bacterial attack. *In* "Biological Transformation of Wood by Microorganisms" (W. Liese, ed.), pp. 74-88. Springer-Verlag, New York.
McLaughlin, D. J. (1982). Ultrastructure and cytochemistry of basidia and basidiospore development. *In* "Basidium and Basidiocarp Evolution, Cytology, Function, and Development" (K. Wells and E. K. Wells, eds.), pp. 37-74. Springer-Verlag, New York.
Nilsson, T., and A. Singh. (1984). "Cavitation bacteria." International Research Group on Wood Preservation, Document No. IRG/WP/1235. Stockholm, Sweden.
Messner, K., and H. Stachelberger. (1982). "Extracellular Osmiophilic Particles in Connection with Brown-Rot." The International Research Group on Wood Preservation, Document No. IRG/WP/1157, Stockholm, Sweden.

Messner, K., and H. Stachelberger. (1984). Transmission electron microscope observations of white rot caused by *Trametes hirsuta* with respect to osmiophilic particles. *Transactions of the British Mycological Society* 83(2):217-221.
Messner, K., and H. Stachelberger. (1985). Transmission electron microscope observations of brown rot caused by *Fomitopsis pinicola* with respect to osmiophilic particles. *Transactions of the British Mycological Society* 83(1):131-137.
Mouzouras, R. (1989). Soft-rot decay of wood by marine microfungi. *Journal of the Institute of Wood Science* 11(5):193-201.
Murmanis, L., T. L. Highley, and J. G. Palmer. (1984). An electron microscopy study of western hemlock degradation by the white-rot fungus *Ganoderma applanatum*. *Holzforschung* 38(1):11-18.
Murmanis, L., J. G. Palmer, and T. L. Highley. (1985). Electron-dense particles in wood decayed by *Ganoderma applanatum*. *Wood Science and Technology* 19: 313-321.
Murmanis, L., T. L. Highley, and J. G. Palmer. (1987). Cytochemical localization of cellulases in decayed and nondecayed wood. *Wood Science and Technology* 21: 101-109.
Murmanis, L., T. L. Highley, and J. G. Palmer. (1988a). The action of isolated brown-rot cell-free culture filtrate, H_2O_2-Fe^{++}, and the combination of both on wood. *Wood Science and Technology* 22:59-66.
Murmanis, L., T. L. Highley, and J. Ricard. (1988b). Hyphal interaction of *Trichoderma harzianum* and *Trichoderma polysporum* with wood decay fungi. *Material und Organismen* 23:271-279.
Nilsson, T., G. Daniel, T. K. Kirk, and J. R. Obst. (1989). Chemistry and microscopy of wood decay by some higher ascomycetes. *Holzforschung* 43(1):11-18.
Otjen, L., and R. A. Blanchette. (1986). A discussion of microstructural changes in wood during decomposition by white-rot basidiomycetes. *Canadian Journal of Botany* 64:905-911.
Otjen, L., and R. Blanchette. (1987). Assessment of 30 white-rot basidiomycetes for selective lignin degradation. *Holzforschung* 41(6):343-349.
Otjen, L., and R. A. Blanchette. (1988). Lignin distribution in wood delignified by white-rot fungi: X-ray microanalysis of decayed wood treated with bromine. *Holzforschung* 42(5):281-288.
Palmer, J. G., L. Murmanis, and T. L. Highley. (1983a). Visualization of hyphal sheath in wood-decay hymenomycetes. I. Brown-rotters. *Mycologia* 75(6):995-1004.
Palmer, J. G., L. Murmanis, and T. L. Highley. (1983b). Visualization of hyphal sheaths in wood-decay Hymenomycetes. II. White-rotters. *Mycologia* 75(6):1005-1010.
Palmer, J. G., L. Murmanis, and T. L. Highley. (1984). Observations of wall-less protoplasm in white- and brown-rot fungi. *Material und Organismen* 19(1):39-48.
Parameswaran, N., and W. Liese. (1982). Ultrastructural localization of wall components in wood cells. *Roh-und Werkstoff* 40:145-155.
Parham, R. A., and R. L. Gray. (1984). Formation and Structure of Wood. *In* "Chemistry of Solid Wood" (R. M. Rowell, ed.), pp. 1-56. American Chemical Society, Washington, D.C.
Postek, M. T., K. S. Howard, A. H. Johnson, and K. L. McMichael. (1980). "Scanning Electron Microscopy: A Student's Handbook." Ladd Research Industries, Inc., Burlington, Vermont.

Reid, I. D. (1985). Biological delignification of aspen wood by solid-state fermentation with the white-rot fungus *Merulius tremellosus*. *Applied and Environmental Microbiology* **50**(1):133-139.

Resch, H., and D. G. Arganbright. (1971). Location of pentachlorophenol by electron microprobe and other techniques in Cellon-treated Douglas fir. *Forest Products Journal* **21**(1):38-43.

Ruel, K., and D. Barnoud. (1985). Degradation of wood by microorganisms. In "Biosynthesis and Biodegradation of Wood Components" (T. Higuchi, ed.), pp. 441-467. Academic Press, New York.

Ruel, K., F. Barnoud, and K.-E, Eriksson. (1984). Ultrastructural aspects of wood degradation by *Sporotrichum pulverulentum*. *Holzforschung* **38**(2):61-68.

Ruel, K., and J. P. Joseleau. (1984). Use of enzyme-gold complexes for ultrastructural localization of hemicelluloses in the plant cell wall. *Histochemistry* **81**:573-580.

Ruel, K., F. Barnoud, and K.- E. Eriksson. (1981). Micromorphological and ultrastructural aspects of spruce-wood degradation by wild-type *Sporotrichum pulverulentum* and its cellulase-less mutant Cel 44. *Holzforschung* **35**(4):157-171.

Ruel, K., F. Barnoud, and J. P. Joseleau, S. C. Johnsrud, and K. E. Eriksson. (1986). Ultrastructural aspects of birch-wood degradation by *Phanerochaete chrysosporim* and two of its cellulase-deficient mutants. *Holzforschung* **40**:Suppl. 5-9.

Ryan, K. G. (1986). Preparation techniques for x-ray analysis in the transmission electron microscope of wood treated with copper-chrome-arsenate. *Material und Organismen* **21**:223-234.

Ryan, K. G., and J. A. Drysdale. (1988). X-ray analysis of copper, chromium, and arsenic within the cell walls of treated hardwoods—new evidence against the microdistribution theory. *Journal of the Institute of Wood Science* **11**(3):108-113.

Saka, S., and D. A. I. Goring. (1985). Localization of lignins in wood cell walls. In "Biosynthesis and Biodegradation of Wood Components" (T. Higuchi, ed.), pp. 51-62. Academic Press, New York.

Schmidt, O., Y. Nagashima, W. Liese, and U. Schmidt. (1987). Bacterial wood-degradation studies under laboratory conditions and in lakes. *Holzforschung* **41**(3):137-140.

Srebotnik, E., and K. Messner. (1988). "Spatial Arrangement of Lignin Peroxidase in Pine Decayed by *Phanerochaete chrysosporium* and *Fomitopsis pinicola*." The International Research Group on Wood Preservation, Document No. IRG/WP/1343, Stockholm, Sweden.

Srebotnik, E., K. Messner, and R. Foisner. (1988a). Penetrability of white-rot degraded pine wood by the lignin peroxidase of *Phanerochaete chrysosporium*. *Applied and Environmental Microbiology* **54**(11):2608-2614.

Srebotnik, E., K. Messner, R. Foisner, and B. Pettersson. (1988b). Ultrastructural localization of ligninase of *Phanerochaete chrysosporium* by immunogold labeling. *Current Microbiology* **16**:221-227.

Thielke, C. (1982). Meiotic divisions in the basidium. In "Basidium and Basidiocarp Evolution, Cytology, Function, and Development" (K. Wells and E. K. Wells, eds.), pp. 75-91. Springer-Verlag, New York.

Unligil, H. H., and S. C. Chafe. (1974). Perforation hyphae of soft-rot fungi in the wood of white spruce [*Picea glauca* (Moench.) Voss.]. *Wood Science and Technology* **8**:27-32.

Wilcox, W. W., and N. Parameswaran. (1974). A method for electron microscopical demonstration of pentachlorophenol within treated wood. *Holzforschung* **28**: 153–155.

Wilcox, W. W., N. Parameswaran, and W. Liese. (1974). Ultrastructure of brown rot in wood treated with pentachlorophenol. *Holzforschung* **28**(6):211–217.

Zabel, R. A., F. F. Lombard, C. J. K. Wang, and F. Terracina. (1985). Fungi associated with decay in treated southern pine utility poles in the eastern United States. *Wood and Fiber Science* **17**(1):75–91.

CHAPTER

10

Changes Caused by Decay Fungi in the Strength and Physical Properties of Wood

Many changes occur in a wide range of strength (mechanical) and physical properties, as wood-inhabiting microorganisms invade and colonize wood. These changes range from drastic effects on wood strength to subtle modifications in properties such as density, hygroscopicity, electrical conductance, acoustics, caloric values, and dimension. These changes and their rates of development in wood vary with wood species, the environmental conditions, and the organisms involved. In many instances these changes are hidden, poorly defined, and may cause, in part, the high variability of many wood properties. A review of the major microbial effects on wood properties will facilitate their detection and help define the roles and interactions of causal agents in the biodeterioration process.

It is important to remember that changes in one property are invariably associated with changes in other properties. For example, a loss in wood weight from decay decreases its caloric value and reduces strength properties. Also, subtle changes in properties such as permeability (aeration), coupled with nutritional factors, may increase decay susceptibility and set the stage for further colonization by other more destructive agents of decay. It should be remembered that most changes in physical properties are simply reflections of the more basic anatomical and chemical changes in wood associated with decay development, reviewed in Chapters 7 and 8.

Since strength is important in most structural uses, it has been the central concern of much wood-decay research. The important changes in wood properties associated with microbial colonization include reduced strength, biomass loss (weight), hygroscopicity, increased permeability, and decreased dimension.

This chapter will emphasize these topics and also review changes in electrical or acoustical properties that may be useful in nondestructive tests to detect early decay in wood products where strength is critical. This chapter places major emphasis on the effects of decay fungi on wood properties (the effects of sapstain and mold fungi are covered in Chapter 14).

Wood Weight Loss (Biomass Loss)

As they grow through wood, most wood-inhabiting microorganisms utilize the various components of the wood cell wall, reducing overall wood weight. Some fungi utilize primarily accessible nutrients in storage tissues or extractives, causing relatively minor weight losses (1 to 3%) and minimal damage. Others attack the more chemically complex components of the wood cell wall, eventually metabolizing these to carbon dioxide and water. Wood weight losses can approach 70% with brown-rot fungi, 96 to 97% for white-rot fungi, and 3 to 60% for soft-rot fungi. Wood weight loss depends on fungal type and wood species evaluated (Fig. 10-1). Weight loss (biomass) is the most commonly used measure of decay capability and is generally expressed as

Weight loss (%) =
  [(Original weight − decayed weight)/Original (oven dry) weight] 100

Weight loss is generally expressed on an oven-dry basis (ODW). Unfortunately, drying wood at high temperatures (100°C) can degrade the wood-cell wall, potentially altering susceptibility to microbial attack. This problem can be overcome by conditioning the wood to an equilibrium moisture content (EMC) (usually 12%) over saturated salt solutions or in a controlled humidity–temperature room. In this method, the weight loss due to decay is the difference between the EMC weights before and after decay fungus exposure. The original ODW necessary for the weight loss percentage determination is calculated from the moisture content of reference blocks kept under the same EMC conditions. However, wood EMC can vary with the degree and type of decay (Fig. 10-2) and become an error source unless accounted for in critical studies. Also, wood is an extremely hygroscopic material, and appreciable errors can be introduced during weighings when small test specimens with high surface:volume ratios are used, unless the EMC conditions are carefully controlled (see Chapter 6). In such cases, the use of critically defined temperature and relative humidity conditions and weighing bottles are necessary for accuracy.

The use of control blocks to account for any changes in EMC conditions between the original and the postdecayed blocks and to monitor any decay chamber effects, such as autoclaving or nutrient migration, are also necessary to achieve accuracy in laboratory decay-determination tests, particularly in those cases in which the weight losses due to decay are low.

Wood weight loss is extremely useful under laboratory conditions, where

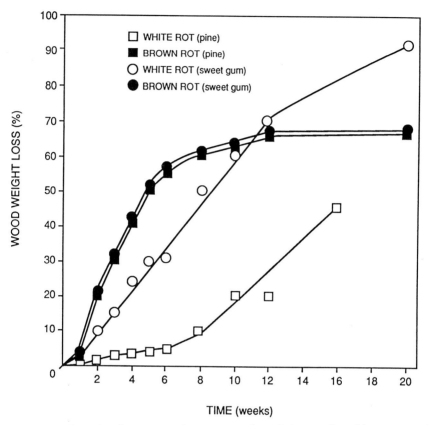

Figure 10-1 Relationships between incubation time and weight loss as affected by rot type and substrate. From the data of Wilcox (1968).

the original wood weight can be obtained, but is of limited use for naturally decayed wood, since the original weight of the wood can be only roughly approximated. Whereas wood weight loss is a useful comparative measure of decay capability (Hartley, 1958), it does not accurately measure the magnitude of decay effects on many other wood properties such as strength (Wilcox, 1978).

Density Loss

Density (wood mass/unit volume at a specific moisture content) and specific gravity (ODW weight/weight of displaced unit volume of water) are also used to measure the effects of microbial attack. The disadvantage of these methods is the difficulty of accurately measuring wood volume due to its sensitive hygroscopic properties and the higher variability, since two wood-property

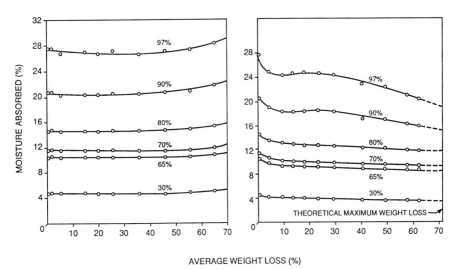

Figure 10-2 Moisture absorbed at various relative humidities at 26.5°C for wood in progressive stages of colonization by (left) the white-rot fungus *Trametes versicolor* and (right) the brown-rot fungus *Postia placenta*. From data of Cowling (1961).

measurements are involved. White-rot fungi cause substantial weight losses with little change in wood volume, whereas brown-rot fungi cause substantial volume reduction as they degrade the wood. Thus, density reductions are not comparable between these two decay types. Low density, measured as abnormal lightness, is often used as a rough decay indicator by lumber graders, since density is closely correlated with some strength properties, such as bending. Density can also be measured indirectly using X-rays, which pass at different rates through sound and decayed wood (Mothershead and Stacey, 1965), but this technique is both expensive and difficult to interpret. Surface density can also be estimated using the Pilodyn, a spring-loaded pin penetration device (Cown and Hutchinson, 1983; Hoffmeyer, 1978); however, readings often vary with wood species and must be corrected for wood-moisture content (Smith and Morrell, 1986).

Strength (Mechanical) Properties

The effects of decay fungi on wood strength have been intensively studied. As fungi grow through the wood, they alter the chemical structure and remove mass, thereby altering the mechanical properties of wood. Wood derives its strength from a combination of highly oriented cellulose microfibrils and encrusting hemicellulose. Any changes in these carbohydrates often cause sharp reductions in wood strength.

As a result of the many obvious deleterious effects of microbial colonization on wood use, a large literature has accumulated on its effects on wood strength. Unfortunately, most of this research examined the effects of a limited microbial flora on a few commercially important wood species using a variety of different measures of wood strength.

Early studies on the effects of decay emphasized heart rots of standing trees and related losses of usable volumes of lumber (Colley, 1921). The effects of these fungi were further quantified by testing beams cut from clear and colonized wood (Scheffer et al., 1941; Hartley, 1958). As the effects of these fungi were quantified, the emphasis shifted to the roles of various fungi associated with the decay of buildings and other wood products (Cartwright et al., 1931). These early studies were dependent on naturally decayed material, but as techniques became more refined, individual fungi were isolated and tested for their ability to cause weight and strength losses under controlled laboratory conditions. One of the first carefully controlled studies of the effects of a single decay fungus on wood strength was performed by Scheffer, who evaluated the effects of *Trametes versicolor* on red gum (1936).

The mechanical properties of wood can be measured by a variety of methods. Wood strength is covered in detail in several textbooks (Panshin and de Zeeuw, 1980; Hoyle, 1972; Bodig and Jayne, 1982). The Committee D-7 on Wood has prepared and periodically updates a publication that describes in detail the standard methods of conducting a series of standard strength tests on wood (ASTM, 1988). The strength properties most often used to measure the effects of decay include modulus of rupture, work to maximal load in bending, maximal crushing strength, compression perpendicular to the grain, impact bending, tensile strength parallel to the grain, toughness, hardness, and shear strength (Brown, 1963; Henningsson, 1967; Kennedy and Ifju, 1962; Mulholland, 1954; Toole, 1969, 1971; Armstrong and Savory, 1959). Of these many strength properties, work to maximal load, toughness, and impact bending are reported to be the most sensitive to detect early decay (Wilcox, 1978).

Most studies of microbial effects on wood strength have used bending tests of small clear specimens that had been exposed to the desired test organism for varying periods. The data collected were then used to compute modulus of elasticity, modulus of rupture, and work to maximal load from stress–strain relationships. Bending tests of simple beams (supported at both ends and center loaded) measure the effects of microbial colonization at the center of the span, and any effects of colonization away from this zone are sharply diminished. Thus, uneven colonization of even small test beams can produce variable results.

Whereas bending tests are useful, many researchers have employed other strength tests using smaller, more easily handled specimens to accelerate the decay process and increase replication. These tests include tension parallel to the grain, radial or longitudinal compression testing, and small-scale tests such as the breaking radius and pendulum tests (Graham and

Safo-Sampah, 1976; Scheffer, 1978). In the latter tests, small wafers are exposed to the test fungus and tested either by dropping a heavy pendulum into the center of the wafer and measuring the force absorbed or by bending the wafer successively around a series of mandrils with smaller diameters until failure is achieved. In both instances, the variation is high, and many specimens must be tested to achieve reasonable accuracy (Sexton, 1988).

Another useful method for rapid strength tests on small beams is use of the toughness-testing machine developed by the U.S.D.A. Forest Products Laboratory (Forest Products Laboratory, 1941). This test measures the kinetic energy absorbed by a swinging pendulum during the sudden breaking of a small beam. The toughness values are measured in inch-pounds, and comparative data have been developed for a number of commercial wood species. Reductions in toughness are a particularly useful and sensitive indicator of the early decay stage. An advantage of the method is that large numbers of test beams can be measured rapidly. A disadvantage is that the measuring unit (inch-pounds) can not be related directly to other strength values. Nevertheless, pendulum or mandril tests are useful methods for the rapid screening of large numbers of microorganisms for rough approximation of decay capability.

Radial and longitudinal compression strength tests (RCS and LCS) can be run on small blocks or samples removed from larger wood members. RCS is especially sensitive to the early stages of decay, particularly for the brown-rot fungi (Kubiak and Kerner, 1963; Smith and Graham, 1983), whereas LCS is more closely related to bending properties (Smith and Morrell, 1987).

Wilcox (1978), in a comprehensive summary of the literature on the effects of fungi on strength properties, reported that toughness or resistance to impact loading was the property most sensitive to the early stage of decay, followed by static bending properties (as work) (Table 10-1, Fig. 10-3). In general, brown rots are perceived to be more damaging because they are associated with greater strength losses at lower weight losses than are white rots. These rapid decreases in strength at early stages of decay are attributed to the fact that brown-rot fungi produce enzymes that diffuse well away from the fungal hyphae. As a result, the relative damaging effect of colonization by a brown-rot fungus extends far beyond the limited occurrence of hyphae present in the incipient stages of decay. White-rot fungi generally produce enzymes that are more closely associated with hyphal tips and, therefore, the strength effects associated with these fungi are less significant in the early stages of attack (<5% weight loss). During the later stages of attack where the fungal mycelia are well distributed through the wood structure, there is little difference in the degree of strength loss between the two fungal groups. A bias in the current literature is the preponderance of strength testing that has been performed using primarily a few brown-rot fungi [e.g., *Postia (Poria) placenta, Gloeophyllum (Lenzites) trabeum*], since they are more commonly associated with decays in structural applications using coniferous woods.

Although the strength effects of decay by brown- and white-rot fungi

Table 10-1

Estimated values for strength losses in softwoods and hardwoods at early stages of decay (indicated by weight loss) by brown-rot and white-rot fungi[a]

Approximate weight loss (%)	Toughness	Impact bending	Static bending				Compression perpendicular (radial)	Compression parallel	Tension parallel	Shear parallel	Hardness
			General bending strength	Work to maximal load	Modulus of rupture	Modulus of elasticity					
Brown rot											
Softwoods											
1	57	20–38	—	—	—	—	—	—	—	2	—
2	—	20–50	5	27	13–50	4–55	18–24	10	23–40	—	—
4	75	25–55	—	—	—	—	25–35	—	—	6	7
6	—	62–72	16	—	61	66	48	25	60	—	—
8	—	78	—	—	—	—	48–60	—	50	15	21
10	—	85	36	—	70	—	66	45	—	20	—
Hardwoods											
1	—	6–27	—	—	—	—	—	—	—	—	—
2	36	31–50	—	54	32	—	6–10	—	56	—	—
4	—	60–70	—	69	49	—	—	—	—	—	—
6	—	80	—	75	61	—	16–25	—	—	—	—
8	—	9–89	13–34	—	—	—	19	—	82	—	—
10	60	70–92	—	—	—	—	—	—	—	—	—

White rot									
Softwoods									
1	55	—	—	—	—	—	—	—	—
2	—	—	—	—	—	—	—	4–38	—
4	75	—	—	—	—	—	—	8–43	—
6	—	—	—	—	—	10–20	—	10–49	—
8	—	—	—	—	—	—	—	14–58	—
10	85	—	—	—	—	32–61	—	20–63	—
Hardwoods									
1	—	21	—	—	4	4	—	—	—
2	—	26	—	28–35	13–14	5	—	22–42	—
4	70	44	—	38	20	—	—	17–44	—
6	75	50	—	45–53	20–27	12–27	14	12–58	18
8	—	—	—	—	—	—	—	14–49	—
10	85	60	—	58	24	35	20	20–50	25

[a] As a percentage of the values for nondecayed samples. From Wilcox (1978). Values obtained from published experimental results and adjusted to equivalent weight-loss levels.

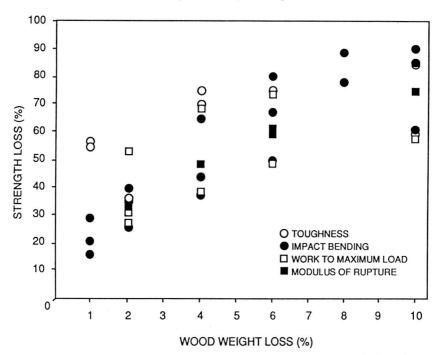

Figure 10-3 Relationship between wood weight loss and strength for selected white-, brown-, and soft-rot fungi. Data from Table 10-1.

have been studied, the effects of other decay types on wood strength remain poorly defined. Soft-rot fungi have been found to also cause large strength losses at low wood weight losses. Some soft-rot fungi (Type 1) also remove sizeable sections of the wood cell wall during cavity formation, forming many failure zones and magnifying the strength losses in very small areas of the wood cell wall. Furthermore, hyphal penetration of multiple cell walls has been noted with some soft-rot fungi (Type 2) and may also create planes of weakness (Morrell and Zabel, 1985). At present, soft-rot-damaged wood, when detectable visually, is presumed to have no residual strength (Hoffmeyer, 1976); however, further research on the enzyme systems associated with soft-rot fungi and their progressive effects on wood strength are needed to better quantify the relationship between types of microbial colonization and strength effects.

One weakness of the many laboratory studies on the effects of various decays on strength is the use of copious quantities of inoculum in an axenic system, generally in the presence of an excess of sugar compounds, to accelerate fungus growth. In natural systems, the fungal colonization originates from spores or hyphal fragments, which must germinate and grow through the wood, generally in the absence of excess nutrients. These fungi must also compete with other microbes that are also attempting to colonize and utilize the substrate. A recent study, using *Stereum sanguinolentum* and *Postia placenta*, indicates that microbial colonization of the wood is relatively rapid, whereas effects on wood properties occur more gradually (Smith and Morrell, 1991). These results suggest that there is a delay between colonization and strength effects.

One area of strength effects that has received less intensive study is the effects of microbial attack on fiber properties. As the fungus attacks the wood, cellulose microfibrils become more susceptible to chemical and physical damage during the pulping process. Although the effects of microbial attack on pulp yield and paper properties have been studied (McGovern *et al.*, 1951; Lindgren and Eslyn, 1961; Rothrock *et al.*, 1961; Christie, 1979), most studies have depended on natural colonization in pulpwood or chip piles. The microflora and its potential effects in a particular study might vary widely with season, wood-moisture content, pulpwood types, and other environmental conditions. Microbial colonization, as measured by wood-density losses with storage time, decreases burst strength by 15% and folging by 24% over a 6-month period (McGovern *et al.*, 1951). Similarly, tear strength declined by 35%, with a 7% decline in density over a 5-month period (Rothrock *et al.*, 1961). The effects of microbial colonization on other pulpwood and pulp properties are discussed in Chapter 13 and 14.

The effects of microbial agents on wood-strength properties vary widely with the microbial agents involved, wood species, and environmental conditions, but the previous studies suggest that most changes are readily measurable before the wood has exceeded a 5% weight loss. These effects highlight the importance of timely and accurate detection of early decay, since substantial reductions in the strength properties in brown rots has already occurred. Improving our knowledge about the types and rates of these strength losses and their detection will enhance greatly the reliability of wood systems and increase the safety of these structures. It may also reveal other property changes highly sensitive to early decay changes, which may improve the effectiveness of early decay detection.

Hygroscopicity

As microbial enzymes degrade the ligno-carbohydrate matrix, they induce changes in the moisture-holding capacity of the wood cell wall. These changes, in turn, alter the EMC and can influence the results of decay studies in which wood weight-loss measurements are made by equilibrating wood at constant temperature–relative humidity conditions (Fig. 10-2). Generally, the EMC of brown-rotted wood is lower than that of sound wood, whereas the EMC of white-rotted wood is higher when induced weight losses exceed 60% (Cowling, 1961). The increase in EMC begins at about 40% weight loss for white-rotted wood, whereas brown-rotters are associated with sharp drops in

EMC at the early stages of decay. The sharp drop in EMC associated with brown-rot fungi probably reflects preferential attack on the amorphous cellulose. Amorphous cellulose retains higher levels of adsorbed water than the crystalline cellulose regions, and its early removal decreases the overall moisture-holding capacity of the wood. The absence of EMC changes in the early stages of white-rot attack probably reflects uniform removal of all wood components, whereas increases in EMC during the later stages of decay by these fungi may reflect selective attack of crystalline cellulose.

Caloric Value

As microbial agents colonize and utilize the wood substrate, they remove and convert wood substance to microbial biomass, carbon dioxide, water, and metabolic waste products. Although microbial biomass will make a slight contribution to the caloric value, the net energy content of the decayed wood declines (Fig. 10-4). Caloric value becomes particularly important wherever

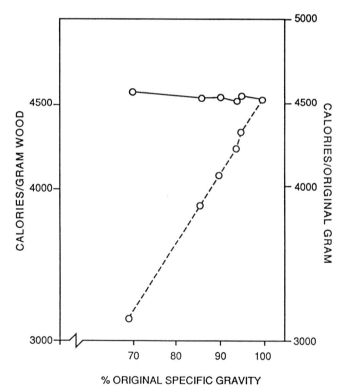

Figure 10-4 Net caloric value of wood decayed to various weight losses on original or final wood-weight basis (calories/final gram of wood = O-----O, whereas calories per original gram = O---O). After Scheffer (1936).

wood is used for heating, since additional biomass must be used to generate the same quantity of heat. In general, the losses in caloric value of decayed wood should be directly proportional to the density reduction (Scheffer, 1936). On a weight basis, conifers, due to their higher lignin content, have a higher caloric value (e.g., pines, 8836 BTUs per lb) than hardwoods (oaks, 8556 BTUs per lb). This higher caloric value usually is masked by the lower density of conifers. On this basis, brown-rotted wood should have a higher caloric value (at equal weight loss) than white-rotted wood, but no confirming data are available. Recently, differential scanning calorimetry has been proposed as a rapid laboratory method for detecting the relative amounts of cellulose, hemicellulose, and lignin remaining in decayed wood (Reh et al., 1986).

Permeability

Although some wood-inhabiting fungi directly penetrate the wood cell wall to move from cell to cell, the majority of the decay organisms initially move through the wood primarily by direct pit penetration. Since pits represent the major limiting factor in fluid flow through fibers and tracheids, the removal of the pit membrane makes the wood markedly more receptive to movement of fluids. As a result of these changes, decayed wood absorbs and desorbs liquids more readily than sound wood.

Excessive absorption can result in sinking of logs during storage, reduces the rate of kiln-drying for some species, and can result in uneven absorptions of preservative treatments. The process of pit removal due to bacterial action in ponded logs is relatively slow (Ellwood and Eckland, 1959); however, over many years, sinker logs can represent a substantial loss in volume. The presence of differing areas of permeability can also influence the rate of lumber drying (Ward, 1986). The bacterial-invaded wood, for example, is markedly wetter than uninfected wood. These materials have differing drying rates from those of the uninfected wood that may be present in the same board (Ward, 1986).

Increased permeability will also increase the absorption of various preservatives, glues, and coatings applied to the wood. The preservative costs are increased, and often serious preservative bleeding or coating retention problems develop. In gluing, dry joints and premature failures may develop. In coatings, color variations and retention problems may develop. The increased permeability of fungal colonized wood has been exploited to improve the treatability of difficult-to-treat species (Lindgren, 1952; Graham, 1954; Schulz, 1956; Lindgren and Wright, 1954). Fungi such as *Trichoderma viride*, a non-decaying scavenger of carbon compounds, were inoculated onto the wood surface and allowed to colonize the freshly peeled wood. Whereas treatability was improved near the surface, the results were inconsistent, and treatment of the heartwood was not markedly improved (Johnson and Gjovik, 1970).

In addition to the obvious effects on end uses, increased permeability may also alter the ecology of the decaying wood. More permeable wood will wet more readily and, as a result, conditions in the wood will be conducive to microbial colonization for longer periods.

Electrical Properties

Wood has much lower electrical conductivity than other construction materials such as steel, and this is one reason for its common use to support electric distribution systems. As the wood is degraded, however, its electrical conductivity increases (Richards, 1954). As a result, electric-resistance type moisture meters can sometimes overestimate the moisture content in decayed wood if it is below the fiber-saturation point.

Measuring changes in electrical properties due to microbial colonization using pulsing electric currents has been proposed to detect early decay (Shigo and Shigo, 1974). In this process, voltages are pulsed through a twisted wire probe that is insulated along its length, but exposed at the tip. As the probe is inserted into a predrilled hole, electrical resistance is measured. Sound wood has a high electrical resistance, whereas decayed or discolored wood registers resistance readings that are 50 to 75% lower than the sound material. The Shigometer has been extensively used to detect decay and discoloration in living trees (Shigo and Shigo, 1974; Shigo et al., 1977; Shigo and Berry, 1975; Shortle et al., 1978; Skutt et al., 1972), but has been less consistent in wood products (Shigo et al., 1977; Shortle, 1982; Piirto and Wilcox, 1978; Zabel et al., 1982; Graham and Inwards, 1980). A disadvantage of the Shigometer is the limitation of accurate use to wood above the fiber-saturation point. Electrical resistance is presumed to increase during decay development owing to the release of ions and organic acids during decay. These changes are easily detected in the standing tree, since wood moisture contents generally exceed the fiber-saturation point; however, moisture contents in wood products vary widely with position in the sample. For example, moisture contents around water-trapping joints will generally be higher than those in zones away from these sites. Furthermore, moisture content in wood products varies with season. Thus, moisture-content variations will minimize the usefulness of a tool that is heavily dependent on moisture. Finally, wood in soil contact or near metal fasteners will often experience inward migration of various salts, which can also influence electrical resistance.

Acoustic Properties

Wood is an excellent transmitter of sound waves and also produces characteristic acoustic emissions as it is stressed mechanically. As wood is colonized by microbial agents, its ability to transmit or emit sounds is generally altered (Pellerin et al., 1986; Noguchi et al., 1986). The alteration in acoustic properties can be exploited to detect various stages of decay. As sound waves move through wood, they will pass around decay pockets or voids. The time of flight (velocity) of a given pulse of sound across the cross section of a timber can be used to determine the internal condition of the wood. Because of the difficulty of measurement and the natural variation in the wood structure, velocity measurements appear to be the easiest to use for detecting advanced decay (>25% wood weight loss) where voids in wood structure are present.

A second approach to measuring changes in acoustic properties is analyzing the characteristics of a sound wave after it has passed through the wood. As a sound wave moves through the wood, the characteristics of the wood (growth rings, knots, checks, etc.) modify the wave pattern. The modified wave represents an acoustic *fingerprint* of the wood. Small changes in the structural integrity of the wood due to microbial activity should be discernable in the resulting fingerprint. At present, the wave-form analysis of wood is still relatively crude; however, progress in other areas of material science suggest that this technique has promise to nondestructively monitor changes in the wood over the course of decay (Pellerin et al., 1986; Ross and Pellerin, 1990).

Summary

- The major effects of decay development on the physical properties of wood are weight and density losses, strength reductions, and changes in permeability, hygroscopicity, electrical conductance, caloric content, acoustics, and dimensions.
- The important changes in wood properties associated with decay development affecting many wood uses are weight loss, strength reductions, and increased permeability.
- When the decay fungi are established and decay conditions optimal, wood-weight loss is nearly linear with time.
- Total wood-weight losses may approach 96 to 97% for white-rot fungi, up to 70% for brown-rot fungi, and range from 3 to 60% for soft-rot fungi.
- Of the many strength properties used to measure the effects of decay, work to maximal load, toughness, and impact bending have proven to be most sensitive for the detection of early decay.
- Of the fungi tested, brown rots cause substantially greater strength losses at lower weight losses than white rots. In brown rots, strength reductions of up to 75% can occur before weight losses of less than 5%.
- The equilibrium moisture content (EMC) of brown-rotted wood is

Summary continues

lower than that of sound wood at all weight losses, whereas for white rots, it is higher after the weight loss exceeds 60%.
- In general, the caloric values of decayed wood are highest in the brown rots and are directly related to weight loss or density reductions.
- Decayed wood has a higher electrical conductivity than sound wood, but moisture content is a significant confounding factor at levels below the fiber saturation point.
- Decayed wood has different patterns of sound-wave transmission than nondecayed wood. These patterns are being studied and have potential as nondestructive tests for detecting early decay.
- Most studies of the physical changes in wood properties associated with decay have utilized a limited number of brown- or white-rot fungi, particularly those isolated from conifers. There is a danger of overgeneralizing from these few fungi and a need to determine the effects of other fungi and decay types on the major use properties of wood.

References

American Society for Testing and Materials (ASTM). (1988). "Standard Methods for Establishing Clear Wood Strength Value." Standard D-2555-88. ASTM Book of Standards Section 4. Philadelphia, Pennsylvania.

Armstrong, F. H., and J. G. Savory. (1959). The influence of fungal decay on the properties of timber. *Holzforschung* 13:84–89.

Bodig, J., and B. A. Jayne. (1982). "Mechanics of Wood and Wood Composites." Van Nostrand Reinhold, New York.

Brown, F. L. (1963). A tensile test for comparative evaluation of wood preservatives. *Forest Products Journal* 13(9):405–412.

Cartwright, K. St. G., W. P. K. Findlay, C. J. Chaplin, and W. G. Campbell. (1931). "The Effect of Progressive Decay by *Trametes serialis* Fr. on the Mechanical Strength of Wood of Sitka Spruce." Great Britain Department of Science and Industrial Research, Forest Products Research Bulletin No. 11.

Christie, R. D. (1979). The use of decayed wood in pulping processes. In "Chip Quality Monograph." Pulp and Paper Technology Series (F. V. Hatton, ed.), pp. 111–128. TAPPI, Atlanta, Georgia.

Colley, R. H. (1921). The effect of incipient decay on the mechanical properties of airplane timber. *Phytopathology* 11:45.

Cowling, E. B. (1961). "Comparative Biochemistry of the Decay of Sweetgum Sapwood by White-Rot and Brown-Rot Fungi." U.S.D.A. Forest Service Technical Bulletin No. 1258, Madison, Wisconsin.

Cown, D. J., and J. D. Hutchinson. (1983). Wood density as an indicator of the bending properties of *Pinus radiata* poles. *New Zealand Journal of Forest Science* 13(1): 87–99.

Ellwood, E. A., and B. A. Eckland. (1959). Bacterial attack of pine logs in pond storage. *Forest Products Journal* 9(9):283-292.

Graham, R. D. (1954). The preservative treatment of Douglas fir post sections infected with *Trichoderma* mold. *Forest Products Journal* 4:164-166.

Graham, R. D., and R. D. Inwards. (1980). Comparing methods for inspecting Douglas fir poles in service. *Proceedings of the American Wood Preservers' Association* 76:283-289.

Graham, R. D., and S. Safo-Sampah. (1976). Rapid agar-stick breaking-radius test to determine the ability of fungi to degrade wood. *Wood Science* 9(2):65-69.

Hartley, C. (1958). "Evaluation of Wood Decay in Experimental Work." U.S.D.A. Forest Products Laboratory General Technical Report 2119, Madison, Wisconsin.

Henningsson, B. (1967). "Changes in Impact Bending Strength, Weight, and Alkali Solubility Following Fungal Attack of Wood." Studia Forestalia Suecica (Stockholm) No. 41, Stockholm, Sweden.

Hoffmeyer, P. (1976). Mechanical properties of soft-rot-decayed Scots pine with special reference to wooden poles. *In* "Soft Rot in Utility Poles Salt-Treated in the Years 1940-1954." pp. 2.1-2.55. Swedish Wood Preservation Institute. Number 117, Stockholm, Sweden.

Hoyle, R. J., Jr. (1972). "Wood Technology in the Design of Structures." College of Engineering, Washington State University, Pullman, Washington.

Johnson, B. R., and L. R. Gjovik. (1970). Effect of *Trichoderma viride* and a contaminating bacterium on microstructure and permeability of loblolly pine and Douglas fir. *Proceedings of the American Wood Preservers' Association* 66:234-240.

Kennedy, R. W., and G. Ifju. (1962). Applications of microtensile testing to thin wood sections. *TAPPI* 45(9):725-733.

Kubiak, M., and G. Kerner. (1963). Die Veranderungen einiger physikalischer Eigerschaffen der Druckfestigkeit und der chemischen Zummensetzing des Buchenholzes im anfangsstadium des Holzabbaues durch *Coniophora cerebella* Pers. und *Stereum hirsutum. Willd. Drev. Vysk* 1963(4):181-193.

Lindgren, R. M. (1952). Permeability of Southern pine as affected by mold and other fungus infection. *Proceedings of the American Wood Preservers' Association* 48:158-174.

Lindgren, R. M., and W. E. Eslyn. (1961). Biological deterioration of pulpwood and pulp chips during storage. *TAPPI* 44(6):419-429.

Lindgren, R. M., and E. Wright. (1954). Increased absorptiveness of moldy Douglas fir poles. *Forest Products Journal* 4:162-164.

McGovern, J. N., J. S. Martin, and A. Hyttiner. (1951). Effect of storage of slash pine pulpwood on sulfate and ground wood pulp quality. *Forest Products Research Society Proceedings* 5:162-169.

Morrell, J. J., and R. A. Zabel. (1985). Wood strength and weight losses caused by soft-rot fungi isolated from treated southern pine utility poles. *Wood and Fiber Science* 17(1):132-143.

Mothershead, J. S., and S. S. Stacey (1965). Applicability of radiography to inspection of wood products. *Proceedings 2nd Annual Symposium on Non-Destructive Testing of Wood.* Spokane, Washington, pp. 307-336. Published by Washington State University at Pullman, Washington.

Mulholland, J. R. (1954). Changes in weight and strength of Sitka spruce associated with decay by a brown-rot fungus. *Forest Products Journal* 4(6):410-416.

Noguchi, M., K. Nishimoto, Y. Imanura, Y. Fujii, S. Okamura, T. Miyauchi. (1986). Detection of very early stages of decay in western hemlock wood using acoustic emissions. *Forest Products Journal* 36(4):35-36.

Panshin, A. J., and C. De Zeeuw. (1980). "Textbook of Wood Technology." 4th Ed. McGraw-Hill, New York.

Pellerin, R. F., R. C. DeGroot, and G. R. Esenther. (1986). Nondestructive stress wave measurements of decay and termite attack in experimental wood units. *Proceedings 5th Nondestructive Testing of Wood Symposium*, Pullman, Washington.

Piirto, D. D., and W. W. Wilcox. (1978). Critical evaluation of the pulsed- current resistance meter for detection of decay in wood. *Forest Products Journal* 28(1):52-57.

Reh, U., G. Kraepelin, and I. Lamprecht. (1986). Use of differential scanning calorimetry for structural analyses of fungally degraded wood. *Applied and Environmental Microbiology* 52(5):1101-1106.

Richards, D. B. (1954). Physical changes in decaying wood. *Journal of Forestry* 52:260-265.

Ross, R. J., and R. Pellerin. (1990). Nondestructive evaluation of wood: Past, present, and future. *Proceedings Symposium on Nondestructive Evaluation of Materials*. Annapolis, Maryland (Conference Draft, Unpublished).

Rothrock, C. W., Jr., W. R. Smith, and R. M. Lindgren. (1961). The effects of outside storage on slash pine chips in the South. *TAPPI* 44(1):65-73.

Scheffer, T. C. (1936). "Progressive Effects of *Polyporus versicolor* on the Physical and Chemical Properties of Sweet-Gum Sapwood." U.S.D.A. Technical Bulletin No. 527, Washington, D.C.

Scheffer, T. C. (1978). Bending tolerance of veneer as related to its toughness. *Forest Products Journal* 29(2):53-54.

Scheffer, T. C., T. R. C. Wilson, R. Luxford, and C. Hartley. (1941). The effect of certain heartrot fungi on the specific gravity and strength of Sitka spruce and Douglas fir. U.S.D.A. Technical Bulletin 779, Washington, D.C.

Schultz, G. (1956). Exploratory test to increase preservative penetration in spruce and aspen by mold infection. *Forest Products Journal* 6(2):77-80.

Sexton, C. M. (1988). "The Development and Use of Breaking Radius and Impact Bending Tests for Measuring Wood Strength as Caused by Basidiomycetes Isolated from Air-Seasoning Douglas fir." M.S. thesis, Oregon State University, Corvallis, Oregon.

Shigo, A. L., and P. Berry. (1975). A new tool for detecting decay associated with *Fomes annosus* in *Pinus resinosa*. *Plant Disease Reporter* 59(9): 739-742.

Shigo, A. L., and A. Shigo. (1974). "Detection of Discoloration and Decay in Living Trees and Utility Poles." U.S.D.A. Forest Service Paper NE-294, Washington, D.C.

Shigo, A. L., W. C. Shortle, and J. Ochrymowych. (1977). "Detection of Active Decay at Ground Line in Utility Poles." U.S.D.A. Forest Service General Technical Report NE-35, Washington, D.C.

Shortle, W. C. (1982). Decay Douglas fir wood: Ionization associated with resistance to a pulsed electric current. *Wood Science* 15(1):29-32.

Shortle, W. C., A. L. Shigo, and J. Ochrymowych. (1978). Patterns of resistance to a pulsed electric current in sound and decayed utility poles. *Forest Products Journal* 28(1):48-51.

Skutt, H. R., A. L. Shigo, and R. M. Lessard. (1972). Detection of discolored and decayed wood in living trees using a pulsed electric current. *Canadian Journal of Forest Research* **2**:54–56.

Smith, S. M., and R. D. Graham. (1983). Relationship between early decay and radial compression strength of Douglas fir. *Forest Products Journal* **33**(6):49–52.

Smith, S. M., and J. J. Morrell. (1986). Correcting Pilodyn pin measurement of Douglas fir for different moisture levels. *Forest Products Journal* **36**(1): 45–46.

Smith, S. M., and J. J. Morrell. (1987). Longitudinal compression as a measure of ultimate wood strength. *Forest Products Journal* **37**(5):49–53.

Smith, S. M., and J. J. Morrell. (1991). Effect of colony size on residual wood strength of Douglas fir sapwood and heartwood. Unpublished report.

Toole, E. R. (1969). Effect of decay on crushing strength. *Forest Products Journal* **19**(10):36–37.

Toole, E. R. (1971). Reduction in crushing strength and weight associated with decay by rot fungi. *Wood Science* **3**(3):172–178.

U.S.D.A. Forest Products Laboratory. (1941). "Forest Products Laboratory Toughness Testing Machine." Report No. 1308, Madison, Wisconsin.

Ward, J. C. (1986). The effect of wetwood on lumber drying times and rates: An exploratory evaluation with longitudinal gas permeability. *Wood and Fiber Science* **18**(2):288–307.

Wilcox, W. W. (1968). "Changes in Wood Microstructure through Progressive Stages of Decay." U.S. Forest Service Research Paper FPL-70, Madison, Wisconsin.

Wilcox, W. W. (1978). Review of literature on the effects of early stages of decay on wood strength. *Wood and Fiber* **9**(4):252–257.

Zabel, R. A., C. J. K. Wang, and F. C. Terracina. (1982). "The Fungal Associates, Detection, and Fumigant Control of Decay in Treated Southern Pine Poles." Final Report EL2768. Electric Power Research Institute, Palo Alto, California.

CHAPTER

11

Colonization and Microbial Interactions in Wood Decay

In this chapter we consider how decay fungi colonize wood and effectively compete with other wood-inhabiting microorganisms. An emphasis is placed on the colonization strategies of decay fungi and their associations and interactions with other wood-inhabiting microorganisms. These topics are a part of the new and expanding discipline of microbial ecology. Topics of related interest to the wood microbiologist include soil microbiology, ruminant digestion, decomposition processes, and microbial waste conversions. The broad outlines of microbial ecology are covered in current articles and textbooks on microbiology and microbial ecology (Pugh, 1980; Atlas and Bartha, 1987; Brock *et al.*, 1984; Campbell, 1983). This chapter provides a background for the next five chapters, which discuss the major decays and discolorations that can develop during processing, and some uses of wood.

First, a brief review of some of the major concepts and terms developed by ecologists to describe organismal roles and interactions in the environment.

Some Ecological Concepts and Terminology

Most organism interactions start with *competition* (interference) for space, nutrients, or energy. The organisms associated with wood decay are heterotrophs, and the competition is primarily nutritional. Wood in the living tree, during its conversion into products, and in its many product forms, represents a rich potential nutrient source or useful habitat for many microorganisms. When wood suddenly becomes available for use after tree wounding or felling or when untreated wood is used in conditions conducive to decay, extreme microbial competition for the substrate is the common rule. *Colonization* is the term used to describe the initial invasion and occupancy of wood by microorganisms. As fungi and other microorganisms invade and utilize wood components (Levy, 1975; Rayner and Todd, 1979, 1982; Rayner and Boddy, 1988a; Swift, 1977, 1984), they create a continuum of changing conditions from the wood surface inward, both radially and longitudinally. The zones of changing conditions created in the wood as a result of these microbial actions and related abiotic factors are called *niches*. A niche might be defined as the physical and biological conditions of a zone in the wood favoring the development of a species or its population. Microorganisms, by the process of natural selection, have adapted to compete for the special nutritional and ecological conditions of niches.

The number of individuals of the same species occupying a *niche* is termed the *population*. Populations of different organisms living and interacting at the same time in the many *niches* composing the *habitat* or *substrate* (e.g., fallen log, stump, pole), make up a *community*. The term substrate is preferred in the case of wood decay where the wood is primarily a nutrient source. The progressive change in species composition and frequency that occurs over time in the various communities of the habitat is termed *succession*. The concept of succession usually implies a predictable direction in communities' development. Among some higher plant and animal communities, succession may lead to rather stable assemblages called *climax communities*. In contrast, the colonization and succession events in the decomposition of wood lead to eventual destruction and, in some cases, total mineralization. Colonization and succession patterns in wood products are not well understood. They appear to be more erratic and vary considerably with the abundance of inoculum available at the time of initial exposure, the wood use, and the presence of any chemical treatment.

When a new wood substrate becomes available for microbial colonization and abiotic conditions are favorable (e.g., moisture content and temperature), severe competition for the available nutrient resource develops. Interactions among competing species range from exclusion to cooperation and coexistence. In small organisms such as bacteria, the conflict is among populations, whereas in larger organisms such as fungi, competition may be between the thalli of individuals. A complex terminology has developed to describe and categorize the many possible interactions. As early as 1879, Anton deBary (the so-called *father of plant pathology*) introduced the term *symbiosis* to describe, in a broad sense, any close interspecific association between two organisms.

Initially, symbiotic relationships were separated into those that were beneficial *(mutualistic symbiosis)* or harmful *(parasitic symbiosis)*. Classic examples of mutualistic symbiosis that received major attention were the lichens and mycorrhiza, where fungi form beneficial associations with algae and plant roots, respectively. Studies of the many interactions among the higher plants and animals, new insights into the complexities of ruminant and termite digestion, and the discovery of antibiotics greatly expanded the categories needed to describe the wide range of interactions. The term symbiosis is now narrowly defined by some microbiologists to describe primarily beneficial associations (Salle, 1961; Brock et al., 1984).

The possible first-order interactions of two microorganisms competing for the same niche in wood are illustrated in Table 11-1. It is important to realize that the effects of the participants on each other are judgmental, so controversy can occur about the grouping of an interaction.

Types of Microbial Interactions in Wood

The principal categories of microbial interaction of direct relevance to wood microbiology include the following.

1. *Mutualism*—both organisms benefit. *Amylostereum chailleti*, a decay fungus, is vectored by a horntail *(Sirex cyaneus)* during oviposition into the dying sapwood of fire-killed or defoliated balsam fir. The fungus can then invade the wood before the arrival of competing fungi. The horntail larvae benefit by ingesting fungal mycelium and acquiring the cellulase enzymes needed for softening and digestion of the wood (Kukor and Martin, 1983).
2. *Commensalism*—one is benefited and the other is unaffected. In the later decay stages, many brown-rot fungi release more simple carbon compounds than they use or store. Some secondary scavenger fungi such as *Trichoderma* spp. or *Mucor* spp. invade in the late decay stage and use the excess sugars.
3. *Amensalism* (antibiosis)—the toxic products of one organism inhibit or retard the growth of another. Prior colonization of logs by *Tricho-*

Table 11-1
An illustration of the types of first-order interactions that may take place between two microorganisms competing for the same niche during the decay of wood

		Effects on species 1 when it competes with species 2		
		No effect	Harmful	Beneficial
Effect on species 2 when it competes with species 1	No effect	0, 0[a] Neutralism	−, 0 Amensalism (antibiosis)	+, 0 Commensalism
	Harmful	−, 0 Amensalism (antibiosis)	−, − Competition	−, + Predation and parasitism
	Beneficial	+, 0 Commensalism	+, − Predation or parasitism	+, + Mutualism

[a]Species 1, left, and species 2, right.

derma spp. has been shown to inhibit subsequent basidiomycete colonization (Hulme and Shields, 1975). A microfungus, *Scytalidium* sp., commonly present in seasoning Douglas fir utility poles, is reported to delay the colonization of *Antrodia carbonica* by an antibiotic (Ricard and Bollen, 1968). This antibiotic association has been proposed as a potential biological-control approach to reduce decay development in utility poles (Ricard, 1976; Morris et al., 1984; Bruce and King, 1986). Some wood-inhabiting bacteria have been shown to produce toxic metabolites that inhibit the growth of white-, brown-, and soft-rot fungi in laboratory tests (Preston et al., 1982).

4. *Parasitism*—one organism obtains its nutrition directly from another, and the relationship is generally harmful to one. Some fungi are pathogens of other fungi and insects. Some white-rot fungi attack and utilize wood-inhabiting fungi, apparently as a nitrogen source (Rayner et al., 1987). Such fungi are called *mycopathogens* and are of special interest as possible biological-control agents.

5. *Predation*—one organism physically captures and consumes part or all of another. Usually the predator is larger than the prey, but some fungi trap nematodes using hyphal coils and digest them as a food source (Thorn and Barron, 1984; Stirling, 1988).

6. *Neutralism*—the organisms occur together with no discernible effect on each other. Bacteria are commonly associated with the gelatinous coating on the outer hyphal wall of some soft-rot fungi. It has been suggested that these bacteria may interfere with the feed-back inhibition of cellulase and accelerate decay.

There are many other interesting types of interaction between organisms. *Synergism* occurs when the combined effect (growth, decay rate, etc.) of both organisms is greater than either alone. The combination of some yeasts and decay fungi has been shown to greatly accelerate decay rates (Blanchette and Shaw, 1978). *Co-metabolism* occurs when one organism modifies a complex substrate, such as a paint polymer, permitting its subsequent digestion by another. Neither organism can completely utilize the substrate alone.

Common Wood Inhabitants during Decay

Whereas fungi play a dominant role in decay development, many other microorganisms are often present and may affect the fungi. They include the bacteria, protozoa, algae, and small animals such as nematodes, mites, and insects. Their roles in decay development range from significant to minor, and, in many cases, are unknown. Insects are important vectors of many stain and decay fungi. Some bark beetles carry nitrogen-fixing bacteria, which can enhance the nutritional quality of the wood (Bridges, 1981). Others harbor fungi that grow in tunnel walls and are consumed by grazing larvae. By tunneling and comminution, insects and other microfauna increase both the aeration and accessibility of the wood to other microorganisms. These roles are common in the later decay stages of logs and stumps in the forest.

Bacteria are ubiquitous in decaying wood and an important food source for protozoa and nematodes. Predator–prey relationships occur among the microfauna in the later decay stages. It is common to find the mycorrhizal roots of forest trees growing inside decayed wood on the forest floor. Decayed wood is a favorable site for nitrogen fixation by bacteria and an important nitrogen source for the invading mycorrhizal feeder roots (Larsen *et al.*, 1982).

Microecology of Wood Decay

It is clear that many organisms directly and indirectly interact during wood decay. However, the microecological aspects of wood decay are poorly understood, and their study has lagged far behind ecological studies of higher plant and animal systems. A major reason is the difficulty of measuring decay development and the microorganisms involved in a nondestructive fashion. The decay process occurs over periods often measured in years in an essentially opaque material that usually must be destructively sampled for analysis. Early studies of the sequences of fungi associated with decay relied primarily on the presence of basidiomata to determine identities and changes in species composition (Spaulding and Hansborough, 1944). Such studies tended to favor the Basidiomycotina, which were better known taxonomically at the time, and neglected less visible organisms, particularly the Deuteromycotina. The appearances of fruiting bodies can simply reflect differences in maturation times and have little or no relationship to when the fungi colonized the material or their impact on the substrate.

The use of cultural characteristics (Campbell, 1938; Davidson *et al.*, 1938; Nobles, 1948; Stalpers, 1978) was an effective way to identify isolated fungi from their vegetative or cultural characteristics. This was an important step in setting the stage for progress in fungal ecology.

Further developments in systematic sampling (Butcher, 1971; Sharp, 1974; Sharp and Levy, 1973) and improved isolation procedures, and advances in cultural taxonomic data have improved our ability to define the identities (or taxonomic descriptions) and relative abundance of the organisms associated with stages of the decay process. Laboratory tests now can determine the decay capability and type, as well as the antibiotic, or mycoparasitic abilities of the isolates. Nevertheless, determining the relative role and importance of individual species in the decay process remains difficult because of the complexity of the system. There is always the underlying question of whether the laboratory results, often from axenic tests, are relevant to decay events in the natural system.

Colonization Strategies

A useful approach for understanding the roles and interactions of microorganisms during wood decay is to consider the characteristics or strategies used to occupy a niche, retain it, or in some cases, capture it from a competing microorganism.

The major microorganism types present at various stages of decay development and their roles can often be traced to nutritional capabilities. Some microorganisms can utilize only the relatively simple carbon compounds that occur in untreated sound wood. Some can utilize hemicelluloses and cellulose but not lignified cell-wall material. Some, such as the white-rot fungi, utilize all cell-wall components.

Rayner and Boddy (1988b) have proposed three roles of wood-inhabiting fungi to explain their appearances and sequences in colonization. These types are listed and discussed with examples as follows.

Opportunists

These fungi are scavengers and appear to have adapted to rapidly exploit a new wood substrate (e.g., broken branches, felled trees, freshly sawed boards, checks in seasoning poles, untreated sapwood). They are analogous to the pioneer species in the colonization of disturbed soil sites by plants. Opportunists are characterized by rapid growth rates and prolific production of airborne spores. A few examples are (1) the molds and sapstain fungi that rapidly invade freshly sawed lumber; (2) the bacteria and microfungi that initially colonize stem wounds; and (3) some decay fungi in wounded stems, slash, stumps, and wood products (e.g., *Schizophyllum commune, Sterum sanguinolentum, Irpex lacteus,* and *Phanerochaete (Peniophora) gigantea.*

Stress Resistors

These fungi have adapted to adverse growth conditions, such as the presence of toxicants or allopathic chemicals, low aeration levels, minimal moisture levels, and high temperatures. A few examples are (1) the molds and soft-rot fungi that tolerate high levels of some toxicants; (2) the heart-rot fungi that tolerate the aeration and restrictive environment of some living stems; (3) the thermotolerant and thermophilic fungi and bacteria that survive in wood-chip piles; (4) the heart-rot fungi with cryptic invasion strategies that may remain dormant for years in the sapwood of small dead branches; and (5) the bacteria that invade ponded logs where low oxygen levels limit fungi.

Combatants

These fungi are able to repel invaders or take the niches of other fungi. A few examples are (1) fungi that produce antibiotics that may repel or assist in invasions (Etheridge, 1961); and (2) fungi that are mycopathogens and parasitize other wood inhabitants (Barnett, 1964). *Lenzites betulina* parasitizes and replaces some *Coriolus* sp. (Rayner *et al.*, 1987).

Many wood-inhabiting microorganisms overlap these categories, or their strategies are as yet unknown. However, these categories are a major step toward understanding the appearances, sequences, and interactions of microorganisms involved in wood colonization and decay development.

Wood-colonization patterns vary with wood types, uses, and exposure conditions. Common colonization patterns are discussed in the next section.

Colonization Patterns of Wood by Fungi

A variety of microbial sequences are reported for various wood products or wood under various exposures. These studies are related primarily to the salvage of trees killed by disease, insects, fire, or windthrow, the storage of pulpwood and logs, the decomposition rates of slash and branches under various silvicultural practices, branch stubs, and the service lives of treated posts and poles. Shigo assembled a comprehensive review of the early literature on the types and sequences of microorganisms occurring during the discoloration and decay of wood, with emphasis on wounded living trees and diseased or dead standing trees (Shigo, 1967). More recently, Käärik (1974, 1975) has summarized the research on fungal successions in various wood substrata with emphasis on slash, logs, stumps, or posts in soil contact.

These studies are now summarized for a range of wood uses and conditions. The topic is discussed further in the subsequent chapters on stem decays, wood storage problems, sapstains, and decays in the major wood products.

Standing Tree

Wood in the living tree represents a hostile environment for most microorganisms, but when branches and roots die or wounds occur, some organisms invade the roots, stem, or branches (Shigo, 1972; Manion, 1975; Good and Nelson, 1962). Heart-rot fungi are believed to invade the central nonliving portions of the stem through wounds or dead branch stubs. These fungi are examples of stress-tolerant organisms. Sapwood decays originate from wounds that destroy the protective bark and reduce the moisture content of the exposed sapwood. Extensive study of microbial sequences in stem wounds and decay development have been made by Shigo and associates (Shigo, 1967, 1972) (see Chapter 12). Many decays in the outer sapwood become heart rots when enveloped later by growth in stem girth. A progression from bacteria to microfungi to decay fungi represents the proposed succession of microorganisms invading stem wounds of living trees.

Standing, Dying, or Dead Trees

Large volumes of standing timber become susceptible to decay when defoliated and killed by insects such as the gypsy moth or spruce budworm. In studies to determine feasible salvage schedules, the sequences of fungi that invade the stems have been determined. Basham (1959) proposed a sequence of microfungi, a weak decayer *Amylostereum chailletii*, and finally a severe decayer, *Hirchioporus abietineus* in spruce budworm-killed balsam fir in eastern Canada. Warren (1989) has recently shown that the appearances of these two decayers may be independent, with *A. chailletii* vectored by a Sirex wood wasp and the *H. abietineus* invasion related to subsequent severe bark beetle damage to the stem.

Seasoning Lumber

Freshly sawn and stacked, untreated lumber represents a rich nutrient source that is suddenly unprotected and available to the first pioneer microorganisms that can invade and exploit the storage materials in the parenchyma cells. Sapstain, a serious source of lumber degrade for many natural-finish wood uses, can develop within a week of the sawing in the absence of proper piling and protective dip treatments. A few stain fungi such as *Ophiostoma (Ceratocystis) picea* may eventually also produce soft rot after prolonged exposure. If the lumber is excessively wet, molds in genera such as *Rhizopus, Penicillium, Aspergillus,* or *Trichoderma* are favored. When lumber drying is restricted or in cases of bulk piling, the stain fungi may be followed by brown-rot decayers such as *Postia placenta*. This often-observed sequence of stainers followed by decayers may not reflect a colonization pattern, but rather differing growth and mycelial development rates. Sapstains are discussed later in Chapter 14.

Untreated Wood in Ground Contact

Merrill and French (1966) reported the major initial colonizers of buried wood to be the microfungi *Fusarium solani* and *Trichoderma viride*, followed by the appearance of soft rot within 6 weeks and pockets of brown rot after 10 weeks. *Sistotrema (Trechiospora) brinkmannii*, a brown-rot fungus, was isolated after 12 weeks. The generalized sequence in this and related studies suggests an initial invasion by some microfungi acting as aggressive pioneers, followed by soft-rotters and later Basidiomycotina decayers.

In the development of decay in untreated wood posts (*Pinus* and *Betula*) in ground contact, Corbett and Levy (1963) reported a similar colonization pattern of the Moniliales group (*Trichoderma* spp., *Penicillium* spp., etc.), Sphaeropsidales (including some soft-rot fungi), and Basidiomycotina.

In 1968, Butcher studied fungal successions over a 13-month exposure period in *Pinus radiata* sapwood stakes in New Zealand. He reported a successional sequence of blue stain to molds above ground and a pattern similar to that shown by Merrill and French (1966) and Corbett and Levy (1963) below the groundline.

In studies of fungal colonization of pine and spruce poles in Sweden over a 4-year period, Käärik (1968) found that fungal isolations varied substantially with soil type and with position below and above the ground zone of the poles.

Many different fungal species were isolated from many zones in the poles, suggesting random initial invasions and severe competition to determine survivors in the succession. In agreement with the other studies, the major decay fungi appeared in wood already occupied by the nondecay fungi and competed effectively with them.

In a 10-year study of the succession of microorganisms into posts of birch and Scot's pine in England, Banerjee and Levy (1971) reported the fungal sequences at the soil line to be sapstain fungi with soft-rot characteristics, followed by Basidiomycotina.

Treated Wood in Ground Contact

The presence of preservative substantially slows and alters microbial colonization and tends to select more chemically tolerant species (Greaves and Savory, 1965; Henningsson and Nilsson, 1976; Nilsson and Henningsson, 1978; Sorkhoh and Dickinson, 1975).

In studies on the sequences and prevalence of fungi isolated from a series of preservative-treated poles representing a wide range of service ages, it was concluded that the probable invasion sequence of fungi at the groundline began with microfungi and soft-rotters followed by primarily white-rot fungi. The incidence of microfungi increased in the older pole age groups. The initial invading fungi were tolerant to wood preservatives, and some tolerant fungi were associated with specific preservative treatments, e.g., *Hermocomis (Cladosporium) resinae* from creosote-treated poles and *Rhinocladiella atrovirens* from penta-treated poles (Zabel et al., 1985).

Other Wood Conditions

Whereas there are numerous reports concerning the major decay fungi associated with many important wood products, none that we know of has studied the sequences of fungi that invade treated wood in its many aboveground uses. Studies on microorganism sequences in chip piles involving bacteria and thermotolerant fungi are discussed in Chapter 13, "Wood Storage Losses."

Succession in Wood Decay

Microbial succession during the decay process, although often postulated, is difficult to unequivocally establish owing to the uncertainty of reliably isolating all the principal microorganisms involved. Important organisms may be missed because of the isolation procedures, types of selective media used, the absence of a critical associate, or even the conditions employed for incubation of the isolated samples. The thallus character and indeterminate growth pattern of many decay fungi present difficulties in comparative enumeration. Prevalence of a single species may simply reflect greater reproductive potential. Isolation frequency can be very misleading, since a single fungus mycelium may dominate several cubic feet of decaying wood, yet appear minor if compared numerically to the frequent isolates of prolific spore-producing scavenger fungi or bacteria. The difficulty of properly identifying each isolate and the tendency to study the roles of the known have also stymied ecological studies of the decay process. The wood-inhabiting bacteria and Deuteromycotina genera are often difficult to identify and hence have been subject to notorious neglect. Important organisms in the various decay processes of wood may still lurk in the unidentified isolates of many earlier studies. As a brief aside, these are some of the reasons that the microbial ecology of wood-inhabiting microorganisms is still largely in a descriptive stage, compared to the highly developed quantitative and statistical approaches to population dynamics and understandings of succession and climax communities in the higher plants and animals.

Successions of microorganisms involved in the decay of various wood materials are exceedingly complex and variable events. It is probable that chance inocula, substrate changes by prior inhabitants, and changes in abiotic conditions, such as wood aeration or moisture, all influence the decay process.

There are several problems with using the term succession to describe the sequences of microorganisms observed in the initial decay of wood. First, as mentioned earlier, *succession* as used in higher plant and animal ecology implies the steady-state condition of communities at the end of a predictable series of changes. With few exceptions, most claims of succession in the wood-decay process start with sound wood and end with the appearance of decay, or when the practical use of the wood product under study is over. Thus, the process being studied is really early colonization. The continuing later stages of decay are rarely studied in either wood products or wood under natural conditions such as decaying logs. Observations of advanced decay in the forest floor or structures indicate that the microfauna play a significant comminuting role. Studies of soil also show that humus development is traceable to recalcitrant lignin and microbial residues that may remain in soil for decades and represent significant carbon reservoirs in the biosphere. Bacteria probably play a key role in the final decomposition of the lignin residues originating from wood. Complete mineralization of wood represents the end of the decay process from the ecologic viewpoint. Second, there is uncertainty about the actual causes of the changes in fungal communities reported in the early stages of decay development. The successional concept implies that sequences of communities modify the substrate to more favorable conditions for subsequent communities. Uncertainty exists about whether the community changes observed in early decay sequences are the cause or simply a direct reflection of changing abiotic conditions under seasonal control, such as moisture content, temperature, or even the phenology of the microorganisms involved.

Research Needs on Ecology of Decay Fungi

Our present limited understanding of organism types, roles, interactions, and sequences (succession) in the decay process of major wood materials and products stresses the need for a new emphasis in research on the ecology of decay fungi. On a broader base, an increased role by mycologists in ecological research and theory development is important to understand the decomposer role of fungi in the biosphere.

There are also immediate and practical applications for information on the sequences and interactions of the microorganisms involved in decay development. Such information is useful to the wood microbiologist and wood-products engineers concerned with controlling decay. Important questions to the wood microbiologist about the development of decay in various wood forms or products include the identities and numbers of the microorganisms involved *(who)*; their time of appearance *(when)*; locations *(where)*; and roles *(why)*.

Answers to these questions can lead to improved decay-prevention practices for some wood uses or suggest decay-vulnerable locations or periods in other wood uses.

Summary

- *Colonization* is the term used to describe the invasion and occupancy of wood by microorganisms.
- Wood-inhabiting microorganisms have adapted to compete for the various nutritional and ecological zones in the wood, termed *niches*.
- When a new wood substrate becomes available and the growth conditions are favorable, severe competition develops among wood-inhabiting microorganisms for the available nutrients.
- Microbial interactions between competing wood-inhabiting microorganisms include *mutualism* (both benefit), *commensalism* (one benefits; one is unaffected), *amensalism* (one poisons or chemically repels the other), *parasitism* (a harmful relationship in which one obtains nutrients from the other), *predation* (one captures and consumes another), and *neutralism* (no discernible effects).
- Common wood inhabitants during decay development, in addition to fungi, are bacteria, protozoa, algae, and small animals such as nematodes, mites, and insects.
- Colonization strategies are a useful way to explain the appearances and sequences of microorganisms that invade wood. *Opportunists* are the pioneers (scavengers) that have adapted to rapidly exploit a new wood substrate; *stress-resistors* are able to withstand adverse growth conditions or tolerate poisons and allopathic chemicals; *combatants*

Summary continues

- are able to repel invaders or capture the niches of other fungi with antibiotics or by acting as mycopathogens.
- The sequences of microorganisms often reported in the colonization of untreated and treated wood in ground contact are microfungi, soft-rot fungi, and Basidiomycotina decayers.
- Microbial succession during the decay process, though often claimed, is difficult to prove because of the uncertainty of reliably isolating, identifying, quantifying, and determining the interactive roles of the many microorganisms often involved.
- Additional research on the microbial ecology of the decay process may lead to future effective biological controls of decay and provide useful information to the wood microbiologist about *when* and *where* various wood products are most vulnerable to attack by destructive biotic agents.

References

Atlas, R. M., and R. Bartha. (1987). "Microbial Ecology Fundamentals and Applications," 2nd Ed. Benjamin/Cummings, Menlo Park, California.

Banerjee, A. K., and J. F. Levy. (1971). Fungal succession in wooden fence posts. *Material und Organismen* 6:1-25.

Barnett, H. L. (1964). Mycoparasitism. *Mycologia* 56:1-19.

Basham, J. T. (1959). Studies in forest pathology II. Investigations of the pathological deterioration of killed balsam fir. *Canadian Journal of Botany* 39:291-326.

Blanchette, R. A., and C. G. Shaw. (1978). Associations among bacteria, yeasts, and basidiomycetes during wood decay. *Phytopathology* 68:631-637.

Bridges, J. R. (1981). Nitrogen-fixing bacteria associated with bark beetles. *Microbial Ecology* 7:131-137.

Brock, T. D., D. W. Smith, and M. T. Madigan. (1984). "Biology of Microorganisms," 4 Ed. Prentice-Hall, Englewood Cliffs, New Jersey.

Bruce, A., and B. King. (1986). Biological control of decay in creosote-treated distribution poles. II. Control of decay in poles by immunizing commensal fungi. *Material und Organismen* 21:165-179.

Butcher, J. A. (1968). The ecology of fungi infected untreated sapwood of *Pinus radiata*. *Canadian Journal of Botany* 46:1577-1589.

Butcher, J. A. (1971). Techniques for the analysis of fungal floras in wood. *Material und Organismen* 6:209-232.

Campbell, R. (1983). "Microbial Ecology," 2nd Ed. Blackwell Scientific, Oxford, England.

Campbell, W. A. (1938). The cultural characteristics of the species of *Fomes*. *Bulletin of the Torrey Botanical Club* 65:31-69.

Corbett, N. H., and J. F. Levy. (1963). Ecological studies on fungi associated with wooden fence posts. *Nature (London)* 198:1322-1323.

Davidson, R. W., W. A. Campbell, and D. J. Blaisdell. (1938). Differentiation of wood-decaying fungi by their reactions on gallic or tannic acid medium. *Journal of Agricultural Research* **57**:683-695.

Etheridge, D. E. (1961). Factors affecting branch infection in aspen. *Canadian Journal of Botany* **39**:799-816.

Good, H. M., and C. D. Nelson. (1962). Fungi associated with *Fomes igniarius* var. *populinus* in living poplar trees and their probable significance in decay. *Canadian Journal of Botany* **40**:615-624.

Greaves, H., and Savory, J. G. (1965). Studies of the microfungi attacking preservative-treated timber, with particular reference to methods for their isolation. *Journal of the Institute for Wood Science* **15**(1965):45-50.

Henningsson, B., and Nilsson, T. (1976). Some aspects on microflora and the decomposition of preservative-treated wood in ground contact. Organismen and Holz. Internationales Symposium Berlin-Dahlem (1975) *Material und Organismen* **3**: 307-318.

Hulme, M. A., and Shields, J. K. (1975). Antagonistic and synergistic effects for biological control of decay. In "Biological Transformation of Wood by Microorganisms" (W. Liese, ed.), pp. 52-63. Springer-Verlag, New York.

Käärik, A. (1968). Colonization of pine and spruce poles by soil fungi after twelve and eighteen months. *Material und Organismen* **3**:185-198.

Käärik, A. (1974). Decomposition of wood. In "Biology of Plant Litter Decomposition," Vol I. (H. Dickinson and G. J. Pugh, eds.), pp. 129-174. Academic Press, London.

Käärik, A. (1975). Successions of microorganisms during wood decay. In "Biological Transformation of Wood by Microorganisms" (W. Liese, ed.), pp. 39-51. Springer-Verlag, New York.

Kukor, J. J., and M. M. Martin. (1983). Acquisition of digestive enzymes by siricid wood wasps from their fungal symbiont. *Science* **220**:1161-1163.

Larsen, M. J., M. F. Jurgensen, and A. E. Harvey. (1982). N_2 fixation in brown-rotted soil wood in an intermountain cedar-hemlock ecosystem. *Forest Science* (2)**28**: 292-296.

Levy, J. F. (1975). Colonization of wood by fungi. In "Biological Transformation of Wood by Microorganisms" (W. Liese, ed.), pp. 16-23. Springer-Verlag, New York.

Manion, P. D. (1975). Two infection sites of *Hypoxylon mammatum* in trembling aspen (*Populus tremuloides*). *Canadian Journal of Botany* **53**(22):2621-2624.

Merrill, W., and D. W. French. (1966). Colonization of wood by soil fungi. *Phytopathology* **56**:301-303.

Morris, P. I., D. J. Dickinson, and J. F. Levy. (1984). The nature and control of decay in creosoted electricity poles. *British Wood Preservers' Association Annual Convention,* 42-53. Cambridge.

Nilsson, T., and Henningsson, B. (1978). *Phialophora* species occurring in preservative-treated wood in ground contact. *Material und Organismen* **13**(40):297-313.

Nobles, M. K. (1948). Studies in forest pathology. VI. Identification of cultures of wood-rotting fungi. *Canadian Journal of Research C, Botanical Science* **26**:281-431.

Preston, A. F., F. H. Erbisch, K. R. Kramm, and A. E. Lund. (1982). Developments in the use of biological control for wood preservation. *Proceedings of the American Wood Preservers' Association* **78**:53-61.

Pugh, G. J. F. (1980). Strategies in fungal ecology. *Transactions of the British Mycological Society* **75**:1-14.

Rayner, A. D. M., and L. Boddy. (1988a). Fungal communities in the decay of wood. *In* "Advances in Microbiological Ecology," Vol. 10. (K. C. Marshall, ed.), pp. 115-166. Plenum Press, New York.
Rayner, A. D. M., and L. Boddy. (1988b). "Fungal Decomposition of Wood. Its Biology and Ecology." John Wiley & Sons, New York.
Rayner, A. D. M., and Todd, N. K. (1979). Population and community structure and dynamics of fungi in decaying wood. *Advances in Botanical Research* **7**:333-420.
Rayner, A. D. M., and Todd, N. K. (1982). Population structure in wood-decomposing basidiomycetes. *In* "Decomposer Basidiomycetes: Their Biology and Ecology" (J. C. Frankland, J. N. Hedger, and M. J. Swift, eds.), pp. 109-128. Cambridge University Press, Cambridge, England.
Rayner, A. D. M., L. Boddy, and G. G. Dowson. (1987). Temporary parasitism of *Coriolus* spp. by *Lenzites betulina*: A strategy for domain capture in wood decay fungi. *FEMS Microbiological Ecology* **45**:53-58.
Ricard, J. L., and W. B. Bollen. (1968). Inhibition of *Poria carbonica* by *Scytalidium* sp., an imperfect fungi isolated from Douglas fir poles. *Canadian Journal of Botany* **46**:643-647.
Ricard, J. L. (1976). Biological control of decay in standing creosote-treated poles. *Journal of the Institute of Wood Science* **7**:6-9.
Salle, A. J. (1961). "Fundamental Principles of Bacteriology." McGraw-Hill, New York.
Sharp, R. F. (1974). Some nitrogen considerations of wood ecology and preservation. *Canadian Journal of Microbiology* **20**:321-328.
Sharp, R. F., and J. F. Levy. (1973). The isolation and ecology of some wood-colonizing microfungi using a perfusion culture technique. *Material und Organismen* **8**(3): 189-213.
Shigo, A. L. (1967). Successions of organisms in discoloration and decay of wood. *International Review of Forest Research* **2**:237-299.
Shigo, A. L. (1972). Successions of microorganisms and patterns of discoloration and decay after wounding in red oak and white oak. *Phytopathology* **62**(2):256-259.
Sorkhoh, N. A., and D. J. Dickinson. (1975). The effects of wood preservative on the colonization and decay of wood by microorganisms. Organism und Holz. Internationales Symposium Berlin-Dahlen 1975. *Material und Organismen* **3**:287-293.
Spaulding, P., and Hansbrough, J. R. (1944). "Decay of Logging Slash in the Northeast." Technical Bulletin 876, U.S.D.A., Washington, D.C.
Stalpers, J. A. (1978). Identification of wood-inhabiting Aphyllophorales in pure culture. Central-bureau voor Schimmelcultures, Baarn, *Studies in Mycology* **16**:1-248.
Stirling, G. R. (1988). Prospects for the use of fungi in nematode control. *In* "Fungi in Biological Control Systems" (M. N. Burge, ed.), pp. 188-210. Manchester University Press, New York.
Swift, M. J. (1977). The ecology of wood decomposition. *Scientific Progress Oxford* **64**:175-199.
Swift, M. J. (1984). Microbial diversity and decomposer niches. *In* "Current Perspectives in Microbial Ecology" (M. J. Klug and C. A. Reddy, eds.), pp. 8-16. American Society for Microbiology.
Thorn, R. G., and G. L. Barron. (1984). Carnivorous mushrooms. *Science* **224**:76-78.
Warren, G. R. (1989). "The Identities, Roles, and Interactions of Tree Fungi Associated with Stem Biodeterioration of Balsam Fir Stressed and Killed after Spruce

Budworm Defoliation in Newfoundland." Ph.D. thesis, State University of New York, College of Environmental Science and Forestry, Syracuse, New York.

Zabel, R. A., F. F. Lombard, C. J. K. Wang, and F. Terracina. (1985). Fungi associated with decay in treated southern pine utility poles in the eastern United States. *Wood and Fiber Science* **17**(1):75–91.

PART 3

CHAPTER
12

Decays Originating in the Stems of Living Trees

Whereas this textbook is primarily about the decays of wood products, a general understanding of the features and origins of the many decays that develop in living trees is important to wood users, wood scientists, and wood engineers for several reasons.

We emphasize, initially, that some of the decays and discolorations that appear commonly in lumber and other wood products in use have their origins in living stems. Undetected incipient decay is a reason for some of the high variability experienced in some wood uses. It is particularly important to detect such decay in roundwood and lumber destined for critical-strength uses such as poles, piling, and ladder rails as early as possible. The physiological and structural responses in the living stem following injury and subsequent decay development can reduce preservative treatability and increase decay susceptibility in products. Information on the mechanisms by which trees resist invading wood-decay fungi may yield clues toward more effective preservatives and decay-prevention practices during the storage, conversion, and use of wood.

Descriptions and general information on the major decays of standing timber are available in various timber-decay and forest pathology textbooks (Hubert, 1931; Cartwright and Findlay, 1958; Boyce, 1961; Sinclair et al., 1987; Manion, 1991). Information on their identities and prevalence in the major timber species in North America has been compiled by Hepting (1971). Descriptions of most of the causal fungi are available in a taxonomic study by Gilbertson and Ryvarden (1986). Monographs on heart rots in living trees have been prepared by Wagener and Davidson (1954) and Shigo and Hillis (1973).

In this chapter, a special emphasis will be placed on types, origins, and developmental patterns of stem decays that are important for the detection and understanding of their various effects on many wood-use properties.

Historical Highlights

As indicated in Chapter 1, forest pathology had its origin in the study of stem decays. In 1884, R. Hartig of Germany first showed that a specific fungus caused a specific decay and a predictable amount of cull. Forest pathology research began in the United States around the 1900s in response to growing fears of a timber famine and deep concerns about extensive decay damage developing in railways and other wood construction (Williams, 1989). For these reasons the study of stem decays was a major concern of the early forest pathologists (Weir and Hubert, 1918; Boyce, 1920, 1932). An emphasis was placed on estimating the extent of internal decay from the locations of external indicators such as basidiomata and decayed knots (Hubert, 1924). Cull-detection manuals were developed for most of the major timber-producing regions, which provided foresters with information on the principal decay indicators and associated cull (Murphy and Rowatt, 1932; Hepting and Hedgcock, 1937; Silverborg, 1954; Lockhard et al., 1950; Roth, 1959; Shigo and Larson, 1969; Kimmey and Hornebrook, 1952; Partridge and Miller, 1974).

Other important studies were the determination of the rates of decay development and pathologic rotations (when decay development begins to exceed annual growth) for the important timber species. The effects of various important decays on wood uses and strength properties were determined for some principal timber species (Scheffer et al., 1941; Stillinger, 1951; Atwell, 1948). Extensive studies demonstrated the close association between fire scars and decay development in southern hardwoods (Hepting, 1935).

More recently, stem-decay studies have focused on the infection process (Haddow, 1938; Etheridge and Craig, 1976; Manion, 1967; Manion and French, 1967; Merrill, 1970). Contributions toward understanding the types and sequences of fungi involved in decay initiation have been made by Davidson and Redmond (1957); Basham (1958); Good (1959); Shigo (1967, 1976, 1984); and recently Rayner and Boddy (1988) and Warren (1989).

Prevention practices for stem decays have been indirect and largely associated with silvicultural operations because of cost limitations. Some of the measures recommended have been sanitation cuttings, early branch pruning (before heartwood forms), wound-prevention practices, fire control, and silvicultural treatments favoring rapid growth and the natural pruning of small branches.

Stem-Decay Types

There are some important differences between the decays that develop in living stems (pathogenic decays) and wood products (saprobic decays). In the living stem, decay development is a disease, and the tree often is able to respond to the initial infection and defend itself. Some tree (host) responses provide useful evidences (symptoms) of the internal decay. In the living tree, the outer sapwood is resistant to decay, and the heartwood less so, whereas heartwood in the wood product is often resistant, and the sapwood is universally susceptible.

Stem decays can be grouped conveniently into *heart rots* and *sap rots* based on the principal stem zones invaded and destroyed. Heart rots are the predominant type of stem decay in mature forests and occupy primarily the physiologically inactive central portion of the stem (heartwood). The sap rots develop in the outer sapwood generally after injury and death of localized bark and xylem tissues.

These positional distinctions are not clear cut and may intergrade in the later stages of decay development. We shall see later that many heart rots in the central zone of stems originated as localized sap rots around wounds and were later enveloped by the radial growth of the tree. Some heart-rot fungi such as *Phellinus (Fomes) pini* and *Inonotus (Poria) obliquus* may invade and kill the outer living sapwood and form cankers.

Stem decays can be also subdivided into those that occupy the top, central, or basal (butt) portions of the tree trunk. Some stem decays are localized, whereas others may occupy entire heartwood zones from the basal roots to the larger branches. Although both brown and white rots are important in standing timber, most are white rots (Gilbertson, 1980).

Stem-Decay Origins

The manner in which decay fungi enter xylem tissues in living stems still remains a perplexing and intriguing question, difficult to study and answer because of the long infection periods, the complex growth patterns of trees, and the many microorganisms involved. The anatomical nature of bark and the high-moisture and low-aeration levels of outer sapwood tissues in uninjured trees are effective barriers to the entry and growth of many decay fungi. During the natural shedding of branches (self-pruning), most dying tissues are effectively sealed form inner living tissues, anatomically, by callus formation and chemically, by infiltration with resins and gums.

Heart Rots

Early forest pathologists (beginning with Hartig in the late 1800s) assumed that all heart-rot infections occurred through the exposed heartwood in branch stubs. This belief, which dominated forest pathology for nearly 50 years, was based on the observed continuity of the central decay columns with heartwood zones in the branch stubs. It was believed that decay fungi were saprobes and could grow only in dead heartwood tissues. The assumptions were made that the basidiospores of decay fungi germinated on the surface of dead branch stubs, grew into the exposed heartwood, and eventually invaded the heartwood zone of the central stem. No experimental evidence to date has verified this infection route; however, evidence that has accumulated slowly in the past few decades indicates that heart-rot fungi enter trees in other ways. In 1938, Haddow showed that *Phellinus (Fomes) pini* on eastern white pine often infected small branches or twigs at the base of pole-sized trees or weeviled tips in larger trees (Fig. 12-3D). From these sapwood zones, decay development could be traced to the central stem. His research clearly established that sapwood injury and small branches (sapwood) were important infection courts for this major heart-rot fungus of conifers, worldwide. Studies of *Echinodontium tinctorium* on western hemlock also showed that small sapwood twigs and branches on the lower stem were the initial infection courts (Etheridge and Craig, 1976). The fungus entered the base of twigs, and remained dormant for years in ray tissues, until the branch tissues were enveloped by the growing stem. Subsequent stem injury or tree stressing appeared to initiate decay development. More recently, extensive studies of stem decays in northeastern hardwoods have shown that many of the apparent heart rots and central discolored zones have their origins as wound-initiated sapwood decays that were compartmentalized and entrapped in heartwood after subsequent stem growth (Shigo and Marx, 1977). Some root-rot pathogens, such as *Heterobasidion (Fomes) annosum*, *Armillaria mellea*, and *Phellinus (Poria) weirii*, are also able to invade the heartwood in the basal section of older trees from roots.

Whereas dead branch stubs are still often posed as probable infection courts for many heart-rot fungi, the evidence is primarily circumstantial, and proof is still lacking. It is clear that small dead branches, sapwood wounds, broken tops, and roots are entry points for many major heart-rot fungi.

Sap Rots

The principal origin of sap rots in the living stem is wounds that expose and kill xylem tissues. Many sap rots in young vigorous stems are localized. Others are transitory and, when contained by compartmentalization, become heart rots when enveloped later by the expanding heartwood. Other sap rots are traceable to drastic injury to stems or injuries on trees of low vigor, where the callus tissue can not enclose the damaged zone, and decay slowly expands in the dead sapwood tissues. Some sap rots develop in the stems of highly stressed and moribund trees when bark tissues die. Localized sap rots are also commonly associated with, or caused by, some stem cankers (e.g., *Eutypella*, *Strumella*, and *Hypoxylon* cankers). Sap rots are a common cause of stem breakage in the forest, since the outer decayed zone is located where bending stresses during storms are at a maximum.

Types of Stem Wounds

Injuries to the bark and outer sapwood are common and occur throughout the life of most trees. Wounds that expose only sapwood contain a boundary of living tissues that may react to the injury, and when small, callus closure and containment are common. Larger wounds or those that expose heartwood close more slowly, and the chances of containment of the infection are much reduced. Some examples of common stem injuries illustrate these wound types.

Injuries that expose heartwood include large, dead branches and branch stubs, broken tops from windstorms, pruning wounds, and large, broken branches or roots from windstorms.

Injuries that expose sapwood arise from fire scars; logging and related mechanical damage; frost cracks; animal injury (i.e., rodent gnawing, bird pecks, deer and elk damage); insect tunnels; sunscald; lightning strikes; bark cankers; bark tears; and abrasions caused by falling trees or broken branches during windstorms.

Stem Tissue Reactions to Wounding

The general observation that stem wounds are common, yet many trees live for centuries, indicates that most timber species have evolved effective defense mechanisms to limit decay. Small sapwood injuries often heal rapidly, particularly in young vigorous trees. Wounding stimulates the production of growth hormones, which in turn, stimulate the formation of callus cells from the cambium at the outer margin of the injury. The callus or wound parenchyma form new phloem and xylem tissues that gradually cover the wound surface with healthy tissue. Tissues internal to the damage also react to isolate the dead and damaged tissues. The isolation of diseased tissue by histological and chemical defenses is a common method many plants have developed to defend themselves against pathogens (Agrios, 1988). Major changes include the formation of tyloses and deposition of resins, gums, and other toxicants (phytoalexins) in the living cells surrounding the injury (Shain, 1979). The term *compartmentalization* has been used by Shigo (1984) to describe this process in injured tree stems and will be discussed in the next section.

Another effective defense developed by some tree species is the infiltration of dying tissues with toxicants. Examples are the infiltration of branch stubs in many conifers and hardwoods with resins or gums and the deposition of highly toxic extractives in the heartwood of some timber species, which creates a hostile ecological niche for most decay fungi. Characteristics of these extractives will be addressed in more detail in Chapter 18.

Compartmentalization and Succession

From long-range studies based on the dissection and analysis of the decay patterns in thousands of stems, Shigo and his associates (Shigo and Marx, 1977; Shortle, 1979) developed a new interpretation of decay development, which involves the ideas of *compartmentalization* and *succession.* They explain the *decay-containment process* in some stems and the eventual formation of serious heart rot in others. The acronym CODIT (compartmentalization of decay in trees) was used to describe the process. The related term *succession* describes the sequence of microorganisms that invade the damaged tissues, modify them, and initiate decay. Four barriers or stem zones restrict the invasion of microorganisms from sapwood wounds (Fig. 12-1). Wall 1 incorporates gums, tyloses, and resins that plug the cells above and below

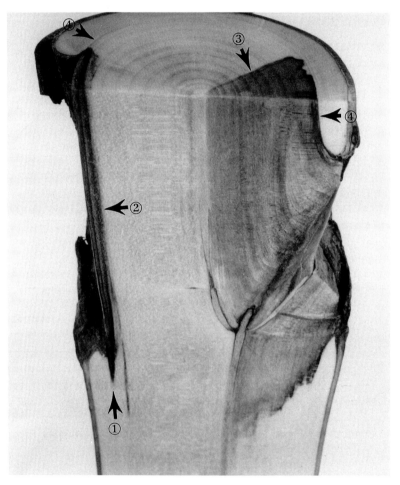

Figure 12-1 An example of the walls (Walls 1, 2, 3, and 4) that formed on red maple *(Acer rubrum)* after two pruning wounds to compartmentalize injury and invasion by wood-inhabiting fungi. The flush-cut wound on the right was not contained, whereas the pruning cut on the left (which retained the branch protection zone) was successfully contained. The four wall zones are described in the text. Courtesy of Dr. A. L. Shigo (1989).

the injury. Wall 2 is the inner wall, consisting of a series of latewood bands of cells where pitting is infrequent, and the cell walls are thickest. Wall 3 includes the radially aligned wood rays, where living cells may react metabolically to the injury and release toxic phenolic compounds. Walls 1, 2, and 3 resist the upward, inward, and lateral spread of microorganisms. Wall 4 represents the new xylem tissue that subsequently develops over the wound surface and protects the new sapwood from infection. Walls 1 and 3 rely on parenchyma cells with limited resources for response. Wall 4 is more intimately tied to the dynamic vitality (cambium) of the tree and is therefore the most effective. It is termed the barrier zone.

Succession, or the sequential involvement of other microorganisms with decay fungi, is the second premise of the expanded concept of stem decay. Based on isolations from many types and ages of stem injuries and inoculation experiments, Shigo (1967) concluded that bacteria were the first wound invaders, followed by a series of microfungi in genera such as *Phialophora, Gliocladium,* and *Ceratocystis.* These microorganisms generally precede the appearance of decay fungi and were assumed to play a role in the conditioning the tissue wound for decay fungus invasion. Whether these organisms are required for infection by decay fungi, or just opportunistic, remains uncertain and difficult to prove. Others showed that the entry of decay fungi into wounded stem tissues may not be immediate, but follows other microorganisms that presumably modify the microecologic conditions of the infection court (Etheridge, 1961; DeGroot, 1965). In contrast, decay fungi appear to be among the first fungi colonizing sweet gum and yellow poplar wounds and are associated with tissue discoloration (Shortle and Cowling, 1978). Manion and French (1969) have shown the *Phellinus tremvlae* first colonizes aspen wounds. Apparently the host defenses in response to a wound inhibit the spore germination of competing microorganisms and stimulate the germination of *P. tremulae* basidiospores.

Rayner and Boddy (1988) believe the presence of microorganisms is not necessary to initiate the host defense reactions and propose that tissue drying and aeration of the injury surface are the initiating factors.

It is clear that much more needs to be known about when, where, and how major decay fungi enter living stems and the roles of associated microorganisms (Manion and Zabel, 1979). Factors and conditions affecting spore germination of decay fungi on branches and injury courts are still largely unknown (Merrill, 1970). Clarification of the roles and sequences of other microorganisms involved in the invasion process may lead to effective *biological* methods for preventing invasion by decay fungi.

Rates of Decay Development

Very little is known about the rates of decay development in single stems, since it is difficult to accurately determine the times of infection. Artificial inoculations of stems have indicated varying annual rates of longitudinal spread, ranging from a few inches to several feet in maples and aspen (Silverborg, 1959). Analysis of decay development in bottomland oaks indicates vertical spreads ranging from 2 to 9 inches annually (Toole and Furnival, 1957).

The rate of decay development in stands varies greatly with species and site. Comparisons of pathologic rotation periods provides some insights into decay-rate differences among species, but there still is the uncertainty of knowing when decay began or whether the rates of decay are uniform. Some examples of pathologic rotations determined for mature timber are as follows: 40–50 yr for aspen in Minnesota (Schmitz and Jackson, 1927); 150 yr for Douglas fir in Washington (Boyce, 1932); 170 yr for eastern white pine in Nova Scotia (Stillwell, 1955); and 250 to 300 yr for Sitka spruce in the Queen Charlotte Islands (Bier et al., 1946). The long pathologic rotation for many timber species clearly illustrates that heart rot develops very slowly in most tree stems.

Recognition of Stem Decays

The detection of internal stem decay from external evidences is difficult and often presumptive at best. Many of the evidences used to detect decay during timber cruising are also useful for roundwood (logs, pulpwood, and poles). Wood users responsible for decay detection during purchasing or inspections should become familiar with regional cull manuals for the species involved. Useful macroscopic evidences of decay (reviewed in Chapter 7) that apply to timber and roughwood are the presence of decayed knots, fungal fruiting bodies, and discolored or decayed wood in the exposed ends. A roughened sawcut often coincides with the exposed decayed zone in the log end (Fig. 12-2). Instruments such as the Shigometer and increment borer are also often very useful for detecting internal decay in stems (see Chapters 10 and 16 for details). Many other useful clues to the field detection of decays are often host or causal fungus specific and are mentioned in regional cull manuals. For example, swollen knots in eastern white pine and beech containing a core of a brown, spongy fungus material are evidences of extensive stem decay by *Phellinus pini* (Fig. 12-3C) and *Inonotus glomeratus*, respectively. Swollen, resin-soaked knots are also presumptive evidence of *Phellinus pini*. A depression below knots in incense cedar, called a *shot-hole cup*, is a reliable indicator of extensive decay caused by the brown pocket-rot fungus, *Postia (Polyporus) amara* (Boyce, 1961). Shigo and Larson (1969) have prepared a photograph guide of many northeastern hardwoods showing that common decays, discolorations, and other defects are often associated with bark evidences of prior wounds.

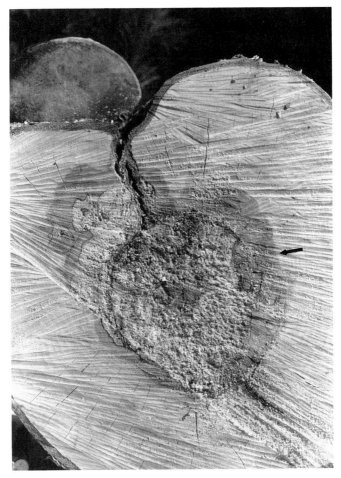

Figure 12-2 A roughened saw cut on the log end and the basidioma are useful indicators of internal decay. Heart rot in a birch log caused by *Phellinus igniarius*. Courtesy of Dr. Manion (1991), with permission of Prentice-Hall, Englewood Cliffs, New Jersey.

Some Common Stem Decays

Information on the identities of the many stem-decay fungi reported as damaging commercially important timber trees in North America has been assembled by Hepting (1971) and the literature sources cited. Detailed descriptions of many of the major decays and their external evidences are also available (Boyce, 1961). The decay fungi reported as causing serious losses in the major timber species in the United States and Alaska are listed in Table 12-1. Several decays that develop in living stems and are later often found in

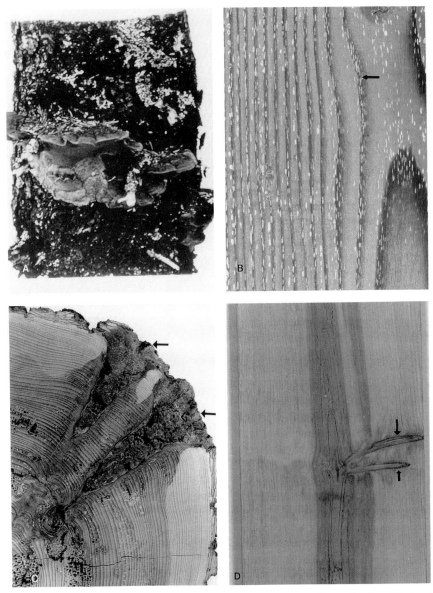

Figure 12-3 Some evidences of the presence of *Phellinus pini* decay in logs or lumber. (A) A typical basidioma of *P. pini* on red spruce; (B) an early stage (firm pocket) of the decay in Douglas fir. Note the clustering of the decay pockets in the earlywood zones; (C) a punk knot indicating extensive decay; (D) buried weeviled tips (arrows) are a common infection court for *P. pini* in eastern white pine and a useful decay indicator. Photo 12-3C courtesy of Dr. A. Shigo, 1989.

roundwood, lumber, and commercial wood products are described briefly in this section.

Major Heart Rots

Phellinus (Fomes) pini is reported to be the major cause of stem decays in conifers in the northern temperate zone. It is particularly severe on Douglas fir in the Northwest and white pine in the East. The common names for this decay are white-pocket rot or red-ring rot, since the small decayed pockets in some hosts are concentrated in the earlywood bands (Fig. 12-3B). In the incipient and early decay stages, the wood is a pink-reddish to purplish color.

In the early stages of the decay, the hemicelluloses and lignin are attacked selectively, and in the later decay stages, cellulose remains in many, small, lenticular-shaped pockets. Wood in the early stages of the red-ring rot is reported to be suitable for general construction uses until the decay pockets are visible (firm-pocket stage). Wood in the firm-pocket stage is also acceptable for low-grade construction lumber and plywood. In advanced stages of decay, the fungus invades the living sapwood and sometimes forms cankers. Infected wood in the outer-sapwood zones is often resin infiltrated. Once the tree is cut and seasoned, the decay fungus dies out rapidly.

External evidences of red-ring rot in standing trees are the presence of swollen knots containing brownish punky masses of fungal material, extensive resin flow from the base of branch stubs, and the presence of characteristic brownish-shelving perennial basidiomata (White, 1953) (Fig. 12-3). Boyce (1961) reported that heart-rot columns extend several feet above the top basidioma or punk knot and about 4 feet below the lowest. Haddow (1938) has shown that small branches or buried weeviled tips are common infection courts on eastern white pine in the East. In studies of the infection process, DeGroot (1965) reported that the weeviled tip preference may reflect the low concentrations of pinosilvin in weeviled tips compared to recently dead branches. As branch stubs age, the pinosilvin levels decrease, and eventually basidiospores of *Phellinus pini* are able to germinate on these substrates; however, heart-rot infections were not obtained.

Echinodontium tinctorium is the major cause of a serious heart rot in the true firs and western hemlock in the Pacific Northwest. The importance of this fungus varies with region and site, with losses as high as 30% of the gross merchantable volume on some poor sites. This fungus is a brown rot and causes a stringy or fibrous-textured decay. Wood in the early decay stage is a light-tan color and incipient decay may extend 2 to 6 feet longitudinally beyond the point of visible decay. In some cases, the decay may extend from the roots to the larger branches. Perennial, hoof-shaped basidiomata form beneath branch stubs (Fig. 12-4A) and are readily recognized by the toothed pore surface and the bright-red context, which gives the fungus its common name *Indian paint fungus*. A single conk on a stem means essentially total stem cull. The earliest appearance of heart rot in stems has been estimated to range from 45 to 75 yr. Infection is believed to originate, some years earlier, at

Table 12-1
A list of some of the common stem-decay fungi associated with major timber species in the United States[a]

Important timber species or genera	Important stem-decay fungi	Appearance	Type of decay	Location
Abies spp. (white, grand, California red firs)	*Echinodontium tinctorium*	Brown, laminated	White rot	Central stem
	Heterobasidion annosum	White, pocket	White rot	Lower stem
	Armillaria mellea	Light brown	White rot	Basal stem
	Fomitopsis pinicola	Brown, cubical	Brown rot	Central stem
Abies balsamea (balsam fir)	*Stereum sanguinolentum*	Brown, stringy	White rot	Central stem
	Scytinostroma galactinum	Yellow, stringy	White rot	Basal stem
	Piloderma bicolor	White, stringy	White rot	Basal stem
Acer saccharum (sugar maple)	*Inonotus glomeratus*	Tan, spongy	White rot	Central stem
	Phellinus igniarius	White, mottled, spongy	White rot	Central stem
	Oxyporus populinus	Yellow, stringy	White rot	Lower stem
	Cerrena unicolor	White, soft	White rot	Central stem
	Ganoderma applanatum	White, mottled	White rot	Lower stem
	Climacodon septentrionalis	White, spongy	White rot	Central stem
Betula alleghaniensis (yellow birch)	*Phellinus igniarius*	White, mottled, spongy	White rot	Central stem
	Inonotus obliquus	White, mottled, spongy	White rot	Central stem
	Stereum murraii	Light brown	White rot	Central stem
	Inonotus hispidus	White, spongy	White rot	Upper stem
	Pholiota adiposa	Brown, mottled	Brown rot	Central stem
	Ganoderma applanatum	White, mottled	White rot	Lower stem
	Fomes fomentarius	White, mottled	White rot	Central stem
Chamaecyparis spp. (Alaska and Port Orford cedars)	*Phellinus pini*	White, pocket	White rot	Central stem
	Fomitopsis pinicola	Brown, cubical	Brown rot	Central stem
Fagus grandifolia (beech)	*Phellinus igniarius*	White, mottled, spongy	White rot	Central stem
	Ganoderma applanatum	White, mottled, spongy	White rot	Lower stem
	Inonotus glomeratus	Light brown, spongy	White rot	Central stem
	Armillaria mellea	Light brown	White rot	Lower stem
	Climacodon septentrionalis	White, spongy	White rot	Central stem
	Cerrena unicolor	White, soft	White rot	Central stem
	Ustilina deusta	White, soft	White rot	Basal stem

Host	Fungus	Rot characteristics	Rot type	Location
Juglans nigra (black walnut)	*Phellinus igniarius*	White, mottled, spongy	White rot	Central stem
	Phellinus everhartii	White, soft-flaky	White rot	Lower stem
	Inonotus hispidus	White, spongy	White rot	Upper stem
Larix occidentalis (western larch)	*Fomitopsis officinalis*	Brown, cubical	Brown rot	Central stem
	Phellinus pini	White, pocket	White rot	Central stem
	Laetiporus sulphureus	Brown, cubical	Brown rot	Central stem
Liriodendron tulipifera (yellow poplar)	*Pleurotus ostreatus*	White, flaky	White rot	Central stem
	Hydnum erinaceus	White, spongy	White rot	Central stem
	Armillaria mellea	Light brown	White rot	Lower stem
Picea spp. (red, white, black)	*Stereum sanguinolentum*	Brown, stringy	White rot	Central stem
	Phellinus pini	White, pocket	White rot	Central stem
	Inonotus tomentosus	White, pocket	White rot	Lower stem
	Scytinostroma galactinum	Yellow, stringy	White rot	Basal stem
	Fomitopsis cajanderi	Brown, cubical	Brown rot	Upper stem
	Fomitopsis pinicola	Brown, cubical	Brown rot	Central stem
	Perenniporia subacida	White, spongy	White rot	Lower stem
Picea sitchensis (Sitka spruce)	*Phellinus pini*	White, pocket	White rot	Central stem
	Fomitopsis pinicola	Brown, cubical	Brown rot	Central stem
	Phaeolus schweinitzii	Brown, cubical	Brown rot	Lower stem
	Postia placenta	Brown, cubical	Brown rot	Central stem
	Neolentinus kauffmanii	Brown, cubical, pocket	Brown rot	Lower stem
	Fomitopsis officinalis	Brown, cubical	Brown rot	Central stem
	Laetiporus sulphureus	Brown, cubical	Brown rot	Central stem
Pinus spp. (white, western, and sugar pines)	*Phellinus pini*	White, pocket	White rot	Central stem
	Phaeolus schweinitzii	Brown, cubical	Brown rot	Lower stem
	Heterobasidion annosum	White, pocket	White rot	Lower stem
	Fomitopsis officinalis	Brown, cubical	Brown rot	Central stem
Pinus spp. (southern pines)	*Phellinus pini*	White, pocket	White rot	Central stem
	Phaeolus schweinitzii	Brown, cubical	Brown rot	Lower stem
	Heterobasidion annosum	White, pocket	White rot	Lower stem
Pinus ponderosa (ponderosa pine)	*Dichomitus squalens*	White, pocket	White rot	Central stem
	Laetiporus sulphureus	Brown, cubical	White rot	Central stem
	Phellinus pini	White, pocket	White rot	Central stem
	Crytoporus volvatus	Gray, soft	White rot	Central stem
	Fomitopsis officinalis	Brown, cubical	Brown rot	Central stem

(continues)

Table 12-1 (continued)

Important timber species or genera	Important stem-decay fungi	Appearance	Type of decay	Location
Populus tremuloides (quaking aspen)	*Phellinus tremulae*	White, mottled, spongy	White rot	Central stem
	Ganoderma applanatum	White, mottled	White rot	Lower stem
	Armillaria mellea	Light brown	White rot	Lower stem
Prunus serotina (black cherry)	*Phellinus prunicola*	White, mottled, spongy	White rot	Central stem
	Phellinus laevigata	White, mottled, spongy	White rot	Central stem
Pseudotsuga menziesii (Douglas fir)	*Phellinus pini*	White, pocket	White rot	Central stem
	Fomitopsis officinalis	Brown, cubical	Brown rot	Central stem
	Fomitopsis rosea	Brown, cubical	Brown rot	Upper stem
	Phaeolus schweinitzii	Brown, cubical	Brown rot	Lower stem
	Postia placenta	Brown, cubical	Brown rot	Central stem
Quercus spp. (white and red oak)	*Hydnum erinaceus*	White, spongy	White rot	Central stem
	Xylobolus frustulatus	White, pocket	White rot	Lower stem
	Pleurotus ostreatus	White, flaky	White rot	Upper stem
	Phellinus everhartii	White, soft, flaky	White rot	Lower stem
	Inonotus hispidus	White, spongy	White rot	Upper stem
	Stereum gausapatum	White, mottled	White rot	Central stem
Sequoia sempervirens (redwood)	*Postia sequoiae*	Brown, cubical, pocket	Brown rot	Central stem
	Ceriporiopsis rivulosa	White, laminated	White rot	Central stem
Taxodium distichum (bald cypress)	*Stereum taxodii*	Brown, cubical, pocket	Brown rot	Central stem
Thuja plicata (western red cedar)	*Phellinus weirii*	Yellow, laminated	White rot	Lower stem
Tsuga heterophylla (western hemlock)	*Phellinus pini*	White, pocket	White rot	Central stem
	Heterobasidion annosum	White, pocket	White rot	Central stem
	Ganoderma applanatum	White, mottled	White rot	Lower stem
	Echinodontium tinctorium	Brown, stringy	White rot	Central stem
	Fomitopsis pinicola	Brown, cubical	Brown rot	Central stem
	Fomitopsis officinalis	Brown, cubical	Brown rot	Central stem

[a]Principal information sources were Hepting (1971) and Gilbertson and Ryvarden (1986, 1987). Names of the fungi were obtained from Farr et al. (1989).

Figure 12-4 The presence of the basidiomata (conks) of some heart-rot fungi indicates extensive internal stem decay. Examples of several important heart-rot fungi are (A) *Echinodontium tinctorium* on white fir; (B) *Phaeolus schweinitzii* associated with roots of eastern white pine; (C) a sterile conk of *Inonotus obliquus* on yellow birch; and (D) *Phellinus igniarius* on beech. Photos 12-4A and 12-4B, courtesy of Dr. P. Manion (1991), with permission of Prentice-Hall, Englewood Cliffs, New Jersey.

the base of small branchlets on shade-killed lower branches. The fungus remains dormant for years, and the branch tissues are slowly enveloped by stem growth. The fungus becomes activated later in response to surface wounds or loss of tree vigor (Etheridge and Craig, 1976). Managerial controls appear to be limited to determining regional pathologic rotations and silvicultural treatments to limit suppression of tree growth.

Phaeolus (Polyporus) schweinitzii causes a serious brown-cubical rot in the roots and the butt log of all conifers species in the north temperate zone, with the exception of cedars, junipers, and cypresses. This fungus is a particularly dangerous heart-rot pathogen, since the early decay stages are barely detectable in the wood, yet associated with sharp reductions in tensile strength and toughness (Scheffer *et al.*, 1941). At the early stage there is a slight yellowing of the wood in vertical spires, and the incipient stage may project several feet above the zone of visible decay. The advanced stage is characterized by a brown-cubical rot with near-total tissue collapse. The decay develops in the roots and butt portions of logs and reaches a height of 2.4 m, although in some cases the decay column may extend 4.5 to 6.0 m above ground. Occasionally young trees are killed by extensive decay in the roots, and windthrow is common in severely diseased stands. The only external evidence of this heart rot is the occasional presence of the characteristic brown, velvety-textured, stipitate basidiomata arising from the forest floor near the base of the infected tree or, rarely, sessile conks attached to basal wounds (Fig. 12-4B). Presumed entry points of the fungus are broken roots or deep basal wounds. Conifer roots previously infected by *Armillaria mellea* may be predisposed to colonization by *P. schweinitzii* (Barrett, 1970). Decayed trees often occur in clusters in pine stands, and the decay may not be detectable until felling. Managerial controls are not known, other than following pathologic rotations and avoiding planting on sites where *A. mellea* is established. Heavy stand stocking to minimize root breakages during wind storms has been suggested as a potential method for decreasing infection.

Fomitopsis officinalis occurs on conifers in North America and Europe. It is important primarily on Douglas fir, sugar pine, ponderosa pine, and western hemlock in western North America and produces a brown-cubical rot in the late stages, which is often characterized by the presence of thick mycelial felts in the shrinkage cracks. Like *P. schweinitzii*, this fungus is dangerous in wood products since the early stage (a faint yellow to brown) is almost imperceptible, and the incipient stage may extend many feet longitudinally beyond the zone of visible decay. Infection is uncertain but is presumed to occur from large broken branches in the upper bole and broken tops, but some infections are also associated with fire scars and logging wounds. Large, chalky white, hoof-shaped perennial basidiomata are found in wound faces or knots, and the presence of a single basidioma indicates total cull.

Phellinus (Fomes) igniarius is an important major cause of heart rot in birches, maples, beech, and oaks in North America. *Phellinus tremulae*, a closely related species, is the major cause of heart rot in aspens. This fungus causes a white-spongy rot in the heartwood, but can also invade and kill living sapwood. The rot, as seen on stem cross sections, is characterized by a yellowish green to brownish black outer invasion zone, which surrounds a core of irregularly mottled, white, spongy wood. The advanced decay often contains fine, concentrically arranged zone lines. The decayed wood is still usable for pulping purposes. The fungus may continue to decay stored wood for some months, but dies out in wood products. Perennial hoof-shaped basidiomata commonly develop in late decay stages at wound margins or the base of branch stubs (Fig. 12-4C). The presence of a single conk generally indicates a cull tree, with decay columns ranging from 10 to 14 feet in length. Wounds are believed to be the principal infection sites. Minimizing stem injuries and following regional pathological rotations are the recommended management practices for limiting damage by this fungus.

Inonotus (Poria) obliquus and *I. (Polyporus) glomeratus* cause a mottled white rot, which is very similar to decays caused by *Phellinus igniarius*. *Inonotus obliquus* is a major cause of heart rot in birches in the northern hemisphere, whereas *I. glomeratus* is important primarily on maples and beech in eastern North America. Both decay fungi form large, black, coal- or clinkerlike abortive basidiomata, which are sterile and perennial on living hosts (Fig. 12-4C). Both fungi also invade the living sapwood and may form stem cankers. The presence of a single sterile conk indicates extensive heart rot and a cull tree. Swollen knots or dead bark on canker faces in beech may partially conceal the sterile conks of *I. glomeratus*. Wounds and cankers appear to be the principal infection courts of these fungi. *Inonotus obliquus* forms a brown resupinate basidioma in the decayed sapwood of dead hosts splitting off the outer sapwood and bark layers to expose the hymenial surface whereas *I. glomeratus* forms large, effused-reflexed basidiomata on the lower surface of logs (Zabel, 1976). Neither fungus continues to decay wood in the product form.

Common Sap Rots

The sap rots are more localized in the stem than most heart rots and are usually associated with dead tissue from large stem wounds such as logging injuries, fire scars, frost cracks, sunscald, and lightning injuries that do not callus over, or injury on trees of low vigor. Most sap rots are white rots and occur more commonly on hardwoods. These fungi are of special importance because of their location in the lower, outer stem zones where lumber grades are high and strength most critical. Sap-rot fungi are primarily facultative pathogens on severely wounded, low-vigor, or moribund trees. Most of the sap-rot fungi in tree stems are also important decayers of forest slash and wood products. Some of the common sap rots that occur on the trunks of living trees are described below. Descriptions of the associated basidiomata are available in the taxonomic literature (Gilbertson and Ryvarden, 1986).

Fomes fomentarius causes a white-mottled rot primarily in beech and the birches. Although it is found commonly on slash and dead stubs, it also causes sapwood decay in moribund and severely injured trees. *Ganoderma applanatum* causes a white-mottled rot in butt log of oaks and other hardwoods.

Ustilina (vulgaris) deusta, an ascomycete, causes a white rot in the basal roots and butt of maples and beech. The decayed wood contains numerous zone lines, which often remain as black, brittle sheets of fungal tissue in badly decayed stumps. Black perithecial stromata often cover the remnants of decayed stumps.

Daedaleopsis (Daedalea) confragosa is a common white-rotter, which is associated with severe upper stem wounds on maples, willow, and other hardwoods. *Cerrena (Daedalea) unicolor* causes a white rot and a canker on maples. Whether this fungus causes the canker or invades the dead tissues is uncertain. *Irpex lacteus* causes a soft-white rot and commonly fruits on dead branches and severe wounds in hardwoods. *Bjerkandera adusta* causes a soft-white rot in the sapwood of beech after the bark has been damaged by the beech scale and Nectria canker. *Amylostereum chailletii* causes a red stain and white rot in balsam fir stressed and killed by the spruce budworm. This fungus is vectored by a wood wasp (*Sirex* sp.) after the tree is damaged by insect defoliation or fire (Warren, 1989). *Hirschioporus abietinus* forms a white-pocket rot in the large wounds and associated dead tissue in conifers.

Host-Specific Stem-Decay Fungi

Some heart-rot fungi appear to be important on only a limited number of hosts. *Postia (Polyporus) amara* causes a brown-cubical pocket rot in incense cedar throughout the range of the species. Decay pockets are localized and surrounded by firm wood. The presence of a basidioma (rare and rapidly consumed by insects) or the characteristic shot-hole cup indicates a rot column throughout the stem heartwood. *Laurilia (Stereum) taxodii* causes a brown-cubical pocket rot in cypress, known as pecky cypress in the lumber trade. The decayed pockets are small, and, as in the case of *Postia amara*, the wood surrounding the pockets appears sound. Both woods species produce very durable heartwood, which performs well in posts and greenhouse benches. Pecky cypress has aesthetically pleasing textural qualities and is a popular paneling product.

Phellinus robiniae (Fomes rimosus) causes a soft-white rot in black locust. The perennial basidiomata are conspicuous and indicate the presence of extensive heartwood decay. Black locust has a very durable heartwood and is often used for fence posts.

Some Colonization Strategies of Stem-Decay Fungi

Stem-decay fungi display a wide variation in host selectivity. Some are limited to the heartwood zones of a few durable species, whereas others have broad host ranges. Some heart-rot fungi develop only in living trees. Some sap-rotters are predominately saprobic fungi in slash or wood products and mostly occupy the dead tissue of severe wounds. Colonization strategies may explain some of these differences (see Chapter 11). Some heart-rot fungi, such as

Stereum taxodii and *Postia amara* appear to have selected for *stress resistance* as a colonization strategy. Tolerance to the toxic heartwood extractives of the host tree would be an important selective trait (Wilcox, 1970). Other heartwood characteristics that restrict fungi are low nitrogen levels and high $CO_2:O_2$ ratios (Highley *et al.*, 1983; Rayner, 1986). Stress resistance to adverse heartwood environmental conditions or unique infection routes, as shown by *E. tinctorium*, may explain the host selectivity of some heart-rot fungi. Many of the fungi that cause severe basal rots are potent pathogens that penetrate root tissues from large inoculum bases in the soil. The stem-decay fungi that primarily colonize damaged sapwood tissues appear to be *substrate opportunists*. Sapwood injury exposes a nutrient-rich substrate, which rapidly dries and aerates to levels favorable for microbial growth (Rayner and Boddy, 1988). It is interesting to note that many of the sap-rotters in living stems are also important slash- and products-rot fungi, whereas many of the heart-rot fungi are unable to survive or compete outside of the unique environmental niche provided by heartwood of the living tree. In large stem wounds, the exposed dead tissues provide a substrate for many opportunistic fungi, similar to sapwood in slash or wood products.

Undoubtedly, there are many other explanations for host selectivity by heart-rot fungi. The many microbial sequences and interactions involved in the infection process in stem decays may be another important cause of host selectivity. Methods for circumventing the tree's containment defenses may also be important. The host-selection process can be viewed as a part of the long term co-evolution of hosts and pathogens in plant diseases.

Summary

- Some decays and discolorations in wood products develop in the living tree. It is particularly important to detect early decay in wood destined for critical-strength uses or uses where decay adversely affects important use properties. A general understanding of the origins and types of decay that originate in the living tree will facilitate its detection in products and encourage more effective wood uses.
- Decay fungi in living stems are pathogens, and their damages are grouped as *heart rots* or *sap rots*, based on the stem zones occupied. Heart rots develop primarily in the physiologically inactive central zone of the stem and may occupy the entire heartwood of a tree. Sap rots develop in the physiologically active outer sapwood, usually after wounding, and are often localized. In contrast to stem decayers, the fungi that decay wood products are saprogens.
- Stem decays may be either white or brown rots. Whereas the white rots predominate, brown rots are a special concern since the incipient stage is often difficult to detect, and is associated with large strength losses.

Summary continues

- Dead branches, wounds, and root tissues appear to be the principal infection courts for many stem-decay fungi. For branches, the tissues invaded are the sapwood. Although exposed heartwood is also a logical entry point, entry in this zone has not been proven experimentally.
- Trees have evolved effective defenses against the invasion of most wood-inhabiting fungi. The defenses involve the impregnation of branch stubs with resins and gums, the development of callus to form new bark and xylem tissues over the wounds, and the isolation of injured tissues.
- The term *compartmentalization* has been coined to describe the *infection-containment process* and utilizes, in part, the natural structure and functioning of the stem. Compartmentalization involves the deposition of gums and resins, the formation of tyloses to restrict longitudinal microbial spread, the successive layers of thick-walled latewood tissues to restrain radial spread, and the release of toxic chemicals (phytoalexins) from radial tissues to reduce tangential spread. Subsequent xylem tissues laid down by callus limit outward spread. The barrier zone formed by the callus is the most effective infection restraint. Compartmentalization effectively isolates the infection of most small wounds, particularly in vigorous trees, and walls it off from the rest of the tree.
- Tissue drying and aeration in the boundary zones of stem wounds have been proposed by Rayner and Boddy (1988) as possible causes of the defense reactions and not necessarily the actions of microorganisms.
- Some heart-rot decays are traceable to compartmentalized infections that became active after they were included in the heartwood or when the outer spread was contained, but the fungus continued to colonize tissues toward the pith.
- *Succession*, or the sequential involvement of bacteria and microfungi with the decay fungi, is believed to be an important part of the infection process. These organisms are presumed to modify the tissue to favor decay fungi.
- The presence of characteristic fruiting bodies, decayed (punky) knots, and discolored or decayed wood in the exposed ends of logs and pulpwood are useful macroscopic evidences of decay in logs and roundwood. A roughened sawcut often coincides with the exposed decay in the log end.
- The major stem decays in the United States and Alaska are briefly described with emphasis on their recognition, decay type, and associated cull.
- Future practical controls of stem decays may rest on better understandings of microbial colonization sequences and interactions in infection courts and the physiologic actions involved in tree responses to injury. Such information could lead to biologic protection strategies

and provide useful information for tree selection for genetic improvements.

References

Agrios, G. N. (1988). "Plant Pathology," 3rd Ed. Academic Press, New York.

Atwell, E. A. (1948). Red-stain and pocket-rot in Jackpine: Their effect on strength and serviceability of the wood. Dept. of Mines & Resources, *Dominion Forest Service Forest Product Laboratory Circular* 63:1-23.

Barrett, D. K. (1970). *Armillaria mellea* as a possible factor predisposing roots to infection by *Polyporus schweinitzii*. *Transactions of the British Mycological Society*, 55:459-462.

Basham, J. T. (1958). Decay of trembling aspen. *Canadian Journal of Botany* 36:491-505.

Bier, J. E., R. E. Foster, and P. J. Salisbury. (1946). Studies in forest pathology IV. Decay of Sitka spruce on Queen Charlotte Islands. *Canadian Department of Agriculture Publication 783, Technical Bulletin* 56:1-35.

Boyce, J. S. (1920) "A Dry-Rot of Incense Cedar." U.S.D.A. Bulletin 871-1-58, Washington, D.C.

Boyce, J. S. (1932). "Decay and Other Losses in Douglas Fir in Western Oregon and Washington." U.S.D.A. Technical Bulletin 286-1-59, Washington, D.C.

Boyce, J. S. (1961). "Forest Pathology," 3rd Ed. McGraw-Hill, New York.

Cartwright, K. St. G., and W. P. K. Findlay. (1958). "Decay of Timber and its Prevention," 2nd Ed. Her Majesty's Stationery Office, London, England.

Davidson, A. G., and D. R. Redmond. (1957). Decay of spruce in the Maritime Provinces. *Forestry Chronicle* 33:373-380.

DeGroot, R. C. (1965). Germination of the basidiospores of *Fomes pini* on pine wood extract media. *Forest Science* 11:176-180.

Etheridge, D. E. (1961). Factors affecting branch infection in aspen. *Canadian Journal of Botany* 39:799-816.

Etheridge, D. E., and H. M. Craig. (1976). Factors influencing infection and initiation of decay by the Indian paint fungus *(Echinodontium tinctorium)* in western hemlock. *Canadian Journal of Forestry Research* 6:299-318.

Farr, D. F., G. F. Bills, G. P. Chamuris, and A. Y. Rossman. (1989). Fungi on Plants and Plant Products in the United States. American Phytopathological Press, St. Paul, Minnesota.

Gilbertson, R. L. (1980). Wood-rotting fungi of North America. *Mycologia* 72:1-49.

Gilbertson, R. L., and L. Ryvarden. (1986 and 1987). "North American Polypores." Vol. 1, 1986, and Vol. II, 1987 Fungiflora, Oslo, Norway.

Good, H. M. (1959). Fungi associated with *Fomes igniarius* in decay of poplar. *Proceedings of the International Botanical Congress, 9th*, 2:136.

Haddow, W. R. (1938). The disease caused by *Trametes pini* (Thore) Fries in white pine *(Pinus strobus* L.). *Transactions of the Royal Canadian Institute* 22:21-80.

Hepting, G. H. (1935). "Decay Following Fire in Young Mississippi Delta Hardwoods." U.S.D.A. Technical Bulletin 494, Washington, D.C.

Hepting, G. H. (1971). "Diseases of Forest and Shade Trees of the United States." U.S.D.A. Forest Service, Agricultural Handbook, Number 386, Washington. D.C.

Hepting, G. H., and G. G. Hedgcock. (1937). "Decay in Merchantable Oak, Yellow

Poplar, and Basswood in the Appalachian Region." U.S.D.A. Technical Bulletin 570-1-20, Washington, D.C.

Highley, T. L., S. S. Bar-Lev, T. K. Kirk, and M. J. Larsen. (1983). Influence of O_2 and CO_2 on wood decay, by heartrot and saprot fungi. *Phytopathology* **73**:630–633.

Hubert, E. E. (1924). The diagnosis of decay in wood. *Journal of Agriculture Research* **29**(11):523–567.

Hubert, E. E. (1931). "An Outline of Forest Pathology." John Wiley, New York.

Kimmey, J. W., and E. M. Hornebrook. (1952). Cull and breakage factors and other tree measurement tables for redwood. U.S.D.A., Forest Service, California Forest and Range Exp. Sta., *Forest Survey Release* **13**:1–28.

Lockhard, C. R., J. A. Putnam, and R. D. Carpenter. (1950). Log defects in southern hardwoods. *U.S.D.A. Handbook* **4**:1–37.

Manion, P. (1967). "Factors Affecting the Germination of Basidiospores of the Heart-Rot Fungus *Fomes igniarius* in the Infection of Aspen." Ph.D. thesis. University of Minnesota, St. Paul, Minnesota.

Manion, P. (1991). "Tree Disease Concepts," 2nd Ed. Prentice-Hall, New York.

Manion, P. D., and D. W. French. (1967). *Nectria galligena* and *Ceratocystis fimbriata* cankers of aspen in Minnesota. *Forest Science* **13**:23–28.

Manion, P. D., and D. W. French. (1969). The role of glucose in stimulating germination of *Fomes igniarius* var. *populinum* basidiospores. *Phytopathology* **59**: 293–296.

Manion, P. D., and R. A. Zabel. (1979). Stem decay perspectives—an introduction to the mechanisms of tree defense and decay patterns. *Phytopathology* **69**(10): 1136–1138.

Merrill, W. (1970). Spore germination and host penetration by heart-rotting hymenomycetes. *Annual Review of Phytopathology* **8**:281–300.

Murphy, T. G., and H. H. Rowatt. (1932). "British Columbia Softwoods: Their Decays and Natural Defects." Department of the Interior, Canadian Forest Service, Bulletin No. 80. Ottawa, Canada.

Partridge, A. D., and D. L. Miller. (1974). "Major Wood Decays in the Inland Northwest." Idaho Research Foundation, University Station, Moscow, Idaho.

Rayner, A. D. M. (1986). Water and origins of decay in trees. *In* "Water, Fungi, and Plants" (P. G. Ayres and L. Boddy, eds.)., pp. 321–341. Cambridge University Press, Cambridge, England.

Rayner, A. D. M., and L. Boddy. (1988). "Fungal Decomposition of Wood. Its Biology and Ecology." John Wiley and Sons, Chichester, United Kingdom.

Roth, E. R. (1959). "Heart Rots of Appalachian Hardwoods." U.S.D.A. Forest Service, Forest Pest Leaflet 38.

Scheffer, T. C., T. R. C. Wilson, R. F. Luxford, and C. Hartley. (1941). "The Effect of Certain Heart-Rot Fungi on Specific Gravity and Strength of Sitka Spruce and Douglas Fir." Technical Bulletin, U.S.D.A., No. 779, Washington, D.C.

Schmitz, H., and L. W. R. Jackson. (1927). Heartrot of aspen with special reference to forest management in Minnesota. *Minnesota Agricultural Experimental Station Technical Bulletin* **50**:1–43.

Schrenk, H. Von. (1900). "Some Diseases of New England Conifers." Bulletin of the U.S.D.A. No. 25, Washington, D.C.

Shain, L. (1979). Dynamic responses of differentiated sapwood to injury and infection. *Phytopathology* **69**:1143–1147.

Shigo, A. L. (1967). Successions of organisms in discoloration and decay of wood. *International Review of Forest Research* **2**:237–249.

Shigo, A. L. (1976). Microorganisms isolated from wounds inflicted on red maple, paper birch, American beech, and red oak in winter, summer, and autumn. *Phytopathology* **66**(5):559–563.

Shigo, A. L. (1984). Compartmentalization: A conceptual framework for understanding how trees grow and defend themselves. *Annual Review of Phytopathology* **22**:189–214.

Shigo, A. L. (1989). "A New Tree Biology: Facts, Photos and Philosophies on Trees and Their Problems and Proper Care," 2nd Ed. Shigo and Tree Associates, Durham, New Hampshire.

Shigo, A. L., and W. E. Hillis (1973). Heartwood, discolored wood, and microorganisms in living trees. *Annual Review of Phytopathology* **11**:197–222.

Shigo, A. L., and E. Larson. (1969). "A Photo Guide to the Patterns of Discolorations and Decay in Living Northern Hardwood Trees." U.S.D.A. Forest Service Research Paper NE-127, Northeastern Forest Experiment Station, Upper Darby, Pennsylvania.

Shigo, A. L., and H. Marx. (1977). "CODIT (Compartmentalization of Decay in Trees)." U.S.D.A. Informational Bulletin 405.

Shortle, W. C. (1979). Mechanisms of compartmentalization of decay in living trees. *Phytopathology* **69**:1147–1151.

Shortle, W. C., and E. B. Cowling. (1978). Development of discoloration, decay, and microorganisms following wounding of sweet gum and yellow poplar trees. *Phytopathology* **68**:609–616.

Silverborg, S. B. (1954). "Northern Hardwoods Cull Manual." State University of New York, College of Forestry at Syracuse University, Bulletin 31:1045, Syracuse, New York.

Silverborg, S. B. (1959). Rate of decay in northern hardwoods following artificial inoculation with some common heartrot fungi. *Forest Science* **5**:223–228.

Sinclair, W. A., H. H. Lyon, and W. T. Johnson. (1987). "Diseases of Trees and Shrubs." Cornell University Press, Ithaca, New York.

Stillinger, J. R. (1951). Some strength and related properties of old-growth Douglas fir decayed by *Fomes pini*. *ASTM Bulletin* **173**:52–57.

Stillwell, M. A. (1955). Decay of yellow birch in Nova Scotia. *Forestry Chronicle* **31**:74–83.

Toole, E. R., and G. M. Furnival. (1957). Progress of heartrot following fire in bottomland red oaks. *Journal of Forestry* **55**:20–24.

Wagener, W. W., and R. W. Davidson. (1954). Heartrots in living trees. *Botanical Review* **20**:61–134.

Warren, G. R. (1989). "The Identities, Roles, and Interactions of the Fungi Associated with Stem Biodeterioration of Balsam Fir Stressed and Killed after Spruce Budworm Defoliation in Newfoundland. Ph.D. thesis, State University of New York, College of Environmental Science and Forestry, Syracuse, New York.

Weir, J. R., and E. E. Hubert. (1918). "A Study of Heart-Rot in Western Hemlock." U.S.D.A. Bulletin 722.

White. L. T. (1953). Studies in forest pathology. X. Decay of white pine in the Timagami Lake and Ottawa Valley areas. *Canadian Journal of Botany* **31**:175–200.

Wilcox, W. W. (1970). Tolerance of *Polyporus amarus* to extractives from incense cedar heartwood. *Phytopathology* **60**:919–923.

Williams, M. (1989). "Americans and Their Forests. A Historical Geography." Cambridge University Press, Cambridge, United Kingdom.

Zabel, R. A. (1976). Basidiocarp development in *Inonotus obliquus* and its inhibition by stem treatment. *Forest Science* **22**(4):431–437.

CHAPTER
13

Biodeterioration of Stored Wood and Its Control

When a living tree is felled and cut into logs or bolts, the exposed wood surfaces rapidly dry, and the sapwood portions become susceptible to fungal colonization. In some cases bark also loses its protective properties and is invaded soon after felling by both insects and fungi. In an ecological sense, the freshly exposed wood tissues provide a new energy-rich substrate for rapid exploitation by opportunistic microorganisms. During warm seasons in the temperate zones and in the tropics, essentially all timber species are subject to serious damage by invading microorganisms within a few months. The recent practices of chipping trees or wood wastes for paper pulp or fuel have many advantages but also provide a greatly expanded, unprotected wood surface for rapid microorganism colonization.

Roundwood (logs, poles, and bolts) and chips are the common raw-material forms for most primary wood-processing operations. In many cases, it is necessary to maintain large inventories to achieve seasonal transportation or purchase-price advantages to ensure continuous operation of large mills. In other cases, there may be seasonal harvesting limitations or the need to air-season the materials before treatments. These storage periods may be as long as several years, in some cases, and can result in serious deterioration losses unless remedial steps are taken.

In this chapter, we review the principal types of storage losses, their causal agents, and the factors affecting their development in stored logs, pulpwood, and chip piles in the United States. An emphasis will be placed on management practices to prevent or reduce losses. Particularly useful sources for additional information on log and chip-wood storage practices are Hajny (1966), Scheffer (1969), Cowling *et al.* (1974), Hulme (1979), and Fuller (1985).

Types of Storage Loss

The principal losses from prolonged wood storage are decays, discolorations (both chemical and biotic), insect damages, and deep checking during drying. There are other subtle losses, difficult to detect and quantify, which are important in some products. Fungal colonization during storage causes variability in wood properties, such as permeability, strength, color, and decay susceptibility in structural uses.

The magnitude of decay losses varies greatly with length of storage, region, wood species, season, and end use. Insect tunnels and associated stains may be very damaging in veneer bolts, but of little concern in pulpwood. Pulp-yield losses from decay and stains are of special concern during pulpwood and chip storages for paper products. End checking may be a minor defect in logs destined for poles or piling but causes serious losses in lumber or veneer. Some conifers are very susceptible to sapstain damage, whereas many hardwoods are nearly immune. White rots, which are more common in hardwoods, may cause modest losses in paper production, whereas the brown rots, which are more common in conifers, may be very damaging. Bark beetles cause serious damage rapidly in hardwood logs and veneer bolts stored in the Gulf region of the United States, but are generally unimportant in northern zones. Particularly severe losses from both insect and fungi can occur rapidly in tropical regions. In northern locations, logs and wood chips may be stored beginning in the winter for a year or so with minimal deterioration losses.

Actual wood losses for the various industries are difficult to measure because wood is continually added and removed from the same storage pile; unit measures often change between the raw material and product; it is difficult to measure large bulk volumes of mixed species of wood; a high percentage of waste products is involved in many wood product conversions; subtle changes affect quality; and there are huge volumes of continually changing raw materials.

Until recently, storage losses were accepted owing to the low cost and abundance of wood and the absence of precise information on the cost of the losses. Some control practices were followed to reduce losses in high-value products, including fall cutting and winter storage, ponding logs, and limiting the storage periods to several months in the South and less than a year in northern regions.

Studies of wood-storage losses in the United States began shortly after the depletion of the eastern forests and the rapid expansion of the pulp and paper industry in the northern states (Humphrey, 1917; Kress, 1925).

General Control Practices

The relative values of the end products generally determine the nature and cost of the feasible control treatments. For example, the high value of a black walnut peeler log for veneer justifies costly control treatments for each log, whereas the low value of pulpwood severely limits treatment possibilities.

The general methods used for protecting green wood during storage are rapid utilization; arranging the major storage period to begin during a cold season; storage under water; water sprinkling or spraying; chemical treatments with fungicides and/or insecticides, applications of end-checking compounds, and conversion to chips.

The actual controls practiced vary greatly with species, periods of storage, end products, region, and the size and sophistication of the industry. For low-cost items with short-term storages, practices are followed to keep the units wet; for long-term storage, the wood is peeled so it can dry out as rapidly as possible. For high-cost items, chemical treatments or water immersions are necessary. Basic control rules are rapid utilization and maintaining the minimal volume of wood in storage that the economics of the operation permits. This is difficult because mills must purchase on anticipated demand. At times, anticipated demands fail to materialize, and wood remains in storage for long periods. It is these times that pose the greatest risks to wood in storage.

Logs and Bolts

Logs of all North American species can be damaged when stored through a summer season (Fig. 13-1). These losses are often severe if late fall- and winter-cut logs are stored more than a year in northern and western zones or 3 months or more in the South. Although these losses are accepted as substantial by the industry and need control, published data documenting the volume and grade losses are meager. In birch bolts stored during a single summer season in the Northeast, visible decay can develop 30 to 45 cm into the bolts from each end (Isaacson, 1958), and in the case of beech bolts, where the bark rapidly loses its protective qualities, decay also develops throughout the bolt from the sides (Scheffer and Zabel, 1951; Scheffer and Jones, 1953). Scheffer and Eslyn (1976) noted relatively shallow decay zones in birch exposed after 3 months in Wisconsin and Michigan, but substantial attack after a full summer of exposure (Table 13-1).

Similar decay rates can occur in the sapwood zones of most other species stored through a summer season unless preventive measures are taken.

Sapstains and ambrosia beetles may become major problems within a month or so of summer storage in *Pinus* spp. and hardwoods and seriously degrade high-quality lumber (Fig. 13-2).

End checking may result in degrade losses in some species during periods of drying stress and solar exposure. A more serious consequence of drying checks is their function as infection courts for decay and stain fungi, greatly prolonging the colonization period of seasoning logs.

General practices followed in most small operations to reduce seasoning losses are as follows:

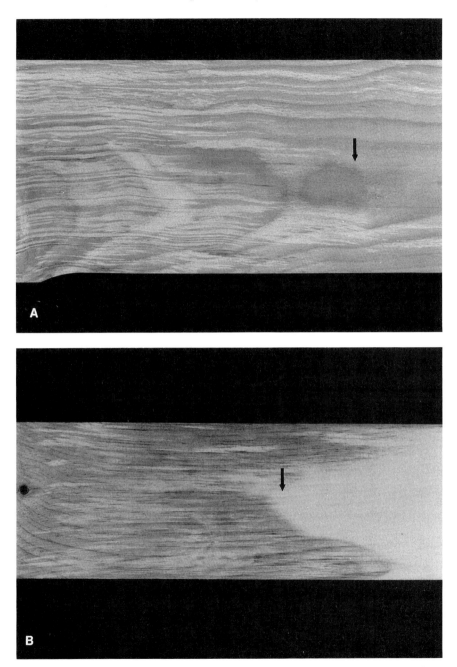

Figure 13-1 Boards sawed from the ends (arrows) of logs showing the penetration of early decay (A, beech) and sapstain (B, eastern white pine) from the ends and bark surface after 2 months of summer storage in northern New York (×0.5).

Table 13-1
Decay development in birch bolts stored for varying periods[a]

Exposure periods	Decay penetration from ends (cm)	
	Average penetration	Deepest penetration
March–July	5	—
March–October	27.5	35
April–October	25	32.5
November–October	32.5	35

[a]Data from Scheffer and Eslyn (1976).

1. Pile logs off the ground on well-drained sites;
2. Facilitate air movement around the piles;
3. Avoid exposing log ends to the south to minimize solar heating and end checking;
4. Utilize rapidly;
5. Initiate storage during colder seasons; and
6. Follow the practice that first logs in are the first logs out for processing.

The other practices followed in large operations for high-value logs are water storage, water spraying or sprinkling, and chemical treatments.

Water Storage

Submerging logs in water (ponding) provides superior protection and has long been used for tide-water mills and where the processing operation is ad-

Figure 13-2 Sapstain and *pinhole* damage from ambrosia beetles that developed in eastern white pine logs after a summer of storage in central New York.

jacent to suitable bodies of water convenient to the processing operation (Fig. 13-3). Water storage provides several years of protection against decay, stains, insect attack, and checking. The presumed protective mechanism is oxygen deprivation of wood-inhabiting fungi because of the low concentrations and slow diffusion rate of oxygen in water. Despite the effectiveness of water immersion to control all storage defects, there are disadvantages. It takes large volumes of water (about 3 to 4 acres of water surface per million bd ft of logs) to store adequate supplies of roundwood for large operations. It is difficult to keep the logs grouped by species, size, and grade for various uses. High-density hardwood logs often sink and may be difficult to recover. Exposed portions of conifer logs are subject to decay, stain, and insect damage unless held under water by baffles. A slight darkening of the wood occurs in some hardwoods, and in smaller ponds during summers, disagreeable odors may develop and permeate the wood. In some cases, serious permeability damage may occur in some conifers and especially the soft pines. Under prolonged or concentrated log-storage conditions, several species of bacteria invade and destroy parenchyma tissue (wood rays and resin canals), increasing the permeability of the wood (Ellwood and Ecklund, 1959; Knuth and McCoy, 1962).

Figure 13-3 A large raft of logs in the Pacific Northwest, assembled for water protection against insects and fungi and transportation advantages.

Water Spraying

Water spraying or sprinkling protects decked logs of all species for an entire warm season (Hansbrough, 1953; Lane and Scheffer, 1960; Carpenter and Toole, 1963). There appears to be no difference between continuous mist sprays and intermittent sprinkling as long as the volumes of water delivered are sufficient to keep the log surfaces saturated at all times. It is important also that the spraying begins while the logs or bolts are still green. Bacteria and molds may develop on the ends and surfaces of the logs, but any wood penetration is shallow and removed as slabs or initial peelings during log processing. For example, longleaf pine logs protected by water spraying for a summer season in the South retained strength properties (toughness) similar to those of fresh-cut logs (Scheld and DeGroot, 1971). The protection mechanism is unknown, but water storage presumably deprives microorganisms of oxygen. Surface microflora (yeasts, bacteria, molds, and algae) may also form physical or chemical (antibiotic) barriers to airborne decay and stain spores, or the parenchyma cells in the bark and sapwood may remain viable longer in water-soaked wood to serve a protective function. The advantages of sprinkling and spraying are the effectiveness and nominal cost. The disadvantages are the need to initiate spraying promptly while the logs are still green, the requirement of large supplies of water, the need to provide drainage to avoid excessively muddy conditions in access to the piles, and the need to assure that run-off does not adversely affect nearby surface water.

Chemical Treatments

Chemical treatments can also limit fungal or insect defects and end checking for many species for several months, even where storage hazards are high. The treatments are applied by spraying or brushing fungicides on the exposed wood on log ends and debarked areas, spraying the bark with insecticides where ambrosia beetles and wood borers are a problem, and applying end dressings to the log ends (after the fungicide treatment) to reduce end checking. It is recommended that log ends and sides be sprayed with fungicides and insecticides in areas south of 40° N latitude when summer storage periods up to 7 weeks are anticipated (Scheffer and Eslyn, 1976). Hardwood logs stored in the northern zones usually require only a fungicidal treatment. End coatings are necessary to reduce checking for longer storage periods. Chemical treatments are particularly useful for small operations where stock piles accumulate rather irregularly and the log values are high. The fungicides used are the antistain chemicals commonly used for lumber dipping (see Chapter 14). The principal insecticide used initially was the gamma isomer of benzene hexachloride. This chemical has been banned recently for wood-protection uses by the Environmental Protection Agency (EPA), and substitutes such as chlorpyrifos or some synthetic pyrethroids are now used. Appropriate state or federal regional agencies should be consulted to determine which insecticides and fungicides are approved currently by EPA or state agencies for local wood-protection uses. End coatings include waxes, hardened gloss oil, coal tar pitches, and asphalt mixtures. Proprietary end-dressing compounds designed to reduce checking are also available.

The principal disadvantages of depending on chemical methods of log protection are (1) the critical need to treat the logs within 24 hr of tree felling to avoid the cases where the rapid growth of fungi in the exposed xylem cells exceeds the subsequent depth of fungicide penetration, and (2) the potential environmental impacts of chemical loss from the treated wood.

Biological Controls

The growth of innocuous molds on exposed wood surfaces that inhibit colonization by damaging decay and stain fungi has long intrigued wood microbiologists. *Trichoderma viride* is a pioneer fungus that rapidly invades exposed wood tissues and forms green masses of spores on the surface. The hyphae that penetrate the wood are colorless and do not adversely affect wood strength or color. Lindgren and Harvey (1952) reported that applying *T. viride* provided considerable protection against stain and decay development in pine bolts stored in the South. Spraying the wood with solutions of ammonium bifluoride appeared to favor the growth of *T. viride* on the log surfaces. Shields and Atwell (1963) have shown that *Trichoderma* inoculations also protect several hardwood species. The erratic nature of the protection has led to no large-scale applications. The potential for bioprotection against both stain and decay fungi has become the subject of renewed interest (Seifert *et al.*, 1988; Benko and Highly, 1990). Biological controls of damaging fungi during log storage remains an intriguing future possibility, warranting additional research.

Poles and Piling

Air-seasoning of poles for periods of several months in the South and years in the Northwest before preservative treatment is practiced by many wood-treating companies. Particular advantages of air-seasoning to reduce pole moisture contents are the low cost and avoidance of the damaging effect of the high temperatures associated with some kiln-drying schedules on pole strength. Practices to hasten air-seasoning are prompt debarking and shaping, stickered piling on foundations, and, sometimes, overhead structures to protect against rainfall. Air-seasoning poles in the South for periods of even a few weeks during periods of wet weather can lead to rapid colonization by stain and decay fungi. The decay fungi enter the pole ends and any deep checks that may form on the pole surfaces. After prolonged storage, zones of early internal decay may develop between the inner sapwood and heartwood. Subsequent preservative treatments, when high temperatures are used or deep penetrations achieved, usually kill the invading fungi in the smaller poles and may mask the internal decay. So-called internal *pretreatment decay* in poles is particularly insidious because it is difficult to detect, may seriously reduce pole-bending strength, and result in erratic preservative treatment. Decay increases wood permeability, and poles exposed to rain just before treatment will be poorly penetrated, whereas those treated after prolonged dry weather will be overtreated. Overtreated poles tend to bleed in service, creating potential environmental hazards. Taylor (1980) made a number of recommendations for limiting pretreatment decay in utility poles. Dip or flood treatments of debarked and shaped poles in concentrated solutions of ammonium bifluoride have been shown to provide protection against decay and stain fungi during air-seasoning periods of up to a year (Panek, 1963). Prompt kiln-drying of southern pine poles before preservative treatment minimizes any decay and related storage problems.

In the Pacific Northwest, Douglas fir poles may be air-seasoned for periods of up to 2 years (Fig. 13-4). Air-seasoning is practiced for economic reasons because of the large size of the poles and the energy costs required for the long kiln schedules necessary to minimize checking.

A study of Douglas fir poles in the eastern U.S. found a high incidence of decay in poles that had been air-seasoned several years before treatment with chromated copper arsenate (CCA) at ambient temperatures (Zabel et al., 1980). One particularly abundant fungus, *Antrodia carbonica*, had previously been found primarily in the Pacific Northwest, suggesting that the fungus was in the pole at the time of installation. Subsequent studies indicated that peeled Douglas fir poles, air-seasoned in the open for more than a year in the Pacific Northwest, are extensively colonized by a variety of decay fungi (Przybylowicz et al., 1987; Morrell et al., 1987). However, no significant strength losses were noted in poles seasoned up to 2 years in the region

Figure 13-4 Douglas fir poles piled for air-seasoning in the Pacific Northwest preparatory to preservative treatment.

(Smith et al., 1987). Chemical treatment with ammonium bifluoride delays and reduces colonization by decay fungi and has been recommended when poles are air-seasoned longer than 1 year (Morrell et al., 1989). Whereas minimizing fungal colonization is important, it is equally critical that air-seasoned poles should receive preservative treatments at temperatures that kill all established decay fungi (Newbill et al., 1988). This is generally considered to require heating of a wood to 65.5°C for at least 75 min (Chidester, 1939).

Pulpwood

Pulp mills require large inventories of raw wood to ensure continuous operation. Initially, most mills were located in northern states and pulpwood was cut in the fall and winter seasons and placed in large pulpwood ricks or jackstraw piles (Fig. 13-5). Wood losses were recognized after prolonged storage, but were overlooked because of the relatively low value of pulpwood. In the 1950s, pulp and paper operations began in the South, and a series of studies was begun to quantify pulpwood storage losses and develop controls for these losses (Pascoe and Scheffer, 1950; Lindgren, 1953; Lindgren and Eslyn, 1961; Mason et al., 1963; Hajny, 1966). Decay and stain fungi are the principal destructive agents in stored pulpwood and reduce both pulp yield (specific gravity) and quality (color changes and losses in paper-strength properties). This damage may also increase bleaching costs and contribute subsequently to slime-control and waste-water problems in the paper mill. The fungi invading stored pulpwood are similar to those that invade forest slash and stored logs. The principal decay fungi invading coniferous pulpwood are *Phanerochaete (Peniophora) gigantea* and *Gloeophyllum (Lenzites) sepiarium;* hardwood pulpwoods are degraded by *Hirschioporus pargamenus, Trametes versicolor,* and *Bjerkandera adusta.* Many members of the genera *Ceratocystis, Ophiostoma, Graphium,* and *Alternaria* also colonize pulpwood and cause stain damage, which can increase bleaching costs.

Some examples of the weight losses associated with several pulpwood species, stored for various periods, in several locations include the following.

1. Southern yellow pine in the southern locations in rick piles, unbarked: in the *summer season,* 2–4% in 2 months, 6–8% in 4 months, 8–10% in 6 months; and in the *winter season,* 0–1.5% in 2 months, 1.5–3% in 4 months, 3–5% in 6 months. Losses for an entire year ranged from 11 to 15% (Lindgren, 1953; Lindgren and Eslyn, 1961). Subsequent studies produced similar results (Volkman, 1966).
2. Jack pine *(Pinus banksianna)* in the Lake States lost 4.5% in 1 year and 9% in 2 years (Pascoe and Scheffer, 1950).
3. Aspen *(Populus tremuloides)* in the Lake States loses 6% per year (Scheffer, 1969).
4. Losses during 6 months of summer storage in the South were 9.6% for pine, 7.2% for oak, and 13.0% for gum (Hajny, 1966).

Estimates of storage losses in pulpwood must also consider the effects of decay type on pulp yields. Brown-rot fungi, such as *Gloeophyllum sepiarium*, selectively attack carbohydrates, and associated increases in alkali solubility substantially decrease pulp yield, magnifying the storage-decay loss. White-rot fungi utilize wood components more or less uniformly and have little effect on yield on a weight basis. Their attack, may, however, require that a larger volume of wood be processed to maintain a given production level. Some white-rot fungi selectively utilize lignin, and these fungi are under intensive study for use in biological pulping.

General approaches to reducing pulpwood storage losses include rapid utilization, rick piling, and procurement schedules that maintain maximal volumes during the colder seasons. Peeling has been reported to substantially extend safe storage periods for aspen in the North. For short storage periods in the South (less than 3 months in summer and 5 in winter), Lindgren (1953) recommended retaining bark, tight piles, and favoring large diameters and long lengths. For storage periods exceeding these time limits, debarking, open piles, favoring small diameters, and splitting large diameter bolts are recommended. In the short storage period, a high initial moisture content is sought to adversely affect microorganisms, and rapid drying is the goal for the longer period.

Other controls, similar to those used for logs, have been used in some larger mills. Chemical treatments, again similar to those used for logs were attempted, but proved to be expensive and often ineffective on peeled pulpwood in the South. Underwater storage is effective, but is cumbersome because of the huge volumes of wood involved. Water sprays were introduced at some mills and proved to be very effective (Chesley *et al.*, 1956; Volkman, 1966). Large-scale experiments demonstrated that spraying could provide 12 months of safe storage at greatly reduced costs compared to conventional dry-wood storage (Djerf and Volkman, 1969). Water spraying has been accepted generally as the most effective and economical way to store both conifer and hardwood roundwood in the South for up to 12 months without appreciable deterioration damage. Minor problems with spraying include loss of parenchyma cells in outer 2.5 cm of sapwood after 4 months of storage and the occasional development of a sour odor. Prolonged storage, however, did not reduce the amounts of extractive by-products such as tall oil, turpentine, or rosins.

As an alternative to spraying, attempts were made to exclude oxygen by sealing large piles of green pulp in polyethylene enclosures (McKee and Daniel, 1966). Many fungi can survive at very low oxygen levels, making it difficult to completely limit growth. Furthermore, the material and handling costs of this technique are prohibitive. Biological protection of pulpwood by inducing *Trichoderma viride* growth on the surface was reported to reduce storage losses in some cases (Lindgren and Harvey, 1952); however, the treatment results were erratic.

Despite the effectiveness of decay-control measures on pulpwood bolts, interest in round wood storage declined with the introduction of chip piling around 1950. Chip storage of pulpwood, because of its handling advantages and relatively low losses, has now become the predominant method of wood storage.

Pulpwood Chips

Chip storage of pulpwood reduces deterioration losses, decreases handling cost, and requires smaller storage areas. It also provides an economic use for slabs, edgings, and other forms of wood waste in large, integrated, wood-conversion operations. Disadvantages include increases in dirt contamination and the loss of by-product yields (tall oil and turpentine) in some pulping operations. Recent developments in mechanized pulpwood harvesting and field chipping have also accelerated the use of chips both for pulp and as fuel. Generally, bolts or wastewood are drum debarked, chipped, and blown into large bins or carried by conveyor belts into huge conical piles (Fig. 13-5).

Safe storage periods for chips of various species in regions of the United States were initially reported as follows: southern yellow pines in the South up to 4 months; conifers in the Northwest up to 3 years; hardwoods in the Northwest up to 1 year; and hardwoods in the Northeast up to 8 months. However, it was soon found that serious deterioration developed in some chip piles after prolonged storage, and factors such as high sapwood proportion, green wood, certain species (e.g., aspen, red alder), and excessive pile height or compaction could reduce safe storage periods. In general, conifer chips can be safely stored for longer periods than hardwood chips. Large-diameter pulpwood contains more heartwood, resulting in more durable chips. Seasoned pulpwood forms drier and more durable chips. Safe storage periods are longer in the cooler northern regions, particularly when the piles are assembled in the colder seasons.

The principal defects that develop in stored chips are chemical decomposition and discolorations (brownish), sapstains, molds, and decays. Temperature in the chip pile is an important factor, determining both the type of chemical damage and location of the microorganisms (Björkman and Haeger, 1963; Eslyn, 1967; Hulme, 1979; Greaves, 1971). Many factors are involved in elevated temperatures, which develop rapidly in large chip piles. The initial temperature rise (up to 7 days) is presumed to result from respiratory activity of living parenchyma cells in the sapwood, which may remain viable for up to 6 months after cutting (Feist et al., 1971). Heat from the respiration of the huge bacterial population that rapidly develops on the chip surfaces and utilizes exogenous simple carbon sources (counts as high as 755×10^6 bacterial cells per gram of wood have been measured) escalates the pile temperature. Temperatures as high as 48.9°C may be reached after 7 days (Springer and Hajny, 1970).

Further temperature rises depend, in part, on pile features. Tall piles with excessive compaction, or accumulated layers of fines reduce air circulation,

Figure 13-5 A large pulp and paper mill in the South displaying the huge piles of chips (left, arrow) and pulpwood bolts (right, arrow) necessary to ensure continuous mill operation. Courtesy of W. S. Fuller, Weyerhauser Paper Company.

and temperatures may rise to 60 to 71°C. At these temperatures, slow heat decomposition of wood begins, and acetic acid is released (Kubler, 1982). The exothermic reaction further elevates the temperature, and the acetic acid increases wood acidity, which may reach a pH of 3. Chips exposed at these acidity and temperature levels turn brown and become friable (Fig. 13-6). Catastrophic losses from chemically degraded chips and occasional fires from spontaneous combustion can occur in piles under these high temperature conditions. The principal heat sources in chip piles and the factors affecting various temperature levels are illustrated in Fig. 13-7.

After the initial rise, temperatures decline in properly constructed and managed piles, but remain higher than external ambient temperatures (Feist

Figure 13-6 A comparison of heat-acid damaged pulpwood chips from improper pile storage with fresh chips. The small pile of chips on the right (2) was exposed to temperatures in excess of 175°F for 3 to 4 weeks, the small pile on the left (3) was exposed to these temperatures for 6 to 8 weeks and had a strong odor of acetic acid. The background chips (1) are fresh. Courtesy of W. S. Fuller, Weyerhauser Paper Company, and permission of *TAPPI* (1985).

et al., 1973a,b). Management practices generally strive to maintain pile temperatures below 60°C.

In an ecological sense, chips in a pile present a drastically different environmental niche to invading microorganisms from that of logs or roundwood during storage. Chips collectively create a huge wood surface that is inoculated with the spores and mycelial fragments of fungi and bacteria during chipping, transport, and piling. The chip surfaces are smeared with extractives and ruptured cell contents, which provide a moist film of simple carbon compounds. Chip piles are colonized rapidly within days by many of the same opportunists that invade roundwood and slash. Respiration and fermentation release considerable heat. The small size and weight of chips leads to compaction in larger piles. The pile temperature rapidly rises, and within a week or so, the pile interior becomes a hostile site where colonization is limited to *stress-resistant* thermotolerant or thermophilic fungi and bacteria. The outer shell of the pile is generally cooler and wetter than the inner zones. Chips in the outer layer of the pile are often stained, molded, or bleached. Common genera of the associated fungi are stainers such as *Ceratocystis, Ophiostoma, Graphium, Aureobasidium, Leptographium,* and *Alternaria,* molds such as

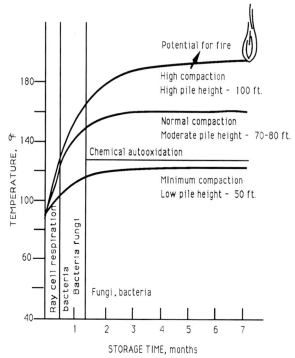

Figure 13-7 The principal heat sources in large piles of pulpwood chips and the various factors affecting the temperature levels and chip damage. Courtesy of W. S. Fuller, Weyerhauser Paper Company, and permission of the *TAPPI* (1985).

Trichoderma, Aspergillus, Gliocladium, and *Penicillium* (Wang, 1965). The bleached chips often reflect the early stages of attack by common white-rot fungi, such as *Phanerochaete gigantea, Hirschioporus (Polyporus) abietinus, Trametes versicolor,* and *Bjerkandera (Polyporus) adusta.* Other decayers such as *Gloeophyllum sepiarium* and *Fomitopsis rosea* are found in localized pockets. Common soft-rot genera include *Chaetomium, Phialophora,* and *Humicola.* The wood-staining fungi (Eslyn and Davidson, 1976), microfungi (Shields, 1969), and thermophilic fungi (Tansey, 1971) in chipwood piles have been studied, although many of their interactions are poorly documented.

Chips in the interior, warmer zones of the pile are characterized by light brown oxidative staining. Bacteria and thermophilic or thermotolerant fungi such as Actinomycetes and soft-rot fungi, predominate in the interior zone. Common thermophilic fungi include *Chaetomium thermophile* and *Talaromyces (Penicillium) duponti.* Major damage to the chips in the interior occurs through a combination of heat decomposition (acetolysis) and soft-rot decay.

As a secondary effect, the fungi colonizing stored chips can also pose health problems from the fungal spores acting as allergens and causing human mycoses (Thörnqvist and Lundström, 1982).

Chemicals were initially evaluated to minimize the microbial activity during chip storage. Eslyn (1973) evaluated a large number of chemicals and reported that dithiocarbamate and dinitrophenol showed particular promise. Propionic acid, which is used to control microorganisms in grain storage, has also been evaluated. Chemicals are sprayed on the chips as the pile is assembled; however, chemical control approaches to reduce chip storage losses are largely impractical owing to high relative costs and environmental concerns.

As an alternative, water spraying was evaluated but provided no benefits over dry storage (Djerf and Volkman, 1969). Current control approaches are to reduce storage times to absolute acceptable minimums and follow pile-management practices to minimize conditions favoring chip deterioration. Current recommendations include limiting pile heights to 15 m (50 ft) and restricting tractor spreading to avoid compaction, which reduces air circulation and sets the stage for disastrous temperature increases (Fuller, 1985). Chips from sources that deteriorate rapidly (hardwoods, such as aspen and alder, or whole-tree chips) should be piled separately to avoid creating heat-generating pockets in the pile. Steps should be taken to avoid layering of fines or sawdust in pile zones. Whole-tree chips, which are particularly subject to heating, should be stored separately in smaller piles (less than 8 m) and for periods not to exceed 4 weeks.

An important step in reducing chip deterioration is to monitor pile condition and temperatures routinely. Rapid temperature rises and the strong odor of acetic acid are signs that a pile is out of control and heading for serious damage. Rapid utilization or ventilation of hot spots in the pile by trenching are then necessary. Prompt remedial measures are needed whenever pile temperatures reach 79°C. Other useful procedures in large operations are reclaim systems, which obtain chips from the base of pile using screw conveyors. These devices permit a first-in, first-out practice, thereby reducing the storage time of individual chip batches. Blending of chips from storage piles of various ages just before use has been proposed as a useful way to avoid quality variations in chips due to differing storage periods (Schmidt, 1990).

A large literature on pulpwood-chip storage has developed on topics such as optimal size and packing of piles; types and sequences of microorganisms that occur in various pile locations and chip types; methods of monitoring weight losses in piles; and the effects of microorganism types and storage periods on paper quality (Hatton, 1979). New information is now accumulating rapidly on the uses of chips for fuel wood (White *et al.*, 1983) and paper (Auchter, 1975); however, many mills have only a limited knowledge of their chip-storage problems. Increasing wood prices will no doubt renew interest in chip quality.

Summary

- Logs, poles, bolts, and wood chips contain substantial amounts of sapwood, which is susceptible to rapid colonization by microorganisms.

Summary continues

- Stored wood is subject to severe damage within months during warm periods unless properly managed.
- The principal losses from prolonged or improper storage of raw wood are decay, chemical and biotic discolorations, insect damage, deep checking, and in special cases, fires from internal combustion in chipwood piles.
- General methods for reducing wood-storage losses are rapid utilization, initiating storage piles during a cold season, storing under water, water spraying or sprinkling, and conversion to chips.
- Visible decay can develop 30 to 45 cm in from each end in hardwood logs stored during the summer season. Sapstains and insect damage in *Pinus* spp. and susceptible hardwoods can seriously degrade high-quality lumber within a month of storage during a warm season.
- Submerging green logs in water (ponding) or continuous spraying protects wood from storage losses for an entire warm season.
- Prompt kiln-drying of southern yellow pine poles before preservative treatment is generally practiced in the South to eliminate pretreatment decay. In the Pacific Northwest, Douglas fir poles should not be air-seasoned for longer than 2 years, and seasoning should be followed by preservative treatments that heat the wood to at least 65.5°C for 75 min.
- Decay and stain attack in stored pulpwood bolts reduces both the yield and quality (brightness and strength) of the pulp. Density losses in southern pine pulpwood stored for a year in the South may range from 11 to 12%.
- Rapid utilization, storage during colder seasons, and continuous water spraying or ponding can reduce pulpwood storage losses.
- The conversion of pulpwood to chips for storage has become common practice because of handling ease, the small storage spaces required, low deterioration losses, and the ability to use other wood wastes. Initial safe storage periods of pulpwood in chip piles were reported as a few months in the South and up to several years in the Northwest. A disadvantage of chipping is the rapid loss of tall oil, turpentine and other valuable by-products.
- Chemical decomposition and discoloration of chips can develop when the piles are too large and become compacted. Respiration of parenchyma cells and fermentation by bacteria rapidly elevate the interior temperature of piles within 1 week of initial assembly. In large or compacted piles, the temperatures reach levels that chemically decompose the wood and release acetic acid.
- The elevated temperatures in chip piles create highly selective conditions for thermophilic and thermotolerant fungi and bacteria. Soft-rot fungi are the principal cause of biotic damage to stored chips.
- Common practices for minimizing wood damage during storage in

Summary continues

> chip piles are rapid turnover, limiting pile heights (< 15 m), avoiding compaction, and limiting accumulation of fines.
> - The development of procurement, storage, and handling systems to minimize deterioration losses will be increasingly important to wood-using industries as wood raw-material costs increase.

References

Auchter, R. J. (1975). Whole-tree utilization: Fact or fancy. *TAPPI* **58**(11):4–5.

Björkman, E., and G. E. Haeger. (1963). Outdoor storage of chips and damage by microorganisms. *TAPPI* **46**(12):757–766.

Benko, R., and T. L. Highley. (1990). "Evaluation of Bacteria for Biological Control of Wood Decay." International Research Group on Wood Preservation Document No. IRG/WP/1426. Stockholm, Sweden.

Carpenter, B. E., and E. R. Toole. (1963). Sprinkling with water protects hardwood logs in storage. *Southern Lumberman* **207**(Aug. 1):25–26.

Chesley, K. G., J. C. Hair, and J. N. Swartz. (1956). Underwater storage of southern pine pulpwood. *TAPPI* **39**(9):609–614.

Chidester, M. S. (1939). Further studies on temperatures necessary to kill fungi in wood. *Proceedings of the American Wood Preservers' Association* **35**:319–324.

Cowling, E. B., W. L. Hafley, and J. Weiner. (1974). Changes in value and utility of pulpwood during harvesting, transport, and storage. *TAPPI* **57**(12):120–123.

Djerf, A. C., and D. A. Volkman. (1969). Experiences with water spray wood storage. *TAPPI* **52**(10):1861–1864.

Ellwood, E. L., and B. A. Ecklund. (1959). Bacterial attack of pine logs in pond storage. *Forest Products Journal* **9**(9):283–242.

Eslyn, W. E. (1967). Outside storage of hardwood chips in the Northeast II. Microbiological effects. *TAPPI* **50**(6):297–303.

Eslyn, W. E. (1973). Evaluating chemicals for controlling biodeterioration of stored wood chips. *Forest Products Journal* **23**(11):21–25.

Eslyn, W. E., and R. W. Davidson. (1976). Some wood-staining fungi from pulpwood chips. *Memoirs of the New York Botanical Garden* **28**(1):50–57.

Feist, W. C., G. J. Hajny, and E. L. Springer. (1973a). Effect of storing green wood chips at elevated temperatures. *TAPPI* **56**(8):91–95.

Feist, W. C., E. L. Springer, and G. J. Hajny. (1971). Viability of parenchyma cells in stored green wood. *TAPPI* **54**(8):1295–1297.

Feist, W. C., E. L. Springer, and G. J. Hajny. (1973b). Spontaneous heating in piled wood chips—contribution of bacteria. *TAPPI* **56**(4):148–151.

Fuller, W. S. (1985). Chip pile storage—a review of practices to avoid deterioration and economic losses. *TAPPI* **68**(8):48–52.

Greaves, H. (1971). Biodeterioration of tropical hardwood chips in outdoor storage. *TAPPI* **54**(7):1128–1133.

Hajny, G. J. (1966). Outside storage of pulpwood chips—a review and bibliography. *TAPPI* **49**(10):97a–105a.

Hansbrough, J. R. (1953). Storage of northern hardwood logs and bolts. *Journal of the Forest Products Research Society* **3**(2):33–35, 92.

Hatton, J. V. (1979). "Chip Quality Monograph." TAPPI Pulp and Paper Technology Series No. 5. Technical Association of the Pulp and Paper Industry, Atlanta, Georgia.

Hulme, M. A. (1979). Outside chip storage—deteriorating during OCS. In "Chip Quality Monograph" (J. V. Hatton, ed.), pp. 213-244. Joint Textbook Committee of the Pulp and Paper Industry, Atlanta, Georgia. Chapter 11.

Hulme, M. A., and D. W. Stranks. (1976). Heat tolerances of fungi inhabiting chip piles. *Wood Science* **8**:237-241.

Humphrey, C. J. (1917). "Timber Storage Conditions in the Eastern and Southern States with Reference to Decay Problems." Bulletin 510. U.S. Department of Agriculture, Washington, D.C.

Isaacson, I. H. (1958). New water fog spray method reduces log losses in storage. *Wood and Wood Products* **63**:1, 25-26, 60-62.

Knuth, D. T., and E. McCoy. (1962). Bacterial deterioration of pine logs in pond storage. *Forest Products Journal* **12**(9):437-442.

Kress, O. (1925). "Control of Decay in Pulp and Pulpwood." Bulletin 1298, U.S. Department of Agriculture, Washington, D.C.

Kubler, H. (1982). Heat release in thermally disintegrating wood. *Wood and Fiber* **14**(3):166-177.

Lane, P. H., and T. C. Scheffer. (1960). Water sprays protect hardwood logs from stain and decay. *Forest Products Journal* **10**:277-282.

Lindgren, R. M. (1953). Deterioration losses of stored southern pine pulpwood. *TAPPI* **36**(6):260-263.

Lindgren, R. M., and W. E. Eslyn. (1961). Biological deterioration of pulpwood and pulp chips during storage. *TAPPI* **44**(6):419-429.

Lindgren, R. M., and G. M. Harvey. (1952). Decay control and increased permeability in southern pine sprayed with fluoride solutions. *Journal of the Forest Products Research Society* **2**:250-256.

Mason, R. R., J. M. Muhonen, and J. N. Swartz. (1963). Water-sprayed storage of southern pine pulpwood. *TAPPI* **46**(4):233-240.

McKee, J. C., and J. W. Daniel. (1966). Long-term storage of pulpwood in sealed enclosures. *TAPPI* **49**:47a-50a.

Morrell, J. J., M. E. Corden, R. D. Graham, B. R. Kropp, P. Przybylowicz, S. M. Smith, and C. M. Sexton. (1987). Basidiomycetes colonization of air-seasoned Douglas fir poles. *Proceedings of the American Wood Preservers' Association* **83**:284-296.

Morrell, J. J., R. D. Graham, M. E. Corden, C. M. Sexton, and B. R. Kropp. (1989). Ammonium bifluoride treatment of air-seasoning Douglas fir poles. *Forest Products Journal* **39**(1):51-54.

Newbill, M. A., J. J. Morrell, and K. L. Levien. (1988). Internal temperature in Douglas fir poles during treatment with ammoniacal copper arsenate or pentachlorophenol. *Proceedings of the American Wood Preservers' Association* **84**: 48-54.

Panek, E. (1963). Pretreatments for the protection of southern yellow pine poles during air-seasoning. *Proceedings of the American Wood Preservers' Association* **59**: 189-202.

Pascoe, T. A., and T. C. Scheffer. (1950). Jack pine pulpwood deterioration in yard storage. *Paper Trade Journal* **131**:20-21.

Przybylowicz, P., B. R. Kropp, M. E. Corden, and R. D. Graham. (1987). Colonization of Douglas fir poles by decay fungi during air-seasoning. *Forest Products Journal* **37**(4):17-23.

Scheffer, T. C. (1969). Protecting stored logs and pulpwood in North America. *Material und Organismen* 4. Bd. Heft 3. 167–199.
Scheffer, T. C., and W. E. Eslyn. (1976). Winter treatments protect birch roundwood during storage. *Forest Products Journal* 26(1):27–31.
Scheffer, T. C., and T. W. Jones. (1953). "Control of Decay in Bolts and Logs of Northern Hardwoods during Storage." Paper 63, U.S.D.A. Northeast Forest Experiment Station, Upper Darby, Pennsylvania.
Scheffer, T. C., and R. A. Zabel. (1951). "Storage of Beech Logs and Bolts in the Northeast." Northeast Forest Experiment Station, Beech Utilization Series, Forest Service, U.S.D.A. Upper Darby, Pennsylvania.
Scheld, H. W., and R. C. DeGroot. (1971). Toughness of sapwood in water-sprayed longleaf pine logs. *Forest Products Journal* 21(4):33–34.
Schmidt, R. L. (1990). The effect of wood chip inventory rotation policies on storage costs, chip quality, and chip variability. *TAPPI* 73(11):211–216.
Seifert, K. A., C. Breuil, L. Rossignol, M. Best, and J. H. Saddler. (1988). Screening microorganisms with the potential for biological control of sapstain on unseasoned lumber. *Material und Organismen* 23:81–95.
Shields, J. K. (1969). Microflora of eastern Canadian chip piles. *Mycologia* 61: 1165–1168.
Shields, J. K., and E. A. Atwell. (1963). Effect of a mold, *Trichoderma viride*, on decay of birch by four storage-rot fungi. *Forest Products Journal* 13:262–265.
Smith, S. M., R. D. Graham, and J. J. Morrell. (1987). Influence of air-seasoning on fungal colonization and strength properties of Douglas fir pole sections. *Forest Products Journal* 37(9):45–48.
Springer, E. L., and G. J. Hajny. (1970). Spontaneous heating in piled wood chips. 1. Initial mechanism. *TAPPI* 53(1):85–86.
Tansey, M. R. (1971). Isolation of thermophylic fungi from self-heating industrial wood chip piles. *Mycologia* 63:537–547.
Taylor, J. A. (1980). Pretreatment decay in poles. *Proceedings of the American Wood Preservers' Association* 76:227–245.
Thörnqvist, T., and H. Lundström. (1982). Health hazards caused by fungi in stored wood chips. *Forest Products Journal* 32(11/12):29–32.
Volkmann, D. (1966). Water-sprayed storage of southern pine pulpwood. *TAPPI* 49:48A–53A.
Wang, C. J. K. (1965). "Fungi of Pulp and Paper in New York." Technical Publication 87, State University of New York, College of Forestry, Syracuse, New York.
White, M. S., M. L. Curtis, R. L. Sarles, and R. W. Green. (1983). Effects of outside storage on the energy potential of hardwood particular fuels: Part 1. Moisture content and temperature. *Forest Products Journal* 33(6):31–38.
Zabel, R. A., F. F. Lombard, and A. M. Kenderes. (1980). Fungi associated with decay in treated Douglas fir transmission poles in the northeastern United States. *Forest Products Journal* 30(4):51–56.

CHAPTER
14

Wood Stains and Discolorations

The natural beauty of wood is a major reason for its widespread use in construction and furniture, and many users pay a premium for clear, defect-free wood. Unfortunately, a number of chemical and biological agents can discolor wood, reducing its aesthetic and, sometimes, structural value. The majority of objectionable discolorations appear in the sapwood during seasoning and storage and are primarily biological in origin.

Stains or discolorations can be defined as abnormal color patterns that develop in wood and adversely affect its value. Most stains are readily delineated from normal color patterns. Recognition of stain depends on familiarity with normal wood-color variations, including heartwood/sapwood differences, color variations associated with buried sapwood, reaction wood, wound zones, and annual-ring variations.

Stains cause significant losses in wood quality worldwide, but the exact cost of these losses is difficult to assess. Scheffer estimated that annual losses due to stain exceeded $50 million in the United States (1973). Stain losses generally increase with increasing sapwood content and become more severe as climate becomes more conducive to fungal growth. Losses associated with stains have become more severe in recent years because of an emphasis on natural finishes and the growth of an export market for clear, light wood timbers.

This chapter will review the historical significance of stains, describe the major types of stains associated with different wood species, and emphasize methods for preventing stain development.

History

For many years, the presence of some stain was accepted by wood users, but the shift to second-growth timbers with higher percentages of stain-susceptible sapwood along with the importation of tropical hardwoods, which tended to stain in transit, encouraged efforts to determine stain causes, to identify the organisms associated with biological stains, and to develop effective prevention measures (Boyce, 1927). Four comprehensive publications summarized the status of sapstain research from 1935 to 1959 and presented effective control strategies, some of which are still practiced (Scheffer and Lindgren, 1940; Verrall, 1945; Verrall and Mook, 1951; Findlay, 1959). Most stain-prevention strategies depend on either rapid processing to dry the wood below the fiber saturation point or surface application of prophylactic biocides to prevent microbial colonization of the wood. Sodium pentachlorophenate and ethyl mercury phosphate were the primary biocides used initially for this purpose, but increasing concerns about the safety of these chemicals stimulated research to identify safer materials. Initially, the protection period was limited largely to the early months of air-seasoning, but protection may be needed for as long as a year with solid-piled unseasoned lumber for export shipments.

Types of Wood Discoloration

The many kinds of wood discoloration can be conveniently grouped in four general categories based on the source of discoloration, as follows:

1. Color changes resulting from enzymatic and chemical activity that develops on the surface or deep in the wood shortly after harvesting or sawing;
2. Color changes caused by contact with chemicals;
3. Color changes associated with the early stages of decay; and
4. Color changes associated with the growth of fungi on the wood surface or deep within the sapwood.

The first two types are commonly referred to as chemical, enzymatic, or oxidation stains, whereas the latter two are biotic stains.

Enzymatic and Chemical Stains

Many lumber discolorations result from chemical changes in cell contents shortly after the wood is exposed to air. These oxidative stains have long been studied (Bailey, 1910), but they remain poorly understood. These reactions are analogous to the browning reactions that occur in freshly cut fruits and may be related to the defense reactions of wounded tissues in living stems.

Kiln-Brown or Coffee Stain   These dark discolorations develop on the surface or deep within ponderosa pine, white pines (eastern, western, and sugar), or western hemlock wood during kiln drying (Hubert, 1931; Barton and Gardner, 1966). The dark brown color is frequently concentrated at the ends of the board, in the vicinity of knots and at the heartwood–sapwood interface. The stain develops in both the sapwood and heartwood and is characterized by longitudinally oriented lenticular-brown discolorations, with occasional white zones or bands between the discolored zones (Fig. 14-1A). Kiln-brown stain is believed to occur as the result of an enzymatic reaction involving a peroxidase and subsequent oxidation or polymerization of a leuco product in a two-step chemical process (Stutz, 1959). High kiln temperatures then cause the polymerization and oxidation that produce the colored compounds (tannins and phlobotannins). Peroxidase activity on phenolic extractives is apparently accelerated at the moisture and oxygen levels that occur in freshly sawn boards during such periods. The stain appears more frequently during warm, humid periods and is exacerbated by bulk piling of freshly sawn lumber. The stain is deeply penetrating and cannot be removed by surface planing (Fig. 14-1B). The risk of stain can be minimized by using mild kiln schedules in which the dry-bulb temperature does not exceed 150°F (65.5°C). Formerly, dip applications of sodium azide and sodium fluoride were found to prevent kiln-brown stain (Stutz, 1959; Stutz et al., 1961; Cech, 1966); however, these compounds are toxic, and safety concerns have largely limited their use. Ammoniacal zinc oxide (Shields et al., 1973) and several alkaline salts (Hulme, 1975) were shown to provide some control. Phosphoric acid and several β-hydroxyquinoline compounds are environmentally acceptable and control brown stain in sugar pine under laboratory conditions (Oldham and Wilcox, 1981). More recently, Hulme and Thomas (1983) have shown that dipping freshly sawn boards in water solutions of 5% sodium sulphite or sodium thiosulfate is an effective and safe control of brown stain in eastern white pine.

A similar stain has also been noted on Douglas fir and develops rapidly under moist, warm conditions (Miller et al., 1983). Water-soluble extractives migrate to the wood surface, where they undergo oxidation to produce a brown, polymerized pigment. The stain usually is close to the surface of the board, but has been observed deeper in bulk-piled boards. Brown stain appears sporadically, making it difficult to collect material for routine study.

Developing a more comprehensive understanding of the nature of oxidative stains might provide more useful insights into the prevention of these stains. One hypothesis for certain oxidative stains suggests that bacteria in the freshly harvested wood either alter wood pH or secrete enzymes that promote the formation of pigmented compounds (Yazaki et al., 1985). Attempts to confirm this effect in the brown-oxidative stain of Douglas fir have, so far, proven negative.

There is also a brown sapstain caused by a *Cytospora* sp., which occurs commonly on several hard pines (jackpine, red pine, and ponderosa pine) and occasionally on white pines (eastern, western and sugar) in northern regions. These biological stains can be confused with the chemical brown stain (Hubert, 1931; Fritz, 1952). Despite the similar brown colors, the *Cytospora* stain is limited to the sapwood and produces numerous dark flecks in the wood. In addition, the whitish margins that characterize the chemical brown

Types of Wood Discoloration / 329

Figure 14-1 Chemical brown or coffee stain on kiln-dried eastern white pine. (A) The appearance on the unplaned board surface and (B) the same stain on a planed board, indicating that the stain develops deep in the wood tissues. Note the intense stain development around a knot.

stain are absent in the fungal colonized wood. An abundance of large, septate, brown hyphae in the *Cytospora*-stained wood cells readily separates the similarly colored stains microscopically. The *Cytospora* stain develops in logs only after prolonged storage, and studies indicate that the stained wood is not weakened (Fritz, 1952).

Oxidative Stains of Hardwoods Many hardwoods develop deep yellow to reddish brown discolorations on the surface when exposed to air immediately after sawing or peeling. These discolorations are especially noticeable on cherry, birch, red alder, sycamore, oak, maple, and sweet gum. This stain develops in red alder, the oaks, birch, and maple during air-seasoning, and is intensified at the point where a sticker contacts the wood, hence the common name *sticker stain*. This stain is absent in wood that is immediately kiln dried and appears less frequently under cool conditions. In 1910, Bailey had already established the probable enzymatic nature of this stain in alder and several birches and reported that it was controlled by immersion in boiling water for several minutes. This method has not proven practical and is rarely used. There are no other reported methods for preventing these stains, although treating stickers with a 4% solution of sodium hydrogen sulphite appears to minimize the intensity of sticker stain.

A related gray stain on several southern oaks also appears to be oxidative (Clark, 1956), and preliminary field studies suggest it can be prevented by dipping boards in a 10% sodium bisulfite water solution and storing, solid-piled for 14 days under cover, before seasoning (Forsyth, 1988).

Mineral Stain or Streak This a puzzling degrading stain that appears in several hardwoods and particularly maples in the Northeast and the Lake States (Scheffer, 1939, 1954). The stain is variable in occurrence and may appear in streaks or as a broad discoloration; moreover, it may be present in small clusters of stems in a stand and be absent elsewhere. Mineral stain appears in both the sapwood and heartwood of lumber, although it develops in the sapwood of the living tree. The stain may reflect tree response to multiple or successive wounds in the sapwood zone; however, little is known about the cause, and there is some confusion about the range of discolorations. Mineral stain on sawed maple generally appears as lenticular-shaped streaks, which range in color from a deep olive to a green-black color. The streaks will often evolve small bubbles of CO_2 when treated with a mineral acid, denoting deposition of carbonates. The stained wood is denser and harder than normal wood and, if heavily stained, it tends to twist and warp badly when dried (Fig. 14-2). The wood also tends to split when nailed and is essentially useless for construction purposes.

Examination of stained versus sound wood indicated that the mineral content of stained wood was about one third greater, and it had a much higher pH than sound wood (Good *et al.*, 1955). Levitin (1970) found that parenchyma cells in mineral-stained wood were filled with numerous brown amber

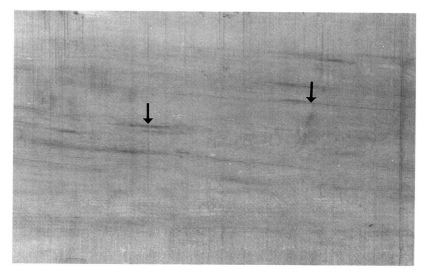

Figure 14-2 Mineral stain on the surface of an air-seasoned sugar maple board showing the characteristic lenticular-shaped streaks of a deep olive to green-black color (arrows). The board also was cupped, warped, and unusually dense.

deposits, that soluble polyphenols were present at higher levels in stained wood, and that condensed polyphenols deposited on the cell walls were not easily removed by solvents or bleaches. He also noted that tannate salts of magnesium (Mg), potassium (K), and calcium (Ca) formed compounds that range from brown to green in color. Based on these findings, he proposed that wounding resulted in enzymatic hydrolysis of phenolic glycosides to polyphenols, which are, in turn, oxidized to form pigmented compounds. A similar yellow-, purple-, or brown-hued pattern of variegated colors may develop in the lower stem of living yellow poplar trees and is commonly termed *blue butt* or mineral stain by lumbermen. This stain is also believed to develop in the sapwood of living trees following wounding and is grouped with the oxidative stains (Roth, 1950).

Iron Stain An intense black stain sometimes develops on the freshly sawn sapwood of some species when the woods come in contact with iron. Iron tannate forms on the wood surface of lumber with high tannin contents, such as the birches, cherry, sweet gum, and oaks, but is easily removed by planing. Iron stain is also common around nails in coniferous lumber. Dipping material in sodium carbonate solutions prevents the discoloration, whereas oxalic acid can be used to remove the stain. Alternatively, the use of nonferrous metals wherever wood contacts machinery in the sawmill can eliminate iron stain; however, this is expensive and largely impractical.

Color Changes Associated with Incipient Decay

As wood is decayed, a number of subtle color changes may also occur, and the wood may develop shades of red, brown, purple, gray, or mottled white (see early decay in Chapter 7). In general, these discolorations may be distinguished from those caused by typical staining fungi by the presence of dark zone lines, textural changes on the sawn surface, or an irregular shape of the color patterns, which may not coincide with annual rings.

Fungal Stains or Molds

Molds

Molds are generally of minor importance in wood and are primarily a factor in very wet wood, such as wood that has been solid-piled or wood that has been covered to restrict aeration. Most molds are airborne, opportunistic fungi with hyphae that are normally colorless, but discolor the wood by forming masses of pigmented spores on the wood surface. Mold discoloration of coniferous woods can most often be removed by brushing or planing the wood surface; however, the discoloration on hardwoods is often deeper and may be more persistent. Common mold fungi and their discolorations include *Aspergillus* spp. (black), *Fusarium* spp. (red or violet), *Gliocladium* spp. (green), *Monilia* spp. (orange), *Penicillium* spp. (green), *Rhizopus* spp. (black), and *Trichoderma* spp. (green). Some molds, such as *Monilia* and *Aspergillus* species, can cause both allergic reactions and worker health problems.

Molds are particularly common on hardwoods when the wood is very wet, but some can develop on wood held for long periods at high relative humidities. Molds tend to enter ruptured cells, vessels (hardwoods), and exposed rays, and spread from cell to cell via the pits. As they attack the pit membranes, these fungi make the wood more receptive to fluids. Treatments with molds have been proposed as a method for improving the permeability of Douglas fir, the spruces, and other difficult-to-treat species (Schulz, 1956; Lindgren and Wright, 1954). *Trichoderma* colonization, however, has also been shown to inhibit colonization of pine pulpwood by decay and stain fungi during storage (Lindgren, 1952; Hulme and Shields, 1972). The principles behind this deterrent effect are now being exploited for possible biological protection of wood from decay fungi.

In addition to their protective and permeability effects, a number of molds have been shown to detoxify preservatives (Brown, 1953; Stranks and Hulme, 1975; Verrall, 1949). These include *Penicillium (cyclopium) aurantiogriseum* on mercury compounds, *Scopulariopsis brevicaulis* on arsenic compounds, *Hormoconis (Cladosporium) resinae* on creosote, and *Trichoderma* sp. on sodium fluoride. The roles of these fungi and the significance of their detoxification abilities on treated wood in the natural environment remains unknown, but under ideal conditions, they could detoxify preservative treatments, permitting decay fungi to colonize the treated wood.

Sapstains

Major sapstains are caused by fungi with pigmented hyphae that grow primarily in the parenchyma tissues of the sapwood (Fig. 14-3A,B). Common sapwood discolorations are shades of blue, black, brown, and gray. Occasional stains of lesser importance are yellow, pink, purple, and green. These latter stains are the result of pigments secreted from the fungal hyphae. The predominance of a bluish discoloration and restriction to the sapwood have resulted in the major stains' being termed *blue stain* or *sapstain*. Fungal stains are worldwide in distribution, but become a more serious problem when species containing high percentages of sapwood are harvested and seasoned under warm, humid conditions, suitable to rapid fungal growth. Sapstain poses major challenges in the southeastern United States because of the ideal climate for fungal growth, coupled with the large percentages of sapwood present in species such as the southern yellow pines, sweet gum, and yellow poplar. White pine in the Northeast and Lake States is similarly affected. In California, similar conditions in the spring and fall make stain prevention difficult during air-seasoning of sugar and ponderosa pine. Sapstain is also a major hazard in western white pine and ponderosa pine in the Inland Empire (Idaho, Montana, Wyoming) during the warmer months. The export of solid-piled green lumber in the Pacific Northwest has led to increasing stain problems in Douglas fir and western hemlock lumber (Cserjesi, 1977).

In all of these cases, fungi rapidly colonize the ray parenchyma to utilize the readily available storage carbohydrates (Fig. 14-3B). As they grow through the wood, the pigmented hyphae of the major fungi discolor the wood. Interestingly, often the hyphae produce dark, melanin-based pigments (Zink and Fengel, 1988, 1989, 1990), but the effect of these darkened hyphae is to color the wood blue. Most stain fungi are in the *Ascomycotina* or *Deuteromycotina*. Several of the important staining fungi in the major timber species in the United States are listed in Table 14-1. Käärik (1980) has assembled a useful, detailed list and described the principal features of the sapwood-staining fungi in temperate zones.

Many stain fungi are specific to a region or wood species. The staining fungi can be placed into two broad groups. Some of the fungi, and particularly those in the genera *Ophiostoma* and *Ceratocystis*, are closely tied with the life cycles of bark beetles and other wood-inhabiting insects (Verrall, 1941; Dowding, 1969, 1970). The spores are sticky and transmitted primarily by insects (vectors) and water splashing or aerosols. These fungi invade and damage wood primarily during log storage and the initial stages of lumber seasoning. The other group of stain fungi, such as *A. pullulans*, *Alternaria alternata*, and *Cladosporium* spp. are general opportunists, whose dry spores are primarily disseminated in air. These fungi invade wood in a wide range of uses when conditions conducive to fungal growth develop.

Like all biological agents, stain fungi require free water, adequate temperature, oxygen, and a food source. Mill operators have long exploited their need for oxygen by flooding or spraying logs to raise the moisture content and exclude oxygen from the wood surface (see Chapter 13). Stain fungi grow at a

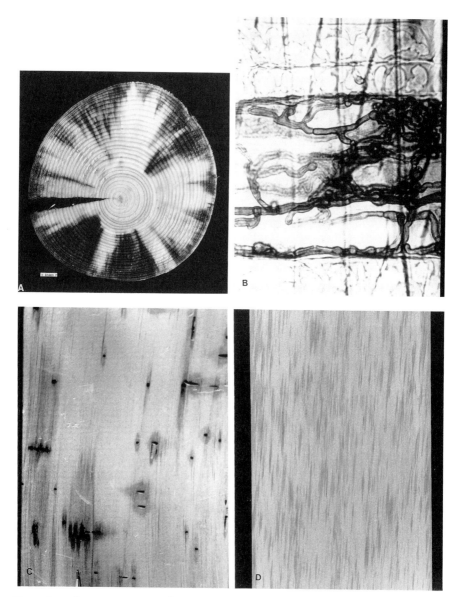

Figure 14-3 Sapstain in eastern white pine. (A) A cross section of pulpwood bolt showing the wedge-shaped pattern of stain development in the sapwood, resulting from surface infections in shallow checks and rapid growth of the fungus in ray tissues. (B) A photomicrograph of the radial section of wood stained by *Aureobasidium pullulans*, indicating preferential colonization of parenchyma cells in the wood rays by the black hyphae (approximately ×600). (C) Ambrosia beetle attack in a log indicating a sapstain and a symbiont stain (restricted to tunnel site and shown by arrows) associated with the insect tunnels. (D) Multiple stain infections (parenchyma cells remain living in stored logs for several weeks), which developed on the surface of a solid-piled board several weeks after heavy inoculum exposure from prior long stain during its sawing. (Photograph A U.S. Forest Service, courtesy Dr. R. DeGroot.)

Table 14-1
Some important sapwood-staining fungi in the United States[a,b]

Conifers	Hardwoods
Ophiostoma (Ceratocystis) ips	*Ophiostoma (Ceratocystis) pluriannulatum*
Ophiostoma (Ceratocystis) piliferum	*Ceratocystis moniliformis*
Ophiostoma (Ceratocystis) piceae	*Lasiodiplodia theobromae (Diplodia natalensis)*
Aureobasidium pullulans	*Ceratocystis coerulescens*
Leptographium lundbergii	*Graphium rigidum*
Alternaria (tenius) alternata	
Cephaloascus fragrans	
Cladosporium spp.	
Lasiodiplodia theobromae	
Phialophora spp.	

[a]Scientific names are based on current listings in Farr *et al.* (1989). In some cases, names that appear commonly in the older literature are placed in parentheses.
[b]From Scheffer and Lindgren (1940); Verrall (1939, 1942); Roff (1973); Davidson (1935, 1942, 1971); Davidson and Eslyn (1976); Zabel (1953); and Käärik (1980).

broad range of temperatures (4 to 30°C), making it difficult to avoid stain development without special measures except in the coolest and driest climates.

Stain Development

Fungal stains most often originate from the germination of spores on the freshly sawn wood surface. These spores are mostly airborne, or carried by insect vectors such as bark or ambrosia beetles. Stain can also develop when hyphae from previously stained stickers grow into the freshly sawn wood. Sawmill machinery can also serve as an inoculum source, inoculating boards for some time after a badly stained log has been sawn. Under favorable staining conditions, a single board may develop scores of distinct stained zones (Fig. 14-3D).

Under favorable conditions, spores germinate within hours of landing and penetrate the wood surface through ruptured tracheids and exposed wood rays. The hyphae then rapidly colonize the parenchyma cells in the wood rays or longitudinal parenchyma surrounding the resin canals by direct pit penetration. Stain fungi can grow up to 0.5 mm in the tangential plane, 1 mm radially, and 5 mm longitudinally over a 24-hr period under ideal conditions (Lindgren, 1942). Movement inward along the rays from the sides is responsible for the wedge-shaped stain forms that are often seen on the crosscut surfaces of logs or pulpwood bolts. The rapid fungal growth rate permits extensive colonization of freshly sawn materials and highlights the importance of stain control at the earliest possible point after sawing. Many times, pigmentation develops 5 to 6 days after hyphal development, making it appear that the stain literally exploded throughout the wood, when in fact it had been there all along. After several weeks, the parenchyma cell walls are often badly eroded, and hyphae occasionally penetrate the wood cell walls directly to move between tracheid and fiber walls in the radial plane. A few stain fungi, after prolonged incubation under ideal moisture conditions, will act as typical soft-rot fungi and attack the S2 layer of tracheid walls. Among these species are *Alternaria alternata*, *Phialocephala dimorphospora* (Wang and Zabel, 1990), and *Ceratocystis picea* (Levy, 1967).

Effects of Fungal Stain on Wood Properties

Stain fungi produce both aesthetic and physical changes in wood value. The dark, melanistic pigments produced by stain fungi consume more bleach when stained wood is pulped, thereby increasing paper-production costs. Conversely, stained wood has aesthetic value to some users who purchase *blue pine* panels, which are actually stained wood. Although the most obvious effect of stain is wood discoloration, some stain fungi also alter other wood properties. Reductions in toughness have been noted for some fungi, but the strength effects are not consistent among stain fungi (Chapman and Scheffer, 1940). For example, *Leptographium lundbergii* has no effect on toughness, whereas toughness losses associated with *Alternaria alternata* may approach 40% (Crossley, 1956). Stained wood is generally not recommended for structural purposes where strength is critical and is not used for structures such as utility poles, glue-laminated timbers, ladders, or piling. Conditions favorable for stain development are also conducive to decay initiation, so decay and stain development in wood are often coupled. In addition to strength effects, stain fungi increase wood permeability by removing pit membranes. As a result, the wood will wet and dry more rapidly. These effects are particularly important in the finishing industry, since stained wood will absorb excessive solution and will often finish unevenly. This material will also absorb water more quickly in service, increasing the development of checks, which provide points of entry for decay fungi (Björkman, 1947). Although stained wood is no more susceptible to decay than sound wood, greater hygroscopicity in stained wood tends to create conditions conducive to fungal growth for longer periods, increasing the risk of fungal decay. Furthermore, stained wood may provide an ideal inoculum source for paint-disfiguring fungi (Zabel and Terracina, 1980).

Sapstain Control

Stain control can be accomplished either through rapid drying to reduce wood moisture content, or dipping or spraying with fungicidal solutions to protect the surface against fungal invasion. Dipping is relatively inexpensive: 1000 bd ft of 1-inch rough lumber absorbs about 15 gal of treating solution, costing a few cents per gallon. Where stain conditions are severe, both fungicidal protection and good drying practices are required.

Drying procedures can include kiln drying or employing proper piling procedures, which promote extensive air-flow around the wood (Fig. 14-4). In the southeastern United States, kiln drying represents the best alternative,

Figure 14-4 A representative lumber pile displaying proper piling procedures. Note the high foundation, the roof covering with overhang, the sticker spacing, and use of a narrow pile to facilitate adequate cross circulation of air during both air seasoning and later storage. Courtesy of the Forest Service, U.S. Department of Agriculture.

although air-seasoning is often used for large timbers such as railroad ties. Decay and stain will undoubtedly occur during air-seasoning and are often referred to as *stack burn*. Air-seasoning is commonly used in the western United States, where drier conditions and wood species containing lower percentages of stain-susceptible sapwood make these practices feasible.

Chemical treatments for preventing stain have been employed since the early 1900s, when water solutions of sodium carbonate or borax were applied to the wood. These chemicals are relatively mobile and provide relatively short-term protection. In the 1930s, significant losses to fungal stains in the southern and western pines led to an extensive chemical evaluation program by the U.S. Department of Agriculture, resulting in the testing and use of chlorinated phenols and organic mercury compounds (Scheffer and Lindgren, 1940; Verrall and Mook, 1951; Zabel and Foster, 1949). These chemicals were traditionally applied to lumber by dipping it in tanks or by passing it through a spray of the chemical. More recently, concerns about worker exposure and environmental contamination have led to the development of elaborate enclosed spray booths that minimize chemical exposure. In the late

1960s, concerns about the safety of mercury compounds resulted in the elimination of this chemical, and most mills depended on sodium pentachlorophenate (penta) or tetrachlorophenate, alone or with borax, for stain prevention. In recent years, the presence of traces of dioxin in some penta products has led to use of alternative chemicals for dipping and spraying in North America, although penta continues to be used in many developing countries.

The restriction of penta has resulted in an extensive evaluation program to identify acceptable alternatives (Cserjesi, 1980; Cserjesi and Roff, 1975; Cserjesi and Johnson, 1982; Drysdale and Preston, 1982; Cassens and Eslyn, 1983; Eslyn and Cassens, 1983; Hulme and Thomas, 1979; Lewis et al., 1985; Presnell and Nicholas, 1990; Unligil, 1979; Miller and Morrell, 1989; Miller et al., 1989). Compounds that show promise include oxine copper, triazoles, methyl thiobenzothiazole, methylene bisthiocyanate, and several quaternary ammonium compounds. In addition, many chemicals have been employed in mixtures to overcome resistant species. In general, all of the acceptable alternatives are more expensive, and none has been as completely effective as penta. Interestingly, although there has been an overwhelming effort to develop acceptable alternative antistain chemicals, studies related to the fundamental nature of fungal stain have been overlooked. The chemical aspects of antistain treatments are covered in more detail in Chapter 19.

Proper handling of wood during harvesting and seasoning is critical for limiting the development of stains. Some important practices are as follows:

1. Rapid utilization, ponding, or spray storage of logs before milling to ensure stain-free logs at the mill;
2. Prompt treatment with antistain chemicals within 24 hr after sawing to prevent germination and growth of fungi beyond the penetration depth of treatment, resulting in internal stain;
3. Maintenance of treatment solutions at recommended levels and frequent replenishment of the treatment solutions to minimize dilution of the fungicide by selective absorption;
4. Protection of dipped lumber from rain to ensure that a sufficient level of chemical remains on the wood and to prevent chemical contamination of the local environment;
5. Use of stickers made only of heartwood or preservative-treated wood; and
6. Use of proper piling practices to maximize air-flow through and over the boards in the pile. Lumber should be far enough off the ground to prevent rain splashing.

In some cases, properly handled logs may still develop some stain. This stain is usually related to poor handling procedures before the log reaches the mill; for example, the log may have been stored in the woods for some time before transport, permitting fungi to colonize the end and debarked areas, or insects to carry fungal spores into the wood.

In addition to chemical control, the possibility of using competitive microorganisms to prevent fungal stain has been explored. Several organisms have been found *in vitro* to inhibit the development of fungal stain and might prove especially useful for protecting lumber under moist, humid conditions (Bernier *et al.*, 1986; Seifert *et al.*, 1988; Benko, 1988). Alternatively, it may be possible to identify the mechanism of pigment formation and develop methods for preventing this process. In this case, the fungus might still attack the wood, but would be unable to produce the discoloring pigment. As an example, fungal pigments form more slowly at lower oxygen levels; however, it is generally not possible to control oxygen in stored lumber. Attempts to develop solvents or chemical treatments to decolor sapstained wood have been unsuccessful.

Summary

- Stains can be chemical or biological in nature. Chemical stains are primarily oxidative, leading to the formation of pigmented compounds in the wood.
- Controlling chemical stains is difficult, although reducing compounds and antioxidants have sometimes proven useful.
- Biological stains include incipient decay, molds, and sapstains. Of these three, sapstains are the most important.
- Stains are becoming more important because of the increased amounts of sapwood, the increased use of wood in decorative situations where appearance is important, and the development of export markets that require shipping for long periods while the wood is still wet.
- In addition to suffering aesthetic damage, stained wood is more permeable, often has substantial loss in toughness, and in some cases the stain may conceal incipient decay.
- Application of chemicals, either through dipping or spraying, can prevent fungal colonization, but care must be taken to ensure that the treatments are properly timed and performed.

References

Bailey, I. W. (1910). Oxidizing enzymes and their relation to "sap stain" in lumber. *Botanical Gazette* **50**:142–147.

Barton, G. M., and J. A. F. Gardner. (1966). "Brown-Stain Formation and the Phenolic Extractives of Western Hemlock [*Tsuga heterophylla* (Raf.)Sarg]." Publication No. 1147, Department of Forestry, Ottawa, Canada.

Benko, R. (1988). "Bacteria as Possible Organisms for Biological Control of Blue Stain." International Research Group on Wood Preservation Document No. IRG/WP/1339. Stockholm, Sweden.

Bernier, R., Jr., M. Desrochers, and L. Jurasek. (1986). Antagonistic effect between *Bacillus subtilis* and wood staining fungi. *Journal of the Institute Wood Science* **10**(5):214–216.

Björkman, E. (1947). On the development of decay in building-timber injured by blue stain. *Sätryck ur Svensk Papperstidning* **50**, 11B.

Boyce, J. S. (1927). Sapstain and decay in Northwest lumber. *Columbia Port Digest* **5**(3):5–6.

Brown, F. L. (1953). Mercury-tolerant *penicillia* causing discoloration in northern white pine lumber. *Journal of the Forest Products Research Society* **3**:67–69.

Cassens, D. L., and W. E. Eslyn. (1983). Field trials of chemicals to control sapstain and mold on yellow poplar and southern pine lumber. *Forest Products Journal* **33**(10):52–56.

Cech, M. Y. (1966). New treatment to prevent brown stain in white pine. *Forest Products Journal* **16**(11):23–27.

Chapman, A. D., and T. C. Scheffer. (1940). Effect of blue stain on specific gravity and strength of southern pine. *Journal of Agricultural Research* **61**(2):125–134.

Clark, J. W. (1956). A gray non-fungus seasoning discoloration of certain red oaks. *Southern Lumberman* **193**(2417):35–38.

Crossley, R. D. (1956). "The effects of five sapstain fungi on the toughness of eastern white pine." Master's thesis, State University of New York, College of Forestry, Syracuse, New York.

Cserjesi, A. J. (1977). Prevention of stain and mould in lumber and board products. *Proceedings of the ACS Symposium No. 43 on Wood Technology: Chemical Aspects*, 24–32.

Cserjesi, A. J. (1980). "Field-Testing Fungicides on Unseasoned Lumber—Recommended Procedure." Technical Report 16, Forintek Canada Corp., Vancouver, B.C., Canada.

Cserjesi, A. J., and E. L. Johnson. (1982). Mold and sapstain control: Laboratory and field tests of 44 fungicidal formulations. *Forest Products Journal* **32**(10):59–68.

Cserjesi, A. J., and J. W. Roff. (1975). Toxicity tests of some chemicals against certain wood-staining fungi. *International Biodeterioration Bulletin* **11**(3):90–96.

Davidson, R. W. (1935). Fungi causing stain in logs and lumber in the southern states, including five new species. *Journal of Agricultural Research* **50**:789–798.

Davidson, R. W. (1942). Some additional species of *Ceratostomella* in the United States. *Mycologia* **31**:650–662.

Davidson, R. W. (1971). New species of *Ceratocystis*. *Mycologia* **63**:5–15.

Davidson, R. W., and W. E. Eslyn. (1976). Some wood-staining fungi from pulpwood chips. *Memoirs, New York Botanical Garden* **28**:50–57.

Dowding, P. (1969). The dispersal and survival of spores of fungi causing bluestain in pine. *Transactions of the British Mycological Society* **52**(1):125–137.

Dowding, P. (1970). Colonization of freshly bared pine sapwood surfaces by staining fungi. *Transactions of the British Mycological Society* **55**(3):399–412.

Drysdale, J. A., and A. F. Preston. (1982). Laboratory screening trials with chemicals for the protection of green timber against fungi. *New Zealand Journal of Forest Science* **12**(3):457–466.

Eslyn, W. E., and D. L. Cassens. (1983). Laboratory evaluation of selected fungicides for control of sapstain and mold on southern pine lumber. *Forest Products Journal* **33**(4):65–68.

Farr, D. F., G. F. Bills, G. P. Chamuris, and A. Y. Rossman. (1989). "Fungi on Plants and

Plant Products in the United States." American Phytopathological Society Press, St. Paul, Minnesota.

Findlay, W. P. K. (1959). Sap-stain of timber. *Forestry Abstracts* **20**(1,2):1–14.

Forsyth, P. G. (1988). "Control of Nonmicrobial Sapstains in Southern Red Oak, Hackberry, and Ash Lumber during Air-Seasoning. M.S. thesis, Mississippi State University, Mississippi.

Fritz, C. W. (1952). Brown stain in pine sapwood caused by *Cytospora* sp. *Canadian Journal of Botany* **30**(4):349–359.

Good, H. M., P. M. Murray, and H. M. Dale. (1955). Studies on heartwood formation and staining in sugar maple, *Acer saccharum* Marsh. *Canadian Journal Botany* **33**:31–41.

Hubert, E. E. (1931). "Outline of Forest Pathology" John Wiley, New York.

Hulme, M. A. (1975). Control of brown stain in eastern white pine with alkaline salts. *Forest Products Journal* **25**(8):36–41.

Hulme, M. A., and J. K. Shields. (1972). Effect of primary fungal infection upon secondary colonization of birch bolts. *Material und Organismen* **7**:177–188.

Hulme, M. A., and J. F. Thomas. (1979). Control of fungal sap stain with alkaline solutions of quaternary ammonium compounds and with tertiary amine salts. *Forest Products Journal* **29**(11):25–29.

Hulme, M. A., and J. F. Thomas. (1983). Control of brown stain in eastern white pine with reducing agents. *Forest Products Research Society* **33**(9):17–20.

Käärik, A. (1980). "Fungi Causing Sap Stain in Wood." The Swedish University of Agricultural Sciences, Report Nr R 114, Document No. IRG/WP/199. Stockholm, Sweden.

Levitin, N. (1970). Lignins of sapwood and mineral stained maple. *Canadian Forestry Service, Bi-Monthly Research Notes* **27**(4):29–30.

Levy, J. F. (1967). Decay and degrade of wood by soft-rot fungi and other organisms. *International Pest Control* (November/December):28–34.

Lewis, D. A., G. R. Williams, and R. A. Eaton. (1985). The development of prophylactic chemicals for the treatment of green lumber. *Record, Annual Convention of the British Wood Preservers' Association* **1985**:14–26.

Lindgren, R. M. (1942). Temperature, moisture, and penetration studies of wood-staining Ceratostomellae in relation to their control. U.S.D.A. Bulletin No. 807, Washington, D.C.

Lindgren, R. M. (1952). Permeability of southern pine as affected by mold growth and other fungus infection. *Proceedings of the American Wood Preservers' Association* **48**:158–174.

Lindgren, R. M., and E. Wright. (1954). Increased absorptiveness of molded Douglas fir posts. *Journal of the Forest Products Research Society* **4**(4):162–164.

Miller, D. J., and J. J. Morrell. (1989). "Controlling Sapstain: Trials of Product Group I on Selected Western Softwoods." Research Bulletin 65. Forest Research Laboratory, Oregon State University, Corvallis, Oregon.

Miller, D. J., D. M. Knutson, and R. D. Tocher. (1983). Chemical brown staining of Douglas fir sapwood. *Forest Products Journal* **33**(4):44–48.

Miller, D. J., J. J. Morrell, and M. E. Mitchoff. (1989). "Controlling Sapstain: Trials of Products Group II on Selected Western Softwoods. Research Bulletin 66. Forest Research Laboratory, Oregon State University, Corvallis, Oregon.

Oldham, N. D., and W. W. Wilcox. (1981). Control of brown stain in sugar pine with environmentally acceptable chemicals. *Wood and Fiber* **13**(3):182–191.

Presnell, T. L., and D. D. Nicholas. (1990). Evaluation of combinations of low-hazard biocides in controlling mold and stain fungi in southern pine. *Forest Products Journal* 40(2):57–61.

Roff, J. W. (1973). Brown mould *(Cephaloascus fragrans)* on wood, its significance and history. *Canadian Journal Forest Research* 34:582–585.

Roth, E. R. (1950). Discolorations in living yellow poplar trees. *Journal of Forestry* 48:184–185.

Scheffer, T. C. (1939). Mineral stains in hard maples and other hardwoods. *Journal of Forestry* 37(7):578–579.

Scheffer, T. C. (1954). "Mineral Stain in Hard Maples and Other Hardwoods." Report 1981. U.S.D.A. Forest Products Laboratory, Madison, Wisconsin.

Scheffer, T. C. (1973). Microbiological degradation and the causal organisms. In "Wood Deterioration and Its Prevention by Preservative Treatments" (D. D. Nicholas, ed,), 31–106. Syracuse University Press, Syracuse, New York.

Scheffer, T. C., and J. T. Drow. (1960). "Protecting Bulk-Piled Green Lumber from Fungi by Dip Treatment." Report 2201, U.S.D.A. Forest Products Laboratory, Madison, Wisconsin.

Scheffer, T. C., and R. M. Lindgren. (1940). "Stains of Sapwood Products and Their Control." Technical Bulletin 714. U.S.D.A., Washington, D.C.

Schulz, G. (1956). Exploratory tests to increase preservative penetration in spruce and aspen by mold infection. *Forest Products Journal* 6(2):77–80.

Seifert, K. A., C. Breuil, L. Rossignol, M. Best, and J. N. Saddler. (1988). Screening for microorganisms with the potential for biological control of sapstain on unseasoned lumber. *Material und Organismen* 23(2):81–95.

Shields, J. K., R. L. Desai, and M. R. Clarke. (1973). Control of brown stain in kiln-dried eastern white pine. *Forest Products Journal* 23(10):28–30.

Stranks, D. W., and M. A. Hulme. (1975). The mechanisms of biodegradation of wood preservatives. *Material und Organismen Symposium Berlin-Dahlem* 3:346–353.

Stutz, R. E. (1959). Control of brown stain in sugar pine with sodium azide. *Forest Products Journal* 9(11):59–64.

Stutz, R. E., P. Koch, and M. L. Oldham. (1961). Control of brown stain in eastern white pine. *Forest Products Journal* 11(5):258–260.

Unligil, H. H. (1979). Laboratory screening tests of fungicides of low toxic hazard for preventing fungal stain of lumber. *Forest Products Journal* 29(4):55–56.

Verrall, A. F. (1939). Relative importance and seasonal prevalence of wood-staining fungi in the southern states. *Phytopathology* 29:1031–1051.

Verrall, A. F. (1941). Dissemination of fungi that stain logs and lumber. *Journal of Agricultural Research* 63(9):549–558.

Verrall, A. F. (1942). A comparison of *Diplodia natalensis* from stained wood and other sources. *Phytopathology* 32(10):879–884.

Verrall, A. F. (1945). The control of fungi in lumber during air-seasoning. *Botanical Review* 11(7):398–415.

Verrall, A. F. (1949). Some molds of wood favored by certain toxicants. *Journal of Agricultural Research* 78(12):695–703.

Verrall, A. F., and P. V. Mook. (1951). "Research on Chemical Control of Fungi in Green Lumber, 1940–1951. Technical Bulletin 1046, U.S.D.A., Washington, D.C.

Wang, C. J. K., and R. A. Zabel. (1990). "Identification Manual for Fungi from Utility Poles in the Eastern United States." American Type Culture Collection. Rockville, Maryland.

Yazaki, Y., J. Bauch, and R. Endeward. (1985). Extractive components responsible for the discoloration of Ilomba wood (*Pycnanthus angolensis* Exell.). *Holz als Roh-und Werkstoff* **43**:359–363.

Zabel, R. A. (1953). Lumber stains and their control in northern white pine. *Journal of the Forest Products Research Society* **3**:1–3.

Zabel, R. A., and C. H. Foster. (1949). "Effectiveness of Stain-Control Compounds on White Pine Seasoned in New York." Tech. Pub. No. 71. N.Y. State College of Forestry at Syracuse University, Syracuse, New York.

Zabel, R. A., and F. C. Terracina. (1980). The role of *Aureobasidium pullulans* in the disfigurement of latex paint films. *Developments in Industrial Microbiology* **21**: 179–190.

Zink, P., and D. Fengel. (1988). Studies on the colouring matter of blue-stain fungi. Part 1. General characterization and the associated compounds. *Holzforschung* **42**(4):217–220.

Zink. P., and D. Fengel. (1989). Studies on the colouring matter of blue-stain fungi. Part 2. Electron microscopic observations of the hyphae walls. *Holzforschung* **43**(6):371–374.

Zink, P., and D. Fengel, (1990). Studies on the colouring matter of blue-stain fungi. Part 3. Spectroscopic studies on fungal and synthetic melanins. *Holzforschung* **44**(3):163–168.

CHAPTER

15

Decay Problems Associated with Some Major Uses of Wood Products

One of the major drawbacks associated with the use of wood products is their susceptibility to biological deterioration. This deterioration often occurs as we inadvertently duplicate natural conditions for decay in our structures. These conditions can sometimes be prevented by proper structural design, but wood often must be exposed to ground contact or periodic wetting. When this is unavoidable, economic realities dictate the use of preservative-treated or naturally durable woods. Despite these efforts, a substantial percentage of the wood in service falls to the agents of decay.

Quantifying decay losses has long stymied researchers. Wood failures occur in buildings, utility poles, railroad ties, bridge timbers, piling, and myriad other unrelated uses. With the exception of utilities and railroads, most wood users lack a systematic method for quantifying their decay losses, and even these groups have a relatively imprecise knowledge of their losses. Quantifying losses in residential structures is particularly difficult since there are no uniform procedures for reporting damage. As a result, many potentially important decay problems may be overlooked.

The replacement of decayed wood alone has been estimated to consume 10% of the timber cut annually in the United States (Boyce, 1961). In 1988, this figure for softwoods amounted to $613 million (Anderson, 1990). Whereas wood decay results in substantial losses, labor costs involved in replacing structures, productivity losses, or liability that stems from poorly maintained wood far exceed the raw value of the wood. The total cost of insect and decay repairs in buildings in California approaches $400 million per year (Brier et al., 1988). Even simple kinds of decay can sometimes cause significant productivity losses. For example, decaying ties decrease the speed at which trains can safely travel, thereby decreasing track use and increasing transit times for trains. These slowdowns are estimated to cost $18.60 per tie per year in main-line track (Anonymous, 1985). When nearly 3000 ties are present in a single mile of track, the cost of decay can rapidly mount. The cost to replace a single utility pole in California approaches $3000, including labor, and may also trigger costly service interruptions. Individual utilities often have 100,000 or more poles within their system and incur rejection rates of 0.3 to 0.4% per year, amounting to $900,000 to $1,200,000 annually in replacement costs. Furthermore, the liability associated with the failure of a wood pole can easily exceed several million dollars if a serious injury is involved. On a more personal note, decay or insect attack in residential homes can markedly reduce the home value.

It is readily apparent that we accept a certain level of decay loss within specific commodities; however, declining supplies of wood and increasing raw-material costs will necessitate a more careful evaluation of wood usage. In this chapter, we review the causes of decay losses in the major commodities where wood is employed and stress the principles and practices used to prevent or minimize these losses.

Decay Hazard

In most interior uses and many structural applications where wood is kept dry, there is no decay hazard and this material will last indefinitely. Decay hazards are related to exterior uses of wood subjected to atmospheric wetting or other moisture sources such as soil contact.

Before we address decay problems associated with specific wood uses, it is important to consider that the risk of decay varies widely with climate and geographic location. This premise is employed in the specifications of the American Wood-Preservers' Association through the incorporation of different levels of chemical protection that users can specify for their particular regions (AWPA, 1990). It is readily apparent that the risk of decay is considerably greater in southern Florida than northern Wyoming and that the degree of exposure has a marked influence on performance. Decay problems are minimal in the dry southwestern United States or at higher elevations, but become quite significant in the southeastern United States. It is, however, less apparent that decay risks can vary widely within closely situated sites. These hazards often necessitate the use of either species with naturally durable heartwood or wood that has been preservative treated.

Types of Decay Hazard

The variations in exposure were used by Scheffer (1971) to develop a climate index for decay hazard for various exterior wood uses above the ground (Fig. 15-1A). This index uses rainfall and temperature data to develop an index rating ranging from 0 (no risk) to 100 (high hazard). These values are then adjusted on the basis of known service records of wood in the various regions. The index establishes three broad hazard zones: *severe decay hazard* (southeastern United States and the Olympic Peninsula), *moderate decay*

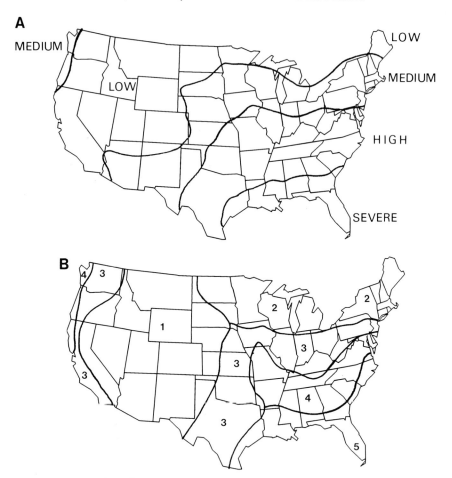

Figure 15-1 Decay hazards for (A) aboveground exposure (from Scheffer, 1971) and (B) utility poles in the United States (from REA, 1973), where 1 represents low decay hazard and 5 represents severe exposure.

hazard (northeastern United States, north and central states, and western Oregon to California), and *low decay hazard* (Southwest, Rocky Mountains, and the eastern Pacific Northwest). Whereas the climate index is useful for predicting performance of wood in non-ground-contact applications, it cannot predict ground-contact performance, since it cannot account for variations in soil type, water-holding capacity, and vegetation. The climate index has, however, been included in the AWPA Standards and is often employed by wood users to assess their relative decay risk.

As an alternative to the climate index, the Rural Electrification Administration (REA) developed a hazard index for utility poles based on inspection data from across the United States (Fig. 15-1B) (REA, 1973). Their system has five decay-hazard categories, and utilities that receive REA loans must utilize these requirements. These zones, based on actual pole failures, probably represent the most comprehensive national data base of wood performance based on climate.

It is interesting that there are relatively few guidelines available to assist in the specifications of wood products; however, this void reflects the wide array of uses in which wood is employed and the absence of an effective mechanism for monitoring wood losses in these many commodities.

Types of Wood Products—Decay Fungi

Almost without exception, the principal wood products decayers are common saprobic fungi that decay fallen timber and slash in the forest.

A wide array of fungi colonize decaying wood in the forest, but the number of species isolated from wood products is relatively limited. This decreased variety may reflect the more restrictive conditions within the wood product as well as the scarcity of extensive, systematic data on fungal colonization of wood. In most instances, the fungi present in wood products probably reflect a close replication of an environmental niche present in decaying wood in the forest ecosystem.

A number of workers have reported on the incidence of fungi in various products (Richards, 1933; Silverborg, 1953; Davidson *et al.*, 1947; Toole, 1973; Cowling, 1957; Eslyn, 1970; Eslyn and Lombard, 1983; Graham and Corden, 1980; Zabel *et al.*, 1980, 1985). The most comprehensive listing was prepared by Duncan and Lombard (1965), which summarized fungi associated with wood-products decays collected by the U.S. Forest Products Laboratory, Madison, Wisconsin, and the Forest Disease Laboratory, Beltsville, Maryland, over a 30-yr period. These identifications tend to favor those organisms that produce large, durable fruiting structures or that are readily cultured and identified from wood, but they do provide a relative guide to the major decayers of wood products in the United States. The Duncan and Lombard results indicate the following.

1. Brown rots constituted 76% of the nearly two thousand samples examined. These results probably reflect the extensive use of coniferous woods in many structures.
2. The 10 most prevalent fungi on conifers were *Neolentinus lepideus*, *Gloeophyllum trabeum*, *G. sepiarium*, *Postia (Poria monticola) placenta*, *Meruliporia (Poria) incrassata*, *Coniophora arida*, *Antrodia (Poria) vaillantii*, *Antrodia (Poria) xantha*, *Coniophora puteana*, and *Antrodia (Poria) radiculosa*.
3. The six most prevalent species on hardwoods were *Gloeophyllum trabeum*, *Trametes (Coriolus, Polyporus) versicolor*, *Antrodia (Poria) oleracea*, *Meruliporia incrassata*, *Xylobolus frustulatus*, and *Schizophyllum commune*.

4. The principal fungi associated with wood decay in buildings were *Meruliporia incrassata, Gloeophyllum trabeum, Tapinella (Paxillus) panuoides, Antrodia vaillantii, Coniophora puteana, Serpula (Merulius) lacrimans, Postia placenta,* and *Antrodia serialis.*

It is interesting to note that fungi with poroid basidiomata (previous family grouping, Polyporaceae) composed 60% of the decay-associated fungi, and those with resupinate basidiomata predominated (previous generic grouping, *Poria*).

One of the shortcomings of the available data on the identities of the wood-inhabiting fungi associated with wood decay is the overemphasis on basidiomycetous decayers. Many members of the Fungi Imperfecti and Ascomycotina have been shown to cause substantial soft rot or white rot in southern pine poles (Zabel *et al.*, 1985), but most surveys of products decayers have excluded these fungi despite their importance in specific environments. Many members of the genera *Phialophora, Daldinia,* and *Xylaria* have been isolated from decaying wood products (Seifert, 1983; Nilsson *et al.*, 1989; Wang and Zabel, 1990). The role of these fungi in the decay of wood materials remains poorly understood, but it is readily apparent that they play important roles under specific environmental conditions.

Decay of Wood Products

Wood is used in a variety of applications, each with its own set of special conditions (niches), which may permit decay development and often a specific fungus flora. Whereas a majority of these problems reflect a failure to control one of the four requirements for decay fungi (moisture, air, favorable temperature, or food source), examining specific problems related to each commodity is helpful for determining controls and delineating future research needs.

Wood Buildings (Homes)

Wood will last for centuries when properly used in well-designed and maintained structures. Examples of this principle can be seen in the temples of Japan, the stave churches of Norway, and early houses in the United States. The Hoxie house, built near Sandwich, Massachusetts, in 1624, remains sound and contains many original timbers, framing, clapboards, and shingles. High foundations, a wide roof overhang, extensive overlapping of roof shingles and siding, and a well-drained building site have all combined to provide this exceptional longevity.

More recently, decay losses in homes appear to be increasing at a substantial rate and have stimulated extensive substitution of wood with aluminum, asbestos shingles, and, more recently, plastics. Increasing decay problems reflect a number of changes in building design and services including:

1. The increased use of slabs or crawl-space foundations in place of basements;
2. Designs that decrease roof pitch, narrow roof overhangs, and extend decorative features such as roof beams, grills, or balustrades beyond the zone protected by the roof overhang;
3. Shifts away from naturally durable wood species and the use of woods with higher percentages of decay-susceptible sapwood;
4. The use of green wood in framing, which, once the wood dried, results in open-butt joints where moisture can penetrate and accumulate inside walls; and
5. Increased water condensation resulting from indoor plumbing, dryers, air-conditioners, and washers. This moisture causes particular problems in buildings that were tightly constructed to minimize heat loss or in those buildings where the vapor barriers were improperly installed.

Many of these problems reflect the decreasing numbers of builders and designers with extensive knowledge of wood properties. Most of these users have been educated in steel and concrete construction and fail to account for the inherent variability of wood. Examples of some of the common design errors in homes and wooden structures that led to serious decay problems have been assembled by Rosenberg and Wilcox (1980).

The principles of decay and insect control in homes and wooden structures have been exhaustively covered in a number of excellent publications (Anonymous, 1969; Biesterfedt et al., 1973; Scheffer and Clark, 1966; Scheffer and Verrall, 1973; Verrall, 1966; Verrall and Amburgey, 1977; Moore, 1979; Rambo, 1988). Amburgey (1971) has prepared a useful annotated bibliography of the prevention and control of decay in homes and other wooden structures. The most important design principle emphasized in these publications is to *exclude* moisture from wood and when this is impossible, the use of durable or preservative-treated wood is essential.

Moisture in buildings can arise in the following ways.

1. Framing with green or wet lumber can temporarily create moisture pockets and permit the growth of molds and stain fungi. Stained wood appears to be more frequently associated with decay damage to sheathing. The use of dry wood in construction can minimize these problems.

2. Direct wood contact with soil or damp masonry results in rapid moisture absorption. Examples include vertical beams in soil contact, grade levels above exterior woodwork, or wood forms left on concrete.

3. Wetting may occur from rain or seepage from roof leaks, splashing from the ground, or water flow through cracks or butt joints (Fig. 15-2). Water can also enter the roof in cold climates through ice dams that form near the roof edge. Although the immediate risk of decay is minimal owing to the cool temperatures, this moisture often remains during warmer periods, creating ideal conditions for decay. Water enters the wood rapidly in liquid form, but

350 / 15. Decay Problems with Uses of Wood Products

Figure 15-2 High-decay-hazard zones in homes are water accumulations in open-butt joints. (A) Paint-peeling and decay in a fascia board where water has seeped into a joint. (B) Advanced decay development in the corners of an untreated window sash. From U.S. Forest Service and Southern Forest Experiment Station, and courtesy of Drs. Arthur Verrall and Terry Amburgey (1977).

must exit slowly as a vapor, leaving the wood wet for long periods after the initial wetting. Appreciable volumes of water may enter joints and cracks during periods of high winds and heavy rainfall. Water then spreads horizontally within walls by gravity flow, capillarity, and condensation of water vapors. Seepage and moisture accumulations can often be detected by rusting nail heads, paint discoloration or peeling, nails backing out, or siding buckling (Fig. 15-3). Damage can be minimized by using proper roof overhangs, installing gutters to minimize splashing, caulking wood joints, flashing horizontally exposed wood, installing drip lines on roof edges, and good paint maintenance.

4. Condensation on wood surfaces has long been recognized as a problem in cool climates where moisture vapor moving through the wood can condense on the cool outer surfaces (Fig. 15-4), but can also be a problem in air-conditioned homes in warmer climates (Anderson, 1972; Duff, 1971). Condensation varies with the volume of vapor diffusing and the magnitude of temperature drop. In cool climates, the condensate may freeze during the

Figure 15-3 Some useful indicators of excessive moisture accumulations in the walls of a home from leaks or condensation, setting the stage for decay development. (A) A rust streak below nails; (B) excessive paint peeling, and (C) paint discolorations, nail pulling, and checks near joints. From U.S. Forest Service and Southern Forest Experiment Station, and courtesy of Drs. Arthur Verrall and Terry Amburgey (1977).

Figure 15-4 Condensation is an insidious source of moisture in some structures. Corners in crawl spaces or zones of restricted air circulation are common locations in homes. From U.S. Forest Service and Southern Forest Experiment Station, and courtesy of Drs. Arthur Verrall and Terry Amburgey (1977).

winter and melt as the weather warms. This condensation may be relatively minor, but can lead to paint blistering, wood staining, compaction of insulation, and corrosion of electrical conduits. More serious condensation can develop in subflooring over damp crawl spaces, especially in the corners of the building. Crawl-space condensation can be minimized by installing adequate vents around the building and by covering the ground with an impermeable plastic barrier.

5. Water leaks from plumbing or inadequate sealing of joints around plumbing fixtures may cause extensive localized decay. Moisture can also come from outside sources, such as sprinklers that spray walls. Control rests mostly with correcting the leak or redirecting the water away from the wood.

6. Biotic sources may also contribute to moisture in wood. Some fungi, such as the notorious building-rot fungi *Meruliporia incrassata* and *Serpula lacrimans*, form rhizomorphs that can transport water from damp soil across foundations and masonry into the wood. These fungi also release a sufficient quantity of water through metabolism to sustain decay in poorly ventilated spaces. These fungi are best controlled by periodic inspection for the pres-

ence of the cable-like rhizomorphs and the removal of wood debris from the soil around the building (Verrall, 1968).

These two building-rot fungi are of special interest and importance. *Serpula (Merulius) lacrimans* is the major damaging building decayer in Europe and occurs also in Canada and the northern United States. It appears to be limited to cool climates and occurs most commonly in crawl spaces or sub-flooring where the ventilation is poor. This fungus is widespread in Europe because of the prevalent practice in older buildings of embedding timbers in masonry. A similar major building decayer is *Meruliporia incrassata*, which occurs commonly in the southern coastal areas of the United States. Outbreaks of this house decay saprobe are often traceable to lumber that was invaded by the fungus during seasoning or storage (Verrall, 1968). Both fungi occur most commonly in poorly ventilated locations and cause a similar appearing brown-cubical rot. They are able to transport water in advance to the wood they colonize and invade the wood via mycelial fans, cords, and in some instances thick cable-like rhizomorphs (Fig. 3-1B). The ability of these fungi to penetrate and survive in mortar makes them particularly difficult to control.

There are numerous simple methods for preventing wood damage in buildings (Table 15-1). Most require good initial design of the structure, followed by a regular maintenance schedule to limit any damage that occurs. It is especially important to maintain paint films or other water-repelling barriers that can minimize the creation of conditions suitable for microbial growth (Feist, 1984; Feist and Mraz, 1978; Rowell and Banks, 1985).

Utility Poles

Decay in utility poles has been heavily researched because of the high degree of reliability demanded of electrical service and the availability of

Table 15-1
Recommendations for preventing decay in buildings

Use bright, kiln-dried lumber
Protect lumber from wetting during construction
Provide adequate roof overhang (>2 ft)
Provide gutters
Use well-drained building sites
Install preservative-treated sill plates >8 in above grade
Install ground covers on soil in crawl spaces
Ventilate crawl spaces (openings 1/160 of surface area)
Flash wood where it is exposed in a horizontal position
Use pressure-treated wood for exposed design features, such as posts or rails
Maintain coatings, such as paints or stains, and recaulk joints regularly
Periodically inspect building for signs of moisture
Use exterior finishes that shed water

utility research support. The utility pole in direct soil contact is exposed to a high decay hazard, but must deliver 30 to 40 years of reliable service. Because of this hazard, most utilities use either naturally durable wood species such as western red cedar, or they pressure-treat less durable woods with preservative. Commercial wood pole species in North America include Douglas fir, lodgepole pine, ponderosa pine, red pine, southern pine, and western red cedar. In most regions, Douglas fir is used for larger size poles, whereas southern pine composes the majority of smaller poles.

Although wood poles have performed well for over 100 years, a certain percentage of poles develop decay problems. Examples of the major decay types and common locations in utility poles are illustrated in Figs. 15-5 and 6. Poles with low preservative retentions may slip through the inspection process and are responsible for some early failures. In a classic study, Lumsden (1960) showed that individual retentions in a pole batch treated to 8 lbs/ft^3 (gauge) ranged from 2 to 14 lbs/ft^3.

The decay problems in poles are generally classified as *internal* or *external* decay. Internal decay can originate in the pole during storage or seasoning before treatment. Such pretreatment decay is particularly insidious in the southern pines, in which the thick sapwood zone is responsible for much of the pole strength. This decay is usually inactivated by the preservative treatment, but the treatment often conceals the damage. Internal decay in larger poles may not be inactivated during the preservative treatment. It also often develops in thin-sapwood species, which are difficult to completely protect with preservative. Poles are normally treated while the inner heartwood remains moist. As the pole seasons in service, deep checks that penetrate beyond the depth of treatment expose the untreated heartwood to attack by decay fungi and insects (Fig. 15-7). Untreated wood can also be exposed when holes or cuts are made after treatment. Internal decay normally occurs at or below the ground line, but can extend for considerable distances above the ground in regions where moisture levels are high, or below the ground in arid regions. In some cases it develops at bolt holes and cross arm connections. Internal decay can cause significant reductions in wood strength in relatively short periods (4 to 10 yr). Internal decay can be minimized by limiting the degree of checking or by improving the depth of treatment in high decay-hazard zones (the ground line). These topics are discussed in more detail in Chapter 19. Internal decay can also occur in species with thicker sapwood, when the wood is treated while the moisture content remains high. This process results in pockets of inferior treatment in which fungal decay can develop.

External decay can develop in several ways. When naturally durable woods are employed, the outer sapwood shell is left on the pole. This zone has no natural decay resistance and will begin to decay both above and below the ground line. This effectively decreases the circumference of the pole, thereby reducing bending strength. Aboveground decay of naturally durable woods can be controlled by full-length preservative treatment before installation or

Figure 15-5 Examples of the major decay types in southern pine utility poles in service. (A) A brown-cubical rot in the outer treated zone; (B) advanced white rot in the untreated center; (C) soft rot as seen on surface of an older creosote-treated pole, the radial development indicated by white arrows; and (D) soft rot as seen on the below-ground surface of a Cellon-treated southern pine pole displaying the typical cracking and exfoliation pattern. From Zabel *et al.* (1982), (A) and (B) with permission of Electric Power Research Institute, Palo Alto, California, and (C) and (D) Wang and Zabel (1990), with permission of American Type Culture Collection, Rockville, Maryland.

by flooding the surface of poles with preservative at 10- to 15-yr intervals in the field. As an alternative, some utilities shave the sapwood off before installation to eliminate the need for chemical treatment.

External decay can also develop below the ground line in preservative-treated poles. Typically southern pine and ponderosa pine are considered to

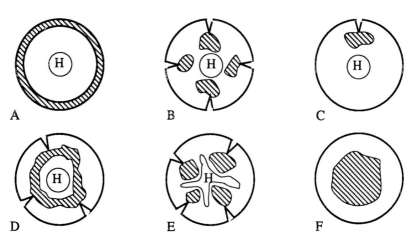

Figure 15-6 Illustrations of the common locations and patterns of decay development in the ground line zone of utility poles. (A) Surface soft rot; (B) and (C) decay pockets associated with checks; (D) internal decay where the decay developed between the outer treated shell and the heartwood and usually originated from the base of deep checks; (E) similar to (D), but the cross section is at a stem nodal zone; and (F) an internal decay hollow that may have resulted from an advanced case of (D) or pretreatment decay that escaped the preservative treatment. From Zabel *et al.* (1982), Wang and Zabel (1990), with permission of Electric Power Research Institute, Palo Alto, California and American Type Culture Collection, Rockford, Maryland.

be susceptible to surface decay; however, Douglas fir treated with pentachlorophenol in liquified petroleum gas or methylene chloride is also susceptible to this damage. Surface decay develops slowly as the outer shell of protective preservative is depleted by leaching or detoxified by pioneering fungi such as *Hormonicus resinae* or *Paecilomyces variotii*. Once these fungi have depleted the preservative below a toxic threshold level, soft-rot fungi capable of degrading the wood structure can invade (Fig. 15-8). Soft-rot damage can develop in pockets near the surface or may be associated with small surface checks. In some cases, soft-rot fungi that are tolerant to a preservative may penetrate the entire treated shell (Zabel *et al.*, 1991). The presence of this damage in the outer shell is critical, since 90% of the bending strength of a pole lies in the outer 5 cm of the surface, and any loss in pole circumference has drastic effects on strength.

The best methods for limiting decay in utility poles are performed before or during preservative treatment. They include adequate seasoning to remove moisture, predrilling of all holes for attachments, and strict adherence to specifications for preservative retention and penetration. In addition, the development of a regular program of inspection and remedial treatment can significantly extend pole service life (Graham and Helsing, 1979). As an example, Bonneville Power Administration once expected as little as 20 years of service life from their Douglas fir poles, but by using a combination of im-

Figure 15-7 A cross-section in the ground-line zone of a treated Douglas fir transmission pole indicating the association between deep checks (arrows) and the development of internal decay. Courtesy of Dr. R. W. Meyer.

proved specifications and maintenance, they now achieve an average service life in excess of 50 years (Lindgren, 1989). In remedial treatments, early decay detection is important since the pole still retains adequate residual strength at this point.

Foundation or Marine Piling

Like wood poles, pilings are subject to the same problems with checking and failure of the treated zone. In most instances, however, pilings are exposed under high-moisture conditions that exclude most fungal or insect attacks. Problems do arise when fluctuating water levels permit oxygen to enter the wood and establish conditions for fungal growth. This can be a particular problem in older buildings supported by untreated foundation piling. As the water table fluctuates or declines, extensive fungal attack can occur. This damage is extremely difficult to correct or prevent because the building rests on top of the piling. Fungal attack can also occur near the tops of marine piling above the water line or wherever joints with caps or stringers act as water traps. In many cases, builders remove the pile top, exposing the untreated wood beyond the treated shell and leading to extensive internal decay. Capping with fiberglass mesh and coal tar creosote will exclude moisture and protect the untreated wood from attack (Newbill and Morrell, 1990). A number

Figure 15-8 Typical soft-rot damage in the ground-line zone of a CCA-treated Douglas fir transmission pole that had been in service for approximately 10 years in a swampy area. Arrows (left to right) identify the pole butt, zone of maximal soft-rot development and surface exfoliation, and the ground line.

of corrective remedial treatments can be applied to decaying pile tops to arrest and prevent decay.

In addition to decay problems, marine pilings are exposed to varying degrees of marine-borer attack. This attack is often concentrated near the mud line or around joints, where the marine borers are more protected. Proper specification for the marine-borer hazards present in the harbor and regular diving inspections can minimize the risk of marine-borer damage.

Railroad Ties, Mine Timbers, and Bridges

Railroad ties, mine timbers, and bridges employ extensive amounts of large-dimension timbers that are susceptible to the development of deep checks extending beyond the depth of preservative treatment. Whereas railroad ties are typically thought to fail most frequently from mechanical failure or failure around the spikes *(spike kill)*, portions of this failure may reflect decay that occurred during air-seasoning (often called *stack burn* by wood treaters) or in service. Other activities, such as drilling for new spikes, exposes untreated wood and may hasten failure. As noted earlier, decaying ties force lower train speed limits, reducing track efficiency.

Bridge timbers represent a slightly different exposure, but many of the problems experienced with ties are similar. Because bridges are structural, however, decay poses significant safety problems. As a result, pretreatment handling is directed to limit the risk of decay and developing checks before treatment. Despite these efforts, the use of large timbers with a portion of the pith virtually assures that some deep checks will develop and that these checks will penetrate beyond the depth of treatment. As a result of these problems and the expense of large-dimension timbers, many bridges are now constructed using either glue-laminated timbers or using many smaller treated timbers in stress-laminated decks. Glue-laminated timbers are dried before treatment, minimizing later check development. In addition, individual laminations can be selected for high strength, improving the properties of the resulting beam. Smaller wood members tend to check less, reducing the risk of check penetration beyond the depth of treatment.

In addition to the problems associated with checking, bridges experience a number of other decay problems, mostly relating to the collection of moisture around wood joints. Particular problems occur in the railings, where the bridge rests on the foundation, and in the deck. Many times, these problems originate where debris collects in joints, creating ideal conditions for fungal growth. Careful design details to minimize debris-collection points, and regular maintenance can minimize this type of damage.

Mine timbers represent a slightly different type of wood exposure, but these products experience many of the same problems found in other large timbers (Eslyn and Lombard, 1983). Mine timbers are often exposed to high moisture conditions and are expected to provide structural support for the tunnel. Initially, these timbers were used untreated, but declining supplies of durable species resulted in substitution of preservative-treated wood. More recently, mine timber use has declined owing to the development of alternative methods for tunnel support.

Cooling Towers

Cooling towers employ extensive amounts of wood in lathing and for support. Basidiomycete attack can occur in portions of the tower above the splash line or in the whole tower when the tower is shut down, but the high moisture conditions present during operation exclude most basidiomycetes. These conditions, however, are ideally suited for soft-rot fungi, which gradually erode the wood surface. In fact, the importance of soft-rot fungi was first noted in cooling-tower slats (Savory, 1954). In most cases, either naturally durable species such as redwood or preservative-treated wood are used in cooling towers. However, soft-rot attack can even occur in these timbers over long periods. Regular inspections and periodic retreatments with preservative sprays can significantly extend cooling-tower service life (Baechler *et al.*, 1966).

Wooden Boats

Wooden boats have long experienced problems with decay. In fact, decay has played significant roles in maritime history, decimating the Spanish

Armada before its planned attack on Britain, and limiting the availability of ships for Britain during the American Revolution (Ramsbottom, 1937). Boat stems and transoms, frameheads, beam ends, and bilges are most susceptible to decay. The use of durable or preservative-treated woods, ventilation of bilge areas, and careful maintenance practices have developed to minimize this damage. The fungi associated with this attack were the subject of much study during and after World War II (Hartley and May, 1943; Davidson et al., 1947; Savory and Packman, 1954; Roth and Hartley, 1957). The use of fungicides in the bilge water was recommended for limiting damage in that area (Scheffer, 1953). Similar problems have also been noted in commercial fishing vessels, where they reduce efficiency and endanger lives (Condon and Graham, 1975).

Pallets and Boxes

Large quantities of wood are employed in pallets and boxes for transporting goods, and a portion of this material is subjected to conditions conducive to fungal growth. In some instances, the material is treated by dipping in preservative at some point during or after assembly. The problems associated with these products for military uses were summarized by Greathouse and Wessel (1954). Chemical treatments for limiting damage have been explored by a number of workers (Blew, 1959; Verrall, 1959, 1965; Verrall and Scheffer, 1969; DeGroot and Stroukoff, 1986), and generally pressure treatment is required to protect wood in tropical environments, whereas dip treatments with water-repellent preservatives are sufficient in the southeastern United States.

Panel Products

Panel products represent an increasingly important segment of the forest products industry. These panel products are employed where the risk of decay is minimal; however, plywood and flakeboards are increasingly used under conditions conducive to fungal attack. Although the presence of resin in the panels slows fungal attack, recent reports suggest that panel products are beginning to experience failures due to decay, particularly when they are exposed under high-moisture conditions. In general, many of the same fungi isolated from solid wood products appear to be common in panel products, but a number of other organisms including *Phanerochaete chrysosporium* and *Schizophyllum commune* are important colonizers of these products. As panel products see increased application, it is likely that a clearer picture will emerge concerning the fungal flora of these products.

A Decay-Control Principle

As we can see from the previous sections, wood products are subject to decay under a variety of uses and environments; however, each of these situations accurately replicates one or more components of the natural environment for that particular decay organism. Thus, if we hope to limit losses due to decay, we must alter our wood uses to eliminate these conditions. It is readily apparent that the prime factor in decay is moisture. Decay prevention depends on designs that keep moisture levels below the fiber saturation point (f.s.p.). Alternatively, naturally durable or preservative-treated woods must be used. In either instance, the structure must be properly maintained to continue to exclude moisture or to maintain an adequate level of preservative protection.

If careful design practices are employed and the wood is properly maintained, wood can last for centuries. As wood becomes increasingly valuable, it will remain for the wood microbiologist to educate designers, builders and users to adhere to these decay-prevention principles.

Summary

- Decay of wood products causes losses amounting to billions of dollars each year.
- Decay of wood products generally results from a failure to control one of the requirements for fungal growth (moisture, oxygen, temperature, or an adequate food source). Moisture is the key factor, and wood kept dry will not decay.
- The risk of decay varies widely with geographic location, application, and wood species. Attempts have been made to quantify these risks into maps describing decay hazards or decay index. These maps, although useful, provide only relative guidelines to decay risk.
- A variety of fungi have been isolated from decayed wood products, but some species are particularly abundant across a variety of wood uses. Unfortunately, previous research has relied heavily on the presumption that non-basidiomycetes are not important wood destroyers. Further research is necessary to better understand the role of non-basidiomycetes in the deterioration process.
- A variety of wood products are subject to decay. Most damage can be limited by using proper design principles, by using naturally durable or preservative-treated wood, and by regular maintenance.

References

Amburgey, T. L. (1971). "Annotated Bibliography on Prevention and Control of Decay in Wooden Structures (Including Boats)." Southern Forest Experiment Station, Forest Service, U.S.D.A., Gulfport, Mississippi.

American Wood Preservers' Association. (1990). "Book of Standards." Stevensville, Maryland.

Anderson, L. O. (1972). "Condensation Problems: Their Prevention and Solution."

U.S.D.A. Forest Service Research Paper FPL 132, Forest Products Laboratory, Madison, Wisconsin.

Anderson, T. (ed.). (1990). "Random Lengths 1989 Yearbook." Random Lengths Publishers, Eugene, Oregon.

Anonymous. (1969). "Wood Decay in Houses: How to Prevent and Control It." Home & Garden Bulletin No. 73, U.S.D.A., Washington, D.C.

Anonymous. (1985). Addressing bad-tie clustering. *Railway Track and Structures* (August):21-25.

Baechler, R. H., B. A. Bendtsen, and H. G. Roth. (1966). Preservative treatment of cooling tower slats. *Heating, Piping, and Air Conditioning* (April):121-123.

Biesterfeldt, R. C., T. L. Amburgey, and L. H. Williams. (1973). "Finding and Keeping a Healthy Home." Southern Forest Experimental Station, U.S.D.A., Forest Service General Technical Report SO-1, New Orleans, Louisiana.

Blew, J. O. (1959). "Preservatives for Wood Pallets." U.S.D.A., Forest Products Laboratory, Report No. 2166, Madison, Wisconsin.

Boyce, J. S. (1961). "Forest Pathology," 3rd Ed. McGraw-Hill Book Company, New York.

Brier, A. N., W. A. Dost, and W. W. Wilcox. (1988). Characteristics of decay and insect attack in California homes. *California Agriculture* 42(5):21-22.

Condon, E. J., and R. D. Graham. (1975). "Wood-Boat Maintenance: Decay and Its Prevention." Extension Service Sea Grant Advisory Program Publication No. 23, Oregon State University, Corvallis, Oregon.

Cowling, E. B. (1957). A partial list of fungi associated with decay of wood products in the United States. *Plant Disease Reporter* 41:894-896.

Davidson, R. W., F. F. Lombard, and R. R. Hirt. (1947). Fungi causing decay in wooden boats. *Mycologia* 39:313-327.

Duff, J. E. (1971). "The Effect of Air Conditioning on the Moisture Conditions in Wood Walls." U.S.D.A., Forest Service, Research Paper SE-78.

Duncan, C. G., and F. F. Lombard. (1965). "Fungi Associated with Principal Decays in Wood Products in the United States." U.S. Forest Service Research Paper WO-4, U.S.D.A., Washington, D.C.

DeGroot, R. C., and M. Stroukoff. (1986). "Efficacy of Alternative Preservatives Used in Dip Treatments for Wood Boxes." U.S.D.A. Forest Products Laboratory Research Paper FPL-RP-481. Madison, Wisconsin.

Eslyn, W. E. (1970). Utility pole decay. Part I. Basidiomycetes associated with decay in poles. *Wood Science and Technology* 4:97-103.

Eslyn, W. E., and F. F. Lombard, (1983). Decay in mine timbers. Part II. Basidiomycetes associated with decay of coal mine timbers. *Forest Products Journal* 33(7/8):19-23.

Feist, W. C. (1984). "The Role of Water Repellents and Chemicals in Controlling Mildew on Wood Exposed Outdoors. U.S.D.A. Forest Products Laboratory Research Note FPL-0247, Madison, Wisconsin.

Feist, W. C., and E. A. Mraz. (1978). Protecting millwork with water repellents. *Forest Products Journal* 28(5):31-35.

Graham, R. D., and M. E. Corden. (1980). "Controlling Biological Deterioration of Wood with Volatile Chemicals." Electric Power Research Institute El-1480, Project 212-1. Palo Alto, California.

Graham, R. D., and G. G. Helsing. (1979). "Wood Pole Maintenance Manual: Inspection and Supplemental Treatment of Douglas fir and Western Red Cedar Poles."

Research Bulletin 24, Forest Research Laboratory, Oregon State University, Corvallis, Oregon.

Greathouse, G. A., and C. J. Wessel. (1954). "Deterioration of Materials—Causes and Preventative Techniques." Reinhold, New York.

Hartley, C., and C. May. (1943). "Decay of Wood in Boats." U.S.D.A. Forest Pathology Special Release No. 8, Washington, D.C.

Lindgren, P. (1989). Bonneville Power Administration's wood pole management program. *Proceedings of the International Conference on Wood Poles*, Fort Collins, Colorado.

Lumsden, G. Q. (1960). Fortified wood preservation for southern pine poles. *Forest Products Journal* **10**(9):456–462.

Moore, H. B. (1979). "Wood-Inhabiting Insects in Houses: Their Identification, Biology, Prevention, and Control." U.S. Department of Housing and Urban Development, Washington, D.C.

Newbill, M. A., and J. J. Morrell. (1990). Marine capping devices in combination with chemical treatments to prevent and arrest pile-top decay. *Holzforschung* **44**(1):73–75.

Nilsson, T., G. Daniel, T. K. Kirk, and J. R. Obst. (1989). Chemistry and microscopy of wood decay by some higher Ascomycetes. *Holzforschung* **43**(1):11–18.

Rambo, G. W. (ed). (1988). "Wood Decay in Structures and Its Control." National Pest Control Association, Dunn Loring, Virginia.

Ramsbottom, J. (1937). Dry rot in ships. *Essex Naturalist* **25**:231–267.

Richards, C. A. (1933). Decay in buildings. *Proceedings of the American Wood Preservers' Association* **29**:389–398.

Rosenberg, A. F., and W. W. Wilcox. (1980). How to keep your award-winning building from rotting. *Wood and Fiber* **14**(1):70–84.

Roth, E. R., and C. Hartley. (1957). "Decay Prevention in Wooden Ships during Storage." Division of Forest Disease Research, U.S.D.A., Forest Service Report NPO 1700S571. Beltsville, Maryland.

Rowell, R. M., and W. B. Banks. (1985). "Water Repellency and Dimensional Stability of Wood." U.S.D.A. Forest Products Laboratory General Technical Report FPL-50, Madison, Wisconsin.

Rural Electrification Administration (REA). (1973). "Pole Performance Study." Staff Report, Rural Electrification Administration, Washington, D.C.

Savory, J. G. (1954). Breakdown of timber by Ascomycetes and Fungi Imperfecti. *Annals of Applied Biology* **41**:336–347.

Savory, J. G., and D. F. Packman. (1954). "Prevention of Wood Decay in Boats." Bulletin Forest Products Report No. 31, London, England.

Scheffer, T. C. (1953). Treatment of bilgewater to control decay in the bilge area of wooden boats. *Journal of the Forest Products Research Society* **3**:72–78.

Scheffer, T. C. (1971). A climate index for estimating potential for decay in wood structures above ground. *Forest Products Journal* **21**(10):25–31.

Scheffer, T. C., and J. W. Clark. (1966). On-site preservative treatments for exterior wood of buildings. *Forest Products Journal* **17**(12):21–29.

Scheffer, T. C., and A. F. Verrall. (1973). Principles for protecting wood buildings from decay. U.S.D.A. Forest Service Research Paper FPL 190. Madison, Wisconsin.

Seifert, K. A. (1983). Decay of wood by the Dacrymycetales. *Mycologia* **75**(6):1011–1018.

Silverborg, S. B. (1953). Fungi associated with the decay of wooden buildings in New York State. *Phytopathology* **43**:20–22.

Toole, E. R. (1973). "Fungi Associated with Decay in Utility Poles." Information Series 15, Mississippi Forest Products Utilization Laboratory, Mississippi State College, State College, Mississippi.

Verrall, A. F. (1959). Preservative moisture-repellent treatments for wooden packing boxes. *Forest Products Journal* **9**(1):1–22.

Verrall, A. F. (1965). "Preserving Wood by Brush, Dip, and Short-Soak Methods." Forest Service, U.S.D.A. Technical Bulletin 1334, Madison, Wisconsin.

Verrall, A. F. (1966). "Building Decay Associated with Rain Seepage." Technical Bulletin 1356. U.S.D.A. Forest Service.

Verrall, A. F. (1968). "*Poria incrassata* Rot: Prevention and Control in Buildings." U.S.D.A. Forest Service Technical Bulletin No. 1385.

Verrall, A. F., and T. L. Amburgey. (1977). "Prevention and Control of Decay in Homes." Superintendent of Documents, U.S. Government Printing Office, Washington, D.C.

Verrall, A. F., and T. C. Scheffer. (1969). "Preservative Treatments for Protecting Wood Boxes." U.S.D.A. Forest Service Research Paper FPL 106, Madison, Wisconsin.

Wang, C. J. K., and R. A. Zabel. (eds). (1990). "Identification Manual for Fungi from Utility Poles in the Eastern United States." American Type Culture Collection, Rockville, Maryland.

Zabel, R. A., F. F. Lombard, and A. M. Kenderes. (1980). Fungi associated with decay in treated Douglas fir transmission poles in the northeastern United States. *Forest Products Journal* **30**(4):51–56.

Zabel, R. A., F. F. Lombard, C. J. K. Wang, and F. C. Terracina. (1985). Fungi associated with decay in treated southern pine utility poles in the eastern United States. *Wood and Fiber Science* **17**(1):75–91.

Zabel, R. A., C. J. K. Wang, and F. C. Terracina. (1982). "The Fungal Associates, Detection, and Fumigant Control of Decay in Treated Southern Pine Poles." Electric Power Research Institute EL-2768, Final Report—Project 1471-1, Palo Alto, California.

Zabel, R. A., C. J. K. Wang, and S. E. Anagnost. (1991). Soft-rot capabilities of some major microfungi isolated from Douglas fir poles in the Northeast. *Wood and Fiber Science* **23**(2):220–237.

CHAPTER
16

Detection of Internal Decay

The reliable detection of early decay is a major problem for many wood users and a principal research goal in wood microbiology. It is important to recognize decay early before serious damage occurs, so corrective actions can be taken or remedial treatments, applied.

Detecting wood decay before significant strength losses occur poses a critical challenge to managers of large wood systems such as utilities, railroads, and marine facilities. Whereas methods for rapidly arresting active decay are now highly effective (Morrell and Corden, 1986), most methods for detecting early decay remain unreliable and relatively unsophisticated (Morrell and Wilson, 1985). Where sophisticated devices have been developed, the need for trained operators to interpret results has limited their usefulness (Wilson, 1988).

Decay recognition from visual (macroscopic) and microscopic features was covered in Chapters 7, 9, and 10. This chapter focuses primarily on nondestructive methods for detecting *internal* decay. The general principles of decay detection are presented in relation to practical approaches to detection in major wood products.

Decay-Detection Difficulties

Decay is difficult to detect since it often occurs internally and in hard-to-reach locations. Sampling must be limited since it further damages the wood; also, there is the inherent difficulty of detecting the beginning stages of decay when the property changes are small and subtle. Complicating the detection of decay are the differences in susceptibility of wood species to decay, the presence of normal wood-growth characteristics, such as knots or annual rings, and considerable variation in rate and pattern of attack by the many fungi that attack wood. Compounding these problems are the differences in the rates at which fungi degrade wood under the wide array of environmental conditions and the fact that subtle changes during the early stages of decay may resemble natural variations in wood quality.

Basic Sampling for Decay

In general, inspection for decay should concentrate on those locations that are most likely to become wet in service. The moisture sources may be seasonal or continuous, and the inspector must carefully observe the structure for signs of wetting. Points of direct contact with the ground, water-trapping joints such as those found in window frames and butt joints, crawl spaces where soil moisture can condense on wood in the foundation, condensation on ceilings or attics, leaks from plumbing, and deep checks are all potential sites for decay. Even when dry, these zones often have tell-tale white deposits where salts have diffused through the wood and precipitated on the surface.

An Ideal Decay-Detection Device

It will be useful to briefly review the features of an ideal decay-detection device. The ideal decay-detection device would separate decay from natural wood defects, particularly at the early stages of decay when visible fungal attack is sparse, but effects on mechanical properties may be quite substantial (see Chapter 10). The device should be inexpensive, simple to use, and require minimal training (i.e., be insensitive to operator variables). Ideally, the device would also assess residual wood strength and be nondestructive. The cost of individual devices may vary, since an expensive device for some (~$10,000) may be quite acceptable to a large utility if it is rapid and highly reliable. Finally, but most importantly, the device must have a high probability of detecting decaying wood at the earliest possible time, while minimizing the amount of sound wood that is labeled as decayed. The latter characteristic becomes particularly important for devices capable of detecting incipient decay, since the inspector is incapable of confirming the device's accuracy in the field.

The ideal decay-inspection device has not been developed yet, and one capable of detecting incipient decay while providing a measure of residual strength is probably beyond the scope of any current research program. However, many tests and devices have been developed that give the inspector a reasonable chance of detecting the intermediate and advanced stages of decay, or estimating residual strength (Fig. 16-1). Unfortunately, wood has undergone substantive changes in properties long before it is visibly decayed (Wilcox, 1978). Losses in some strength properties approach 60% before the wood has lost 5% of its weight. At this early stage, a well-trained wood anatomist can detect the damage microscopically, but the wood appears sound visually to the average field inspector. In addition to strength losses,

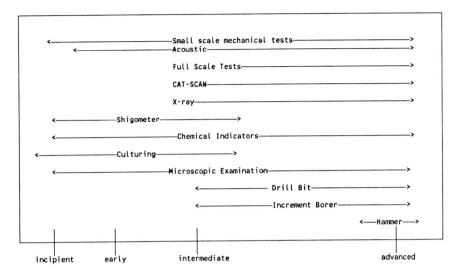

Figure 16-1 Applicability of various wood-inspection techniques to the detection of incipient, early, intermediate, or advanced stages of internal decay.

previous studies also suggest that the efficacy of certain remedial treatments, particularly fumigants, is reduced in decayed wood because of reduced binding (Morrell and Corden, 1986).

Because of the absence of an ideal device for decay detection, inspectors routinely employ several methods. Current inspection methods can be classified as physical, mechanical, electrical, and acoustic.

Physical Decay Detection

Sounding

Sounding with a hammer is perhaps the simplest and most widely used physical wood-inspection technique (Graham and Helsing, 1979; Inwards and Graham, 1980; Goodell and Graham, 1983). In this test, the experienced inspector can generally detect advanced decay where large voids are present, but cannot detect earlier stages of decay when control measures would be most effective. Sounding provides the basis for more sophisticated acoustic-inspection devices, but is not sufficiently sensitive to early decay.

Boring

Boring the wood with a drill bit readily detects soft zones or large internal voids. Torque release is usually evident in the later decay stages. In addition, a trained inspector can examine the shavings for the evidences of decay, thereby improving the ability to identify decaying poles. The inspection hole can also be used to determine the thickness of the remaining sound shell and can be used to apply remedial internal treatments. Whereas boring the wood has these advantages, it inspects only a small portion of the wood pole, leaving the possibility that a decay pocket will be missed. In addition, each inspection hole removes a small area of the wood cross section. Although relatively minor in terms of overall strength, continued inspection (at 10-yr intervals) can reduce strength significantly if all the holes are drilled in the same plane of wood. Proper pole plugging is also necessary to avoid creation of water-entrapment zones.

Visual Examination of Increment Cores

In this procedure, cores are removed with an increment borer, and the intact internal wood tissues can be studied visually or with a hand lens for decay features such as pulled fibers, zone lines, variations in color patterns, core compactions, wet zones, and obvious intermediate and late decay stages. The locations of any decay provide an estimate of the residual sound wood in the outer radius. Cores can be collected aseptically and cultured for the presence of decay fungi to check the reliability of the decay criteria used and the major associated decay fungi. The increment borer hole can be used to apply remedial internal treatments. Also, any torque release during the boring provides additional information on the wood condition.

Mechanical Decay Detection

The primary concern of a wood-inspection program is determining the amount of residual strength remaining in a given structure. Although some acoustic tests provide a relative measure of residual strength as modulus of rupture (MOR) or modulus of elasticity (MOE), using statistical population estimates, none of the decay-inspection devices currently available in the United States directly tests the mechanical strength of the wood.

Compression Tests

Small-scale mechanical tests of cores, plugs, or veneers removed from wood in service represent one approach to measuring residual wood strength. Longitudinal compression has been shown to be well correlated with MOR of small, clear beams and to predict the strength of western red cedar poles (Smith and Morrell, 1987). Radial compression of similar cores detects very early stages of decay (Smith and Graham, 1983); however, this technique measures the weakest growth ring and invariably measures earlywood properties, which are not well correlated to bending properties. Although both of these techniques show some promise as tools for selecting high-strength wood (Smith and Morrell, 1989), they require the removal of multiple samples from a given structure to develop a reliable strength estimate. Also, both techniques require the use of a mechanical test device in the field and an operator with some degree of training in wood properties. A more sensitive, but less useful property that could be tested on wood samples is toughness. Whereas tests have been developed for evaluating toughness of veneers using breaking radius (Safo-Sampah and Graham, 1976), it is doubtful that these tests could be adapted for field use.

Penetration Resistance

The Pilodyn is a spring-loaded, pin penetration device that has been extensively used in other countries to detect surface decay or to measure specific gravity in tree-improvement programs (Taylor, 1981; Friis-Hansen, 1980; Hoffmeyer, 1978; Cown, 1978; Cown and Hutchinson, 1983). The measurements are made on the wood surface. The Pilodyn has also been used to measure pole hardness in relation to various chemical treatments (Pierce, 1983). Attempts to use the Pilodyn to detect internal decay have been unsuccessful, but the device provides a means for measuring the severity of surface soft-rot attack. Whereas the Pilodyn is relatively nondestructive, one problem that complicates its use is the need to make moisture-content corrections for each reading (Smith and Morrell, 1986).

Drilling Torque Release

A second small-scale test method utilizes a power drill and a sensor that detects changes in resistance to drilling. This device reflects the fact that drilling through decayed wood is easier. An experienced inspector can also detect these changes by listening for sound changes (torque release) as the drill bit penetrates the decayed zone. The meter merely places a printed output in the hands of the inspector and provides a record that permits more detailed examination of the results. The inspector must still determine the nature of the defect (i.e., decay, ring-shake) using other methods.

Pick or Splinter Test

The pick test is a simple method for detecting surface decay in poles and timber. In practice, a sharp screwdriver or awl is driven in the wood at an acute angle and bent back to snap a small piece of wood from the surface. The break characteristics of the splinter removed are then examined. Brash breaks reflect low strength and the possible presence of decay, whereas splintery breaks reflect sound wood (Fig. 16-2). The pick test measures toughness and is fairly sensitive to early decay (Scheffer and Verrall, 1979; Wilcox, 1983; Morrell *et al.*, 1986). Drawbacks of the pick test include the relatively large sample removed and the inability to accurately assess internal condition.

Extensiometer

The pole extensiometer was developed in Europe and is a mechanical method for estimating bending strength. It consists of a lever frame that attaches to a tripod that braces the lever against the ground, and an exten-

Figure 16-2 A pick or splinter test being conducted on a board by Dr. T. C. Scheffer. (A) The splinter break indicates sound wood; (B) the brash break suggests the possible presence of decay. Photograph M124-721 courtesy of the U.S. Forest Service, Forest Products Laboratory, Madison, Wisconsin.

siometer fixed between two screws affixed to the pole surface (Morris and Friis-Hansen, 1984). A screw in the tripod turns to mechanically force the lever up and the pole away from the tripod, giving the pole a bending moment between two points. A dial in the lever records the force required to produce a given extension, which yields a predicted failure load in kilojoules (kNm). Whereas the pole extensiometer test is nondestructive, the technique has not been widely used, possibly owing to the set-up time per test, the inability to test specific sites for decay, or the difficulty of testing poles in line.

Vibration

Vibrational testing does not strictly measure mechanical properties; it does measure the ability of the wood to transmit energy. The wood is struck by a rubber mallet, and the vibrational characteristics of the wood are measured (Murphy *et al.*, 1987; Murphy and Taylor-Wilson, 1988). Sound wood vibrates more rapidly than decayed samples, and these differences can be used to determine residual wood strength. The vibrational technique does not determine the location or nature of damage and is not applicable to wood that contains hardware such as wires or connectors.

The best method for assessing wood strength is to test the structure to failure. Obviously testing structures to failure to determine their wood strength is not practical, but limited full-scale testing of occasional structures does have a place in any large-scale, decay-detection program. Full-scale testing of some units provides a measure of reliability for the decay-detection methods employed and provides a valuable training exercise and reference materials for inspectors. Wood users should consider routine testing of occasional selected structures to train their inspectors. In practice, the inspectors would inspect the wood using their approved methods, then make an assessment on wood condition and treatment recommendations. The structure would then be tested to failure and dissected to determine the nature and location of any defects present. In this way, the inspectors learn the accuracy of each device for detecting specific wood defects.

Electrical Decay Detection

A number of inspection methods have been developed that use the electrical properties of wood and the moisture often associated with wood in service.

Moisture Meters

The simplest electrical-inspection device is the moisture meter, which measures wood resistance in ohms between two metal pins driven in the wood (James, 1988). Since dry wood (<20% moisture content) will not decay, determining wood moisture content can provide a relative measure of decay risk. Moisture meters are often used by building or wood-pole inspectors to ensure that posttreatment moisture requirements are met. Wood moisture contents often vary seasonally. In addition, the zone sampled by the moisture meter is limited by the length of the pins, which range from 2.5 to 7.5 cm long. Thus, the moisture meter is of little use for detecting internal decay or the conditions with which this damage is often associated. Further, the moisture meter is sensitive to temperature (James, 1988) and the presence of inorganic salt-type wood preservatives (James, 1980). Currently available moisture meters represent a useful tool for the inspector, but cannot detect decay or determine its severity.

Shigometer

The Shigometer measures changes in electrical resistance of wood using a twisted wire probe inserted into a hole drilled in the wood. In principle, the Shigometer measures changes in electrical resistance caused by the release of ions as fungi degrade the wood (Shigo and Shigo, 1974). There is considerable debate about the usefulness of this instrument, which some equate with the moisture meter (Piirto and Wilcox, 1978). The Shigometer appears to be very useful for detecting decay in living trees, where wood moisture contents are uniformly above the fiber saturation point (Shigo and Shigo, 1974), but has produced inconsistent results in timber products where moisture contents vary more widely (Shigo *et al.*, 1977; Shortle *et al.*, 1978; Shortle, 1982; Thornton, 1979). A limitation of the device is that, at the time of the test, the wood must be above the f.s.p. for useful readings. In addition, several field trials indicate that the Shigometer requires a higher degree of experience to interpret the patterns of resistance changes than other currently used wood-inspection devices (Inwards and Graham, 1980). The Shigometer is used commercially to inspect utility poles, but its application has declined as users have seen little improvement over existing decay-detection methods.

X-ray

X-ray inspection was once commercially used to detect internal voids in wood and has the ability to detect small changes in wood density related to decay or other defects (Mothershead and Stacey, 1965). In a comparison of methods of decay detection in poles, x-ray radiographs appeared to be effective primarily when the wood was in the intermediate to advanced decay stages (Zabel *et al.*, 1982). X-ray units have a high initial expense, require safety training for operators, and generally require fairly sophisticated analyses. In addition, many users are concerned about potential health risks associated with ionizing radiation. As a result, x-rays are primarily used to detect marine-borer infestations in marine test panels, insect attack in museum test pieces, or for measuring wood density in relation to wood quality (Hoag and McKimmy, 1988; Yamamoto and Fujii, 1987).

Tomography

Computer-aided axial tomography scanning (CAT scan) was developed in Europe to produce three-dimensional maps of internal wood conditions (Hattori and Kanagawa, 1985; Morris and Friis-Hansen, 1984; Lindgren, 1987). One such device is attached around a structure and moves upward while performing a scan. The data is collected and analyzed to develop the internal map. CAT scans represent a potentially useful tool, but the time required to collect and process the data, as well as the initial equipment costs, currently make this technique impractical for field purposes. Improvements in data processing and microcircuitry will ultimately make this method practical, but high initial costs will probably prevent extensive use of this technique.

Acoustic Decay Detection

Acoustic detection of decay and other wood defects has received much recent attention, as techniques have been refined for rapid data processing under field conditions.

Acoustic Emissions

Acoustic emission refers to elastic waves produced by deformation and failure processes in stressed materials, as a result of the presence of small defects (Knuffel, 1988). Acoustic emission has been used to evaluate incipient decay in western hemlock specimens that were stressed in bending (Noguchi et al., 1986). Acoustic emission appears to be useful for detecting defects in small samples that can be easily manipulated, but most structures such as poles present formidable challenges to the use of acoustic emission since they are already in tension and compression. In addition, any equipment attached to the structure, or weather factors such as changes in wind velocity and direction during the test, can affect the results. Thus, large wood structures do not appear to permit the controlled test conditions required for acoustic emission analysis.

Stress-Wave Timers

The use of stress-wave timers, which measure the time required for a sound wave to travel through a material, represent a major advance in the ability of wood users to detect internal defects. In principle, a sound wave is induced at one point on the wood and a transducer picks up that signal at a second point. The time between sending and receiving can be correlated with wood MOE and density (Agi, 1983; Bucur, 1983; Ross and Pellerin, 1988). These data can also be related to the presence of internal voids or other defects, since the wave will travel faster through sound wood (Pellerin et al., 1985). Whereas stress-wave timers can be used to detect the presence of defects (Wilcox, 1988), they cannot distinguish between active decay, voids, ring shakes, or other wood defects. One device, the Pol-Tek, was developed for inspecting the ground line of wood poles. Several field trials indicated that this device was less sensitive than conventional inspection methods (Inwards and Graham, 1980). In addition, the device was unable to distinguish between defects that did not adversely affect strength and more serious defects, such as decay or insect attack.

Wave-Form Analysis

The recent improvements in data processing have led a number of investigators to explore advanced sonic inspection techniques. In addition to velocity measurements, these devices attempt to analyze portions of the resulting sound wave (Anonymous, 1987). As a sound wave passes through a material, it is modified by the characteristics of that material. Nondecayed wood transmits sound waves more efficiently and with less modification than does decayed wood. These differences can be analyzed and incorporated with velocity measurements to develop estimates of residual wood strength (Wilson, 1988). These types of devices provide residual strength values, which are based on previous data collected from a large population of the same material (poles, piling, timbers) that were destructively tested and used as reference data (Bodig, 1986). Whereas such devices represent a step forward in wood inspection, there are lingering questions concerning the performance of the currently available system. Sonic devices cannot detect the presence of decay, nor can they determine the exact location of any damage. Thus, they can provide an estimate of residual strength for a line of poles, but no estimate of future risk of increased damage, or the decay damage in individual poles.

As sonic technology improves, wave-form analysis should eventually replace velocity measurement. At present, the complexity of the wave after it has been modified by the wood is too great to provide useful information; however, defining the components of the wave can permit more detailed analysis. Despite this anticipated improvement, it is doubtful that sonic testing will be capable of detecting incipient decay or determining whether this decay is active.

Laboratory Decay Detection

Although most laboratory techniques are not currently directly applicable for field use, they can provide valuable information for inspection programs and be used to develop more accurate decay-detection devices.

Culturing

Culturing involves the isolation, growing, and identification of the fungi inhabiting the wood sample. The process normally takes 4 to 6 weeks and requires the services of a trained microbiologist as well as laboratory facilities. Culturing can determine the reliability of the decay criteria used in inspection programs and provide valuable clues as to when and where some fungi entered the poles (Zabel *et al.*, 1980,1991; Eslyn, 1970; Graham and Corden, 1980). It can authenticate uncertain decay diagnoses and indicate when preservative treatments are approaching threshold levels and remedial treatments are necessary (Zabel and Wang, 1988). Culturing can determine whether viable decay fungi are present (Ricard and Mothershead, 1966; Graham and Helsing, 1979; Morrell *et al.*, 1987). Culturing provides a useful measure of decay risk, and the presence of viable fungi indicates that favorable decay conditions are present. Identification of the fungi isolated is an important feature of the culturing process and can provide useful information on the decay hazard and residual preservative effectiveness (Wang and Zabel, 1990; Stalpers, 1978; and Nobles, 1965). Culturing cannot distinguish between active decay and the presence of viable, but inactive fungi. In addition, culturing may artificially inflate the importance of fungi that sporulate heavily or those that grow well under the cultural conditions used. Because of the delay between inspection and cultural results, culturing probably is not feasible for routine wood inspection, but can provide a valuable supporting role.

Microscopy

Light microscopy has been effectively used to detect incipient decay and study the progressive stages of degradation (Wilcox, 1970,1978; Bravery, 1971). Light microscopy detects decay at very early stages, but does not appear to be practical for field use. Extensive training is required for this time-consuming procedure. Light microscopy can be useful along with culturing to identify the features of decay that can be detected by simpler, more practical methods in the field.

Chemical Indicators

The use of chemical indicators for detecting decay has been proposed by a number of researchers (Lindgren, 1955; Cowling and Sachs, 1960; Eslyn, 1979; Line, 1982; Gilbertson *et al.*, 1975). In general, these indicators are based on changes in pH within the wood as the fungus alters wood structure or releases metabolic acids. In principle, indicators can be used to detect various stages of fungal attack; however, in practice, indicators have shown a wide degree of variability. Eslyn (1979) reviewed the use of nine indicators that showed some promise for detecting specific fungal agents in southern pine cores. More recently, a fluorescent dye, acridine-orange, has been used to detect attack by brown-rot fungi (Krahmer *et al.*, 1982). Subsequent studies have shown that the results of this technique were characterized by extensive operator variation. Currently available indicators have proved to be of limited use, but presage the development of a rapid, simple, and practical decay-detection device for field use.

Lectins

A new, recently identified class of fungal indicators are the plant-derived lectins. These compounds are highly specific for sugars and have been used with great success as probes in cell biology (Leiner, 1976). Several lectins are specific for components of the fungal cell wall. One lectin, wheat germ agglutinin (WGA), is highly specific for fungal chitin, which is a cell wall component of most higher fungi (Morrell *et al.*, 1985). The lectin can be visualized in the wood by coupling to a fluorescent or metal compound and observing it under a properly equipped microscope. Preliminary studies indicate that lectins improve the ability to detect fungi at very early stages of attack (Morrell *et al.*, 1985; Krahmer *et al.*, 1986). A wide array of nondematiaceous (dark-pigmented) fungi can be detected using WGA (Morrell *et al.*, 1986). This technique, however, can not distinguish between living and dead fungi.

Serological Tests

Serological techniques based on the development of antibodies that are specific for the presence of a compound or class of compounds also show promise for decay detection (Goodell *et al.*, 1988a,b; Daniel *et al.*, 1988; Srebotnik and Messner, 1990; Garcia *et al.*, 1987). The use of serological techniques to study wood decay was proposed over 20 years ago (Ricard and Mothershead, 1966); however, the methodology has recently seen dramatic improvements. The presence of fungal hyphae or compounds specific to the decay process can be detected by coupling an antibody to a fluorescent compound or by using enzyme-linked immunosorbent assays (ELISA). The latter technique is extensively used as a diagnostic tool for the detection of viruses, such as that causing acquired immune deficiency syndrome. It could provide a high degree of reliability if a single factor common to wood decay by fungi could be identified. Unfortunately, the decay types differ widely, making identification of a single factor at present difficult. If an enzyme unique to decay fungi could be identified, serological detection of fungal decay could prove very useful. This technique would be practical only when the fungus was actively degrading the wood, since the enzymes produced are normally present at low levels and are relatively ephemeral. Thus, inspections during warm summer months when decay fungi are more active would be more likely to detect decay using such techniques. The ideal serological system will probably use an antibody specific for chitin (most higher fungi contain chitin) and a second antibody specific for an enzyme or chemical agent typically produced by wood-decay fungi.

Analytical Techniques

Although chemical and cultural decay-detection methods have been available for some time, recent advances in analytical techniques have permitted the use of more sophisticated chemical methods for decay detection. Infrared (IR) spectroscopy has been evaluated for detection of incipient decay in a number of conifers and hardwoods (Gibson *et al.*, 1985; Nicholas and Schulz, 1987); it can detect relatively small shifts in the ratios of sugars and phenols in the wood. A preliminary study indicated that this technique could detect incipient decay by brown-rot fungi through the presence of a specific transmittance peak produced at the early stages of fungal attack; however, the technique was not so useful for detecting attack by white-rot fungi (Gibson *et al.*, 1985). The peak correlated with brown rot has been tentatively identified as a key lignin breakdown product (Holt, 1986). Infrared spectroscopy has been improved by the use of Fourier transformations (FTIR), which can increase the precision of analysis, permitting the detection of the more subtle shifts in IR spectra (Nicholas and Schultz, 1987). At present, the equipment required for IR detection of decay limits this procedure to the laboratory, but continued development of portable microprocessors may permit extension of this technique to the field. In addition, further studies are required to identify characteristics of decay by white-rot and soft-rot fungi which can be used for IR spectroscopy. This search is compounded by fact that these groups of fungi generally utilize wood breakdown products at rates that approximately correspond to the rate of degradation, leaving few breakdown products available for detection. In addition, the technique must be refined to distinguish between active and inactive decay, since incipient decay before treatment may be inactivated, but still produce a positive response with the IR.

In addition to these laboratory techniques, a number of other analytical methods may have application to decay detection. Infrared photo acoustic spectroscopy (FTIR-PAS) is a relatively recent development for examining solid wood samples (Kuo *et al.*, 1988). In conventional FTIR, the samples are ground, extracted and compressed in a potassium bromide (KBr) pellet. The process is fairly slow and not amenable to field sampling. FTIR-PAS permits solid-state examination of relatively thick sections (400 μm). Although it has not yet been studied for detecting decay, the use of solid samples would permit more rapid analysis.

Scanning wood using microwaves has been proposed for detecting defects in lumber (Martin *et al.*, 1987); however, the sensitivity of such a device in thicker wood sections and its ability to distinguish between the various defects present are unknown.

Differential scanning calorimetry has also been proposed for detecting degraded wood (Reh *et al.*, 1986; Baldwin and Streisel, 1985). This technique requires oven temperatures of approximately 600°C and appears to be feasible only for laboratory studies.

One technique that might have some applicability to decay detection, when used in conjunction with indicators, could measure the optical characteristics of the wood. Changes in the luster and color of wood as it decays have been noted by a number of researchers (Zabel *et al.*, 1982). If the characteristics of color as wood decayed could be quantified, then optical measurements could be used to detect decay. Indicators could also be used to enhance the chances of detecting the damage. Successful identification of color characteristics would require a high degree of background work, since different decay fungi may cause different color changes, particularly between wood species. In addition, the method would have to be capable of detecting decay in the presence of preservative treatments that alter the wood color. Although field-practical optical devices for this purpose are not currently available, it is likely that devices used for other applications could be modified with relatively little effort.

Olfactory Decay Detection

As wood is degraded by fungi and insects, certain volatile chemicals are emitted. Whereas these chemicals cannot be detected by humans, they can be detected analytically. The current cost of instrumentation and the degree of training required for these purposes are beyond the means of most inspection agencies or utilities. However, some animals have the ability to detect these emissions. The use of sniffer dogs to detect decay has been proposed in Europe, and dogs have been used in the United States to detect termite infestations in buildings (Morris and Friis-Hansen, 1984). At present the use of dogs for decay detection remains largely an interesting idea with many uncertainties as to its feasibility under the wide range of conditions and treatments for wood in use.

As the fundamental aspects of decay are unraveled, it is likely that more sophisticated and accurate methods for detecting this damage will be developed.

Summary

- The reliable detection of incipient and early decay is a major problem in many wood uses and remains an important research goal in wood microbiology.
- The detection of internal decay in its early stages is difficult because of its occurrence in locations that are difficult to access. Detection of decay poses major challenges because of the sporadic patterns of deterioration, the range of effects, and the wide variety of natural defects present in wood.
- Points where moisture can collect, such as deep checks, crawl spaces, soil contact, around window frames, or leaking pipes represent the most likely zones for detecting decay and should be investigated first.
- The wide variety of inspection techniques can be classified as chemical, mechanical, electrical, and acoustic. No universally effective inspection method has been found, and decay detection often requires the use of more than one device.
- Mechanical methods such as drilling or boring are most frequently used for detecting internal decay, but efforts are underway to develop improved electric and acoustic detection devices. In addition, efforts to use serological techniques and other chemical indicators for detecting the presence of fungi or their degradation products are being studied.

References

Agi, J. J. (1983). Structural evaluation of marine structures. *Proceedings, Specialty Conference on Port Modernization, Upgrading and Repairs.* American Society of Civil Engineers. New Orleans, Louisiana.

Anonymous. (1987). Innovative wood-pole-management program uses NDE technique. *Electrical World* (March, 1987):39–43.

Baldwin, R. C., and R. C. Streisel. (1985). Detection of fungal degradation at low weight loss by differential scanning calorimetry. *Wood and Fiber Science* **17**(3): 315–326.

Bodig, J. (1986). Practical definition of wood pole strength: Relation to construction standards. *Proceedings, The Wood Pole Conference.* Portland, Oregon.

Bravery, A. F. (1971). The application of scanning electron microscopy in the study of timber decay. *Wood Science and Technology* **19**:35–46.

Bucur, V. (1983). An ultrasonic method for measuring the elastic constants of wood increment cores bored from living trees. *Ultrasonics* **83**:116–126.

Cowling, E. B., and I. B. Sachs. (1960). Detection of brown rot with osmium tetroxide stain. *Forest Products Journal* **10**(11):594–596.

Cown, D. J. (1978). Comparison of the Pilodyn and torsiometer methods for the rapid assessment of wood density in living trees. *New Zealand Journal of Forestry Science* 8(3):384-391.

Cown, D. J., and J. D. Hutchinson. (1983). Wood density as an indicator of the bending properties of *Pinus radiata* poles.*New Zealand Journal of Forestry Science* 13(1): 87-99.

Daniel, G. F., T. Nilsson, and B. Petterson. (1988). "Immunolabelling Studies on the Detection of Enzymes During the Degradation of Wood by *Phanerochaete chrysosporium.*" International Research Group on Wood Preservation, Document IRG/WP/1364. Stockholm, Sweden.

Eslyn, W. E. (1970). Utility pole decay. Part 2. Basidiomycetes associated with decay in poles. *Wood Science and Technology* 4:97-103.

Eslyn, W. E. (1979). Utility pole decay. Part 3. Detection in pine by color indicators. *Wood Science and Technology* 13:117-126.

Friis-Hansen, H. (1980). "A Summary of Tests and Practical Experiences with the Pilodyn Wood-Testing Instrument." International Research Group on Wood Preservation, Document No. IRG/WP/282. Stockholm, Sweden.

Garcia, S., J. P. Latge, M. C. Prevost, and M. Leisola. (1987). Wood degradation by white-rot fungi: Cytochemical studies using lignin peroxidase-immunoglobulin-gold complexes. *Applied and Environmental Microbiology* 53(10):2384-2387.

Gibson, D. G., R. L. Krahmer, and R. C. DeGroot. (1985). Early detection of brown-rot decay in Douglas fir and southern yellow pine by infrared spectrophotometry. *Wood and Fiber Science* 17(4):522-528.

Gilbertson, R. L., F. F. Lombard, and E. R. Canfield. (1975). "Gum Guaiac in Field Tests for Extracellular Phenol Oxidases of Wood-Rotting Fungi and Other Basidiomycetes" USDA Forest Service Research Paper FPL 269. Madison, Wisconsin.

Goodell, B. S., and R. D. Graham. (1983). A survey of methods used to detect and control fungal decay of wood poles in service. *International Journal of Wood Preservation* 3(2):61-63.

Goodell, B. S., G. Daniel, J. Jellison, and T. Nilsson. (1988a). "Immunolocalization of Extracellular Metabolites from *Poria placenta.*" International Research Group on Wood Preservation, Document No. IRG/WP/1361. Stockholm, Sweden.

Goodell, B. S., J. Jellison, and J. P. Hosli. (1988b). Serological detection of wood decay fungi. *Forest Products Journal* 38(3):59-62.

Graham, R. D., and M. E. Corden. (1980). "Controlling Biological Deterioration of Wood with Volatile Chemicals." Electric Power Research Institute, Report EL-1480. Palo Alto, California.

Graham, R. D., and G. G. Helsing. (1979). "Wood Pole Maintenance Manual: Inspection and Supplemental Treatment of Douglas Fir and Western Red Cedar Poles." Research Bulletin 24. Forest Research Laboratory, Oregon State University, Corvallis, Oregon.

Hattori, Y., and Y. Kanagawa. (1985). Non-destructive measurement of moisture distribution in wood with a medical CT-scanner. *Journal of the Japanese Wood Research Society* 31(12):974-982.

Hoag, M., and M. D. McKimmy. (1988). Direct scanning x-ray densitometry of thin wood sections. *Forest Products Journal* 38(1):23-26.

Hoffmeyer, P. (1978). The Pilodyn instrument as a non-destructive tester of the shock resistance of wood. *Proceedings, Non-Destructive Testing Symposium*, Vancouver, Washington.

Holt, K. (1986). "Chemical Characterization of Breakdown Products Involved in Detection of Incipient Decay in Wood. M. S. thesis, Oregon State University, Corvallis, Oregon.

Inwards, R. D., and R. D. Graham. (1980). Comparing methods for inspecting Douglas fir poles in service. *Proceedings, of the American Wood Preservers' Association* **76**:283-286.

James, W. L. (1980). "Effects of Wood Preservatives on Electric Moisture-Meter Readings." U. S. Forest Service, Forest Products Laboratory Research Note FPL-0106. Madison, Wisconsin.

James, W. L. (1988). "Electric Moisture Meters for Wood." U. S. Forest Service, Forest Products Laboratory, General Technical Report FPL-GTR-6. Madison, Wisconsin.

Knuffel, W. E. (1988). Acoustic emission as strength predictor in structural timber. *Holzforschung* **42**(3):195-198.

Krahmer, R. L., R. C. DeGroot, and E. C. Lowell. (1982). Detecting incipient brown rot with fluorescence microscopy. *Wood Science* **15**(2):78-80.

Krahmer, R. L., J. J. Morrell, and A. Choi. (1986). Double-staining to improve visualization of wood-decay hyphae in wood sections. *International Association of Wood Anatomists Bulletin* **7**(2):165-167.

Kuo, M., J. F. McClelland, S. Luo, P. L. Chien, R. D. Walker, and C. Y. Hse. (1988). Applications of photoacoustic spectroscopy for wood samples. *Wood and Fiber Science* **20**(1):132-145.

Leiner, I. E. (1976). Phytohemagglutinins (Phytolectins). *Annual Review of Plant Physiology* **27**:291-319.

Lindgren, O. (1987). "Computerized Axial Tomography: A Nondestructive Method for Three-Dimensional Wood Density/Moisture Content Measurements." International Research Group on Wood Preservation, Document No. IRG/WP/2285. Stockholm, Sweden.

Lindgren, R. M. (1955). "Color Test for Early Storage Decay in Southern Pine." U.S.D.A. Forest Service, Forest Products Laboratory, Report Number 2037. Madison, Wisconsin.

Line, M. A. (1982). "Toward a Colour Assay of Wood Degradation." International Research Group on Wood Preservation, Document No. IRG/WP/2180. Stockholm, Sweden.

Martin, P., R. Collet, P. Barthelemy, and G. Roussy. (1987). Evaluation of wood characteristics: Internal scanning of the material by microwaves. *Wood Science and Technology* **21**:361-371.

Morrell, J. J., and M. E. Corden. (1986). Controlling wood deterioration with fumigants: A review. *Forest Products Journal* **36**(10):26-34.

Morrell, J. J., and J. B. Wilson. (1985). Detecting and controlling decay of wood utility poles in service. *Proceedings, Northwest Public Power Association Engineering and Operations Conference,* Portland, Oregon.

Morrell, J. J., D. G. Gibson, and R. L. Krahmer. (1985). Effect of fluorescent-labeled lectins on visualization of decay fungi in wood sections. *Phytopathology* **75**: 329-332.

Morrell, J. J., R. L. Krahmer, and L. C. Lin. (1986). "Use of Fluorescent Coupled Lectins as Probes for Studying Fungal Degradation of Wood." International Research Group on Wood Preservation, Document No. IRG/WP/1288. Stockholm, Sweden.

Morrell, J. J., S. M. Smith, M. A. Newbill, and R. D. Graham. (1986). Reducing internal and external decay of untreated Douglas fir poles: A field test. *Forest Products Journal* **36**(4):47-52.

Morrell, J. J., M. E. Corden, R. D. Graham, B. R. Kropp, P. Przybylowicz, S. M. Smith, and C. M. Sexton. (1987). Basidiomycetes colonization of Douglas fir poles during air-seasoning and its effect on wood strength. *Proceedings of the American Wood Preservers' Association* **83**:284-296.

Morris, P. I., and H. Friis-Hansen. (1984). "Report on a Field Demonstration of Methods for Detecting Defects in Wood Poles." International Research Group on Wood Preservation, Document No. IRG/WP/2232. Stockholm, Sweden.

Mothershead, J. S., and S. S. Stacey. (1965). Applicability of radiography to inspection of wood products. *Proceedings, Second Symposium on Non-Destructive Testing.* Spokane, Washington.

Murphy, M. W., and J. Taylor-Wilson. (1988). Vibration analysis: An alternative to sonic testing. *Proceedings, The Wood Pole Conference,* Portland, Oregon.

Murphy, M. W., D. E. Franklin, and R. A. Palylyk. (1987). "A Non-Destructive Testing Technique for Wood Poles." International Research Group on Wood Preservation, Document No. IRG/WP/2293. Stockholm, Sweden.

Nicholas, D. D., and T. P. Schultz. (1987). "Detection of Incipient Brown-Rot Decay in Wood by Fourier Transform Infrared Spectroscopy." International Research Group on Wood Preservation, Document Number IRG/WP/2275. Stockholm, Sweden.

Nobles, M. F. (1965). Identification of cultures of wood-inhabiting hymenomycetes. *Canadian Journal of Botany* **43**:1097-1139.

Noguchi, M., K. Nishimoto, Y. Imamura, Y. Fujii, S. Okumura, and T. Miyauchi. (1986). Detection of very early stages of decay in western hemlock wood using acoustic emissions. *Forest Products Journal* **36**(4):35-36.

Pellerin, R. F., R. C. DeGroot, and G. R. Esenther. (1985). Nondestructive stress wave measurements of decay and termite attack in experimental wood units. Proceedings of the 5th Nondestructive Testing of Wood Symposium. Pullman, Washington

Pierce, B. (1983). Hardness-penetration tests on three species and five treatments. Unpublished report. Arizona Public Service.

Piirto, D. D., and W. W. Wilcox. (1978). Critical evaluation of the pulsed current resistance meter for detection of decay in wood. *Forest Products Journal* **28**: 52-57.

Reh, U., G. Kraepelin, and I. Lamprecht. (1986). Use of differential scanning calorimetry for structural analysis of fungally degraded wood. *Applied and Environmental Microbiology* **52**(5):1101-1106.

Ricard, J. L., and J. S. Mothershead. (1966). Field procedure for detecting early decay. *Forest Products Journal* **16**(7):58-59.

Ross, R. J., and R. F. Pellerin. (1988). NDE of wood-based composites with longitudinal stress waves. *Forest Products Journal* **38**(5):39-45.

Safo-Sampah, S., and R. D. Graham. (1976). Rapid agar-stick breaking-radius test to determine the ability of fungi to degrade wood. *Wood Science* **9**(2):65-69.

Scheffer, T. C., and A. F. Verrall. (1979). (rev.). "Principles for Protecting Wood Buildings from Decay." U.S.D.A. Forest Service Research Paper FPL-190, Madison, Wisconsin.

Shigo, A. L., and A. Shigo. (1974). "Detection of Discoloration and Decay in Living

Trees and Utility Poles." U. S. D. A. Forest Service Research Paper NE 294. Upper Darby, Pennsylvania.

Shigo, A. L., W. C. Shortle, and J. Ochrymowych. (1977). "Detection of Active Decay at Groundline in Utility Poles." U. S. Forest Service General Technical Report NE-35. Upper Darby, Pennsylvania.

Shortle, W. C. (1982). Decaying Douglas fir wood: Ionization associated with resistance to a pulsed electric current. *Wood Science* **15**(1):29–32.

Shortle, W. C., A. L. Shigo, and J. Ochrymowych. (1978). Patterns of resistance to a pulsed electric current in sound and decayed utility poles. *Forest Products Journal* **28**(1):48–51.

Smith, S. M., and R. D. Graham. (1983). Relationship between early decay and radial compression strength of Douglas fir. *Forest Products Journal* **33**(6):49–52.

Smith, S. M., and J. J. Morrell. (1986). Correcting Pilodyn measurement of Douglas fir for different moisture levels. *Forest Products Journal* **36**(1):45–46.

Smith, S. M., and J. J. Morrell. (1987). Longitudinal compression as a measurement of ultimate wood strength. *Forest Products Journal* **37**(5):49–53.

Smith, S. M., and J. J. Morrell. (1989). Comparing full-length bending strength and small-scale test strength of western red cedar poles. *Forest Products Journal* **39**(3):29–33.

Srebotnik, E., and K. Messner. (1990). "Immunogold Labelling of Size Marker Proteins in Brown-Rot-Degraded Pine Wood." International Research Group on Wood Preservation. Document No. IRG/WP/1428. Stockholm, Sweden.

Stalpers, J. A. (1978). Identification of wood-inhabiting Aphyllophorales in pure culture. *Centralbureau Voor Schimmelcultures, Baarn, Netherlands, Studies in Mycology* **16**:1–248.

Taylor, F. W. (1981). Rapid determination of southern pine specific gravity with a Pilodyn tester. *Forest Science* **27**(1):59–61.

Thornton, J. D. (1979). Detection of decay in wood using a pulsed-current resistance meter (Shigometer). I. Laboratory tests of the progression of decay of *Pinus radiata* D. Don sapwood by *Poria monticola* Murr. and *Fomes lividus* (Kalch.) Sacc. *Material and Organismen* **14**(1):15–26.

Wang, C. J. K., and R. A. Zabel. (1990). "Identification Manual for Fungi from Utility Poles in the Eastern United States." American Type Culture Collection, Rockville, Maryland.

Wilcox, W. W. (1970). Anatomical changes in wood cell walls attacked by fungi and bacteria. *Botanical Review* **36**(1):1–28.

Wilcox, W. W. (1978). Review of the literature on the effects of early stages of decay on wood strength. *Wood and Fiber* **9**(4):252–257.

Wilcox, W. W. (1983). Sensitivity of the "pick test" for field detection of early wood decay. *Forest Products Journal* **33**(2):29–30.

Wilcox, W. W. (1988). Detection of early decay stages of wood decay with ultrasonic pulse velocity. *Forest Products Journal* **38**(5):68–73.

Wilson, J. B. (1988). An overview of non-destructive testing. *Proceedings, The Wood Pole Conference*, Portland, Oregon.

Yamamoto, K., and T. Fujii. (1987). Application of soft x-ray microdensitometry to 2-dimensional quantitative evaluations of density decreases by wood decay. *Journal of the Japanese Wood Research Society* **33**(12):974–982.

Zabel, R. A., and C. J. K. Wang. (1988). Utility pole decay problems in the eastern

United States: Changing viewpoints. *Proceedings, The Wood Pole Conference,* Portland, Oregon.

Zabel, R. A., F. F. Lombard, and A. M. Kenderes. (1980). Fungi associated with decay in treated Douglas fir transmission poles in the northeastern United States. *Forest Products Journal* **30**(4):51-56.

Zabel, R. A., C. J. K. Wang, and F. C. Terracina. (1982). "The Fungal Associates, Detection, and Fumigant Control of Decay in Treated Southern Pine Poles." Final Report, Electric Power Research Institute. EPRI EL-2768. Palo Alto, California.

Zabel, R. A., C. J. K. Wang, and S. E. Anagnost. (1991). Soft-rot capabilities of major microfungi isolated from Douglas fir poles in the Northeast. *Wood and Fiber Science* **23**(2):220-237.

CHAPTER

17

Paint Mildew and Related Degradative Problems

Paints and other coatings applied to the surfaces of many building materials provide important protective and decorative benefits. A large multibillion dollar industry has developed, worldwide, to provide the specialized coatings and application techniques needed for buildings and other materials (bridges, ships, automobiles, machinery, etc.). Coats of paint are particularly useful in most exterior uses of wood and, in addition to a decorative function, protect the wood surface from weathering damage (see Chapter 2) and facilitate the shedding of rain water. The coatings used on buildings are often called architectural paints since they are primarily applied to building materials, including wood. They represent nearly one half of the coverings volume sold, annually.

This chapter will cover the microflora of paint, the factors affecting mildew development, and general control practices. Emphasis will be on paint coverings over wood in exterior home uses. Brief mention will be made of mold and related biodeterioration problems in home and industry.

Comprehensive coverages of the principles and practices of coatings technology are available in textbooks, coatings journals, and governmental publications (Weismantel, 1980; Martens, 1981; Lambourne, 1987). General reviews of biodeterioration problems in coatings on wood have been assembled (Ross and Hollis, 1976; Brand and Kemp, 1973; Bravery, 1988).

Hess's classic manual, *Paint Film Defects,* is a useful source of information on the nature, cause, and control of paint defects (Hamburg and Morgans, 1979).

Types of Paint Biodeterioration

Microorganisms (fungi and bacteria) are the principal biotic destructive agents of paints under some use conditions or locations. Paint disfigurement results primarily from the copious mycelial growth of pigmented fungi and the development of spore clusters on the paint surface. The general term used by industry for these growths is *mildew*. Other terms are *disfigurement* or *defacement*, which are really preferable to avoid confusion in the minds of many with the mildew diseases of agricultural crop plants. The aesthetic properties of the paint film are often further reduced by the accumulations of dirt and airborne detritus on the sticky exposed mycelia or spore clusters. Physical and chemical degradation (cracking, chipping, and exfoliation) of the paint film may also occur from the digestive and growth activities of microorganisms. Digestion of the primer coat by bacteria lodged at the wood–paint adhesion zone, when the film remains wet, is also believed to be an important contributor to loss of adhesion and peeling. Waterborne paints are degraded primarily by bacteria during the liquid storage phase, and gas formation, foul odors, and losses in viscosity are some of the paint properties affected (Opperman, 1986). It has been estimated that millions of dollars are now spent annually by the paint industry for the various chemicals needed to protect liquid paint (stored in cans) and paint films from microbial damage.

Paint Types and Compositions

A general idea of the chemical ingredients and structure of paint is necessary to understand its degradation by microorganisms. Paint is a suspension of solid particles in a liquid, which after application to a material, dries, hardens, and forms a durable, continuous surface film. The principal components of paint can be grouped by function into four general categories.

1. The *vehicle* or *binder* provides the matrix or major physical structure to the paint film. Common current vehicles are alkyd, acrylic, polyvinyl acetate, polyurethane, epoxy, and other resins or polymers.
2. Pigments (TiO_2) and extenders (calcium carbonate, talc, mica) provide color and bulk the voids in the paint matrix. Semicolloidal suspensions (emulsions) of the resins in water form the latex type of paints.
3. Various additives such as thickeners (hydroxy-ethyl cellulose), emulsifiers, driers, plasticizers, and biocides confer special properties or application features to the paints.
4. A solvent (organic or water) is the carrier for the paint solids.

The coatings applied to homes for exterior and interior uses can be conveniently grouped into two general types: (1) water-thinned paints (emulsions of resins, polymers, and pigments in water); and (2) solvent-thinned paints (various drying oils, polymers, and pigments in organic solvents such as turpentine).

Both types of paint contain ingredients subject to microbial attack. In the older solvent-thinned paints, vehicles such as linseed oil, soya oil, safflower oil, and dehydrated castor oil are degraded by fungi or bacteria with esterase enzymes. In the water-thinned paints, resins such as acrylic and polyvinyl polymers are resistant to microbial attack, but thickeners such as ethylated cellulose are degraded by some fungi and bacteria.

The use of water-based latex paints has increased steadily in the past few decades because of their ease of application and rapid drying. In general, they are softer films and more prone to microbiological attack than are oil-based paints. The light colors in the latex paints have enhanced the visibility of mildew.

The structure of paint films also affects their susceptibility to microbial damage. Internal voids in the film provide zones for water accumulation and penetration routes for fungi. It is important that all the voids around the binder matrix be occupied by pigment particles. This value is termed the *critical pigment volume concentration.* A low porosity reduces the susceptibility of latex paints to fungal invasion. Chalking paints have been formulated so unbound pigment particles on the surface can slowly slough off and cleanse the surface. Design of paint films specifically to minimize microbial damage represents a neglected research area of potential promise (Ludwig, 1974).

Paint Microflora

It is surprising to note that until the late 1940s, most paint disfigurements were believed to be unavoidable detritus accumulations. One of the earliest reports of fungi as the major culprits of paint mildew was made by Goll *et al.* (1952). They developed techniques to microscopically study the surfaces of many paint samples and found that most disfigurements were actually small patches of pigmented hyphae and clusters of spores. Comprehensive studies of the microflora on paint film soon followed (Klens *et al.*, 1954; Rothwell, 1958; Drescher, 1958; Eveleigh, 1961; and Smith, 1977) and are summarized by Brand and Kemp (1973) and Ross and Hollis (1976). Whereas many genera of opportunistic microorganisms were reported, only a few were consistently obtained in large numbers. The major fungi were *Aureobasidium (Pullularia) pullulans, Cladosporium* spp., *Alternaria* spp. and *Phoma glomerata. Aureobasidium pullulans* was the predominant mildew agent and was found in all regions of the United States and subsequently worldwide. Of the other genera, *Phoma glomerata* was found primarily in the Pacific coast region and *Alternaria* spp. in the eastern region of the United States. The principal bacteria were species of *Flavobacterium* and *Pseudomonas* (Ross, 1964). *Pseudomonas aeruginosa* was isolated commonly from spoiled latex paints in liquid phase (storage), and this may account for its prevalence in latex films.

In recent studies of microorganisms isolated from the waterborne paint formulations and films, Jakubowski *et al.* (1983) reports that the recent changes in paint formulations and manufacturing practices are now creating favorable niches for other fungi and bacteria. In addition to *Pseudomonas* spp., various yeasts (*Torula* and *Saccharomyces*) were found to be important spoilage agents in stored latex paints. Also, whereas *A. pullulans* continued to be the predominant fungus associated with mildew on exterior latex paints, *Alternaria* spp. were also judged to be important.

Characteristics and Growth Features of *A. pullulans*

Some detail on the nature and capabilities of *A. pullulans* is necessary to explain its apparent ubiquity on paint surfaces, worldwide. As the accepted major causal agent of paint mildew, *A. pullulans* has been studied intensively, and a large literature has been assembled on its taxonomy, physiology, and role in paint films (Brand and Kemp, 1973).

Classification *Aureobasidium pullulans* is grouped in the Hyphomycetes of the Deuteromycotina and is characterized by pigmented hyphae (usually black) and reproduction by blastospores (Fig. 17-1). It is *dimorphic* and may assume a yeast or mycelial form depending on the substrate. As a *heterokaryon*, *A. pullulans* is extremely variable, and isolates can assume a wide range of phenotypic forms. The fungus occurs commonly in the soil, wood, and on plant surfaces (phylloplane). This fungus forms an extracellular carbohydrate (pullulan), which facilitates tenacious adherence to surfaces.

Physiology *Aureobasidium pullulans* thrives over broad temperature and pH ranges. It can utilize many simple carbon compounds and displays resistance to some toxicants. The fungus resists desiccation and actinic exposure. It adheres tenaciously to substrates, appears to be well adapted to surfaces, and can utilize detritus and paint extractives whenever favorable moisture levels and temperatures occur.

Growth Patterns in Latex Paint Films *Aureobasidium pullulans* grows primarily in the hyphal form on paint surfaces. Small hyaline and pigmented hyphae grow on the surface and into any air bubbles or voids in the paint film (Fig. 17-2A). Large discrete clusters of blastospores (conidia) and thick-walled chlamydospores commonly develop (Fig. 17-1A).Copious amounts of a brownish hyphal exudate forms on the film surface and embeds the hyphae. After drying, the hyphae adhere tenaciously to the paint surface and may cause the exfoliation of small paint fragments (Fig. 17-2C).

Roles in Nature It is perhaps significant to note other known and suspected roles in nature of this interesting organism. *Aureobasidium pullulans* is a well-known causal agent of sapstain in lumber (Käärik, 1974). Schmidt and French (1976) have shown this fungus to be responsible for discolorations that develop in shingles and shakes. Plant pathologists recognize the organism as a common phylloplane resident of leaf and bud surfaces (Baker and

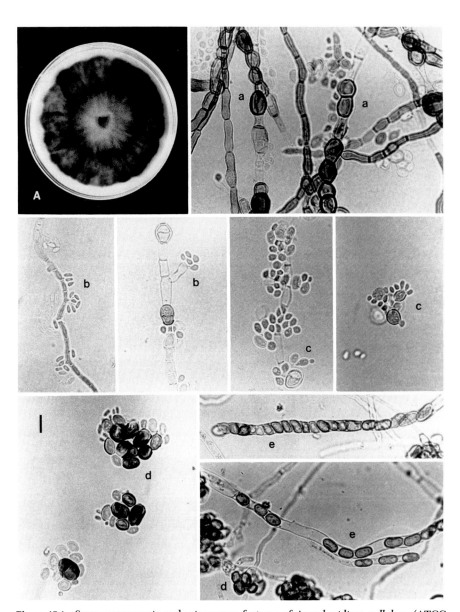

Figure 17-1 Some macroscopic and microscope features of *Aureobasidium pullulans* (ATCC 66656) (A) Colony on malt extract agar at 3 weeks; (a) hyphae with melanized or pigmented spherical cells; (b) hyaline conidigenous cells and conidia; (c) germinating conidia; (d) pigmented conidia; and (e) endoconidia. All ×600. Bar = 10 μm. Published with permission of American Type Culture Collection and the courtesy of Dr. Chun J.K. Wang.

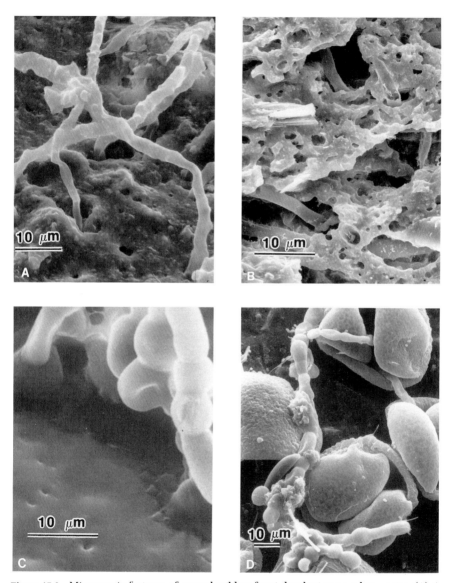

Figure 17-2 Microscopic features of several mildew fungi developing on a latex paint. (A) *A. pullulans* growing on the surface of a paint and penetrating into air holes; (B) *Cladosporium sphaerospermum* hyphae growing in the cavities and voids of a paint film prepared by freeze-drying and fracturing; (C) *A. pullulans* hyphae tenaciously adhering to a paint surface after sample drying; and (D) *A. pullulans* hyphae associated with a pine pollen grain on a paint surface. (Figures 2C and D from Zabel and Terracina, 1980, permission of Society for Industrial Microbiology).

Cook, 1974), and this fungus has also been named as a probable symbiont in lichens (Ahmadjian, 1967).

Factors Affecting Mildew Development

Both waterborne and organic-solvent paints may be severely damaged by microorganisms depending on factors such as geographic location, the material covered, the type of exposure, detritus sources, and the paint components.

Geographic Location Paint mildew problems are more severe in coastal and tropical regions where high humidities and temperatures prevail.

Type of Surface In general, mildew develops more commonly on paints on wood than on metals or masonry. Wood substrates may provide nutrients from their water-soluble extractives that diffuse into the paint film and can be a significant moisture and inoculum source. Also, film integrity and adhesion may be affected by the dimensional changes occurring with fluctuations in moisture content. Paint coats on redwood and the cedars are less susceptible to mildew development than those on the southern yellow pines, Douglas fir, and hemlock.

Type of Exposure On exterior surfaces of homes, mildew is often heaviest in locations where shade trees or shrubbery restrict rapid surface drying and on north-facing walls. Paint surfaces protected from direct rainfall accumulate more detritus and mildew. Ceilings over open garages and porches may be subject to severe mildew owing to condensation caused by rapid temperature drops in the evening in humid regions. Unsightly mildew may develop on painted surfaces in bathrooms and window units because of condensation, particularly during the colder seasons. Poorly ventilated closets and damp basement rooms may also accumulate mildew and other molds during periods of high humidity. The spores and fungal fragments from mildew and mold development may be an overlooked significant source of allergens in some buildings.

Detritus Sources The walls and ceilings of food-processing plants, paper mills, breweries, and dairies often present severe mildew problems attributable to the combination of excessive moisture and detritus accumulations on the surfaces. In severe cases, mold (*Penicillium* spp., *Aspergillus* spp., and *Trichoderma* spp.) may also develop and require specially treated paints and frequent washings to control.

Detritus that accumulates on paint surfaces includes a variety of nutritive sources for mildew fungi such as plant exudates (aphid secretions) and pollen. It has been shown that small amounts of pollen (Fig. 17-2D) and simple sugars simulating natural conditions, stimulate mildew development on latex paints (Zabel and Terracina, 1980).

Nutrient Sources   The nutritive status of paint ingredients varies greatly between and within paint types. Spoilage problems during paint storage occur primarily in waterborne paints. Vehicles consisting of natural oils are readily degraded by some paint-inhabiting microorganisms, but most of the synthetic resins are resistant. Some pigments, such as zinc oxide and barium metaborate, reduce mildew growth, whereas titanium dioxide appears to have no effect. Some paint additives, such as ethylated cellulose, can be degraded by bacteria during paint storage and can also be a minor nutrient source for some fungi on paint films.

Free moisture on the paint film appears to be the critical environmental requisite for mildew development. However, little basic information exists on the amounts and state of water necessary to initiate mildew growth on paints. Hill and April (1971) have postulated that there is critical moisture level to sustain mildew development in and on paint. It has been suggested that prolonged relative humidities above 75% are necessary for mildew formation. At these high humidities, however, slight decreases in temperature lead to a dew point and free water accumulations on the film. In controlled studies of temperature, relative humidity, and moisture, it has been reported that mildew first began to appear on latex films over wet wood blocks at between 30 and 60% moisture content (Zabel and Terracina, 1980).

Postulated Roles of A. pullulans *in Latex Paints*   Surprisingly, despite the importance of *A. pullulans* as a mildew agent, its nutritional sources are uncertain. It is a common observation in the laboratory that *A. pullulans* cannot sustain growth on a latex film under axenic conditions.

Studies on the succession of microorganisms on test paint panels have reported that bacteria and other fungi must precede and modify the nutritive status of the film before the appearance of *A. pullulans*, which then remains as the final dominant or climax species (Winters *et al.*, 1975; Schmitt, 1974). Surface isolation difficulties and whether the microorganism sequences are fortuitous or obligatory raise uncertainties about these claims. Co-metabolism of some paint ingredients with other microorganisms has been proposed (Horvath and Esposito, 1979).

Similarities between *A. pullulans* growth patterns on paints, plastic greenhouse roofs (Durrell and Goldsbury, 1970), and cedar shingles, and its role in nature as a major phylloplane resident suggest it is uniquely adapted to utilize nutrients on these hostile sites. In all cases, the fungus must adhere to a surface, withstand desiccation and high UV radiation, and rely on detritus as the major nutrient source. It is conspicuous and outlasts competitors, so it appears as the dominant paint inhabitant. Many other roles or sequences may occur among the many types of paint and paint-inhabiting microorganisms.

General Control Practices

The basic principles of mildew control are to minimize moisture and detritus accumulations on paint films and to use the more resistant paint treated with mildewcides where mildew hazards are severe. An analysis of the various factors affecting mildew development indicates a range of actions or decisions the homeowner can take to reduce the problem. Periodic washing with disinfectants (sodium hypochlorite) is sometimes helpful. Surface cleaning and disinfection is necessary when repainting over a previous mildewed coat. Buildings designed to facilitate rapid water shedding and promote good aeration can be useful. Steps can often be taken to reduce excessive condensation in bathrooms, windows, and other high-humidity zones. Proper installation of insulating materials in ceilings and walls can reduce the moisture levels in walls. The selection of paints containing the more *nondegradable* ingredients and acceptable film hardness is recommended in all uses where mildew problems can develop.

However, these steps are insufficient in many cases. The most reliable practice has been the addition of effective biocides (mildewcides) to the paint during formulation. These biocides provide protection to the paint during both storage and use. Two widely used pigments, barium metaborate and zinc oxide, are known to be effective inhibitors when used at proper concentrations. Preservatives that have proven to be remarkably effective as mildewcides are the phenyl mercury compounds and chlorinated phenols.

Recent environmental restrictions on uses of mercury compounds (phenyl mercury acetate) and environmental concerns about the chlorinated phenols have stimulated a major search for effective replacement compounds. A wide range of proprietary compounds are now marketed as mildewcides for various paint types and uses. A few examples of their toxic constituents are quaternary ammonium compounds, zinc dimethyl dithiocarbamate, isothiazoline compounds, substituted pyridines such as trichlorodicyanopyridine, triazines, copper-8-hydroxyquininololate, tributyltin oxide, tetraethyl-thiuram disulphide, and tetrachloropyridine-4-methylsulfone.

Ideal mildewcides must be highly effective against mildew fungi, be compatible with other paint components, possess limited solubility, produce no color changes, be nonvolatile, be safe in handling and use, and be environmentally acceptable.

Mildewcide Evaluations

In the development of potential mildewcides, agar-plate toxicity assays are used for the preliminary screening. Promising compounds are then evaluated by chamber test procedures developed by the ASTM (1974). Painted wood panels are exposed in a closed chamber at a temperature of 30°C and relative humidities ranging from 90 to 100%. A bed of loam soil is seeded with selected test fungi and placed below the test panels. Air circulation is provided to ensure a constant source of airborne propagules of the mildew fungi. The effectiveness of a compound is determined by the percentage of mildew coverage compared with that of controls. Weatherometers have also been developed so paint films may be exposed to schedules of simulated rainfalls and UV exposure before or intermittent with the chamber exposures. An accelerated laboratory procedure for effectively growing *A. pullulans* on fresh latex paints on wood axenically was developed to permit the direct testing of mildewcides in the laboratory (Zabel and Horner, 1981).

Potentially effective compounds are tested further in selected homes or extensive panel field tests. Painted wood panels are exposed in severe mildew hazard locations such as Miami, Florida, or New Orleans, Louisiana. Vertical exposures of the panels facing north are generally used. The panels are mounted on offset racks to minimize washing from one panel to the other. Ineffective mildewcides can generally be eliminated after a 1-year exposure. Compounds that appear promising in the laboratory may prove ineffective in the field tests.

Some Research Considerations

The search for more effective and lasting mildewcides that are nontoxic to mammals will continue. A potentially useful procedure of incorporating toxicants, chemically or by encapsulation, into the binder or vehicle has been reported by Pittman and Lawyer (1982). Such a process could reduce leaching losses and deliver the toxicant only when needed at the time of microbial attack. Paint chemists and formulators have demonstrated remarkable skill in designing paints for a wide range of diverse purposes. The use of nondegradative ingredients and moisture control of films may be the basic key to mildew control. Designing paint films to reduce the size and frequency of inner voids and to minimize free water entry from surfaces could be useful. Film designs might be sought to facilitate passage of water vapor out and repel free water ingress or to minimize surface detritus accumulations. Given the current state of knowledge on the factors affecting mildew development, it is probable that bioresistant paints will be achieved when this property assumes a higher priority.

Related Degradative Problems of Microorganisms on Surfaces

In addition to the stains and molds that form on wood during its seasoning and structural uses (discussed in Chapter 14), the growth of microorganisms on machine surfaces during paper manufacture and other industrial and home surfaces can cause serious biodeterioration problems. All these problems can be traced to conditions in which adequate moisture and food substances (simple carbon compounds) accumulate. The offending fungi and bacteria involved are often the same opportunist microorganisms that rapidly exploit any new wood resource. These additional problems are briefly mentioned and referenced here because of their close relationship with lumber molds and paint mildew.

Slime Formation during Pulp and Paper Manufacture

The formation of masses of bacteria and fungi on piping and machine surfaces during the pulping and papermaking processes are commonly known as *slime*. Fragments of slime that loosen and get into the pulp suspension cause paper breaks, degrading holes or specks on the sheet, odors, production disruptions, and reductions in machine efficiency. Slime is caused by various species of bacteria and/or fungi that develop in recesses, corners, and other zones of low circulation in the system. Leathery and gelatinous slimes are often caused by bacteria. Semisolid slimes are traceable to mixtures of bacteria and fungi. Stringy slimes are generally caused by fungi.

Slime formation varies considerably with season of the year, the kind of pulp, type of mill system, and mill practices. Slimes are generally most severe in the spring and fall when bacterial counts are highest in the intake water. General slime problems are more severe on groundwood or semimechanical than on chemical pulps and on sulphite than on kraft mills. Slime problems also are more severe where large volumes of water are recirculated or where paper finishes include clays and starch. Many of the slime-forming microorganisms are traceable to the incoming mill water and additives. Chlorination of intake water and bleaching agents used during pulping reduce slime formation and can minimize pipe-plugging problems with iron bacteria. Control approaches for slime begin with good housekeeping practices, weekly cleanups, and periodic flushing of the mill system with hot-toxicant detergent solutions. In cases where chlorination of intake water and good housekeeping are inadequate, the use of slimicides is necessary. Some current slimicides are the carbamates (disodium ethylene *bis*-dithiocarbamate, sodium dimethyldithiocarbamate), methylene *bis*-thiocyanate,3,5-dimethyl-1,3,5-2H-tetrahydrothiadiazine-2-thione, and 1,2,-dibromo-3-nitrilopropionamide. Slimicides must have low retentions in paper products and must be essentially nontoxic to higher organisms, and nonhazardous to fish and wildlife.

Various companies and consultants listed in the pulp and paper journals generally determine and provide the specialized slimicide treatments necessary for a mill. These services involve a study of the mill system; determination of sampling points; periodic sampling of slurry and paper; cultural determinations of the frequency and type of microorganisms causing the slime; and decisions on the type of slimicide and concentration to use. A comprehensive review of the history and controls of slime formation in pulp and paper mills has been assembled by Ross and Hollis (1976). Slime-measuring units are available also to provide information on the type and rate of slime formation. Recent trends toward alkaline pulping processes and closed mill systems to conserve fiber and energy have favored the formation of anaerobic bacteria in white water tanks or storage chests, particularly where circulation is reduced. Additional problems with the development of anaerobes are pipe corrosion, disagreeable odors, and in some cases, the production of explosive hydrogen and methane gases (Robichaud, 1991). Aeration by agitation, ventilation, and use of biocides are the recommended preventative measures.

Molds within the Home

In the home, the growth of fungi (molds) on surfaces in poorly ventilated areas where moisture accumulates are unsightly and can result in musty odors, allergies, and in some cases, disease.

Mold Growth on Inert Materials

In cases of prolonged high humidity or condensation, some molds such as species of *Penicillium, Aspergillus,* and *Alternaria* grow on glass surfaces from a food base and may etch and damage the surface (Subramanian, 1983). This may become a serious problem on optical instruments in tropical zones. The growth of molds on electronic equipment under prolonged high-humidity conditions can damage plastic materials and disrupt the circuitry (Inoue, 1988). Favorable conditions for the growth of fungi and bacteria sometimes occur on the surfaces and filters in air-conditioning units. Allergies and some diseases are reported to be associated occasionally with air-conditioning and cooling units (Banaszak *et al.*, 1970; Dondero *et al.*, 1980), and their causes and controls have been reviewed by Ager and Tickner (1983).

Summary

- Fungi and bacteria are the principal biotic agents responsible for the destruction and disfigurement of paints.
- Mildew, a major surface disfigurement, is caused by the copious mycelial growth of pigmented hyphae on the surface.
- Paints are suspensions of solid particles in a liquid, which after application to a surface, chemically react to form a durable, continuous film over the surface. Paints are often grouped as those with the solids and other additives in an organic solvent (solvent-thinned paints) or water (waterborne paints).
- Waterborne paints are generally more prone to mildew damage than are solvent-thinned paints.
- *Aureobasidium pullulans* is the major mildew agent and occurs on paints worldwide in humid regions. Other principal mildew fungi are *Cladosporium* spp., *Alternaria* spp., and *Phoma glomerata*.
- *Aureobasidium pullulans*, with its tolerance to a wide range of growth conditions, resistance to desiccation and actinic exposure, and tenacious adherence to substrates, appears to be well adapted to util-

Summary continues

ize detritus and simple carbon compounds that accumulate on paint surfaces when favorable moisture levels and temperatures occur.
- Mildew development occurs on paints primarily in humid coastal and tropical regions. It develops more commonly on painted wood than on metals or masonry. On the exterior of homes, it occurs more commonly on shaded zones, north-facing walls, or locations where shade trees or shrubbery restrict rapid drying of water on the paint surface.
- Inside the home, mildew develops principally on painted surfaces in bathrooms and window units owing to condensation. In food-processing plants, paper mills, dairies, and other industries where high humidities and airborne detritus occur, severe mildew and surface mold problems may occur on ceilings and walls and require specially treated paints and frequent washings to control.
- Free moisture on the paint film appears to be the critical growth requisite for mildew development.
- General control practices are to minimize moisture and detritus accumulations on paint films, and where mildew hazards are severe, to use the more resistant paints treated with mildewcides. A wide range of proprietary compounds are available as mildewcides for various paint types and uses.
- The design of paint films specifically to minimize microbial damage represents an area of potential promise, and it is probable that bioresistant paints can be achieved when this property assumes a higher priority.
- The growth of molds and bacteria on surfaces in pulp and paper mills results in the formation of slimes, which can degrade the paper, produce odors, and reduce machine efficiencies. Related mold accumulations in air-conditioning and cooling units can cause allergies and some diseases.

References

Ager, B. P., and J. A. Tickner. (1983). The control of microbiological hazards associated with air-conditioning and ventilation systems. *Annals of Occupational Hygiene* **27**(4):341–358.

Ahmadjian, V. (1967). "The Lichen Symbiosis." Blaisdell, Waltham, Massachusetts.

American Society for Testing Materials (ASTM). (1974) "Resistance to Growth of Mold on the Surface of Interior Coatings in an Environmental Chamber." General Services Administration, Federal Test Method 627.11.

Baker, K. F., and R. J. Cook. (1974). "Biological Control of Plant Pathogens." W. H. Freeman Co., San Francisco, California.

Banaszak, E. G., W. H. Thiede, and J. N. Fink. (1970). Hypersensitivity pneumonitis due to contamination of an air conditioner. *The New England Journal of Medicine* **283**(6):271–276.

Brand, B. G., and H. T. Kemp. (1973). "Mildew Defacement of Organic Coatings. A Review of Literature Describing the Relationship between *Aureobasidium pullulans* and Paint Films." Federal Society Paint Technology, Philadelphia, Pa.

Bravery, A. F. (1988). Biodeterioration of paint—a state-of-the-art comment. In "Biodeterioration 7," (D. R. Houghton, R. N. Smith, and H. O. W. Eggins, eds.). Elsvier Applied Science, New York. 466-485.

Dondero, T. J., R. C. Rendtorff, G. F. Mallison, R. M. Weeks, J. S. Levy, E. W. Wong, and W. Schaffner. (1980). An outbreak of Legionnaire's disease associated with a contaminated air-conditioning cooling tower. *The New England Journal of Medicine* 302(7):365-370.

Drescher, R. F. (1958). Microbiology of paint films. IV. *American Paint Journal* (March):1-12.

Durrell, L. W., and K. L. Goldsberry. (1970). Growth of *Aureobasidium pullulans* on plastic greenhouse roofing. *Mycopathologia et Mycologia Applicata* 42:193-196.

Eveleigh, D. E. (1961). The disfiguration of painted surfaces, with special reference to *Phoma violacea*. *Annals of Applied Biology* 49:403-411.

Goll, M., H. D. Snyder, and H. A. Birnbaum. (1952). New developments in the diagnosis of paint mildew. *Official Digest* 24:149-152.

Hamburg, H. R., and W. M. Morgans (eds.). (1979). "Hess's Paint Film Defects, Their Causes and Cure" 3rd Ed. Chapman and Hall, London.

Hill, D. O., and G. C. April. (1971). Influence of moisture transport on fungal growth. *Journal of Paint Technology* 43(560):81-88.

Horvath, R. S., and M. M. Esposito. (1979). Co-metabolism of paint by *Aureobasidium pullulans* BH-1-ATCC 34621. *Current Microbiology* 2:169-170.

Ionue, M. (1988). The study of fungal contamination in the field of electronics. In "Biodeterioration 7" (D. R. Houghton, R. N. Smith, and H. O. W. Eggins, eds.), pp. 580-584, Elsevier Applied Science, New York.

Jakubowski, J. A., J. Gyuris, and S. L. Simpson. (1983). Microbiology of modern coatings systems. *Journal of Coatings Technology* 55(705):49-53.

Käärik, A. (1974). "Sapwood-Staining Fungi." The International Research Group on Wood Preservation. Document No. IRG/WP/125. Stockholm, Sweden.

Klens, P. F., G. Leitner, and H. D. Snyder. (1954). A field method for detecting mildew on paint. *American Paint Journal* 38(39):76-82.

Lambourne, R. (ed.). (1987). "Paint and Surface Coatings: Theory and Practice." Ellis Horwood Ltd., Chichester, United Kingdom.

Ludwig, L. E. (1974). Formulation of mildew-resistant coatings. *Journal of Paint Technology* 46 (No. 594):31-39.

Martens, C. R. (1981). "Waterborne Coatings." Van Nostrand Reinhold, New York.

Oppermann, R. A. (1986). The anaerobic biodeterioration of paint. In "Biodeterioration 6" *Proceedings of the 6th International Biodeterioration Symposium, CAB*, International Mycology Institute, Washington, D. C.

Pittman, C. V., and K. R. Lawyer (1982). Preliminary evaluations of the biological activity of polymers with chemically bound biocides. *Journal of Coating Technology* 54(690):41-46.

Robichaud, W. T. (1991). Controlling anaerobic bacteria to improve product quality and mill safety. TAPPI 74(2):149-153.

Ross, R. T. (1964). Microbiological deterioration of paint and painted surfaces. *Developments in Industrial Microbiology* 6(14):149-163.

Ross, R. T., and C. G. Hollis (1976). Microbiological deterioration of pulp wood, paper, and paint. *In* "Industrial Microbiology" (B. M. Miller and W. Litsky, eds.), pp. 309–354. McGraw-Hill, New York.

Rothwell, F. M. (1958). Microbiology of paint films. II. Isolation and identification of microflora on exterior oil paints. *Official Digest* **30**:368–379.

Schmidt, E. L., and D. W. French (1976). *Aureobasidium pullulans* on wood shingles. *Forest Products Journal* **26**(7):34–37.

Schmitt, J. A. (1974). The microecology of mold growth. *Journal of Paint Technology* **46**:59–64.

Smith, R. A. (1977). Fungi surviving in a latex paint film after surface disinfection. *Developments in Industrial Microbiology* **18**:565–569.

Subramanian, C. V. (1983). "Hyphomycetes—Taxonomy and Biology." Academic Press, New York.

Weismantel, G. D. (ed.). (1980). "Paint Handbook." McGraw-Hill, New York.

Winters, H., I. R. Isquith, and M. Goll. (1975). A study of the ecological succession in biodeterioration of a vinyl acrylic paint film. *Developments in Industrial Microbiology* **17**:167–171.

Zabel, R. A., and W. E. Horner (1981). An accelerated laboratory procedure for growing *Aureobasidium pullulans* on fresh latex films. *Journal of Coatings Technology* **53**(675):33–37.

Zabel, R. A., and F. Terracina (1980). The role of *Aureobasidium pullulans* in the disfigurement of latex paint films. *Developments in Industrial Microbiology* **21**:179–190.

CHAPTER
18

Natural Decay Resistance (Wood Durability)

A first principle of decay prevention is the use of naturally durable or preservative-treated woods in those high hazardous-decay conditions in which the wood cannot be kept dry by structural design or handling practices. This chapter discusses the types and features of durable woods, and the next chapter considers similar aspects of preservative-treated woods.

The benefits of using naturally durable woods have long been known (Graham, 1973). Phoenician boat builders routinely employed naturally durable cedars of Lebanon or oaks, and their overuse contributed to the decline of natural forests along the Mediterranean. These same builders clearly understood the differences between sapwood and heartwood, noting that sapwood was susceptible to ravages of decay and marine-borer attack. It is interesting to note that these builders utilized wood in a discriminatory manner similar to that of homeowners in the twentieth century, but had no understanding of the nature of decay.

Although supplies of naturally durable woods have declined, interest in these species continues for many reasons. Decay resistance is an important tree property that has been neglected to date in tree-improvement programs. Knowledge of the naturally toxic compounds in the durable woods may lead to the development of more effective wood preservatives and other wood-protective chemicals. The use of naturally durable woods has also become an important option where there are concerns about the environmental safety of wood preservatives.

In this chapter the emphasis will be on the factors affecting decay resistance and the distribution and properties of the major responsible agents, the heartwood extractives.

The term *durability* is used commonly in the lumber trade to designate decay resistance. Actually, wood durability is a broader term and reflects the resistance of wood to other deteriorating factors, such as insects, marine borers, and weathering, as well as decay fungi. In this chapter we shall limit our discussion to decay resistance, but use the terms as synonyms.

An extensive literature has accumulated over the years on the many facets of decay resistance in woods. Many reports are of a practical type, reporting service tests of various local species for fence post and local construction use. Brief reviews of decay resistance in commercial wood species are presented in both the U.S. Forest Service's *Wood Handbook* (1987) and Panshin's and deZeeuw's *Textbook of Wood Technology* (1980). A comprehensive review of the topic, placing some emphasis on the nature and toxicity of the wood extractives, has been assembled by Scheffer and Cowling (1966).

Variations in Decay Resistance

Wood consists of several natural polymers and a wide range of cell-wall extractives, which are primarily localized in the heartwood. Heartwood durability, as with that of any natural product, is characterized by wide variability both between species and between individual trees of the same species (Scheffer and Cowling, 1966). This variation reflects both the genetic potential of a tree and the environmental conditions under which the tree is grown. The heartwood durability of a species may vary dramatically, as with the differences exhibited between highly durable old-growth redwood and moderately durable second-growth timber of the same species (Clark and Scheffer, 1983).

Species Variations

Great variations in decay resistance occur among species ranging from a few months of service for some susceptible species to 40 to 50 years of service for a few highly durable species such as black locust or western red cedar in high-decay-hazard uses. The decay-resistance ratings of the heartwoods of many commercial domestic woods, based on extensive service records, post-farm experiments, and laboratory evaluations, have been summarized by Scheffer and Cowling (1966) and are listed in Table 18-1. The decay-resistance rankings of many exotic species are available also (Scheffer and Cowling, 1966; Clark, 1969).

Stem Position Variations

The sapwood of all wood species demonstrates high susceptibility to decay regardless of durability status of the heartwood. There are some exceptions. White oak sapwood in the transition zone between recently formed heartwood and innermost sapwood is more decay resistant than recently formed sapwood. Sapwood in the vicinity of wounds where prior injuries have been walled off is more decay resistant than surrounding sapwood (Shigo, 1965). In those species in which colored heartwood is not formed (spruces, fir, etc.), the inner stem tissues are somewhat more resistant to decay than newly formed sapwood. Sometimes in local retail markets, durable species,

Table 18-1
Relative decay resistance of some commercial wood species commonly used in North America[a,d]

Resistant and very resistant	Moderately resistant	Slightly or nonresistant
Bald cypress (old growth)	Bald cypress (new growth)	Alder
Catalpa	Douglas fir	Ashes
Cedars	Honey locust[b]	Aspens
Black cherry	Western larch	Basswood
Arizona cypress	Swamp chestnut oak	Beech
Junipers	Eastern white pine	Birches
Black locust[c]	Longleaf pine	Cottonwood
Red mulberry[c]	Slash pine	Elms
Bur oak	Tamarack	Hackberry
Chestnut oak		Hemlocks
Oregon white oak		Hickories
Post oak		Maples
White oak		Red or black oak[b]
Osage orange[c]		Pines[b]
Redwood		Spruces
Sassafras		Sweet gum[b]
Black walnut		Sycamore
Pacific yew[c]		Willows
Western red cedar		Yellow poplar

[a]From Scheffer and Cowling (1966).
[b]These species have shown a higher decay resistance than most of the other woods in their respective categories.
[c]These woods have exceptionally high decay resistance.
[d]Common names are from Little (1979), which provides the scientific names.

such as northern white cedar, white oak, or black locust, containing large proportions of sapwood are sold in roundwood form for posts. In these cases, durable wood is limited to the heartwood portion of the product. A small cedar post consisting largely of sapwood will offer no more protection than a post of a decay-susceptible species.

In general, decay resistance increases from the cambium to the sapwood-heartwood interface (Fig. 18-1). In many species, durability is highest near the sapwood–heartwood interface and declines toward the pith (Zabel, 1948; Scheffer and Hopp, 1949; Scheffer et al., 1949; Gardner and Barton, 1958; Gardner, 1960; Behr, 1974; Hillis, 1985, 1987). This decline is believed to reflect either biological detoxification, natural oxidation of heartwood extractives, or continued polymerization of extractives to produce less toxic compounds (Anderson et al., 1963). Microbial activity also may reduce heartwood durability with age (Jin et al., 1988). Decay resistance varies with stem height, with the most durable wood occurring near the base of the tree. Variations in the decay resistance of western red cedar appears to be well correlated with distribution of the wood extractive *thujaplicin* (Nault, 1988). Durability

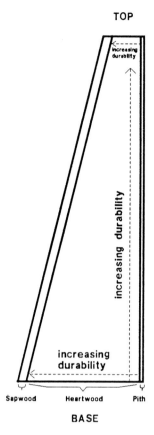

Figure 18-1 Effect of position in the tree on the durability of heartwood in decay-resistant species.

reflects both the natural genetic potential of a tree and the environmental stresses to which the tree is subjected.

Factors Affecting Durability

The wide variation in durability among and within species reflects a number of factors. As stated, extractive content plays the chief role in durability and will be addressed at length later in this chapter. A number of other factors also influence durability.

Lignification

As discussed in Chapter 6, lignification was the important step in higher plant evolution that provided stiffness to stem tissues facilitating stem aerial development and protection against destruction by microorganisms. Lignin type and amount can have a significant influence on the rate and types of decay found in wood. Lignin content can vary widely between species and within individual portions of the same tree, rendering some portions of the wood, such as the primary cell wall and middle lamella, more resistant to microbial attack. Hardwood lignins are composed of both syringyl and guaiacyl phenyl propane units, whereas coniferous lignin contains only guaiacyl residues. Hardwood lignins also differ from coniferous lignins in the types of linkages between phenyl propane units.

Many studies have noted the resistance of softwoods to soft-rot attack; however, this effect can be negated by partial delignification (Zainal, 1976). Lignin type, content, and pattern of deposition appear to play critical roles in inception of soft-rot attack, and relatively minor changes in lignin content can produce large decreases in decay resistance. These effects suggest that the pattern of lignin deposition in the wood cell wall may be as critical as lignin type in determining natural decay resistance to some fungi.

Wood may be naturally durable for a number of other reasons, depending on the environment and organisms to which it is subjected. For example, many species exhibit extreme hardness or contain large quantities of silica or calcium carbonate (Taniguchi et al., 1986). These species are often resistant to marine-borer attack, which require that the animal chew into the hard wood (Bultman et al., 1977; Southwell and Bultman, 1971). Crystals may also alter the moisture-holding capacity, making it more difficult to wet, thereby limiting the microorganisms that can colonize the wood. Hardness creates similar difficulties for sawmills processing these species. Whereas hardness represents one means of protection against insects or marine borers, the most common mechanisms of natural durability involve the production of toxic chemicals during heartwood formation.

Growth Characteristics

In addition to chemical differences, some silvicultural factors may also influence durability. Many users have suggested that wood cut as the tree begins a growth flush and produces leaves or needles may contain higher levels of readily accessible carbohydrates, rendering the wood more susceptible to microbial attack. Wood harvested during dormant periods would be expected to contain lower levels of these compounds and should, therefore, be less susceptible to attack. No evidence of this effect has been noted, although cutting during cooler dormant periods undoubtedly reduces the risk of staining and decay. Similarly, limited studies of southern pine suggest that fertilization using high-nitrogen-content fertilizers can adversely influence resistance to disease and may affect durability. In theory, fertilization should increase tree growth, producing a wider band of decay-susceptible sapwood. Furthermore, higher nitrogen levels are often correlated with increased susceptibility to fungal attack (Merrill and Cowling, 1965). Conversely, rapid growth might increase the levels of carbohydrates available for synthesis of toxic extractives, although this effect has never been noted.

One factor related to natural durability that has become very important is the decay resistance of second-growth timber. Until recently, naturally durable species were primarily cut from old-growth forests; however, these stocks are nearly depleted, and we now harvest many trees from second-growth stands. Preliminary results suggest that the wood from these second-growth trees lacks the natural durability found in the old-growth trees. This effect has been noted with southern pines, bald cypress, and most recently, coastal redwood (Clark and Scheffer, 1983). Interestingly, this effect appears to be limited to the highly durable species. For example, there is no significant difference in decay resistance of old-growth and second-growth Douglas fir heartwood, a moderately decay-resistant species (Scheffer and Englerth, 1952).

As we move to a largely managed, second-growth forest, we may find it necessary to alter our reliance on naturally durable woods, or identify methods for encouraging the development of these woods at earlier points within a rotation age. Increased attention should be also paid to decay resistance as an important trait to be selected for and enhanced in tree-breeding programs.

Miscellaneous Factors

Whereas many growth characteristics may affect durability, one factor that is poorly correlated with decay resistance is wood density. Many woods exhibit high density and are naturally durable, but other dense woods are rapidly degraded. As an example, white oak is dense and durable, but woods with similar density, such as beech or maple, exhibit no natural resistance to decay fungi. Conversely, several light woods, including western red cedar and coastal redwood, are among our most durable species.

Once a tree is harvested and processed, handling procedures can also affect durability. For example, heat treatments of wet wood can either volatilize or denature wood extractives, decreasing natural durability (Scheffer and Eslyn, 1961). Exposure to gamma radiation can also adversely affect natural durability (Scheffer, 1963). Exposure to excessive wetting can lead to leaching of water-soluble extractives, also reducing natural durability (Johnson and Cserjesi, 1980). Although this is generally a slow process, exposures under certain conditions, such as excessively high acidity or under nutrient regimens that encourage the growth of detoxifying microbes, can also reduce durability.

Decay Resistance and Heartwood Formation

Natural durability is related primarily to the presence of toxic extractives that form and are deposited in the heartwood as the sapwood dies. Sapwood has little or no natural durability.

As normal sapwood ages and is buried farther into the tree, the parenchyma cells gradually die. These cells contain higher levels of extractives that form as carbohydrates in the parenchyma are transported from living cells to cells near the heartwood–sapwood transition zone, where they are converted to a variety of phenolic compounds. The nature of the conversion to heartwood remains unclear. Several researchers have suggested that toxic metabolites are transported to the sapwood–heartwood boundary, and their accumulation to lethal levels induces cell death and heartwood formation. Alternatively, ethylene production has been shown to stimulate heartwood formation in cell tissue cultures. Further improvements in *in vitro* techniques for cell tissue culture should permit more detailed studies of heartwood formation.

The phenolic extractives produced in the parenchyma diffuse into the surrounding tracheids, fibers, or vessels, where they are absorbed into the wood cell walls. Some species produce an array of compounds toxic to fungi, insects, or marine borers. In some species, phenolic extractive production is extremely limited or absent. These species are often considered to produce no heartwood, although large portions of the cells are no longer living. These species exhibit no noticeable heartwood durability (Table 18-1). Heartwoods are generally colored, but similar coloration can be produced by factors such as wounding, mineral stain or bacterial attack.

In this chapter we shall limit the discussion of heartwood extractives that contribute to durability to representative types of compounds in four major groups: *polyphenols, terpenoids, tropolones,* and *tannins* (Table 18-2). Heartwood extractives are covered in greater detail by Scheffer and Cowling (1966) and Hillis (1987).

Polyphenols

Polyphenols include the stilbenes and flavenoids and are, by far, the most common heartwood extractives, occurring in the heartwood of nearly all species. Stilbenes are synthesized via the shikimic acid pathway (Kindl, 1985) and are common in the heartwood of *Pinus, Eucalyptus,* and *Maclura* species (Hart, 1981). One stilbene, pinosylvin, and its monomethyl ether are the principal protective agents in pine heartwoods. Isolated stilbenes are toxic to bacteria, fungi, and insects, although their ability to protect wood from these agents varies (Hart, 1981; Hart and Hillis, 1974; Schultz et al., 1990). Flavenoids include many important phytoalexins (compounds produced by plants in response to microbial attack) (Grisebach, 1985) and heartwood extractives such as quercitin and taxifolin, which are found in Douglas fir (Kennedy, 1956).

Terpenoids

Terpenoids are derived from the condensation of C-5 isoprenoid units and can vary from relatively volatile isoprene monomers to the polymer, rubber (Croteau and Johnson, 1985). Monoterpenes, such as pinene, consist of two isoprene units. Terpenoids have created the large naval stores industry,

Table 18-2
Examples of heartwood extractives that inhibit fungal attack[a]

Class of compound	Chemical name
Terpenoids	Carvacrol
	p-Methoxycarvacrol
	p-Methoxythymol
	Thymoquinone
	Sugiol
	Totarol
	Ferruginol
	Chamic Acid
	Chaminic Acid
	1-Citronellic acid
Flavonoids	Quercetin
	Robinetin
	Taxifolin
	Dihydrobinetin
	Homoferreirin
	Ougenin
Stilbenes	Pinosylvin monomethylether
	Pinosylvin dimethylether
	3,5,4'-Trihydroxystilbene
	2,3,4,5'-Tetrahydroxstilbene
	3,4,3',5'-Pentahydroxystilbene
	4-Hydroxystilbene
	Pterostilbene
Tropolones	α,β, or γ-Thujaplicin
	α or β-Thujaplicinol
	Pygemaein
	β-Dolabrin
	Nootkatin

[a] As summarized by Scheffer and Cowling (1966).

which plays an important role in the forest industry. Turpentine is a mixture of terpenes and tall oil, an important by-product of the Kraft pulping process, consisting of mixtures of diterpenes and fatty acids. Terpenoids are synthesized by a variety of plant tissues, such as the resin ducts in conifers. Highly resinous wood exhibits some natural durability.

Tropolones

Terpenoid compounds are important precursors for more toxic extractives, such as the *tropolones*. Of the heartwood extractives, the chemicals that have received the most attention for their decay resistance are the tropolones. These seven-membered rings are specific to members of the Cupressaceae and include several isomers of thujaplicin. Tropolones are believed to derive from terpenoids, which are in turn synthesized from acetyl coenzyme A (CoA). The biochemistry of tropolones has not been studied in great detail, but monoterpenes such as carane and thujane are believed to give rise to two seven-members rings, thujic acid and β-thujaplicin, respectively.

Tannins

Tannins are important components of the wood and bark of many trees. Tannins from vegetables were originally used to cure animal skins, because of their ability to precipitate and stabilize proteins in the skin. Tannins are proanthocyanidins, which are highly complex polymers (Hillis, 1985) (Fig. 18-2). These compounds have been extensively studied in the tanning industry and are coming under increasing scrutiny as replacements for oil-derived components of adhesives. Tannins, because of their ability to precipitate proteins, have also been evaluated as potential wood preservatives; however, their water solubility and relatively low toxicity have limited application (Laks, 1987). The durability of oaks is believed to be associated with their high tannin content (Zabel, 1948). These compounds arise along synthetic pathways that are similar to those for lignin and stilbenes, but the systems are poorly understood.

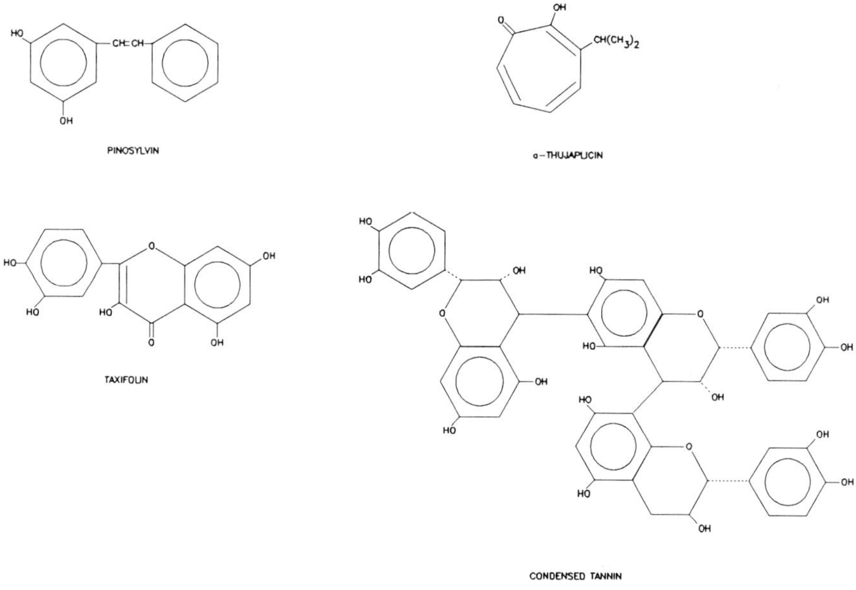

Figure 18-2 Examples of proposed structures for pinosylvin, α-thujaplicin, taxifolin, and a condensed tannin.

Evaluating Natural Durability

In general, natural durability has been evaluated by exposing wood samples to the decay agents for various periods and rating the resultant degree of degradation. These tests are sometimes supplemented by extracting the wood and evaluating the toxicity of various fractions against specific organisms. Field trials (post farms) of natural durability were prevalent beginning in the 1920s, as scientists began to evaluate the properties of many local wood species as alternatives to chestnut and the cedars (Humphrey, 1916; White, 1922). These tests reflected a desire to identify woods with properties similar to existing naturally durable species. Hawley *et al.* (1924) first established the toxic nature of heartwood extractives and posed them as the reason for decay resistance.

Laboratory assays to evaluate natural durability began in the 1940s in an attempt to further explain the nature of durability and identify the toxic compounds (Anderson *et al.*, 1963; Scheffer, 1957; Southam and Erlich, 1943; Kennedy, 1956; Zabel, 1948; Scheffer *et al.*, 1949). In most cases, warm water, ethanol, or other solvents were used to remove extractives from the wood. These extractives were then tested for activity against a variety of decay and non-decay fungi. In addition, these extracts were often tested for activity against common wood-destroying insects (McDaniel, 1989). Most tests were performed in petri dishes or decay chambers using nutrient agar. Although such tests provide a relative guide to chemical toxicity, they cannot evaluate more subtle effects such as variation in deposition of extractives in the wood or interactions between different extractives that must also play roles in natural wood durability.

Much of the laboratory research was performed to identify chemicals that could be synthesized as *natural preservatives* under the premise that such chemicals would be inherently safer and more effective than other biocides. In fact, many of the chemicals responsible for natural wood durability are as toxic as or more toxic than existing wood preservatives. The major advantage of employing naturally durable woods over artificial wood preservation is the elimination of the need for treating facilities to deliver chemical into the wood. Natural durability, however, can never completely replace the need for wood pressure-treated with chemicals, since some hazards, such as marine exposures, are too severe for adequate performance of even the best naturally durable wood species.

Improved chemical assay methods using such techniques as radioisotope labeling, C-13 nuclear magnetic resonance (NMR) spectroscopy, and ion magnetic spectroscopy have enhanced the study of natural wood durability, but methods for studying *in situ* deposition of toxic extractives are still lacking. The use of tissue-culture techniques may further our knowledge of the synthesis of heartwood extractives, but it will be difficult to study subsequent deposition processes using these methods. Developing an improved understanding of the nature and distribution of extractives related to durability could be especially useful for identifying new approaches to depositing chemicals in wood and protecting it. This approach would be particularly intriguing because of evidence that the decay-resisting extractives in some very durable woods are distributed in the wood in a manner more resistant to leaching than fungicides that are applied artificially.

As concerns about the use of artificial wood preservatives increase, however, we may come to depend increasingly on natural durability for wood used in some locations, such as water reservoirs, even under high decay hazards. This increased demand comes at a time when supplies of durable species are declining and when concerns are being raised about decreased durability of second-growth timber. These trends suggest that renewed efforts must be made to identify methods for improving the genetic capabilities to produce durable heartwood and to develop silvicultural practices that favor maximum production of this wood in the shortest period.

Summary

- A first principle of decay prevention is the selection of naturally decay-resistant or preservative-treated wood for uses where decay-hazardous conditions cannot be avoided.
- Decay resistance varies widely among species. Some species such as cypress and the cedars are very durable, whereas others, such as beech and maple, are susceptible to decay.
- Decay resistance in the durable woods also varies with stem position. The sapwood of all wood species is generally susceptible to decay. Decay resistance is highest near the sapwood–heartwood interface and decreases toward the pith for many durable species. Also, durability generally decreases with stem height, and the most durable wood occurs at the base of the trunk.
- Durability also varies within a species, and some genotypes are much more durable than others. Old-growth timber in durable woods such as cypress and redwood is more decay resistant than second-growth timber.
- The nature and amount of the toxic extractives in the heartwood appears to be the major factor affecting decay resistance. The primary toxic heartwood extractives are the polyphenols, terpenoids, tropolones, and tannins.
- The use of durable woods as replacements for preservative-treated materials may increase in some environmentally sensitive areas. This suggests increased attention should be paid to the growing of decay-resistant species in timber management.
- An increased understanding of the mechanisms of natural decay resistance may lead to the development of improved treatment methods and lead to effective replacement wood preservatives.

References

Anderson, A. B., T. C. Scheffer, and C. G. Duncan. (1963). The chemistry of decay resistance and its decrease with heartwood aging in incense cedar (*Libocedrus decurrens* Torrey). *Holzforschung* **17**(1):1–5.

Behr, E. (1974). "Decay Resistance of Northern White Cedar from a Wet Site in Lower Michigan." Res. Rep. 229. Michigan State University Agricultural Experiment Station, East Lansing, Michigan.

Bultman, J. D., R. H. Beal, and F. F. K. Ampong. (1977). Natural resistance of some tropical African woods to *Coptotermes formosanus* Shiraki. *Forest Products Journal* **29**(6):45–51.

Clark, J. W. (1969). "Natural Decay Resistance of Fifteen Exotic Woods Imported for Exterior Use." U.S.D.A. Forest Products Laboratory Research Paper FPL 103. Madison, Wisconsin.

Clark, J. W., and T. C. Scheffer. (1983). Natural decay resistance of the heartwood of coast redwood *Sequoia sempervirens* (D. Don) Engl. *Forest Products Journal* **33**(5):15–20.

Croteau, R., and M. A. Johnson. (1985). Biosynthesis of terpenoid wood extractives. In "Biosynthesis and Biodegradation of Wood Components" (T. Higuchi, ed.), pp. 379–439. Academic Press, New York.

Gardner, J. A. F. (1960). The distribution of dehydroquercitin in Douglas fir and western larch. *Forest Products Journal* **10**:171–173.

Gardner, J. A. F., and G. M. Barton. (1958). The extraneous components of western red cedar. *Forest Products Journal* **8**:189–192.

Graham, R. D. (1973). History of wood preservation. In "Wood Deterioration and Its Prevention by Preservative Treatments" (D. D. Nicholas, ed.), pp. 1–30. Syracuse University Press, Syracuse, New York.

Grisebach, H. (1985). Biosynthesis of flavonoids. In "Biosynthesis and Biodegradation of Wood Components" (T. Higuchi, ed.), pp. 291–324. Academic Press, New York.

Hart, J. H. (1981). Role of phytostilbenes in decay and disease resistance. *Annual Review of Phytopathology* **19**:437–458.

Hart, J. H., and W. E. Hillis. (1974). Inhibition of wood-rotting fungi by stilbenes and other polyphenols in *Eucalyptus sideroxylon*. *Phytopathology* **64**:939–948.

Hawley, L. F., L. C. Fleck, and C. A. Richards. (1924). The relation between durability and chemical composition in wood. *Industrial and Engineering Chemistry* **16**:699–700.

Hillis, W. E. (1985). Biosynthesis of tannins. In "Biosynthesis and Biodegradation of Wood Components" (T. Higuchi, ed.), pp. 325–347, Academic Press, New York.

Hillis, W. E. (1987). "Heartwood and Tree Exudates." Springer-Verlag, New York.

Humphrey, C. J. (1916). Laboratory tests on the durability of American woods—I. Flask tests on conifers. *Mycologia* **8**(2):80–93.

Jin, L., B. J. van der Kamp, and J. Wilson. (1988). Biodegradation of thujaplicins in living western red cedar. *Canadian Journal Forest Research* **18**:782–786.

Johnson, E. L., and A. J. Cserjesi. (1980). Weathering effect on thujaplicin concentration in western red cedar shakes. *Forest Products Journal* **30**(6):52–53.

Kindl, H. (1985). Biosynthesis of stilbenes. In "Biosynthesis and Biodegradation of Wood Components" (T. Higuchi, ed.), pp. 349–377. Academic Press, New York.

Kennedy, R. W. (1956). Fungicidal toxicity of certain extraneous components of Douglas fir heartwood. *Forest Products Journal* **5**:80–84.

Laks, P. E. (1987). Flavonoid biocides: Phytoalexin analogues from condensed tannins. *Phytochemistry* **26**(6):1617-1621.

Little, E. L. (1979). "Checklist of Native and Naturalized Trees of the United States (including Alaska)." U.S.D.A. Forest Service, Agriculture Handbook No. 541, Washington, D. C.

McDaniel, C. A. (1989). Major termiticidal components of Port Orford cedar, *Chamaecyparis lawsoniana* (A. Murr.) Parl. *Material und Organismen* **24**(1):1-15.

Merrill, W. B., and E. B. Cowling. (1965). Effect of variation in nitrogen content of wood on rate of decay. *Phytopathology* **55**:1067-1068.

Nault, J. (1988). Radial distribution of thujaplicins in old growth and second growth of western red cedar (*Thuja plicata* Donn). *Wood Science and Technology* **22**:73-80.

Panshin, A. J., and C. deZeeuw. (1980). "Textbook of Wood Technology." 4th Ed. McGraw-Hill, New York.

Scheffer, T. C. (1957). Decay resistance of western red cedar. *Journal of Forestry* **55**:434-442.

Scheffer, T. C. (1963). Effect of gamma radiation on decay resistance of wood. *Forest Products Journal* **13**(5):268.

Scheffer, T. C., and E. B. Cowling. (1966). Natural resistance of wood to microbial deterioration. *Annual Review of Phytopathology* **4**:147-170.

Scheffer, T. C., and G. H. Englerth. (1952). Decay resistance of second-growth Douglas fir. *Journal of Forestry* **50**(6):439-442.

Scheffer, T. C., and W. E. Eslyn. (1961). Effect of heat on decay resistance of wood. *Forest Products Journal* **11**:485-490.

Scheffer, T. C., and H. Hopp. (1949). "Decay Resistance of Black Locust Heartwood." U.S. Department of Agriculture Technical Bulletin No. 984. Washington, D.C.

Scheffer, T. C., G. H. Englerth, and G. G. Duncan. (1949). Decay resistance of seven native oaks. *Journal of Agricultural Research* **78**:129-152.

Schultz, T. P., T. F. Hubbard, Jr., L. H. Jin, T. H. Fisher, and D. D. Nicholas. (1990). Role of stilbenes in the natural durability of wood: Fungicidal structure-activity relationships. *Phytochemistry* **29**(5):1501-1507.

Shigo, A. L. (1965). The pattern of decays and discolorations in northern hardwoods. *Phytopathology* **55**:648-652.

Southam, C. M., and J. Ehrlich. (1943). Effects of extract of western red cedar heartwood on certain wood-decaying fungi in culture. *Phytopathology* **33**(6):517-524.

Southwell, C. R., and J. D. Bultman. (1971). Marine-borer resistance of untreated woods over long periods of immersion in tropical waters. *Biotropica* **3**(1):81-107.

Taniguchi, T., H. Harada, and K. Nakato. (1986). Characterization of inorganic deposits in tropical woods by a new method of isolation. *Journal of the Japan Wood Research Society* **32**(4):227-233.

U.S.D.A. (1987). "Wood Handbook: Wood as an Engineering Material." U.S.D.A. Agricultural Handbook 72, Washington, D. C.

White, D. G. (1922). Relative durability of untreated native woods. *Southern Lumberman* (September) **2**:46-47.

Zabel, R. A. (1948). "Variation in the Decay Resistance of White Oak." New York State College of Forestry Technical Publication 68. Syracuse, New York.

Zainal, A. S. (1976). The effect of a change in cell-wall constituents on decay of Scots pine by a soft-rot fungus. *Material und Organismen* **11**:295-301.

CHAPTER
19

Chemical Protection of Wood (Wood Preservation)

Although the proper handling and use of wood in well-designed structures can minimize damage from the biodeterioration agents, some uses require special chemical treatments to obtain economic service lives. This chapter reviews briefly the treatments and chemicals currently used to protect wood used under decay-hazardous conditions, placing special emphasis on fungal damage and environmental aspects. Two comprehensive treatments of this topic are also available (Nicholas, 1973; Hunt and Garrett, 1967).

A Brief Early History

Preventing decay has long been a concern of wood users, preceding by hundreds of years the discovery of fungi as the causal agents of decay. Wooden ships played a major role in the early explorations, discoveries, and development of world commerce. Even the earliest ship builders (*circa* 1000 BC) understood the need to use wood from durable species, and there are reports of their attempts to protect wood by daubing the surface with toxic oily extracts from cedar (Graham, 1973). A variety of surface barrier treatments were used to reduce the ravages of marine borers, but only copper sheathing proved effective. Unfortunately, nothing protected wood from decay in some uses, and the search for effective treatments and new sources of durable woods accelerated in the 1800s. Durable wood was sorely needed, not only for new ships, but also to replace decaying wood in men-of-war of the feuding nations of Europe.

The huge fleets necessary to control the far-flung colonial empires of England, Spain, France, and the Netherlands placed tremendous pressure on locating new supplies of wood for construction and repairs. Shortages of naturally durable timbers (e.g., oaks, cedars) forced builders to use less durable species that further accelerated the decay problem. The shortage of durable woods stimulated a frantic search for effective methods to prolong wood service life. Of the many advances at this time, the use of creosote (a coal-tar distillate) by Moll around 1836 and the patenting of the full-cell process for pressure-treating wood by Bethell in 1839 had a major influence on the development of a wood-preservation industry. It is interesting to note that these advances preceded, by nearly 50 years, Hartig's discovery that fungi caused wood decay. The early history and major subsequent developments in wood preservation are available in several sources (Van Groenou et al., 1951; Graham, 1973; and Baechler and Gjovik, 1986).

The chemical protection of wood has developed into a major worldwide industry. In most countries, associations have developed among the wood scientists, preservative producers, and treaters of the region to develop specifications for preservatives, standards for treatments, and promote research. The American Wood Preservers' Association (AWPA) was formed in 1904 in the United States for these purposes. It publishes, annually, a *Proceedings*, which is the principal source of information on wood preservation and new developments, nationwide. The AWPA has also developed and maintains a set of accepted standards for both preservatives and treatments that are widely used in the wood preservation industry. The International Research Group on Wood Preservation (IRGWP), formed in 1969, now plays a major international role in dissemination of information on the destructive agents of wood and stimulation of research on wood preservation.

Treatment Choice and the Biological Hazard

Although methods for protecting wood have expanded and improved, the damages of wood-destroying agents and requirements for their control have remained relatively constant. Selection of the preservatives and treating methods requires consideration of the type of product used, the degree of biotic hazard, the length of time the wood must be protected, and the financial merits of such treatments.

For example, wood subjected to high decay hazard, such as that in ground contact in tropical zones, requires the most effective preservatives and treatments available. Wood subjected to repeat shock loads (for example, railroad ties), which cause rapid strength losses, can generally be treated with lower levels of chemical, even though the decay hazard is high, since the wood will fail mechanically before serious decay damage occurs. In sawmill operations, the producers are primarily interested in preventing fungal and insect attack for relatively short periods during log storage and lumber seasoning until wood dries below the fiber saturation point (f.s.p.). Users of wood in cooling towers, in other adverse decay conditions, or in applications where termites are also a threat are concerned with protecting the wood for 30 or more years. It is important to stress that use requirements and environmental considerations now generally dictate the type of wood preservative used and the treatment method.

Treatment results are normally expressed in terms of chemical loading and depth of preservative penetration. For short-term protection, chemical loadings are normally expressed on the basis of solution absorbed per unit volume of wood. Typically, gallons of chemical per thousand board feet (bd ft) of wood is employed. Penetration of chemical is not considered to be important for short-term protection.

Results for chemical treatments for longer-term protection are normally expressed as weight of chemical per cubic foot of wood treated (lbs chemical/ ft^3 or wood kg/m^3) or retention in an assay zone. Preservative penetration is essential for long-term performance and is usually measured visually or through the use of indicators specific for the treatment chemical.

Short-Term Wood Protection

Freshly cut wood is susceptible to invasion by decay and stain fungi until it dries below the f.s.p. Whereas rapid processing and kiln drying can limit this damage, unforeseen delays do occur. The degradation of logs and unseasoned timber (pulpwood, chips, poles, posts, etc.) are major problems in areas where temperature and humidity conditions favor fungal growth or where the species being harvested contain high percentages of sapwood. These deterioration problems were considered in some detail in the earlier chapters on log and chipwood storage (Chapter 13) and sapstains and wood discolorations (Chapter 14), and will be only briefly summarized here as related to other treatment methods and preservatives.

To prevent stain damage, most sawmills treat freshly sawn high-grade lumber by dipping or spraying with a fungicidal solution to coat the wood surface. This thin, prophylactic barrier prevents fungal spores from germinating on the wood surface and provides only short-term (5 to 6 months) protection. The protective barrier is usually removed in the final planing process.

The effectiveness of potential antisapstain chemicals was extensively evaluated on southern pine sapwood in the 1930s (Lindgren and Scheffer, 1939). Ethyl mercury phosphate and a sodium salt of pentachlorophenol (NaPenta) became the most commonly used chemicals for sapstain control. In the 1960s, concerns about the safe use of mercury compounds left NaPenta as the mainstay of the sapstain-control industry.

NaPenta had the advantage of low-cost, broad-spectrum toxicity, and solubility in water for ease of application. Until recently, these properties made it difficult for other chemicals to compete, but increasing environmental concerns about the safety of NaPenta have stimulated the search for safer fungicides. The major concern with NaPenta has been the presence of contaminants, known as dioxins. Recent rulings by the U.S. Environmental Protection Agency (EPA) have reduced the allowable dioxin levels in NaPenta. It is important to note that there are many isomers of dioxins and that the dioxins reported in NaPenta do not include the highly toxic tetrachlorodibenzodioxin found in some herbicides such as Agent Orange. Many countries have now banned the use of NaPenta or the importation of lumber dipped in the chemical.

It is not within the scope of this chapter to outline all of the substitute chemicals currently being used, but several that merit attention include copper-8-quinolinolate, 2(thiocyanomethylthiobenzothiazole), a number of quaternary ammonium compounds, and 3-iodopropynyl butylcarbamate (Smith et al., 1985; Miller and Morrell, 1989). In general, these compounds lack the broad-spectrum toxicity of NaPenta, but present much lower environmental risks.

Although the chemicals used to prevent sapstain are changing, it is unlikely that we can completely divorce ourselves from their use during the air-seasoning of high-quality lumber. Increasing kiln capacity to eliminate air-seasoning is expensive and impractical for large-dimension material. All too often, delays in processing lead to the staining of green lumber before kiln drying. It is hoped that the current interest in alternatives for sapstain control will continue and lead to development of effective, environmentally acceptable chemicals.

Long-Term Wood Protection

Although short-term decay and sapstain damage can be minimized by proper handling of wood and by dipping or spraying with fungicidal solutions, preventing degradation for long periods requires more effective treatments.

Wood preservation is divided into nonpressure and pressure treatments (for more detailed references see Wilkinson, 1979; Levi, 1973; AWPA, 1990; U.S. Government Standard TTW-571g; Hunt and Garrett, 1967; or Nicholas, 1973).

Nonpressure Treatments Dipping, brushing, spraying, and soaking require minimal equipment and are often used by homeowners for limited amounts of wood or by farmers who have a source of inexpensive timber. Brushing or spraying produces only a thin shell of preservative treatment that provides short-term protection to wood exposed out of ground contact, but is of minimal value for long-term protection needs or when the wood is exposed in soil. As expected, dipping or soaking provides absorption of preservative solution deeper into the wood, particularly if the wood is dry. Manufacturers of wood door frames and windows make extensive use of a dip and low-vacuum treatment to provide a barrier for wood exposed in above-ground uses. The species used are easily penetrated with liquids, particularly at the end grain of joints where moisture absorption is high and the decay normally occurs. Many farmers soak air-dry posts in oil-borne preservatives for periods of 1 to 7 days before use. Some wood users further utilize this simple treatment by dipping peeled, green posts first in a water-soluble preservative such as copper oxide and then in a second such as chromium trioxide. These compounds react in the wood to form a water-insoluble precipitate. This process permits the wood to be treated while it is still green.

Nonpressure treatments can improve wood service life, but the high variation in preservative penetration and distribution, along with the limited protection periods these treatments provide, makes them less attractive where wood in costly structures must be exposed under adverse conditions.

Pressure Treatments For a long, reliable service life for wood used in decay-hazard conditions, pressure-treatment with effective preservatives is mandatory. When performed properly, pressure-treatment results in deeper, more uniform treatment of the wood. There are three basic processes that we shall consider pressure treatments. The *thermal* process uses the natural development of small pressure differences inside the wood to force solution into the wood. In this process, dry wood is placed into an open tank, and preservative solution is added to the tank and heated to 150 to 230°F (65.5 to 110°C) for periods ranging from 26 to 48 hr. After the heating period, the preservative solution is withdrawn, and a cooler preservative solution is pumped in. The cool solution results in a pressure differential in the wood that draws preservative into the wood, increasing uptake. The thermal process is used most extensively to treat western red cedar and lodgepole pine posts. Because of the high temperature of the baths, the thermal process is generally used with oilborne chemicals, which are less likely to evaporate.

The two remaining pressure treatments, the full- and empty-cell processes, require more elaborate equipment and result in deeper, more uniform penetration than the other processes (Fig. 19-1). The full-cell or Bethell process, developed in 1839, represented a quantum leap in treatment technology. In this process, wood is placed in a treating cylinder (retort), a vacuum is drawn to remove as much air as possible from the wood, and the preservative is added to the retort. The pressure is gradually raised to a maximum of 150 to 200 pounds per square inch (psi) (1050–1408 kPa) and held until gauges on the outside of the retort indicate that a sufficient amount of solution has been forced into the wood (Fig. 19-2). This level, called *gauge retention*, depends on the volume of wood being treated, as well as on the specified retention and penetration values required by specifications for a given commodity. Once this target value has been achieved, the pressure is released, and the preservative solution is withdrawn. At this point, the pressure release causes air in the wood to expand and force outward a certain amount of preservative. The amount of preservative released from the wood is called the *kickback*. Additional periods of heating in solution, steaming, and vacuum may also be used to remove surface deposits and reduce subsequent bleeding of preservative once the wood is placed in service. The full-cell process results in a maximal uptake, or retention, of preservative for a given depth of penetration and is used for treating marine piling with creosote and all wood with waterborne solutions. In the latter preservative type, the concentration of chemical in the water can be changed to achieve the desired retention.

Figure 19-1 Commercial treating cylinders range in size from 1.5 to 3.0 m in diameter and up to 42 m long.

In the empty-cell or Rueping process, no vacuum is used. The solution is added and the pressure is raised and held until the desired amount of chemical is forced into the wood. The pressure is then released and air that was trapped in the wood expands outward, forcing excess preservative from the wood. The Rueping process results in larger kickback of preservative at the end of the treatment cycle than the Bethell process, producing a lower retention for a given depth of penetration (Fig. 19-2). In the other empty-cell process (Lowry process), a low pressure (30 psi, 211 kPa) is applied before the solution is added to the cylinder, trapping additional air in the wood and increasing the amount of kickback. The empty-cell processes are used to treat utility poles and lumber with oilborne chemicals for terrestrial uses where a clean surface is desirable.

In addition to the treatments described, a number of variations have been developed to improve treatment with certain chemicals or wood species. One involves the application of preservative in liquified petroleum gas (LPG). In this process, the solution is forced into the wood by pressure, the excess solution is withdrawn, and the pressure is released. Once the pressure is decreased, the LPG volatilizes, leaving a clean, paintable surface, and the solvent is recovered for reuse. This process has been used for treatment of laminated timbers and utility poles with pentachlorophenol, but certain

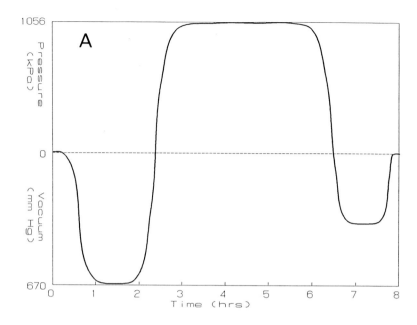

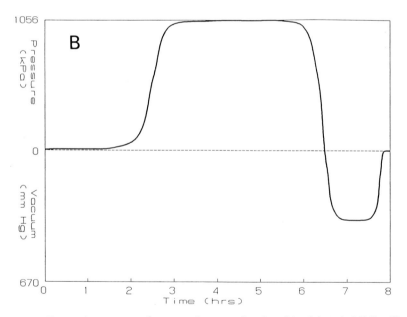

Figure 19-2 Vacuum/pressure conditions used in typical cycles of the (A) Bethel (full-cell), (B) Rueping (empty-cell), and (C) Lowry (empty-cell) processes. *(Figure continues)*

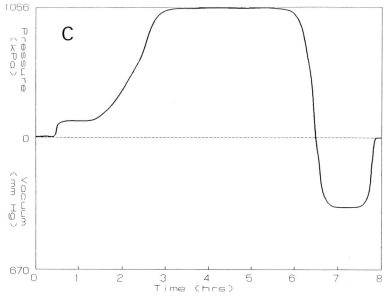

Figure 19-2 *(continued)*

problems with surface decay and concerns about the safety of this highly flammable solvent during treatment have limited its use (Lew and Wilcox, 1981).

Major Wood Preservatives

Numerous chemicals have been evaluated for their ability to protect wood, but only a select few have proven effective and are used in large quantities. These compounds can be divided conveniently into oil- and waterborne chemicals. The oilborne chemicals have the advantage of resistance to leaching and ease of penetration into the wood owing to decreased surface tension. Most waterborne chemicals do not require costly solvents and leave the wood surface clean and paintable (Hartford, 1973). There is no definitive data, but some workers have suggested that the type of carrier improves the chemical performance (Arsenault, 1973). An example is a comparison of the use of pentachlorophenol (PCP) in LPG or heavy oil. In heavy oil, PCP has performed well, but LPG treatments with the same chemical are less effective and highly susceptible to surface decay (Lew and Wilcox, 1981). Compositions or formulas for the major preservatives are indicated in Fig. 19-3.

Organic (Oilborne) Preservatives

Four oilborne preservatives are or have been used in the United States for pressure treatment of wood.

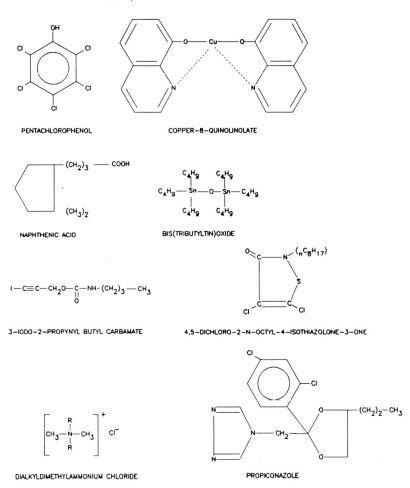

Figure 19-3 Chemical structures of some commonly used wood preservatives.

Creosote Creosote is among our oldest, most reliable preservatives, and ushered in the era of effective wood protection. Creosote is a complex mixture of over 200 polynuclear aromatic hydrocarbons (PAH) produced as a by-product of the high-temperature carbonization (900–1175°C) of coal to produce coke for the steel-making process and from distillation of the tars and pitches formed.

The composition of creosote is variable, depending on the distillation ranges used. The AWPA preservative standards specify boiling-point ranges for various fractions rather than specific chemicals. The heavy-oil fractions of creosote have proven to be the most effective toxic component. Because of its complexity, the toxic mode of creosote has never been fully determined; however, it protects wood against insects, fungi, and most marine borers

(Webb, 1980). Commonly used retentions for wood uses in ground contact range from 8 to 12 lb/ft^3, depending on severity of exposure. Poles treated with this mixture of chemicals have lasted 80 years with no loss in effectiveness. Creosote was once among our most important wood preservatives, but recent emphasis on clean, dry wood products free of surface deposits and concerns about the environmental safety of PAHs in general have relegated this preservative to applications where direct human contacts are minimal (Webb, 1987). Nevertheless, creosote is still used to protect 20 to 25% of the wood treated each year, primarily for utility poles, marine piling, and railroad ties (Micklewright, 1987). Creosote can also be mixed with pentachlorophenol, copper naphthenate, naphthalene, and petroleum for specific uses. The majority of these blends are available only on a special-customer basis. Creosote–petroleum blends are widely used to protect railroad ties and bridge timbers.

As a by-product of coke ovens, the creosote supply was dictated by the steel industry, which experienced wide fluctuations in production. As a result, periodic creosote shortages arose. In addition, the composition of creosote varied widely as some producers removed some of the more toxic fractions as distillate for other uses. At present, creosote specifications are tightly controlled, but the supply and quality variations stimulated a search for other preservatives. Creosote is a restricted-use pesticide, and users of the chemical must be licensed by their respective states.

Pentachlorophenol The search for a clean, effective preservative led to the development of PCP, one of our first synthetic pesticides, in the 1930s (Rao, 1977). It is easily synthesized by successive chlorinations of phenol. It functions as a biocide by uncoupling oxidative phosphorylation. Pentachlorophenol, and its water soluble sodium salt, exhibited broad-spectrum toxicity to fungi and insects, were relatively inexpensive, and could be supplied in large quantities. As a result, PCP rapidly replaced creosote in many applications. Concern over the presence of dioxins and related environmental problems have led to reduced use of this chemical. The EPA's recent decision to list wastes created during the synthesis and application of the chemical as acutely hazardous has provided additional incentive to develop effective alternative preservatives. At present, PCP is a restricted-use pesticide and users must be licensed by the appropriate state agencies.

Copper Naphthenate Copper naphthenate, a copper complex of naphthenic acids derived from the oil-refining process, has recently been promoted as a PCP replacement. This chemical, which is 20 times less toxic to humans than penta, has been successfully tested over a 40-year period and appears to be a promising "new" chemical, as its cost becomes more competitive with other preservatives. It has been used extensively for many years in the retail industry as a brush-on preservative for home and marine use. In cases where the green color of the preservative is objectionable, the colorless but less effective zinc salt is used.

Tributyltinoxide Tributyltinoxide (TBTO) is used extensively in Europe to protect window and door frames and in many marine paints as an antifoulant for barnacles. In the United States, TBTO has been used primarily for treating millwork, particularly where rapid removal of the solvent following treatment is desired. It has not performed well in ground contact, where fungi have been shown to degrade this chemical. Recent reports about the damage to shellfish from TBTO leaching from antifouling paints may lead to further restrictions on the use of this chemical.

Copper-8-Quinolinolate This chemical, also known as Cu-8 or oxine copper, has the lowest mammalian toxicity of any wood preservative currently used for wood treatment. It is the only preservative approved for direct food contact by the U. S. Food and Drug Administration. Although Cu-8 is used to protect food crates and picnic tables, its high cost and questions about its performance in direct ground contact have largely limited its use.

Waterborne Chemicals

Waterborne chemicals have been used in the United States for over 100 years, beginning with mercuric chloride ($HgCl_2$) in the Kyanizing process (Hunt and Garrett, 1967). Water reduces solution costs and leaves wood surfaces clean and paintable. Another important advantage is the reduction in release of volatile hydrocarbons as atmospheric pollutants during the treatment process.

Chromated Copper Arsenate At present, only chromated copper arsenate (CCA), ammoniacal copper zinc arsenate (ACZA), chromated zinc chloride (CZC), and acid copper chromate (ACC) are used to any extent. Developed in India in the 1930s, CCA has rapidly grown to become one of the most widely used preservatives for wood protection. In the United States, this wood preservative is formulated from chromium trioxide, copper oxide, and arsenic pentoxide. Copper and arsenic are excellent broad-spectrum fungicides. Arsenic is a competitive inhibitor for phosphorus in adenosine-triphosphate (ATP) synthesis and provides insect and marine-borer protection. Chromium has strong affinity for and complexes with wood lignin to limit leaching (Hartford, 1973). As a result, CCA is extensively used to protect lumber from decay fungi, insects, and most marine borers. Various formulations of CCA are used in other countries.

Unlike oilborne chemicals that are deposited on the wood cell wall but do not appear to be chemically bound, CCA is strongly fixed to the lignin component by the reduction of Cr^{6+} to Cr^{3+}. This fixation, which is strongly influenced by wood pH and temperature, also appears to help retain arsenic and copper in the wood (Hartford, 1986).

Ammoniacal Copper Arsenate (ACA) Ammoniacal copper arsenate and its recently developed replacement, ammoniacal copper zinc arsenate (ACZA), find extensive use in the treatment of western wood species. The ammonia appears to improve preservative penetration. Unlike CCA, ACZA does not have chromium to hasten fixation; nevertheless, this chemical appears to react strongly with wood components to provide excellent long-term protection (Johnson and Gutzmer, 1984; Wilcox, 1987).

Like PCP and creosote, CCA, ACA, and ACZA are restricted-use pesticides that can only be applied by certified applicators who have passed a state-administered test on pesticide safety and usage.

Chromated Zinc Chloride (CZC) and Acid Copper Chromate (ACC) These compounds are primarily used to protect wood in cooling towers and other non-ground-contact applications. Chromated zinc chloride and its predecessor, zinc chloride, were used initially to protect railroad ties and other timbers, but these chemicals had a tendency to leach from the wood and did not perform well in adverse environments. The excellent performance of the other waterborne chemicals has sharply reduced demand for these chemicals.

Boron Boron compounds are widely used outside North America for protecting wood from fungi and insects. These compounds diffuse with moisture and can completely penetrate wood. Borates are normally applied by dipping, followed by a 4 to 8 week wet-storage period. Although borates are very safe, their application is limited by high susceptibility to leaching in most applications where preservative treatment is required.

Other Compounds Several attempts have been made to develop waterborne formulations of penta and creosote. Waterborne ammoniacal penta formulations have been developed and are approved for above-ground use only. At present, there appear to be no commercial users of this chemical. There have also been reports of waterborne creosote; however, commercialization of this chemical has not occurred in North America (Krzylewski, 1986; Watkins, 1977; Cookson and Greaves, 1986).

Miscellaneous Compounds

In addition to the initial preservative treatments, a number of chemicals have been developed to provide supplemental or remedial protection to wood in service (Graham and Helsing, 1979). These chemicals can be divided into two broad areas: those protecting against surface degradation and those that can penetrate farther into the wood to control internal decay.

External treatments are applied as liquid or pastes. Liquid preservatives are most often applied when cuts or borings are made in preservative-treated wood. The preservative provides a shallow barrier against damage. A variety of chemicals can be used for this purpose, although 2% copper naphthenate is most commonly used. Pastes are generally used to protect the ground-line zone of large wood structures from soft-rot attack. These formulations often include an oil-soluble chemical to provide surface protection and a water-soluble component that penetrates for a short distance into the wood. Groundline treatment chemicals include PCP, copper naphthenate, creosote, sodium dichromate, sodium fluoride, and sodium octaborate tetrahydrate. Environmental concerns have resulted in a gradual shift to pastes containing copper naphthenate, sodium fluoride, or sodium borates. These treatments are extensively used in the electric utility industry to provide supplemental protection to the ground-line zone of wood poles, particularly with southern pine or western red cedar.

Internal treatments include void treatments and fumigants. Void treatments are applied by drilling holes into wood voids and pouring or forcing a set quantity of chemical into the void. Void treatments normally include an oilborne preservative and an insecticide. Until recently, void treatments incorporated creosote or PCP and chlordane, but environmental restrictions have mandated formulation changes. Chloropyrifos has replaced chlordane, and copper naphthenate has been proposed as a replacement for the oilborne preservative. Void treatments are presumed to function by producing a barrier in the void to prevent further insect or fungal attack, but their effectiveness has never been fully proven.

Fumigants are agricultural chemicals applied as liquids through steep-angled holes drilled into the wood. The holes are plugged, and the chemical volatilizes to move throughout the wood as a gas. Fumigant movement through wood up to 4 m from the point of application has been reported (Helsing *et al.*, 1984). Four fumigants are registered for decay control; chloropicrin (96% trichloronitromethane), Vapam (32.1% sodium *n*-methyldithiocarbamate), Vorlex (20% methylisothiocyanate in chlorinated C-3 hydrocarbons), and methylisothiocyanate. Fumigants are highly effective fungicides that rapidly move through the wood to eliminate most decay fungi within 1 year after application and provide protection against renewed invasion for periods ranging from 7 to 20 years (Zabel *et al.*, 1982; Helsing *et al.*, 1984; Morrell, 1989; Wang *et al.*, 1989). Despite their excellent performance, fumigants are effective in ground contact only when there is an existing preservative shell to protect the wood surface (Morrell *et al.*, 1986). Fumigants are widely used in North America to protect wood utility poles, bridge timbers, laminated beams, and marine piling from internal decay.

Fumigants are highly effective, but their volatility and high toxicity have led some users to question their safety. One alternative to fumigant treatment is fused borate rods. Boron can be formed into water-soluble, glass-like rods for application. The rods are inserted in the same holes used for fumigant application. Once applied, moisture in the wood solubilizes the boron, which moves across and down through the wood. Although not yet widely used in North America, fused borate rods are widely used in Europe for controlling decay in aboveground structures.

Nonchemical Methods for Improving Wood Performance

Wood varies widely in its degree of treatability. For example, sapwood is generally treatable, whereas heartwood, because of the high percentage of aspirated pits, tyloses, or extractives, is difficult to treat. In addition, geographic source has a marked effect on the treatability of certain woods. For example, Douglas fir from the coastal regions is far easier to treat than the same species grown farther inland. As a result of the treatment variables, a number of preparative steps have been developed to improve the treatment of wood, and, thus, enhance performance (Graham et al., 1969; Helsing and Graham, 1976; Graham and Helsing, 1979) (Fig. 19-4).

Incising uses steel teeth to punch numerous small holes into the wood surface, increasing the amount of end-grain exposed to the preservative solution, improving the depth and uniformity of preservative treatment. Incising is required for treatment of many thin-sapwood species. *Through-boring* or *radial drilling* are also used to increase the amount of cross section exposed in decay-sensitive zones and are extensively used to improve treatment in the groundline zone of electric utility poles (Graham et al., 1969). *Kerfing* is a different approach to improving the performance of large timbers and poles. A saw kerf is cut to the pith center, normally from the pole base to about the ground-line zone, and acts to relieve normal wood-drying stresses. Kerfing reduces the development of deep checks, which are common colonization sites, and significantly reduces the degree of internal decay in utility poles (Helsing and Graham, 1976).

In addition to specific pretreatments, simple processes including drying the wood before treatment, cutting the wood to length, and drilling all holes before treatment can ensure that the envelope of treatment remains intact.

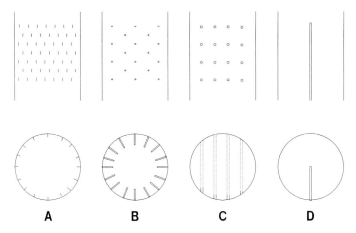

Figure 19-4 (A) Incising, (B) radial drilling, (C) through-boring, and (D) kerfing can improve preservative treatment of wood.

These processes are covered in detail elsewhere (Nicholas, 1973; Hunt and Garrett, 1967; Wilkinson, 1979).

Environmental Considerations

Within the last decade or so, environmental considerations have become a major preoccupation for commercial wood treaters and major users of treated wood products. This is particularly true of users of PCP, inorganic arsenicals, and creosote, since these chemicals are now restricted-use pesticides. In 1977, the U.S. EPA listed these three chemicals for potential revocation of their pesticide registrations. This process, called "rebuttable presumption against registration and reregistration" (RPAR), reviews supporting data for each pesticide provided by industry and the major users to determine whether the benefits of registration outweigh the risk of continued chemical use. In this process, the three chemicals were declared restricted-use pesticides, and users must now be licensed by the state. In addition, the wastes generated during the production of these chemicals and the subsequent treatment process are now more heavily regulated (Wise, 1986). The latter development has resulted in considerable expense to established treaters, who must control existing waste problems while dealing with environmental contamination that resulted from previous handling practices. This damage means that some treating plants would qualify as EPA toxic-waste sites if they were closed. Some plants estimate that complying with environmental regulations costs them about $1 million per year.

The increased environmental restrictions have stimulated a search for safer, more environmentally acceptable wood preservatives. This effort was spearheaded by the Electric Power Research Institute (EPRI) through research at Michigan Technological University's Institute of Wood Research (Preston, 1986; Preston et al., 1983). Unfortunately, the process of preservative development takes many years, and none of the potential replacement chemicals identified in these tests is commercially available. The immediate need to substitute safer chemicals has also stimulated renewed interest in previously identified chemicals such as copper naphthenate, Cu-8, and TBTO. Of these chemicals, copper naphthenate appears to be the closest to commercial use.

Although the current spate of environmental restrictions appears to have stabilized, the wood-treating industry will very likely come under increasing scrutiny. Looming problems are the safe disposal of used treated-wood products and soil contamination in sensitive sites. As a result, chemicals that are currently acceptable may find themselves in the same position as that of PCP, creosote, or inorganic arsenicals. This possibility emphasizes the need to continue developing and testing new preservatives and develop other control approaches.

Wood Preservative Development and Testing

Unlike agricultural chemicals, which need protect a plant from attack by a limited number of pathogens for only a short time, preservatives must protect wood against a wide array of organisms for periods of 30 years or more. Therefore, the process for developing a new wood preservative is considerably more involved and time consuming.

Since field trials of a new preservative would require decades, a number of accelerated, short-term tests have been developed as a guide to long-term performance (Behr, 1973; Baines, 1984; Fahlstrom, 1975; Halbrisky and Ifju, 1968; Duncan, 1958; McKaig, 1986; Smith, 1967). An interesting history of the developments in the evaluation of wood preservatives and of accelerated tests has been assembled by Colley (1953).

Petri Plate Tests

Most chemicals are initially tested for their ability to control decay fungi in petri plate tests (Richards, 1923). This test method varies widely between laboratories. It can include exposing the test fungus to a medium containing known concentrations of the test chemical, or soaking filter paper in the test chemical and placing this on the surface of a previously inoculated plate. The presence of the fungus and its growth rate are used as a measure of chemical effectiveness. These results are then compared with results of similar tests on accepted preservatives. Whereas this method provides a relative measure of toxicity, the growth of fungi on artificial media is markedly different from growth in wood. For example, the addition of excess sugar can largely negate the protective value of some preservatives. In addition, some chemicals that have low water solubility or those that are strongly fixed to wood cannot migrate into the media to inhibit the test fungus (Dost and Scheffer, 1983). As a result, these chemicals will not perform well in these tests, even though they may be very effective wood preservatives. Also, the short-term toxicity of volatiles can give misleading information on the long-term activity of a compound.

Soil-Block Test

A second, and more common method for testing preservative involves the soil-block test developed initially by Leutritz (1946) and later modified and improved (Duncan, 1953, 1958; AWPA, 1990). In this method, a block is treated with the test chemical at a given solution strength to produce a desired target retention. After weighing to determine the amount of chemical absorbed, the blocks are dried, and reweighed, and exposed to one of four selected test fungi (*Postia placenta, Trametes versicolor, Neolentinus* (*Lentinus*) *lepideus,* or *Gloeophyllum trabeum*), which were previously inoculated onto wood wafers on top of a sterile moist soil bed in a 227- or 454-ml bottle. The blocks are exposed to the test fungus for 3 months, then removed from the bottles, oven-dried, and reweighed. This weight is compared to the initial treated dry weight to determine the weight loss due to fungal exposure. The fungi used in this test were selected because of their known tolerances to specific toxicants: *P. placenta* has a high tolerance to copper; *T. versicolor* is tolerant to arsenic; *G. trabeum* is tolerant to PCP; and *L. lepideus* is resistant to creosote. The results following exposure of the treated blocks are compared to those obtained following exposure of untreated blocks to the same test fungi and similarly treated blocks not exposed to the test fungi. In many cases, blocks treated with a known wood preservative are included for comparison. The chemical retention in the block is plotted against the wood weight loss, and the point where weight loss due to test variables intersects with weight loss from fungal attack is termed the threshold (Fig. 19-5). This level can then be compared to *thresholds* for known chemicals and can be used to develop target retentions for field tests. Information on the stability of the preservative being tested can also be gained by exposing a matching set of test blocks to various leaching and weathering exposures. There are many variations on the soil-block test, but the most common substitutes malt agar in place of the soil. There remain, however, some questions concerning the effects of sugars present in the media on decay rates.

Soil-block tests provide a relative measure of preservative performance against white and brown fungi, but do not test the resistance to soft-rot organisms, which are known to be less sensitive to wood preservatives. At

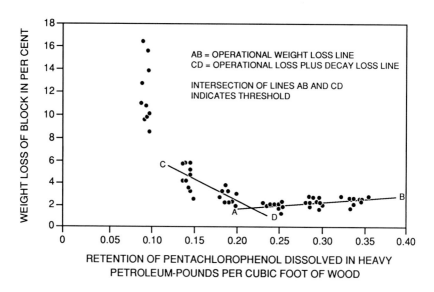

Figure 19-5 Relationship between chemical loading and wood weight loss in a soil-block test. Threshold values represent the point where fungal-associated weight losses equal those associated with method variables.

present, there is no standard test for resistance to soft-rot attack, although the use of soil or vermiculite burial tests, which increase the wood moisture content to levels typical for soft-rot attack, appear most popular (Nilsson, 1973; Zabel et al., 1985).

Soil-block tests provide a relative guide to decay resistance, but this method cannot adequately represent the associations between various fungi in a natural environment.

Decay Cellars

The fungal cellar is another accelerated method for testing fungal associations without establishing field tests in tropical countries. In this method, small wood stakes (0.5 × 1.0 × 6.0 inches) are treated to selected retentions with the test chemical and exposed in a soil bed, which is usually situated in a greenhouse. The temperature in this facility is usually maintained above 28°C, and the soil moisture levels are kept high. These conditions simulate tropical conditions and can produce rapid degradation within 1 year (Preston, 1986).

Termite Tests

Preservatives must protect against insects, as well as fungal attack. In most cases, the chemical is tested against termites since they cause the most widespread insect-related wood damage. In most tests, a set number of termite workers is placed in a chamber with the preweighed, treated wood for a given time (30 days). In some cases, the termite workers are given a choice of untreated or treated wood. At the conclusion of the test, the living workers are counted, the block is rated for degree of damage, and the wood is reweighed to determine the weight loss due to termite attack. These results are compared to those from similar tests using untreated wood or wood treated with accepted wood preservatives. These tests can also be performed with certain beetles. In general, laboratory insect evaluations are difficult, and there are many questions concerning the validity of these procedures.

Field Stake Tests

In general, laboratory tests cannot substitute for well-designed field tests, but most chemical producers have only a limited time in which the patent for a prospective new preservative remains valid. As a result, there is a need to accelerate field testing. To accomplish this task, the tests can be performed under more decay-hazardous conditions, or the specimens can be scaled down to the point where the surface:volume ratio allows decay to proceed most rapidly. Both approaches are used, and most field decay tests are performed on small specimens under tropical or subtropical conditions. Whereas these approaches provide rapid information on performance, it is important that the chemicals be tested under less severe conditions to see whether they might perform well in temperate climates.

In most cases, field tests are performed using stakes of the selected species (in the United States, southern yellow pine is most often used) treated to specific retentions. These stakes are half buried vertically in soil and inspected at regular intervals that vary with the decay hazard of the test site (Fig. 19-6). At each inspection, the stakes are pulled from the ground and visually rated from 10 (sound) to 0 (destroyed) for both termite attack and fungal deterioration. The results of the test chemicals are compared to those obtained with other stakes treated with an accepted wood preservative. In this method, the performance of stakes treated with lower retention of the test chemical can be used as indicators of future performance of the more heavily treated test chemicals (Colley, 1984; Hartford and Colley, 1984). Stake tests can last from 3 to 20 years depending on the test site and chemical retentions used. The most extensive stake tests in the United States are operated by the U.S. Forest Products Laboratory and evaluate a variety of pressure, dip, and brush-on treatments (Blew, 1948; Gjovik and Gutzmer, 1989). Test installations of larger treated units such as pole-diameter posts, closely simulate service tests, and have provided much useful information on pole species, wood preservatives, and treatment effectiveness (Ochrymowych and Kressbach, 1983) (Fig. 19-7).

Wood Assembly Tests

Tests have been developed to evaluate the performance of preservatives when wood is not in direct ground contact. These tests simulate door frames or joinery, where the decay hazard is far lower than in ground contact. Most aboveground tests have used an L-joint design with a tight, water-trapping joint to accelerate decay. These tests are evaluated in a manner similar to that used in stake testing, although the test takes longer to produce substantial decay. To overcome this problem, some laboratories have removed small samples from test members at regular intervals to determine changes in wood permeability related to fungal attack or have cultured the wood to determine whether fungal colonization has occurred.

Marine-Borer Tests

Whereas most testing methodology involves protecting wood from fungi and insects, some chemicals may also be promising marine-borer control agents. Testing these chemicals against terrestrial decay agents provides little information on marine performance. Although most marine testing is done in the ocean, potential marine biocides can also be tested under controlled laboratory conditions to determine the toxicity to certain marine borers. This task is difficult owing to the need to maintain marine conditions. Ocean testing involves the exposure of small wood samples treated to selected retentions of the test chemical at sites where marine-borer hazards have been previously documented (ASTM, 1981; Roe and Hochman, 1959; Johnson and Gutzmer, 1984). The specimens are attached to racks that can be easily retrieved for inspection. At each inspection, the specimens are rated for the degree of and type of marine-borer attack. Where appropriate, the specimens

Figure 19-6 New preservatives are generally evaluated in small wood samples (A) partially buried in soil or (B) submerged in the ocean. The wood is inspected annually for insect and fungal damage or marine borer attack. Stake photo courtesy P. G. Forsyth, Oregon State University.

Figure 19-7 A section of a large field installation of hundreds of treated pole-diameter posts evaluating various pole species, wood preservatives, and treatment methods. The test site was originated by Bell Telephone Laboratories at Chester, New Jersey, and maintained by Western Electric until 1984. Currently the pole-test site is operated by Bell Communications Research, Inc. Specialized equipment permits lifting the pole stubs a few feet so the ground-line zones can be critically inspected periodically for decay and termite attack. Photograph courtesy of John Kressbach, Bell Communications Research, Inc.

can be X-rayed to detect internal shipworm attack. Although the ASTM standard for the testing of wood preservatives in the marine environment list two specimen sizes, a variety of test dimensions are employed ranging from thin (6 mm) coupons to full-size test piling. Very small specimens are easily handled and produce rapid results, but they may not truly represent exposure conditions because they expose a large surface area relative to the wood volume and expose higher amounts of end grain, which will lose chemical more rapidly. Use of larger panels (2.5 × 10 × 45–60 cm) can overcome many of the limitations of small sample size and are most commonly used for marine tests. Large specimens are difficult to handle, expensive to install, require diving inspections that may be less accurate, and take longer to test.

Potential wood preservatives must also be evaluated for possible environmental hazards. Ideally, the preservative should have low mammalian toxicity, be safe to handle, be nontoxic to fish, and be easily disposed. In fact, most chemicals do not fit the ideal profile, but proper testing can identify the potential risks and allow registration of the chemical to be based on sound scientific knowledge, and not conjecture or public pressure. It is not within the scope of this chapter to discuss environmental testing.

Preservative Resistance (Tolerance)

There are large variations in the capacity of many microorganisms to resist or tolerate exposure to toxicants. Resistance and tolerance to toxicants are used synonymously in the literature. The development of resistance by bacteria to antibiotics (e.g., penicillin, streptomycin) and insects or fungi to pesticides (e.g., DDT and benomyl) is a serious problem in both medicine and agriculture (Dekker, 1976). Resistance arises both by selection among the various genotypes in a species and by occasional mutations. Favorable mutations can rapidly accumulate in bacterial and fungal populations, and produce huge numbers of propagules that have short life cycles. Resistance most commonly occurs with heavy usage of those chemicals with single-site toxic action and in cases in which the threshold between treatment levels and toxic levels to the microorganism is narrow. Resistance is less common in the older inorganic pesticides that have toxic action at multiple sites and where the treatments of surfaces or inert materials (wood) permit the use of higher concentrations. Most wood preservatives can protect wood for many years against a wide array of fungi, insects, and marine borers, but a limited number of organisms have developed either high tolerances to some of these chemicals (probably to similar compounds in nature) or in a few cases, evolved the ability to degrade the compound. Some of the reported tolerances of several decay and other common wood-inhabiting fungi to wood preservatives or their toxic components are listed in Table 19-1. Information on the tolerances of fungi to chemicals has several important uses. In the selection of test fungi for evaluating new wood preservatives or treatments, tolerant fungi are selected to increase the rigor of test. Also, the selection of a preservative and its retention for a wood product in some special uses must consider the common decay fungi and their tolerances. A new and potentially valuable use of the tolerant fungi is the degradation of the preservatives in toxic sites (Madhosingh, 1961; Duncan and Deverall, 1964; Cserjesi, 1967). Such fungi are promising candidates for disposal of wastes generated during the synthesis and application of these chemicals and provide an environmentally sound approach for hazardous-waste disposal (see Chapter 20).

Table 19-1
The wood-preservative and related chemical-toxicity tolerances reported for several decay and other wood-inhabiting fungi[a]

Preservative or toxicant	Fungus[b]	Role	Reference(s)
Creosote	Neolentinus lepideus	Brown rot	Zabel (1954)
	Irpex lacteus	White rot	Cowling (1957)
	Antrodia radiculosa	Brown rot	Hudson (1952)
	Hormoconis resinae	Stainer	Marsden (1954)
	Paecilomyces variotii	Soft rot	Kerner-Gang (1975)
Pentachlorophenol	Gloeophyllum trabeum	Brown rot	Zabel (1954)
	Irpex lacteus	White rot	Duncan (1953)
	Chaetomium globosum	Soft rot	Savory (1955)
	Cephaloascus fragrans	Stainer	Cserjesi (1967)
	Meruliporia incrassata	Brown rot	DaCosta and Kerruish (1964)
Copper naphthenate	Meruliporia incrassata	Brown rot	Cowling (1957)
	Postia placenta	Brown rot	Zabel (1954)
	Fomitopsis cajanderi	Brown rot	Cowling (1957)
	Wolfiporia cocos	Brown rot	Cowling (1957)
Chromated copper arsenate	Meruliporia incrassata	Brown rot	DaCosta and Kerruish (1964)
	Phialophora malorum	Soft rot	Daniel and Nilsson (1988)
Arsenic	Gloeophyllum trabeum	Brown rot	Cartwright and Findlay (1958)
Chromated zinc chloride	Coniophora puteana	Brown rot	Cartwright and Findlay (1958)
Sodium fluoride	Trichoderma spp.	Molds	Verrall (1949)
Phenyl mercury oleate	Xylobolus frustulatus	White rot	Cowling (1957)
Mercury	Penicillium cyclopium	Molds	Brown (1953)

[a]Data adapted in part from Levi (1973).
[b]The names used are from Farr et al. (1989).

The methods by which many microbes detoxify chemicals remain poorly understood; however, a brief discussion of several of the detoxification agents can provide a guide to these processes.

Decomposition of Creosote

Creosote is a highly complex mixture of organic chemicals, but several organisms have developed the ability to utilize this mixture as a sole carbon source, including the marine bacterium, *Pseudomonas creosotensis* (O'Neill *et al.*, 1960; Drisko and O'Neill, 1966; Stranks and Hulme, 1975; Seesman *et al.*, 1977), and the imperfect fungus, *Hormocomis (Cladosporium) resinae*. Almost a third of the fungi colonizing creosoted wood have the ability to degrade creosote (Kerner-Gang, 1976). One fungus, *H. resinae*, is the most commonly isolated fungus from creosoted southern pine utility poles (Zabel *et al.*, 1982) and can be grown on creosote as a sole carbon source (Christensen *et al.*, 1942; Marsden, 1954). Although this species commonly colonizes creosoted wood, it does not cause wood degradation, and its effect on longevity of the treated wood remains uncertain. A number of other fungi exhibit tolerance to creosote, including the basidiomycete, *Neolentinus lepideus*, which is used in the soil-block test. More recently, the soft-rot fungus, *Paecilomyces variotii* and the yeast *Candida albicans* were shown to be capable of degrading creosote (Kerner-Gang, 1976).

The ability of microorganisms to degrade creosote has recently been employed in land-farming schemes to detoxify waste products produced during the wood-treating process (Wise, 1986). Since creosote is virtually removed from the soil, this method offers considerable advantages over disposal in toxic-waste sites. In the latter instance, the waste is only stored, and can be returned to the original producer at a later date.

Decomposition of PCP

This chemical is a potent inhibitor of microbial oxidative phosphorylation, but a number of organisms have developed the ability to decompose PCP in soil. Savory (1955) was among the first to suggest that soft-rot fungi were more resistant to PCP than were basidiomycetes, comparing *Chaetomium globosum* and *Trametes versicolor*, respectively. Exposure of PCP-treated blocks to *Trichoderma* sp. has also been shown to increase weight losses caused by *Gloeophyllum trabeum*, suggesting some preservative modification or interaction between the two fungi (Duncan and Deverall, 1964). Similarly, *T. viride* and *Coniophora puteana* removed approximately 63% of the original PCP in treated red pine blocks (Unligil, 1968); however, other studies suggest that soil mineralization or immobilization of preservative might also account for some of these losses (Luetritz, 1965).

Several species, including *Trichoderma harzianum* and *T. virgatum*, have been shown to degrade PCP, generally by methylation of PCP to pentachloroanisole (Cserjesi, 1967; Cserjesi and Johnson, 1972). The latter compound is much less toxic than PCP, and a number of other species were subsequently shown to utilize this method of PCP detoxification (Stranks and Hulme, 1975).

In addition to fungi, bacteria may also play a role in PCP detoxification in soil contact, and a number of isolates have been shown to use PCP as a sole carbon source (Chu and Kirsch, 1972). In some cases, the normal soil microflora can significantly reduce the levels of labeled PCP in as little as 24 hr (Kirsch and Etzel, 1973).

The use of bacterial degradation has been proposed as one method for eliminating PCP from contaminated soil (Edgehill and Finn, 1983; Pignatello *et al.*, 1983). Several systems are currently available for this process (Wise, 1986).

As expected in such a complex system, the environmental conditions can play an important role in the degree and rate of decomposition. Several studies suggest that the addition of exogenous nutrients can increase the preservative resistance of some fungi (Schmidt and Ziemer, 1976), but may slow degradation by others (Bumpus *et al.*, 1985). Thus, each system must be carefully analyzed before attempts are made to use biological degradation to remove toxic wastes.

Decomposition of Inorganic Arsenicals

Unlike creosote and penta, the inorganic arsenicals pose a different problem for organisms attempting to utilize the treated wood. Generally, these chemicals have strong interactions with the wood and may not be as accessible to microbial activity. Furthermore, the lack of carbon in these formulations makes these compounds less useful as an energy source. Since the organism cannot use these chemicals directly as an energy source, it must develop methods for inactivating them or enhance alternative physiologic pathways that avoid the toxicant. The former method requires far less "effort" and is the more common method for detoxifying the inorganic arsenicals.

The literature on arsenical resistance by wood-decay fungi is limited (DaCosta, 1972; Madhosingh, 1961; Chou et al., 1973; Young, 1961). Examination of CCA-treated pine blocks colonized by *Postia placenta*, a copper-tolerant fungus, indicated that the hyphae absorbed an average of 3 to 4% of the total metal ions on a mycelial dry-weight basis in a pattern that closely followed the original concentrations of copper, chromium, or arsenic. In some instances, no copper or arsenic could be detected in the wood cell wall, suggesting complete removal by the test fungi (Chou et al., 1973).

In general, the strong wood–chemical interactions coupled with low energy value of the inorganic arsenicals probably limits the number of organisms capable of degrading these compounds, although a number have developed tolerances to high levels.

Summary

- The early use of creosote (by Moll in 1836) and the development of pressure treatments (by Bethell around 1839) for the preservation of wood preceded, by nearly 50 years, Hartig's discovery that fungi caused wood decay.
- Critical steps in selecting the preservative and treatment type necessary for a wood product are estimating the decay hazard, judging the length of time the wood must be protected, and weighing the relative financial merits of various treatments.
- Short-term treatments, involving dipping or spraying of fungicides onto lumber or log surfaces, are available to protect logs and lumber from sapstains and early decay during storage or lumber seasoning.
- Long-term treatments are necessary for wood used when conditions conducive to decay prevail. These treatments are often subdivided into nonpressure and pressure treatments. The nonpressure treatments include dipping, brushing, spraying, and soaking the wood in a preservative. This treatment type provides only a thin shell of treated

Summary continues

wood and provides useful protection to wood exposed above ground and exposed to intermittent wetting. It is of minimal value to wood in ground contact and not recommended. Pressure treatments are necessary to protect wood exposed to decay conditions in costly structures.
- Pressure treatments use combinations of pressure, vacuum, and elevated temperatures to force the preservative solutions deep into the wood. Long-term protection requires that the penetration exceed the depth of subsequent check formation.
- The major wood preservatives are oilborne: creosote, pentachlorophenol, copper naphthenate, copper-8-quinolinolate, and tributyltin oxide; and water soluble: chromated copper arsenate, ammoniacal copper arsenate, chromated zinc chloride, and acid copper chromate.
- Remedial treatments such as ground line pastes, void treatment, or fumigants are useful for arresting decay that occurs on treated wood in service.
- During the last decade, environmental considerations have led EPA to place creosote, pentachlorophenol, and arsenical compounds in a restricted-use category. This has resulted in a major search for new wood preservatives.
- Wood preservatives are tested and compared with other preservatives by petri-plate and soil-block tests, and the exposure of treated stakes in decay cellars, field-stake tests, or wood-assembly tests.
- Tolerance or resistance to wood preservatives has developed among some of the common decay fungi. The resistant strains of decay fungi are generally used as test fungi in the evaluation of new wood preservatives. Information on the tolerant decay fungi associated with a specific wood use is useful in decisions on what preservative to use and the retention level.
- Some of the tolerant or resistant fungi are able to degrade the preservative. This group may provide an economical and environmentally acceptable way to dispose of some toxic wastes.
- Major research needs are the development of effective, economical, and environmentally acceptable wood preservatives and the exploration of biological control possibilities to ensure the future protection of the enormous amounts of wood that must be used under conditions conducive to decay.

References

American Society for Testing and Materials. (1981). "Standard Methods of Accelerated Evaluation of Wood Preservatives for Marine Services by Means of Small-

Size Specimens." ASTM Standard Designation D 2481-70. Philadelphia, Pennsylvania.

American Wood Preservers' Association. (1990). "Book of Standard." Stevensville, Maryland.

Arsenault, R. D. (1973). Factors influencing effectiveness of preservative systems. In "Wood Deterioration and Its Prevention by Preservative Treatments" (D. D. Nicholas, ed.), pp. 121–278. Syracuse University Press, Syracuse, New York.

Baechler, R. H., and L. R. Gjovik. (1986). Looking back at 75 years of research in wood preservation at the U.S. Forest Products Laboratory. *Proceedings of the American Wood Preservers' Association* **82**:133–147.

Baines, E. F. (1984). Preservative evaluation using a soil bed and static bending strength to measure performance. *Proceedings of the American Wood Preservers' Association* **80**:67–79.

Behr, E. A. (1973). Decay test methods. In "Wood Deterioration and Its Prevention by Preservative Treatments" (D. D. Nicholas, ed.), pp. 217–246. Syracuse University Press, Syracuse, New York.

Blew, J. O. (1948). Comparison of wood preservatives in stake tests. *Proceedings of the American Wood Preservers' Association* **43**:88–119.

Brown, F. L. (1953). Mercury-tolerant *Penicillia* causing discoloration in northern white pine lumber. *Journal of the Forest Products Research Society* **3**(4):67–69.

Bumpus, J. A., M. Tien, D. Wright, and S. D. Aust. (1985). Oxidation of persistent environmental pollutants by a white-rot fungus. *Science* **228**:1434–1436.

Cartwright, K., St. G. and W. P. K. Findlay. (1958). Decay of Timber and Its Prevention. Her Majesty's Stationery Office, London.

Chou, C. K., J. A. Chandler, and R. D. Preston. (1973). Uptake of metal toxicants by fungal hyphae colonizing CCA-impregnated wood. *Wood Science and Technology* **7**:206–211.

Christensen, C. M., F. H. Kaufert, H. Schmitz, and J. L. Allison. (1942). *Hormodendrum resinae* (L.), an inhabitant of wood impregnated with creosote and coal tar. *American Journal of Botany* **29**:542–558.

Chu, J. P., and E. J. Kirsch. (1972). Metabolism of pentachlorophenol by an axenic bacterial culture. *Applied Microbiology* **23**:1033–1035.

Colley, R. H. (1953). The evaluation of wood preservatives. *Bell System Technical Journal* **32**:120–169 and 425–505.

Colley, R. H. (1984). What is log-probability index? *Proceedings of the American Wood Preservers' Association* **80**:247–256.

Cowling, E. B. (1957). The relative preservative tolerances of 18 wood-destroying fungi. *Forest Products Journal* **7**(10):355–359.

Cookson, L. J., and H. Greaves. (1986). Comparative bioassays of two high-temperature and two low-temperature-derived creosotes. *Holzforschung* **40**(1):59–64.

Cserjesi, A. J. (1967). The adaptation of fungi to pentachlorophenol and its degradation. *Canadian Journal of Microbiology* **13**(3):1243–1249.

Cserjesi, A. J., and E. L. Johnson. (1972). Methylation of pentachlorophenol by *Trichoderma virgatum. Canadian Journal of Microbiology* **18**(1):45–49.

DaCosta, E. W. B. (1972). Variation in the toxicity of arsenic compounds to microorganisms and the suppression of the inhibitory effects by phosphate. *Applied Microbiology* **23**:46–53.

DaCosta, E. W. B., and Kerruish, R. M. (1964). Tolerance of *Poria* spp. to copper-based wood preservatives. *Forest Products Journal* **14**:106–112.

Daniel, G. F., and T. Nilsson. (1988). Studies on preservative-tolerant species. *International Biodeterioration* 24:327-335.

Dekker, J. (1976). Acquired resistance to fungicides. *Annual Review of Phytopathology* 14:405-428.

Dost, W. A., and T. C. Scheffer. (1983). *Aspergillus* bioassay of wood treated with waterborne arsenical salts. *Material und Organismen* 18(2):135-140.

Drisko, R. W., and T. B. O'Neill. (1966). Microbial metabolism of creosote. *Forest Products Journal* 16:31-34.

Duncan, C. G. (1953). "Soil-Block and Agar-Block Techniques for Evaluating Oil-Type Preservatives: Creosote, Copper Naphthenate and Pentachlorophenol." Division of Forest Pathology, Bureau of Plant Industry, Special Release No. 37. Forest Products Laboratory, Madison, Wisconsin.

Duncan, C. G. (1958). "Studies of Methodology of Soil-Block Testing." U.S. Forest Products Laboratory Report 2119.

Duncan, C. G., and F. J. Deverall. (1964). Degradation of wood preservatives by fungi. *Applied Microbiology* 12(1):57-62.

Edgehill, R. U., and R. K. Finn. (1983). Microbial treatment of soil to remove pentachlorophenol. *Applied and Environmental Microbiology* 45(3):1122-1125.

Fahlstrom, G. B. (1975). A small stake test procedure for accelerated evaluation of wood preservatives. *Proceedings of the American Wood Preservers' Association* 71:216-220.

Farr, D. F., G. F. Bills, G. P. Chamuris, and A. Y. Rossman. (1989). "Fungi on Plants and Plant Products in the United States." A. P. S. Press, American Phytopathological Society, St. Paul, Minnesota.

Gjovik, L. R., and D. I. Gutzmer. (1989). "Comparison of Wood Preservatives in Stake Tests (1987 Progress Report)." USDA Forest Products Laboratory. Research Note FPL-RN-02. Madison, Wisconsin.

Graham, R. D. (1973). History of wood preservation. *In* "Wood Deterioration and Its Prevention by Preservative Treatments" (D. D. Nicholas, ed.), pp. 1-30. Syracuse University Press, Syracuse, New York.

Graham, R. D., and G. G. Helsing. (1979). "Wood Pole Maintenance Manual: Inspection and Supplemental Treatment of Douglas Fir and Western Red Cedar Poles." Oregon Forest Research Laboratory Bulletin. 24.

Graham, R. D., D. J. Miller, and R. H. Kunesh. (1969). Pressure treatment and strength of deeply perforated Pacific Coast Douglas fir poles. *Proceedings of the American Wood Preservers' Association* 65:234-242.

Halabisky, D. D., and G. Ifju. (1968). Use of respirometry for fast and accurate evaluation of wood preservatives. *Proceedings of the American Wood Preservers' Association* 64:215-223.

Hartford, W. H. (1973). Chemical and physical properties of wood preservatives and wood-preservative systems. *In* "Wood Deterioration and Its Prevention by Preservative Treatment" (D. D. Nicholas, ed.), pp. 1-120. Syracuse University Press, Syracuse, New York.

Hartford, W. H. (1986). The practical chemistry of CCA in service. *Proceedings of the American Wood Preservers' Association* 82:28-43.

Hartford, W. H., and R. H. Colley. (1984). The rationale of preservative evaluation by field testing and mathematical modeling. *Proceedings of the American Wood Preservers' Association* 80:257-269.

Helsing, G. G., and R. D. Graham. (1976). Saw kerfs reduce checking and prevent internal decay in pressure-treated Douglas fir poles. *Holzforschung* **30**(6): 184–186.

Helsing, G. G., J. Morrell, and R. D. Graham. (1984). Evaluations of fumigants for control of internal decay in pressure-treated Douglas fir poles and piles. *Holzforschung* **38**(5):277–280.

Hudson, M. S. (1952). *Poria radiculosa*, a creosote-tolerant organism. *Forest Products Journal* **2**(2):73–74.

Hunt, G. M., and G. A. Garrett. (1967). "Wood Preservation," 3rd Ed. McGraw-Hill, New York.

Johnson, B. R., and D. I. Gutzmer. (1984). "Marine Exposure of Preservative-Treated Small Wood Panels." USDA Forest Products Laboratory Note, FPL-0248.

Kerner-Gang, W. (1975). Effects of microorganisms on creosote. International Symposium Berlin-Dahlem. *Organismen und Holz* **3**:319–330.

Kirsch, E. J., and J. E. Etzel. (1973). Microbial decomposition of pentachlorophenol. *Journal of Water Pollution Control Federation* **45**:359–363.

Krzyzewski, J. (1986). Formulating an arsenical creosote system—possible wood preservative. *Proceedings of the American Wood Preservers' Association* **82**:118–124.

Leutritz, J., Jr. (1946). A wood–soil contact culture technique for laboratory study of wood-destroying fungi, wood decay, and wood preservation. *Bell System Technical Journal* **25**:102–135.

Luetritz, J. (1965). Biodegradability of pentachlorophenol as a possible source of depletion from treated wood. *Forest Products Journal* **15**:269–272.

Levi, M. P. (1973). Control methods. *In* "Wood Deterioration and Its Prevention by Preservative Treatment" (D. D. Nicholas, ed.). pp. 183–216. Syracuse University Press, Syracuse, New York.

Lew, J. D., and W. W. Wilcox. (1981). The role of selected Deuteromycetes in the soft-rot of wood treated with pentachlorophenol. *Wood and Fiber* **13**(4):252–264.

Lindgren, R. A., and T. C. Scheffer. (1939). Effect of blue stain on the penetration of liquids into air-dry southern pine. *Proceedings of the American Wood Preservers' Association* **35**:325–336.

Madhosingh, C. (1961). Tolerance of some fungi to a water-soluble preservative and its components. *Forest Products Journal* **11**:20–22.

Marsden, D. H. (1954). Studies on the creosote fungus, *Hormodendrum resinae*. *Mycologia* **46**:181–183.

McKaig, P. A. (1986). "Factors Affecting Decay Rates in a Fungus Cellar II. International Research Group on Wood Preservation IRG/WP/2259. Stockholm, Sweden.

Micklewright, J. T. (1987). Wood preservation statistics, 1985. A report to the wood preserving industry of the United States. *Proceedings of the American Wood Preservers' Association* **83**:438–447.

Nicholas, D. D. (ed.) (1973). "Wood Deterioration and Its Prevention by Preservative Treatments." Syracuse University Press, Syracuse, New York.

Miller, D. J., and J. J. Morrell. (1989). "Controlling Sapstain Trials of Product Group I on Selected Western Softwoods." Forest Research Laboratory, Oregon State University, Research Bulletin 65, Corvallis, Oregon.

Morrell, J. J. (1989). The Use of Fumigants for Controlling Decay of Wood: A Review of Their Efficacy and Safety. International Research Group on Wood Preservation IRG/WP/3525. Stockholm, Sweden.

Morrell, J. J., S. M. Smith, M. A. Newbill, and R. D. Graham. (1986). Reducing internal and external decay of untreated Douglas fir poles: A field test. *Forest Products Journal* **36**(4):47-52.

Nilsson, T. (1973). "Studies on Wood Degradation and Cellulolytic Activity of Microfungi." Studies Forestalia Suecica Nr. 104 Stockholm, Sweden.

Ochrymowych, J., and J. N. Kressbach. (1983). Effectiveness of the pentachlorophenol/methylene chloride preservative system as determined by field performance of treated pole-diameter posts. *Proceedings of the American Wood Preservers' Association* **79**:46-52.

O'Neill, T. B., R. W. Drisko, and H. Hochman. (1960). "A Creosote-Tolerant Marine Bacterium." U.S. Naval Civil Engineering Laboratory Technical Note N-398. Port Hueneme, California.

Pignatello, J. J., M. M. Martinson, J. G. Steiert, R. E. Carlson, and R. L. Crawford. (1983). Biodegradation and photolysis of pentachlorophenol in artificial freshwater systems. *Applied and Environmental Microbiology* **46**(5):1024-1031.

Preston, A. F. (1986). Development of new wood preservatives: The how and the what. *Proceedings of the Wood Pole Conference*, Portland, Oregon, pp. 78-97.

Preston, A. F., P. McKaig, and P. J. Walcheski. (1983). The use of laboratory, fungus cellar, and field tests in the development of wood preservatives. *Proceedings of the American Wood Preservers' Association* **79**:207-212.

Rao, K. R. (ed.). (1977). "Pentachlorophenol." Plenum, New York.

Richards, C. A. (1923). Methods of testing relative toxicity of wood preservatives. *Proceedings of the American Wood Preservers' Association* **19**:127-130.

Roe, T., and H. Hochman. (1959). Use of small panels in screening potential wood preservatives. *Symposium on Treated Wood for Marine Use*. Special Technical Publication 275. American Society for Testing Materials, Philadelphia, Pennsylvania.

Savory, J. G. (1955). The role of microfungi in decomposition of wood. Records Annual Convention, British Wood Preservers' Association.

Schmidt, O., and B. Ziemer. (1976). Hemmstaff-Toleranz von Pilzen. *Material und Organismen* **11**(3):215-230.

Seesman, P. A., R. R. Colwell, and A. Zachary. (1977). Biodegradation of creosote/naphthalene-treated wood in the marine environment. *Proceedings of the American Wood Preservers' Association* **73**:54-60.

Smith, R. S. (1967). Carbon dioxide evolution as a measure of attack by fungi and its application to testing wood preservatives and sapstain preventives. *Annals of Applied Biology* **59**:473-479.

Smith, R. S., A. Byrne, and A. J. Cserjesi. (1985). "New Fungicidal Formulations Protect Canadian Lumber." Special Publication No. SP-25. Forintek Canada Corp., Vancouver, B.C., Canada.

Stanlake, G. J., and R. K. Finn. (1982). Isolation and characterization of a pentachlorophenol-degrading bacterium. *Applied and Environmental Microbiology* **44**(6):1421-1427.

Stranks, D. W., and M. A. Hulme. (1975). The mechanism of biodegradation of wood preservatives. *Organism und Holz* **3**:345-353.

Unligil, H. H. (1968). Depletion of pentachlorophenol by fungi. *Forest Products Journal* **18**:45-50.

U.S. Government Printing Office. (1974). Federal Specification TT-W-571g. "Wood Preservation: Treating Practices." Washington, D. C.

Van Groenou, H. B., H. W. L. Rischen, and J. Van Den Berge. (1951). "Wood Preservation during the Last 50 Years." A. W. Sijthoff's Uitgeversmaatschappij N. V., Leiden, Holland.

Verrall, A. F. (1949). Some molds on wood favored by certain toxicants. *Journal of Agricultural Research* 78(12):695-703.

Wang, C. J. K., F. C. Terracina, and R. A. Zabel. (1989). "Fumigant Effectiveness in Creosote- and Penta-Treated Southern Pine Poles." Final Report Electric Power Research Institute Report No. EPRI EL-6197. Palo Alto, California.

Watkins, J. B. (1977). Timber Preservation. Australian Patent No. 514897.

Webb, D. A. (1987). Creosote—recent developments and environmental considerations. *Proceedings of the American Wood Preservers' Association* 83:11-18.

Webb, D. A. (1980). Creosote, its biodegradation and environmental effects. *Proceedings of the American Wood Preservers' Association* 76:65-69.

Wilcox, W. W. (1987). Effectiveness of ammoniacal copper zinc arsenate wood preservative as indicated by a soil-block analysis. *Forest Products Journal* 37(7/8): 62-63.

Wilkinson, J. G. (1979). "Industrial Timber Preservation." Associated Business Press, London, United Kingdom.

Wise, E. D. (1986). RCRA surface impoundment options for creosote and pentachlorophenol wood-treating facilities. *Proceedings of the American Wood Preservers' Association* 82:13-19.

Young, G. Y. (1961). "Copper Tolerance of Some Wood-Rotting Fungi. USDA Forest Products Laboratory Report 2223. Madison, Wisconsin.

Zabel, R. A. (1954). Variations in preservative tolerance of wood-destroying fungi. *Forest Products Research Society* 4:166-169.

Zabel, R. A., F. F. Lombard, C. J. K. Wang, and F. Terracina. (1985). Fungi associated with decay in treated southern pine utility poles in the eastern United States. *Wood and Fiber Science* 17(1):75-91.

Zabel, R. A., C. J. K. Wang, and F. C. Terracina. (1982). "The Fungal Associates, Detection, and Fumigant Control of Decay in Treated Southern Pine Poles. Report EL 2768. Electric Power Research Institute, Palo Alto, California.

PART
4

CHAPTER
20

Some Trends in Wood Microbiology Research and a New Emphasis (Biotechnology)

Like many of the biological sciences, wood microbiology is in a period of rapid change. Powerful new analytical tools, techniques, and advances in other disciplines have opened up new insights into and understandings about how wood-inhabiting microorganisms damage wood and how fungi function.

Wood-colonizing microorganisms are generally considered to be pests requiring control, but these same agents are important in the recycling of carbon and nitrogen. Without this natural decomposition, large quantities of these organic elements would be permanently lost, we would soon be inundated in piles of debris, and photosynthesis would cease.

Now special attention has been placed on understanding the decay process as a major biologic event, considering the microecological factors affecting the colonization of wood by sequences of wood-inhabiting microorganisms (Rayner and Boddy, 1988), and considering the environmental impacts and alternatives to the traditional chemical approaches to decay control (Eriksson et al., 1990). A new emphasis is emerging on the potential useful roles fungi may play in the wood-using industries as effective biopulping or biobleaching agents, as efficient degraders or biotransformers of wastes and toxicants (Hammel, 1989), or as biological control agents against decay and stain fungi. Recent advances in biotechnology now offer the possibility of modifying fungal genotypes to carry out some of these roles more effectively.

In this final chapter, we first review some important societal changes affecting future wood supplies and then summarize some of the major research needs and opportunities that may lead to more effective and environmentally acceptable prevention and control of decays and discolorations in wood products. We close with a section on how biotechnology may affect wood-using industries and make fungi useful, beneficial agents.

Changes in Forests and Wood Supplies

We are currently in a period of declining and changing wood supplies worldwide. As the population continues to expand, the wise, efficient use of wood will become increasingly important. Wood is a major renewable industrial material for construction and serves as an important fuel source in many regions. Wood currently accounts for 25% of the value of the major industrial materials in the United States, and demand is projected to increase by 25% by 2030 (National Research Council, 1990).

Changing views on forest values and shifts from timber production to multiple-use management schemes that favor protection, recreation, and watershed uses in many areas will further reduce the timber base in a period of increasing demand for wood. As the volume of available old-growth forests declines, there will be a shift to alternative species and to smaller-diameter stems containing increasing percentages of sapwood and juvenile wood. This material will be increasingly employed in modified wood products, such as particle board and laminated stock.

Reduced timber availability will place added emphasis on improved utilization of existing resources for a variety of applications, some of which remain to be developed. The eventual decline in petrochemical resources will further pressure the wood resource, which is one of our major renewable fiber and chemical resources. Microorganisms can be employed in a number of processes that protect the resource from deterioration or enhance recovery of various products from the material.

New Decay Control Approaches

Since many major industrial uses of wood involve its modification, understanding how fungi decay wood has great potential for developing more efficient processes and minimizing decay and stain damage.

Information on the chemical, physical, and ultrastructural aspects of decay and the physiology of fungi is basic to developing new decay-control approaches as alternatives to current poisoning (chemical) treatments. Examples of potentially rewarding topics are

1. Identifying the small, diffusible oxidizing agent in brown rots that is responsible for the rapid depolymerization of crystalline cellulose. If known, its production or functioning may be blocked or exploited;
2. Determining methods that disrupt the matrix that surrounds the hyphal tips of decayers to potentially block enzyme transfer and the decay process;
3. Developing chemicals that inhibit cell-wall synthesis of decay fungi; and
4. Identifying critical biochemical pathways in protein synthesis that can be disrupted.

Chemicals that disrupt critical syntheses in fungi (a separate life form from animals) may offer minimal human threat. Many fungi are able to decay wood; however, only a few types have been studied thoroughly. Novel methods by which fungi digest lignocellulose alone or in concert with other organisms may await discovery. There are thousands of decay fungi, yet our knowledge of the chemistry of decay rests on the study of only a few dozen species.

Many, if not most, laboratory studies of fungi are conducted in axenic cultures. In nature, many microorganisms are involved sequentially in the colonization and alternation of wood during the decay process. Studying the microecological aspects of the decay process may reveal cohorts of organisms that more efficiently decay wood, or sequences of organisms and interactions that suggest potent new biological-control agents or ways to accelerate decay.

Studies of the types and sequences of fungi involved during wood colonization and decay also may reveal effective ways to modify wood to impart attractive colors or patterns or improve properties such as penetrability.

Wood as Substrate for Mushroom Production

The ability of fungi to colonize wood and produce edible reproductive structures has been exploited for centuries in Asia for the production of mushrooms like shiitake (*Lentinulus edodes*) and the oyster mushroom (*Pleurotus ostreatus*) (Leatham, 1982; Zadrazil, 1974). These white-rot fungi are inoculated into freshly cut hardwood logs, which are incubated in cool, moist environments for up to 2 years (Leatham, 1982; San Antonio, 1981). Once they begin fruiting, the fungus produces mushrooms for 1 to 2 years before it exhausts the nutrients in the log. At that point there is little value left in the log, although there have been some suggestions to use this material as animal feed, since the fungi have solubilized the wood polymers to make them more digestible (Kirk, 1983).

Although the majority of shiitake and other wood-using mushrooms are produced in Asia, the need to develop markets for underutilized hardwoods has stimulated the development of shiitake production in the United States. The recent health consciousness and the growth of the gourmet food industry have also stimulated the development of a U.S. market. Recent studies in New Zealand suggest that high yields of shiitake mushrooms can also be attained using coniferous sawdust.

The use of logs to produce mushrooms converts relatively low-value material into a high-value product with a ready market. As with any commodity, the producer must always contend with biological problems that include drying of the logs and contamination by competing microorganisms. The latter problem can be particularly vexing, since certain shiitake strains exhibit tolerance to competing species, such as *Trichoderma;* however, this resistance can be overcome, and producers must continually screen for new, resistant strains. Conversely, the problems of *Trichoderma* inhibition of basidiomycetes can be exploited to prevent wood decay.

Microbial Generation of Feedstocks

The use of microorganisms to convert wood into a more accessible substrate for animal consumption or the use of fungi to degrade lignin to produce valuable by-products have both been discussed, but little research has been conducted in these areas (Kirk, 1983). Similarly, fermentation of glucose and other hexoses to ethanol is commercially feasible using spent sulfite liquor as the substrate (Eriksson, 1990).

The brown-rot fungi would appear to be ideally poised for these applications, since they can readily depolymerize cellulose and modify the lignin to make it more accessible to fungal enzymes. The use of a controlled brown-rot pretreatment could make wood more usable for ruminants. Unfortunately, such pretreatments have received little attention (Kirk and Shimada, 1985).

A second application of this technology could involve the modification of lignin-based preparations by brown-rot fungi to produce glues or resins. Currently produced resins are derived from petroleum, and their availability may be sharply reduced in future years. Brown-rot fungi modify lignin by demethylation and, in the process, create reactive sites that, if controllable, could be used to develop new types of polymers. Substitution of biomodified lignin for even a portion of the currently used resin would result in significant savings, while reducing the need to import oil.

The use of fungi to produce chemical feedstocks from wood or wood waste has also been proposed; however, the cost of production coupled with the availability of less expensive substitutes have largely limited research in this area. Renewed oil shortages would undoubtedly alter the economics, making biological production of feedstocks more feasible.

Biological Control of Fungal Stain and Decay

Most attempts to utilize wood-inhabiting fungi have involved their ability to degrade wood, but some organisms have the potential to protect wood from attack by other fungi. The use of fungi to protect wood from decay was advanced in the early 1960s by a number of researchers (Kallio, 1971; Hulme and Shields, 1972; Nelson, 1969; Ricard and Bollen, 1967) based on findings in control of agricultural pathogens (Baker and Cook, 1974). In the biological-control scheme, a control agent is introduced into the wood either through a freshly cut stump, or by inoculation into a hole drilled in the wood. In practice, control of established pathogens is extremely difficult, and most strategies seek to prevent colonization or protect the host from attack. Thus, *biological protection* is a more appropriate term for this strategy. The control agent then colonizes the wood to eliminate any existing decay fungi and prevent subsequent fungal invasion. The use of bioprotection is predicated on two critical points; the fungus must be capable of completely colonizing the substrate without damaging the wood, and the protection must be long-term. These two requirements have severely limited the list of potential bioprotectants (Preston et al., 1982).

The first bioprotection of wood decay fungi in forest products was reported by Ricard and Bollen (1967), using *Scytalidium* sp. to inhibit *Antrodia carbonica* in Douglas fir poles. Although the effectiveness of this agent was debated, the findings stimulated additional research. More recently, a bioprotection formulation, Binab AB, has been marketed in Europe for protecting Scots pine (Ricard, 1976; Bruce and King, 1986; Morris et al., 1984). This wood species appears to be colonized by a limited number of fungi, including *Neolentinus lepideus*, a brown-rot fungus that is particularly sensitive to the bioprotectants. Where this fungus dominates, the biocontrol agents should perform well, but there is considerable debate concerning the long-term effectiveness of this formulation (Morris et al., 1984; Morris and Dickinson, 1981; Bruce and King, 1983; Bruce et al., 1983). Recent tests on southern pine and Douglas fir suggest that this bioprotectant cannot completely control the numerous decay fungi associated with these species (Morrell and Sexton, 1990). In addition, the agent was unable to completely eliminate decay fungi already established in the wood, nor did it perform well against white-rot fungi. Brown-rot fungi are an important component of many decay systems, but recent studies have shown that white-rot fungi are far more common in coniferous woods than previously thought (Zabel et al., 1982; Graham and Corden, 1980; Eslyn, 1970). Whereas agriculture can deal with a small percentage of incomplete protection (as yield loss in disease control), the presence of small amounts of decay that can subsequently enlarge to destroy additional wood cannot be tolerated in a large wood structure. As a result, bioprotection does not appear to be feasible without the use of supplemental treatments that alter the ecology of the wood to favor growth or activity of the bioprotectant. For example, chemical pretreatments to eliminate any competing fungi may provide an edge to the bioprotectant, which is applied some years after chemical treatment.

One area in which bioprotection may have immediate application is the prevention of fungal stains. Stains are normally limited by prophylactic chemical treatments, but increasing environmental concerns have led to a search for safer methods of stain control. Biological protection may be especially appropriate for this application since the protective period is relatively short, and the protection can be delivered by surface colonization of the bioprotection agent. Preliminary testing suggests that bioprotectants can limit fungal staining, but variations in performance have been noted (Benko, 1988; Seifert et al., 1988).

Biotechnology in Pulp and Paper

The use of microbes in the field of pulp and paper technology has received extensive study, probably because the net gains from their application are so easily measured (Kirk *et al.*, 1983; Kirk and Chang, 1990). The pulp and paper industry is much more process oriented than other forest-products industries and is therefore more willing to consider new technologies. Furthermore, the economies of scale in a large paper mill make small returns on investments quite feasible because of the total mill capacity. Incorporation of biotechnology has been proposed from the treatment of wood chips all the way through to the treatment of pulping effluents.

The use of fungi that selectively utilize lignin before chemical or mechanical pulping processes has received tremendous attention, as paper companies seek to reduce their chemical and energy costs while increasing pulp yields (Kirk, 1983; Kirk and Shimada, 1985; Setliff *et al.*, 1990). These savings can be particularly significant in thermomechanical pulping, where small decreases in wood weight due to fungal attack can substantially decrease energy consumption during the pulping process (Kirk and Cowling, 1984). Since mechanical pulping is an energy-intensive process, these savings can become significant. In this process, a delicate balance must be maintained between lignin modification and carbohydrate utilization, which decreases both paper strength and yield; however, some studies have considered incorporating residual fungal mycelium in the pulp to increase yields. Further studies are underway to reduce treatment times, which currently last up to 18 days (Eriksson and Kirk, 1985).

Biological pretreatment has tremendous potential in the pulp and paper industry, but there are no commercially viable processes in use. In general, the process of delignification requires the addition of energy, either by the degradation of cellulose or hemicellulose, or by the addition of sugars (Kirk *et al.*, 1984). At present, the degree of control required for stable commercial pretreatments has not yet been determined, although each month brings new advances in our understanding of lignin decomposition. Most of these advances have been performed on a relatively limited number of white-rot fungi. For example, most of the studies in the United States have concentrated on *Phanerochaete chrysosporium* or *Trametes versicolor*, and only 25 to 30 of the thousand or more white-rot species have been researched on decay capabilities to any extent (Kirk and Shimada, 1985). This lack of depth suggests that more efficient species for bioconversion processes remain undiscovered.

In addition to the few species involved, a number of substantial hurdles loom before the successful implementation of bioconversion strategies, including the difficulty of scaling up from a basic laboratory process to a large-scale commercial endeavor, the need to prevent contamination of the mixture by competing fungi, and the slowness of delignification in conifers. Whereas all of these present a challenge, recent progress suggests that these hurdles will be overcome in the near future.

A second application of biotechnology to the pulp and paper industry addresses concerns about dioxin by-products produced during chlorine-bleaching processes. Recent rulings have sharply limited the permissible levels of dioxins present in pulp-mill effluent. Alternative chemical bleaching processes are possible, and biobleaching has also been proposed. Although not yet commercial, biobleaching has shown promise in hardwood pulps, where the combination of reduced lignin content and the white color of the fungal hyphae in the pulp resulted in decreases in Kappa number and increased brightness.

Biodecolorization is a third area in which fungi, particularly white rotters, immobilized on filter media, can be used to remove pigmented compounds from pulp-mill effluent. In this process, chlorinated compounds are substantially reduced, decreasing the need for further treatment.

Other potential areas for process improvement within the pulp and paper industry include pretreatments of hemicellulose for ethanol products, the use of microfungi for improving strength properties of paper, the use of brown-rot fungi to create highly reactive, modified lignins that could be used to produce modified polymers, or the use of enzymes within paper mills to reduce the build-up of slime within pipes (Eriksson, 1990). These provide just a few examples of many opportunities for utilizing microorganisms in the pulp and paper industry.

Another new area in the bioconversion field is the use of white-rot fungi to solubilize coal. This approach has the advantage of making the coal easier to handle without diminishing energy content. Recent studies indicate that Leonardite coal can be solubilized by exposure to cell-free fractions of *Trametes versicolor,* a common white-rot fungus (Pyne et al., 1987; Cohen et al., 1987), and studies are progressing in this area.

Biotechnology in Chemical Waste Management

Each year, millions of pounds of chemicals are hauled to hazardous-waste sites, where they are catalogued and stored. These chemicals remain the responsibility of the producer until they are destroyed. Unfortunately, the technology to safely eliminate toxic wastes remains elusive. Chemicals can be incinerated in high-temperature kilns; however, the licensing requirements for such kilns are extensive. The ability of white-rot fungi and some bacteria to decompose complex structures such as lignin may provide one alternative to incineration or continued hazardous-waste storage.

The ability of some fungi to detoxify preservatives is well known (Lyr, 1962; Duncan and Deverall, 1964; Cserjesi, 1967; Chu and Kirsch, 1972; Seesman et al., 1977; Stranks and Hulme, 1975), but only recently have these same organisms been evaluated for their ability to completely eliminate toxic wastes (Brown et al., 1987; Stanlake and Finn, 1982; Hammel et al., 1986). In addition, many natural microbial populations become more tolerant when continuously exposed to low levels of the toxin, and this trait has been exploited to develop chemically resistant microbial populations (Edgehill and Finn, 1983). Recent results have indicated that bioelimination can result in the conversion of such highly toxic wastes as DDT to carbon dioxide (Bumpus and Aust, 1987; Bumpus et al., 1985).

Although the results to date have been encouraging, and biological agents have been used in a number of soil-farming schemes to purify contaminated soil (Dean-Ross, 1987), many details need to be resolved before bioelimination becomes a proven technology. At present, the rates of elimination are very slow, and the chemicals tested still remain in the solution at low levels (i.e., complete elimination has not been achieved). In addition, care must be taken to ensure that the toxic wastes are completely converted to carbon dioxide, not just converted to less toxic intermediates. We must also decide the appropriate levels of residual toxin we are willing to live with. In some instances, incomplete detoxification can not be tolerated, although the levels required for site remediation continue to change.

In general, laboratory bioelimination studies have concentrated on the use of pure cultures of white-rot fungi and mixed cultures of bacteria in the presence of low levels of relatively pure chemicals. Many bacteria are capable of utilizing complex waste products either as sole carbon sources, or in conjunction with externally applied sugars. The effectiveness of these organisms when challenged by higher levels of combinations of chemicals often found in hazardous-waste sites remains unclear.

In practice, bioremediation agents must compete with an existing microbial flora, which may also be degrading the compounds present. The ecology of bioremediation sites appears to vary widely and, as a result, is poorly understood. Most commercial remediation efforts employ the microbial flora already present at the site. Further improvements in bioremediation will most likely depend on the development of more detailed information on the characteristics of specific sites, the organisms present in chemically contaminated soils and the physiology of these organisms in relation to chemical decomposition.

The Future

It is apparent that there are numerous potential uses for wood-inhabiting fungi within our current technology, including biopulping, bioelimination, biomass production, and biocontrol; however, the technologies to fully implement most of these approaches appear to be some years away. Nevertheless, the potential savings through reduced energy consumption, decreased environmental hazards, increased wood utilization, and longer service life for many forest products indicate that research outlays in this area are well worth the investment.

One of the areas not addressed has been biotechnology or the modification of existing organisms to improve certain physiological characteristics, such as lignin-degrading capability (Alic and Gold, 1985; Tien et al., 1987). This technology is well-developed in other fields, but it is relatively new to the study of wood deterioration; however, studies are underway to select improved delignifiers, and the technology is likely to be applied to the bioelimination field. As with other fields, there will be considerable debate about the dangers of releasing modified organisms into the natural environment, and it would appear that the most likely path for rapid implementation of biotechnology in wood products would involve the selection of existing strains of fungi to accomplish the desired task. As we become more sophisticated in our approach to the manipulation of microbes, the use of altered microbes can then be contemplated.

As these technologies are developed, the information concerning the physiology and ecology of the wood inhabiting fungi will provide an improved understanding of the roles of wood-inhabiting fungi and should ultimately lead to further applications of these fungi in industrial processes.

Summary

- Wood-degrading fungi may have many potential uses for improving wood utilization in a time of declining and changing forest resources.
- High-value mushrooms can be grown on logs of underutilized hardwood species.
- Some fungi may be useful for improving the digestibility of wood components for ruminants.
- Specific organisms can be used to inhibit stain or decay fungi.
- White-rot fungi have excellent potential for modifying lignin in wood before pulping to reduce energy consumption. These same fungi may also be used for pulp bleaching and decolorization of process wastewater.
- White-rot fungi may also be used for detoxifying certain hazardous wastes and, in combination with other organisms, represent an important component in remediation of hazardous-waste sites.
- The number of wood-degrading organisms screened for potential users is limited. More useful capabilities may exist in the large number of less thoroughly researched fungi. Genetic engineering in the future may permit the development of superior fungal genotypes to carry out various wood modifications, fermentations, or bioremedial processes. An exciting future lies ahead for wood microbiology.

References

Alic, M., and M. H. Gold. (1985). Genetic recombination in the lignin-degrading basidiomycete *Phanerochaete chrysosporium*. *Applied and Environmental Microbiology* **50**(1):27-30.

Baker, R., and R. J. Cook. (1974). "Biological Control of Plant Pathogens." Freeman Press, San Francisco, California.
Benko, R. (1988). "Bacteria as Possible Organisms for Biological Control of Blue Stain." International Research Group on Wood Preservation Document No. IRG/WP/1339, Stockholm, Sweden.
Brown, J. F., Jr., D. L. Bedard, M. J. Brennan, J. C. Carnahan, H. Feng, and R. E. Wagner. (1987). Polychlorinated biphenyl dechlorination in aquatic sediments. *Science* **236**:709–712.
Bruce, A., and B. King. (1983). Biological control of wood decay by *Lentinus lepideus* (Fr.) produced by *Scytalidium* and *Trichoderma* residues. *Material und Organismen* **18**(3):171–181.
Bumpus, J. A., and S. D. Aust. (1987). Biodegradation of DDT [1,1,1-trichloro-2,2-*bis*-(4-chlorophenyl) ethane] by the white-rot fungus *Phanerochaete chrysosporium*. *Applied and Environmental Microbiology* **53**(9):2001–2008.
Bumpus, J. A., M. Tien, D. Wright, and S. D. Aust. (1985). Oxidation of persistent environmental pollutants by a white-rot fungus. *Science* **228**:1434–1436.
Cohen, M. S., W. C. Bowers, H. Aronson, and E. T. Gray, Jr. (1987). Cell-free solubilization of coal by *Polyporus versicolor*. *Applied and Environmental Microbiology* **53**(12):2840–2843.
Cserjesi, A. J. (1967). The adaptation of fungi to pentachlorophenol and its biodegradation. *Canadian Journal of Botany* **13**:1243–1249.
Chu, J. P., and E. J. Kirsch. (1972). Metabolism of pentachlorophenol by an axenic bacterial culture. *Applied Microbiology* **23**:1033–1035.
Dean-Ross, D. (1987). Biodegradation of toxic wastes. *ASM News* **53**(9):490–492.
Duncan, C. G., and F. J. Deverall. (1964). Degradation of wood preservatives by fungi. *Applied Microbiology* **12**(1):57–62.
Edgehill, R. U., and R. K. Finn. (1983). Microbial treatment of soil to remove pentachlorophenol. *Applied and Environmental Microbiology* **45**(3):1122–1125.
Eriksson, K.-E. L. (1990). Biotechnology in the pulp and paper industry. *Wood Science and Technology* **24**:79–101.
Eriksson, K.-E. L., R. A. Blanchette, and P. Ander. (1990). "Microbial and Enzymatic Degradation of Wood and Wood Components." Springer-Verlag, New York.
Eslyn, W. E. (1970). Utility pole decay. Part II. Basidiomycetes associated with decay in poles. *Wood Science and Technology* **4**:97–103.
Graham, R. D., and M. E. Corden. (1980). "Controlling Biological Deterioration of Wood with Volatile Chemicals." Final Report Project 212-1. Electric Power Research Institute, Palo Alto, California.
Hammel, K. E. (1989). Organopollutant degradation by ligninolytic fungi. *Enzyme and Microbial Technology* **11**(November):776–777.
Hammel, K. E., B. Kalyanaraman, and T. K. Kirk. (1986). Oxidation of polycyclic aromatic hydrocarbons and dibenzo-(*p*)-dioxins by *Phanerochaete chrysosporium* ligninase. *Journal of Biological Chemistry* **261**(36):16948–16952.
Hulme, M. A., and J. K. Shields. (1972). Interactions between fungi in wood blocks. *Canadian Journal of Botany* **50**(6):1421–1427.
Kallio, T. (1971). Protection of spruce stump against *Fomes annosus* (Fr.) Cooke by some wood-inhabiting fungi. *Acta Forstalia Fennica* **117**:19.
Kirk, T. K. (1983). Degradation and conversion of lignocellulose. In "The Filamentous Fungi. Volume 4. Fungal Technology" (J. E. Smith, D. R. Berry, and B. Kristiansen, eds.), pp. 266–295. Edward Arnold, London.

Kirk, T. K., and Hou-Min Chang. (eds.) (1990). "Biotechnology in Pulp and Paper Manufacture. Applications and Fundamental Investigations." Butterworth-Heinemann, Boston.

Kirk, T. K., and E. B. Cowling. (1984). Biological decomposition of solid wood. In "The Chemistry of Solid Wood" (R. M. Rowell, ed.), pp. 455-487. Advances in Chemistry Series, Vol. 207. American Chemical Society, Washington, D. C.

Kirk, T. K., and M. Shimada. (1985). Lignin biodegradation: The microorganisms involved and the physiology and biochemistry of degradation by white-rot fungi. In "Biosynthesis and Biodegradation of Wood Components" (T. Higuchi, ed.), pp. 579-605. Academic Press, New York.

Kirk, T. K., T. W. Jeffries, and G. F. Leatham. (1983). Biotechnology: Applications and implications for the pulp and paper industry. TAPPI 66(5):45-51.

Kirk, T. K., M. Tien, and B. D. Faison. (1984). Biochemistry of the oxidation of lignin by Phanerochaete chrysosporium. Biotechnical Advances 2:183-199.

Leatham, G. F. (1982). Cultivation of shiitake, the Japanese forest mushroom, on logs: A potential industry for the United States. Forest Products Journal 32:29-35.

Lyr, H. (1962). Detoxification of heartwood toxins and chlorophenols by higher fungi. Nature (London) 4836:289-290.

Morrell, J. J., and C. M. Sexton. (1990). Evaluation of a biocontrol agent for controlling Basidiomycete attack of Douglas fir and southern pine. Wood and Fiber Science 22:10-21.

Morris, P. I., and D. J. Dickinson. (1981). "Laboratory Studies on the Antagonistic Properties of Scytalidium spp. to Basidiomycetes with Regard to Biological Control." International Research Group on Wood Preservation IRG/WP/1130, Stockholm, Sweden.

Morris, P. I., D. J. Dickinson, and J. F. Levy. (1984). The nature and control of decay in creosoted electricity poles. Record Annual Convention British Wood Preservers' Association, 42-55.

National Research Council. (1990). "Forestry Research: A Mandate for Change." National Academy Press, Washington, D. C.

Nelson, E. E. (1969). Occurrence of fungi antagonistic to Poria weirii in a Douglas fir forest soil in western Oregon. Forest Science 15(1):49-54.

Preston, A. F., F. H. Erbisch, K. R. Kramm, and A. E. Lund. (1982). Developments in the use of biological control for wood preservation. Proceedings of the American Wood Preservers' Association 78:53-61.

Pyne, J. W., Jr., D. L. Stewart, J. Frederickson, and B. W. Wilson. (1987). Solubilization of Leonardite by an extracellular fraction from Coriolus versicolor. Applied and Environmental Microbiology 53(12):2844-2848.

Rayner, A. D. M., and L. Boddy. (1988). "Fungal Decomposition of Wood. Its Biology and Ecology." John Wiley, New York, N.Y.

Ricard, J. L. (1976). Biological control of decay in standing creosote-treated poles. Journal of the Institute of Wood Science 7(4):6-9.

Ricard, J. L., and W. B. Bollen. (1967). Inhibition of Poria carbonica by Scytalidium sp., an imperfect fungus isolated from Douglas fir poles. Canadian Journal of Botany 46:643-647.

San Antonio, J. P. (1981). Cultivation of the shiitake mushroom. Horticultural Science 16:151-156.

Seesman, P. A., R. R. Colwell, and A. Zachary. (1977). Biodegradation of creosote/

naphthalene-treated wood in the marine environment. *Proceedings of the American Wood Preservers' Association* **73**:54-60.

Seifert, K. A., C. Breuil, L. Rossignol, M. Best, and J. H. Saddler. (1988). Screening of microorganisms with the potential for biological control of sapstain on unseasoned lumber. *Material und Organismen* **23**:91-95.

Setliff, E. C., R. Marton, S. G. Granzow, and K. L. Eriksson. (1990). Biomechanical pulping with white-rot fungi. *TAPPI* **73**(8):143-147.

Stanlake, G. J., and R. K. Finn. (1982). Isolation and characterization of a pentachlorophenol-degrading bacterium. *Applied and Environmental Microbiology* **44**(6):1421-1427.

Stranks, D. W., and M. A. Hulme. (1975). The mechanisms of biodegradation of wood preservatives. *Organismen und Holz* **3**:345-353.

Tien, M., P. J. Kersten, and T. K. Kirk. (1987). Selection and improvement of lignin-degrading microorganisms: Potential strategy based upon lignin model–amino acid adducts. *Applied and Environmental Microbiology* **53**(2):242-245.

Zabel, R. A., C. J. K. Wang, and F. E. Terracina. (1982). "The Fungal Associates, Detection, and Fumigant Control of Decay in Treated Southern Pine Poles." Report EL 2768. Electric Power Research Institute, Palo Alto, California.

Zadrazil, F. (1974). The ecology and industrial production of *Pleurotus ostreatus, Pleurotus florida, Pleurotus cornucopia,* or *Pleurotus eryngii. Mushroom Science* **9**:621-652.

Author Index

Abad, A. R., 230, 232, 235
Abbot, E. G. M. 83
Adaskaveg, J. E., 230
Adler, E., 152, 206
Ager, B. P., 395
Agi, J. J., 373
Agrios, G. N., 104, 153, 287
Ahmadjian, V., 390
Aho, P. E., 109
Ainsworth, G. C., 81, 73
Alexopoulos, C. J., 69
Alic, M., 453
Allen, T. C., 30
Allison, J. L., 434
AWPA, 44, 345, 413, 427
Amburgey, T. L., 29, 349, 350, 351, 352
Ampong, F. F. K., 403
Anagnost, S. E., 356
Ander, P., 12, 182, 195, 202, 208, 210, 445
Anderson, L. O., 344
Anderson, T., 351
Anderson, A. B., 401, 408
Anonymous, 344, 349, 373
April, G. C., 391
Arganbright, D. G., 239
Armstrong, F. H., 252
Aronson, H., 451
Arsenault, R. D., 419
ASTM, 252, 392, 431
Atlas, R. M., 266
Atwell, E. A., 284, 313
Auchter, R. J., 321
Aust, S. D., 435, 452

Baechler, R. H., 110, 359, 413
Baecker, A. A., 216

Bailey, I. W., 183, 327, 330
Bailey, P., 208
Baines, E. F., 427
Baker, A. F., 28
Baker, W. L., 29
Baker, K. F., 387
Baker, R., 448
Baldwin, R. C., 377
Banaszak, E. G., 395
Banerjee, A. K., 274
Banks, W. B., 353
Bar-Lev, S. S., 98, 301
Barghoorn, E. S., 43, 183
Barnard, F., 195, 230
Barnes, H. M., 27
Barnett, H. L., 90, 271
Barrett, D. K., 298
Barron, G. L., 269
Bartha, R., 266
Barthelemy, P., 377
Bartnicki-Garcia, S., 60
Barton, G. M., 26, 327, 401
Basham, J. T., 273, 284
Bauch, J., 328
Bavendamm, W., 93, 178
Baxter, D. V., 178
Beal, R. H., 29, 33, 403
Becker, C. E., 63
Beckett, A. 60, 228
Bedard, D. L., 451
Behr, C. T., 33
Behr, E. A., 33, 97, 401, 427
Bendtsen, B. A., 359
Benko, R., 313, 339, 449
Bernier, R. Jr., 339
Berry, P., 260
Best, M., 313, 339, 449

Betterman, A., 212
Bier, J. E., 290
Biesterfelt, R. C., 349
Bills, G. F., 82, 296, 434
Birnbaum, H. A., 386
Björkman, E., 317, 337
Blaisdell, D. J., 178, 213, 270
Blanchette, R. A., 11, 12, 182, 195, 202, 208, 210, 229, 230, 231, 232, 233, 235, 239, 269, 445
Bland, D. E., 228
Blew, J. O., 360, 430
Boddy, L., 12, 267, 269, 271, 272, 284, 289, 301, 445
Bodig, J., 252, 373
Bollen, W. B., 269, 448, 449
Borden, J. H., 35
Boutelje, J. B., 18
Bowers, W. C., 451
Boyce, J. S., 98, 178, 283, 284, 290, 291, 293, 327, 344
Boyer, P. D., 122
Boyle, P. J., 43
Bracker, C. E., 228
Brand, B. G., 384, 386, 387
Bravery, A. F., 183, 374, 384
Brennan, M. J., 451
Breuil, C., 313, 339, 449
Bridges, J. R., 269
Brier, A. N., 344
Brock, T. D. 82, 90, 116, 266, 268
Brown, W., 195, 207
Brown, F. L., 24, 252, 332, 434
Brown, J. F. Jr., 451
Bruce, A., 269, 449
Buchanan, R. E., 82, 83
Bucur, V., 373
Buller, A. H. R., 182
Bultman, J. D., 403
Bumpus, J. A., 435, 452
Butcher, J., 11, 83, 270, 273
Byrne, A., 415

Caldwell, J., 101
Calum, J. S., 82, 116
Campbell, A. H., 170
Campbell, W. G., 176, 252
Campbell, W. A., 10, 64, 178, 213, 270
Campbell, R., 266
Canfield, E. R., 375
Cano, R. J., 82, 116
Carlson, M. C., 11, 231
Carlson, R. E., 435

Carnahan, J. C., 451
Carolin, V. M., 29
Carpenter, R. D., 284
Carpenter, B. E., 312
Cartwright, K. St. G., 10, 93, 94, 178, 185, 252, 283, 434
Casida, J. E., 30
Cassens, D. L., 338
Cavalcante, M. S., 216
Cease, K. R., 232, 375
Cech, M. Y., 328
Chafe, S. C., 236
Chamuris, G. P., 82, 296, 434
Chandler, J. A., 436
Chang, H.-M., 195, 210, 450
Chaplin, C. J., 252
Chase, W. W., 101
Chen, C.-L., 210
Chesley, K. G., 316
Chidester, M. S., 95, 96, 106, 315
Chien, P. L., 377
Chin, A. A., 212
Chin, P. P. S., 26
Choi, A., 375
Chou, C. K., 232, 239, 436
Christensen, C. M., 434
Christian, M. B., 37
Christie, R. D., 257
Chu, J. P., 135, 151
Clark, J. W., 330, 349, 400, 404
Clarke, M. R., 328
Cochran, V. W. 90
Cohen, M. S., 451
Collett, R., 377
Colley, R. H., 10, 252, 427, 430
Colwell, R. R., 451
Condon, E. J., 360
Cook, R. J., 390, 448
Cooke, W. B. 97
Cookson, L. J., 423
Cooney, D. G., 104
Corbett, N. H., 10, 178, 273
Corden, M. E., 11, 176, 184, 314, 315, 347, 365, 367, 374, 449
Core, H. A., 136, 144, 227
Corner, E. J. H., 68, 80
Côté, W. A., 136, 140, 142, 143, 144, 145, 149, 158, 156, 154, 195, 202, 227
Coulson, R. N., 28
Cowling, E. B., 10, 109, 176, 183, 186, 195, 196, 197, 199, 207, 230, 251, 257 289, 347, 375, 400, 401, 403, 405, 406, 434, 450

Cown, D. J., 251, 369
Craig, H. M., 284, 286, 298
Crandell, B. S., 101
Crawford, R. L., 195, 210, 216, 435
Crossley, R. D., 336
Crossley, A., 236
Croteau, R., 405
Cserjesi, A. J., 333, 338, 404, 415, 433, 434, 435, 351
Cundell, A. M., 44
Curtis, M. L., 321

DaCosta E. W. B., 434, 436
Dale H. M., 330
Daniel, G. F., 83, 180, 215, 216, 231, 233, 236, 237, 239, 348, 375, 434
Daniel, J. W., 316
Daniels, A. S., 10
Darrah, J. A., 97
Datta, A., 212
Davidson, R. W., 10, 64, 94, 101, 178, 213, 270, 283, 320, 335, 347, 360
Davidson, A. G., 284
Dawes, C. J., 226
Day, A. C., 136, 144, 158, 227
de Zeeuw, C., 136, 252, 400
Dean-Ross, D., 453
DeGroot, R. C., 33, 260, 289, 293, 312, 373, 375, 376
Dekker, R. F. H., 208, 209
Dekker, J., 433
Desai, R. L., 328
Desrochers, M., 339
Deverall, F. J., 433, 435, 451
Dickinson, D. J., 269, 274, 449
Dietrichs, H. H., 216
Dill, I., 230
Djerf, A. C., 316, 321
Dondero, T. J., 395
Dost, W. A., 344, 427
Dowding, P., 333
Dowson, G. G., 269, 272
Drescher, R. F., 386
Drisko, R. W., 434
Drysdale, J. A. 83, 237, 239, 338
Duff, J. E., 351
Duncan, C. G., 10, 102, 105, 178, 186, 347, 401, 427, 433, 434, 435, 351
Durrell, L. W., 391

Eades, H. V., 99

Eaton, R. A., 216, 236, 238, 337
Ebeling, W., 29
Ecklund, B. A., 102, 183, 238, 259, 311
Edgehill, R. U., 435, 452
Effland, M. J., 201, 203, 215, 230
Ehrlich, J., 408
Eichorn, S. E. E. 116
Ellis, W. D., 11, 24
Ellwood, E. L. 102, 183, 238, 259, 311
Emerson, R. 104
Endeward, R., 328
Englerth, G. H., 404, 408
Enoki, A., 215
Erbisch, F. H., 269, 449
Erickson, H. D. 27
Eriksson, K.-E. L., 11, 12, 97, 182, 195, 202, 205, 206, 208, 210, 230, 445, 448, 450, 451
Esau, K., 137
Esenther, G. R., 30, 260, 373
Eslyn, W. E., 105, 108, 176, 201, 203, 215, 230, 257, 308, 310, 312, 315, 317, 320, 321, 335, 338, 347, 359, 374, 375, 404, 449
Esposito, M. M., 391
Etheridge, D. E., 271, 284, 286, 289, 298
Etzel, J. E., 435
Evans, R. C., 116
Evans, H. J., 109
Eveleigh, D. E., 208, 386
Evert, R. F. 116

Fahlstrom, G. B., 427
Faison, B. D., 450
Farr, D. F. 82, 296, 434
Farrell, R. L., 11, 210, 212, 214, 230, 232, 235
Feist, W. C., 23, 24, 317, 318, 353
Feng, H., 451
Fengel, D., 136, 155, 195, 333
Findlay, W. P. K., 10, 93, 94, 178, 252, 283, 327, 434
Fink, J. N., 395
Finn, R. K., 435, 451, 452
Fisher, R. C., 35
Fisher, T. H., 405
Fleck, L. C., 408
Flemming, R. M., 10
Fogarty, W. M., 102
Foisner, R., 230, 231, 232
Forest Products Laboratory, 253

Forney, L. J., 231
Forsyth, P. G., 330
Foster, R. C., 228
Foster, C. H., 337
Foster, R. E., 290
Franklin, D. E., 370
Frederickson, J., 451
French, D. W., 178, 273, 284, 289
Freyfeld, E. E., 93
Fries, E. M., 80
Friis-Hansen, H., 369, 370, 372, 377
Fritz, C. W., 328, 330
Fujii, Y., 261, 373
Fujii, T., 372
Fuller, W. S., 306, 319, 320, 321
Funatsu, M., 204
Furniss, R. L., 29, 39
Furnival, G. M., 290
Furtado, S. E. J., 83, 238
Fuse, G., 215

Garcia, S., 232, 375
Gardner, J. A. F., 327, 401
Gareth-Jones, E. B., 83, 238
Garraway, M. O., 116
Garrett, G. A., 415, 422, 426
Geyer, H., 43
Gibbons, N. E., 82, 83
Gibson, D. G., 375, 376
Gilbertson, R. L., 81, 230, 283, 285, 296, 299, 375
Gjovik, L. R., 260, 413, 430
Glasser, W. G., 151
Glasser, H. R., 151
Glenn, J. K., 212, 213
Gokhale, A. A., 98, 101
Gold, M. H., 210, 212, 213, 453
Goldsbury, K. L., 391
Goldstein, I. S., 26
Goll, M., 386, 391
Good, H. M., 97, 98, 101, 272, 294, 330
Goodell, B. S., 367, 375
Goring, D. A. I., 146, 156, 228
Graham, R. D., 11, 28, 41, 176, 184, 252, 253, 259, 260, 314, 315, 347, 356, 360, 367, 368, 369, 372, 373, 374, 399, 412, 413, 423, 424, 425, 449
Granzow, S. G., 450
Gray, D. R., 35
Gray, R. L., 227
Gray, E. T., Jr., 451
Greathouse, G. A., 360

Greaves, H., 83, 183, 238, 274, 317, 423
Green, F., 232
Green, R. W., 321
Griffin, D. M., 93
Griffin, D. H., 90, 93, 102, 116
Grisebach, H., 405
Grove, S. N. 63
Grunewald, A., 230
Gutzmer, D. I., 423, 430
Gyuris J., 386

Haddow, W. R., 284, 286, 293
Haeger, G. E., 317
Hafley, W. L., 306
Hair, J. C., 316
Hajny, G. J., 306, 315, 317, 318
Halabisky, D. D., 97, 427
Hale, M. D., 236, 238
Halliwell, G., 204
Hamburg, H. R., 384
Hammel, K. E., 445, 451
Hansborough, J. R., 270, 312
Harada, H., 145, 227, 403
Hart, J. H., 405
Hartford, W. H., 419, 422, 430
Hartig, R., 8, 176, 182
Hartley, C., 14, 101, 250, 252, 284, 298, 360
Harvey, A. E., 109, 270
Hatton, J. V., 321
Hattori, Y., 372
Hawksworth, D., 73, 74
Hawley, L. F., 10, 408
Hayashi, K., 204
Heath, I. B., 60, 228
Hedgecock, G. G., 10, 284
Hedges, J. I., 229, 230, 232
Helsing, G. G., 41, 356, 367, 374, 423, 424, 425
Henderson, R., 110
Henningsson, B., 252, 274
Hepting, G. H., 283, 284, 291, 296
Higgins, N. C., 95
Highley, T. L., 83, 95, 110, 123, 196, 198, 199, 202, 207, 208, 230, 232, 233, 234, 235, 239, 301, 313
Higuchi, T., 11, 136, 195, 210
Hill, C. L., 41
Hill, D. O., 391
Hillis, W. E., 283, 401, 405, 407
Hirt, R. R., 182, 347, 360
Hoag, M., 372
Hochman, H., 430, 434

Hoffman, P., 230
Hoffmeyer, P., 251, 256, 369
Hollis, C. G., 384, 386, 394
Holt, J. G., 82
Holt, D. M., 11, 83, 216, 238
Holt, K., 376
Hon, D. N-S., 23
Hopp, H., 170, 401
Horezcko, G., 43
Hornebrook, E. M., 284
Horner, W. E., 393
Horvath, R. S., 391
Howard, N. O., 93
Howard, K. S., 226
Hoyle, R. J., Jr., 252
Hse, C. Y., 377
Hubbard, T. F., Jr., 405
Hubert, E. E., 10, 178, 182, 283, 284, 327, 328
Hudson, M. S., 434
Hughes, S. J., 66
Hulme, M. A., 105, 269, 306, 317, 328, 332, 338, 434, 435, 448, 451
Humphrey, C. J., 10, 104, 307, 408
Hunt, G. M., 415, 422, 426
Huntgate, R. E., 109
Hutchinson, J. D., 251, 369
Hyat, M. A., 226
Hyttiner, A., 257

Ifju, G., 23, 97, 252, 427
Illman, B. L., 207
Imanura, Y., 261, 373
Ingraham, J. L., 90
Inoue, M., 395
Int'l Union of Biochem., 121
Inwards, R. D., 260, 367, 372, 373
Isaacson, I. H., 308
Isquith, I. R., 391
Iwasake, T., 204

Jackson, L. W. R., 290
Jakubowski, J. A., 386
James, W. L., 371
Jamison, G. M., 23
Jayne, B. A., 252
Jellison, J., 375
Jennison, M. W., 110
Jensen, K. F., 98, 99, 100, 101
Jin, L., 401
Johnson, A. H., 226

Johnson, E. L., 338, 404, 435
Johnson, W. T., 283
Johnson, B. R., 260, 423, 430
Johnson, M. A., 405
Johnsrud, S. C., 230
Jones, S. C., 29, 33
Jones, T. W., 308
Joseleau, J. P., 228, 230
Jurasek, L., 339
Jurgenson, M. F., 109, 270
Jutte, S. M., 184, 234, 236, 314, 374

Käärik, A., 272, 273, 333, 335, 387
Kallio, T., 448
Kalnins, M. A., 43
Kalyanaraman, B., 451
Kanagawa, Y., 372
Kaufert, F. H. 95, 434
Keilich, G., 208
Kelman, A., 213
Kemp, H. T., 384, 386, 387
Kenderes, A. M. 184
Kendrick, B., 66
Kennedy, R. W., 252, 405, 408
Kerner, G., 253
Kerner-Gang, W., 434, 435
Kerr, A. J., 228
Kerruish, R. M., 434
Kersten, P. J., 215, 453
Kimmey, J. W., 284
Kindl, H., 405
King, K. W., 204
King, B., 216, 269, 449
Kirchman, D. 44
Kirk, T. K., 11, 98, 180, 195, 198, 199, 201, 202, 203, 210, 211, 212, 213, 214, 215, 236, 301, 348, 446, 447, 448, 450, 451, 453
Kirsch, E. J., 435, 451
Kistler, B. R., 180
Klens, P. F., 386
Klingstrom, A. 109
Knoll, 226
Knuffel, W. E., 373
Knuth, D. T., 83, 183, 311
Knutson, D. M., 328
Koch, P., 328
Koehler, C. S., 35
Koenigs, J. W., 11, 28, 207
Kofoid, C. A., 41
Kohlmeyer, E., 43
Kohlmeyer, J., 43

Kollman, F. F. P., 140, 154
Kraeplin, G., 230, 258, 377
Krahmer, R. L., 375, 376
Kramm, K. R., 269, 449
Kress, O., 307
Kressback, J. N., 430, 432
Krishna, K., 29
Kropp, B. R., 184, 314, 315, 374
Krzylewski, J., 423
Kubiak, M., 253
Kubler, H., 318
Khne, H., 24
Kukor, J. J., 268
Kunesh, R. H., 525
Kuo, M., 377
Kuwahara, M., 212

Laks, P. E., 407
Lamb, M. V., 93
Lambert, D. H., 9
Lambourne, R., 384
Lamprecht, I., 258, 377
Lane, C. E., 43
Lane, P. H., 312
Larsen, M. J., 98, 109, 232, 270, 301
Larson, E., 284, 290
Latge, J. P., 232, 375
Lawyer, K. R., 393
Leatham, G. F., 447
Leathers, T. D., 230, 232, 235
Leisola, M., 232, 375
Leitner, G., 386
Lentz, P. L.
Lessard, R. M., 260
Leukens, U., 24
Leutritz, J., 427, 435
Levi, M. P., 215, 232, 239, 415, 434
Levien, K. L., 315
Levitin, N., 330
Levy, C. R., 10
Levy, J. F., 83, 102, 183, 236, 238, 267, 269, 270, 273, 274, 336, 449
Levy, J. S., 395
Lew, J. D., 419
Lewis, D. A., 338
Lidstrom, M. E., 82
Liese, W., 9, 11, 83, 178, 179, 183, 185, 186, 200, 208, 228, 230, 238, 239
Lilly, V. G., 90
Lin, L. C., 375
Lindenmayer, A., 53
Linder, D. H., 43, 183

Lindgren, R. M., 10, 105, 93, 257, 259, 313, 315, 316, 327, 332, 335, 336, 337, 414, 375
Lindgren, P., 357, 372
Line, M. A., 375
Little, E. L., 401
Livingston, B. E., 98
Ljungdahl, L. G., 97
Lockhard, C. R., 284
Loewus, F. A., 11, 195
Logan, A. F., 228
Lombard, F. F., 11, 184, 236, 274, 314, 347, 348, 359, 360, 374, 375, 429
Lovrien, R. E., 232, 335
Lü, H., 58
Lowell, E. C., 375
Ludwig, L. E., 386
Lumsden, G. Q., 354
Lund, A. E., 29, 269, 449
Lundström, H., 178, 320
Luo, S., 377
Luxford, R., 252, 284, 298
Lyon, H. H., 283
Lyons, J. W., 26
Lyr, H., 206, 451

MacLean, J. D., 25, 107
Madhosingh, C., 433, 436
Madigan, M. T., 82, 90, 116, 266, 268
Maguire, A., 209
Maitland, C. C., 204
Mallison, G. F., 395
Mandelo, M., 208
Manion, P., 181, 197, 272, 283, 284, 289, 291
Margulis, L., 54
Marion, J. E., 28
Marsden, D. H., 434
Martens, C. R., 384
Martin, M. M., 268
Martin, J. S., 257
Martin, P., 377
Martinson, M. M., 435
Marton, R., 450
Marx, H., 11, 286, 287
Mason, R. R., 315
Maudlin, J. K., 29, 33
May, C., 360
McClelland, J. F., 377
McCoy, E., 83, 183, 311
McDaniel, C. A., 408
McGovern, J. N., 257

McKaig, P., 426, 427
McKee, J. C., 316
McKimmy, M. D., 372
McLaughlin, D. J., 58, 60, 228
McMichael, K. L., 226
Meinholtz, D. C., 207
Menzies, R. J., 43
Merrill, W., 9, 109, 178, 180, 273, 284, 289, 403
Messner, K., 230, 231, 232, 234, 375
Meyer, R. W., 357
Meyers, S. P., 44
Miller, B. D. 176
Micklewright, J. T., 421
Miller, D. J., 328, 338, 415, 425
Miller, D. L., 284
Mims, C. W., 69
Miniutti, V. P., 24
Mitchell, R., 43, 44
Mitchoff, M. E., 337
Miyauchi T., 261, 373
Montencourt, B. S., 208
Mook, P. V., 327, 337
Moore, W. E., 198
Moore, H. B., 30, 33, 34, 35, 37, 39, 349
Moore-Landecker, E., 67, 69
Morgan, F. D., 38
Morgans, W. M., 384
Morohoshi, N., 151
Morrell, J. J., 41, 102, 105, 203, 251, 253, 255, 257, 314, 315, 338, 357, 365, 367, 368, 369, 374, 375, 415, 424, 449
Morris, P. I., 269, 370, 372, 377, 449
Mothershead, J. S., 251, 372, 374, 375
Mouzouras, R., 236
Mraz, E. A., 353
Muhonen, J. M., 315
Mulholland, J. R., 252
Murmanis, L., 230, 232, 233, 234, 235, 239
Murphy, M. W., 370
Murphy, T. G., 284
Murray, P. M., 330

Nagashima, Y., 83, 238
Nakato, K., 403
National Research Council, 4, 446
Nault, J., 401
Nelson, C. D., 272
Nelson, E. E., 448
Nester, M. T., 82
Nester, E. W., 82
Newbill, M. A., 315, 357, 369, 424

Newcomb, M. D., 110
Nicholas, D. D., 12, 183, 338, 376, 405, 426
Niederpruem, D. J., 102
Nilsson, T., 10, 11, 83, 178, 180, 215, 216, 231, 233, 236, 237, 238, 239, 274, 348, 375, 429, 434
Nishimoto, K., 261, 373
Nobles, M. K., 10, 64, 270, 374
Noguchi, M., 261, 373
Norkrans, B., 204
Nutman, F. J., 184

O'Neill, T. B., 434
Obst, J. R., 180, 215, 229, 232, 236, 348
Ochrymowych, J., 260, 272, 430
Okamura, S., 261, 373
Oksbjerg, E. 109
Oldham, N. D., 328
Oldham, M. L., 328
Oppermann, R. A., 385
Oteng-Amoako, A. 28
Otjen, L., 11, 230, 231

Packman, D. F., 360
Painter, P. R., 90
Palmer, J. G., 230, 232, 233, 234, 235
Panek, E., 314
Pankratz, H. S., 231
Panshin, A. J., 136, 252, 400
Parameswaran, N., 228, 230, 239
Parham, R. A., 227
Parmeter, J. R., 101
Parrish, F. W., 209
Partridge, A. D., 284
Pascoe, T. A., 315
Pearsall, N. N., 82
Peck, E. C., 163
Pellerin, R. F., 261, 373
Périé, F. H., 213
Peterson, B., 206, 231, 232, 233, 375
Pierce, B., 369
Pignatello, J. J., 435
Piirto, D. D., 260, 371
Pinion, L. C., 203, 215
Pittman, C. V., 393
Popek, R. A., 176
Postek, M. T., 226
Presnell, T. L., 338
Preston, R. D., 215, 239
Preston, A. E., 11
Preston, A. F., 269, 338, 426, 429, 449

Preston, R. D., 436
Prevost, M. C., 232, 375
Proctor, P., 185
Przybylowicz, P., 184, 314, 374
Pugh, G. J. F., 266
Putnam, J. A., 284
Pyne, J. W., Jr., 451

Quayle, D. B., 41
Quinn, V. P., 152

Raju, P. N., 109
Rambo, G. W., 349
Ramsbottom, J., 360
Rao, K. R., 421
Raper, J. R., 69
Raper, C. A., 69, 71
Raven, P. H., 116
Ray, D. L., 43
Rayner, A. D. M., 11, 12, 267, 269, 271, 272, 284, 289, 301, 445
REA, 346
Reddy, C. A., 231
Redmond, D. R., 284
Reese, E. T., 11, 26, 195, 208, 209
Reh, U., 259, 377
Reichelt, H., 93
Reid, I. D., 100, 230
Rendtorff, R. C., 395
Resch, H., 239
Reynolds, E. S., 44
Rhoads, A. S., 182
Ricard, J. L., 239, 269, 374, 375, 448, 449
Richards, C. A., 10, 347, 408, 427
Richards, D. B., 260
Rischen, H. W. L., 413
Robards, 226
Roberts, E. E., 82
Robichaud, W. T., 395
Roe, T., 430
Roff, J. W., 99, 335, 338
Röhr, M., 230
Rosenberg, A. F., 349
Ross, J. D., 27
Ross, R. T., 384, 386, 394
Ross, R. J., 261, 373
Rossell, S. E., 83
Rossignol, L., 313, 339, 449
Rossman, A. Y., 82, 296, 434
Roth, H. G., 359
Roth, E. R., 284, 331, 360

Rothrock, C. W., Jr., 257
Rothwell, F. M., 386
Roussy, G., 377
Rowatt, H. H., 284
Rowell, R. M., 11, 353
Rue, J. D., 10
Ruel, K., 195, 228, 230
Rumbold, C. T., 10
Runeckles, V. C., 11, 195
Ruska, 226
Rutland, P. J., 83
Ryan, K. G., 237, 239
Ryvarden, L., 8, 283, 296, 299

Sachs, I. B., 176, 375
Saddler, J. H., 313, 339, 449
Safo-Sampah, S., 253, 369
Saka, S., 156, 228
Salisbury, P. J., 290
Salle, A. J., 268
San Antonio, J. P., 447
Sarko, A., 147
Sarles, R. L., 321
Savory, J. G., 10, 178, 183, 203, 215, 252, 274, 359, 360, 434, 435
Schacht, H., 17
Schaffner, W., 395
Scheffer, T. C. 10, 95, 96, 98, 99, 101, 176, 182, 252, 253, 258, 259, 284, 298, 306, 308, 310, 312, 315, 326, 327, 330, 335, 337, 345, 346, 349, 360, 369, 400, 401, 404, 405, 406, 408, 414, 427
Scheld, H. W., 312
Schmid, R., 179, 185, 186
Schmidt, O., 83, 216, 238, 435
Schmidt, U., 83, 238
Schmidt, R. L., 321
Schmitt, J. A., 391
Schmitz, H., 95, 290, 434
Schultz, T. P., 376, 405
Schulz, G., 259, 332
Seesman, P. A., 451
Seidler, R. J., 10, 109
Seifert, K. A., 100, 215, 313, 339, 347, 449
Selby, K., 204
Sell, J., 24
Setliff, E. C., 450
Sexton, C. M., 72, 184, 253, 315, 374, 449
Shafizadeh, F., 26
Shain, L., 287
Sharp, R. F., 270
Shaw, R. A., 269

Shaw, C. G., 239
Shenfelt, R. D. 30
Shields, J. K., 269, 313, 320, 332, 448
Shigo, A., 153, 176, 260, 371
Shigo, A. L., 11, 153, 176, 260, 272, 283, 284, 286, 287, 288, 289, 290, 292, 371, 372, 400
Shimada, M., 212, 448, 450
Shortle, W. C., 153, 260, 287, 289, 372
Siau, J. F., 136
Sietsma, J. H., 59
Siggers, P. V., 104
Sigman, D. S., 122
Silverborg, S. B., 284, 290, 347
Simeone, J. B., 39
Simpson, S. L., 386
Sinclair, W. A., 283
Singh, A. P., 11, 216, 238
Sjöström, E., 195
Skaar, C., 95, 136, 160
Skutt, H. R., 260
Smith, C. A., 27
Smith, R. S., 97, 98, 101, 415, 427
Smith, S. M., 184, 251, 253, 257, 268, 314, 315, 369, 374, 424
Smith, D. W., 82, 90, 116, 266, 268
Smith, A. H., 68
Smith, R. A., 386
Smith, W. R., 257
Snell, W. H., 10, 93, 98
Snyder, T. E., 33
Snyder, H. D., 386
Sorkhoh, N. A., 274
Southam, C. M., 408
Southwell, C. R., 403
Sparrow, F. K. 73
Spaulding, P. 182, 270
Springer, E. L., 317, 318
Srebotnik, E., 231, 232, 234, 375
Stacey, S. S., 251, 372
Stachelberger, H., 230, 233
Stahmann, M. A., 60
Stalpers, J. A., 64, 270, 374
Stanier, R. Y., 90
Stanlake, G. J., 451
Steiert, J. G., 435
Steiger, F., 43
Stewart, D. L., 451
Stillinger, J. R., 284, 290
Stirling, G. R., 269
Stoker, R. S., 176
Stranks, D. W., 105, 332, 434, 435, 451
Streamer, M. K. E., 206

Streisel, R. C., 377
Stroukoff, M., 360
Stryer, L., 116
Stutz, R. E., 328
Subramanian, C. V., 395
Sussman, A. S., 73
Sutton, B. C., 73
Swartz, J. N., 315, 316
Swift, M. J., 60, 267

Tabak, H. H., 97
Taber, W. A., 102
Tanaka, H., 215
Taniguchi, T., 403
Tansey, M. R., 320
Taylor, F. L., 176
Taylor J. A., 314
Taylor, F. W., 369
Taylor-Wilson, J., 370
Terracina, F. C., 236, 260, 274, 337, 347, 348, 355, 356, 372, 374, 377, 389, 391, 424, 429, 449
Thacher, D. G., 98, 101
Thiede, W. H., 395
Thielke, C., 228
Thomas, R. J., 183
Thomas, J. F., 328, 338
Thompson, G. H., 35
Thompson, W. C., 27
Thorn, R. G., 269
Thörnqvist, T., 320
Thornton, J. D., 372
Tickner, J. A., 395
Tien, M., 195, 212, 214, 435, 450, 452, 453
Timell, T. E., 136, 146, 148, 149, 157, 195, 202
Tocher, R. D., 328
Todd, N. K., 11, 12, 267
Toole, E. R., 97, 98, 252, 290, 312, 347
Turner, R. D., 40

U.S. Navy, 40
U.S. Government, 415
Unligil, H. H., 236, 338, 435
USDA, 400

Valii, K., 210
Vallander, L., 230
van der Kamp, B. J., 98, 101, 401
VanDenBerge, J., 413

VanGroenou, H. B., 413
Vaughn, D. B., 10, 64
Verrall, A. F., 11, 327, 332, 333, 335, 337, 349, 350, 351, 352, 360, 369, 434
Vessal, M. I., 204
Vestal, M. R., 183
Vogels, G. D., 97
Volc, J., 232
Volkman, D. A., 315, 316, 321

Wagner, R. E., 451
Walchi, O., 24
Walcheski, P. J., 426
Walker, R. D., 377
Wang, C. J. K., 64, 176, 236, 260, 274, 320, 336, 347, 348, 355, 356, 372, 374, 377, 388, 424, 429, 449
Wangaard, F. F., 27
Ward, O. P., 102
Ward, J. C., 259
Wardrop, A. B., 141
Warren, G. R., 273, 284, 300
Warüshi, W., 210
Watkins, J. B., 423
Webb, W. E., 35
Webb, D. A., 421
Webster, J., 69, 71
Weeks, R. M., 305
Weesner, F. M., 29, 34
Wegener, G., 136, 155, 195, 283
Weiner, J., 306
Weir, J. R., 284
Weismantel, G. D., 384
Weliky, K., 229, 230, 232
Wessel, C. J., 360
Wessels, J. G. H., 59
Wheelis, M. L., 90
Whitaker, R. H., 53
White, J. H., 182
White, L. T., 293
White, D. G., 408
White, M. S., 321
Wilcox, W. W., 11, 182, 239, 250, 252, 253, 255, 260, 301, 328, 344, 349, 366, 369, 371, 373, 374, 419, 423
Wilkinson, J. G., 415, 426
William, G. R., 337
Williams, L. H., 37, 349
Williams, M., 284
Wilson, L. F., 33
Wilson, M. M., 28
Wilson, T. R. C., 252, 284, 298
Wilson, J. B., 365, 373
Wilson, B. W., 451
Wilson, J., 401
Winandy, J. E., 27
Winters, H., 391
Wise, L., 10
Wise, E. D., 426, 435,
Wissing, A., 28
Wolter, K. E., 208
Wong, E. W., 395
Woodcock, C., 147
Worrall, J. J., 101
Wright, E., 259, 332
Wright, D., 435, 452
Wu, L. C., 60

Yamamoto, K., 372
Yazaki, Y., 328
Young, H. E., 152
Young, G. Y., 436

Zabel, R. A., 64, 149, 176, 184, 202, 203, 234, 236, 255, 260, 274, 289, 299, 308, 314, 336, 337, 347, 348, 355, 356, 372, 374, 377, 389, 391, 393, 407, 408, 424, 429, 434, 449
Zachary, A., 451
Zadrazil, F., 447
Zainal, A. S., 403
Zeikus, U. G., 101, 210
Ziemer, B., 435
Zimmerman, M. H., 136
Zink, P., 333

Subject Index

Absorption, active transport, 125
Activation energy, 119, 120
Adenosine diphosphate, *See* ADP
Adenosine triphosphate, *See* ATP
ADP, 118
Adsorption, 159
Aerobic respiration, 125, 129
 biochemical pathways, 126
Agaricaceae, 80
Agaricales, 79
Alternaria alternata, 187, 333
Ambrosia beetles, 36
 damage, 310
American Wood Preservers' Association, 9
Amylopectin, 149
Amylostereum chailletii, 268, 273, 300
Anabolism, 117
Anaerobic respiration, 129-130
Animal feedstocks, wood, 448
Annual rings, 138
Anobiidae, 36
Antibiosis, 268
Antrodia carbonica, 79, 176, 177, 235, 269, 314
Armillaria mellea, 286
Aphyllophorales, 79-80
Arthropoda, 29
Ascogonium, 71
Ascomata, typical types, 67
Ascomycotina, 57,67,74,76
Aspergillus niger, 63
ATP, 118,119
Aureobasidium pullulans, 23, 66, 77, 96, 333, 334, 386, 391
 growth, features, 387
 microscopic features, 388
 roles in nature, 387
 structural features, 389

Bacillus polymyxa, 54, 83, 102
Bacteria
 cell wall erosion, 238
 classification, 82
 energy sources, 82
 morphology, general, 54
 morphological features, 82-83
 reproduction, 54
 role in ecosystems, 84-85
 roles in wood decomposition, 215
 wood-inhabiting, 83-84
Bacterial damage, 44
Bacterial wetwood, 101, 130
Basidioma, 70
Basidiomycotina, 57, 68, 77
 microscopic features, 65
Basidiospores, 68
Basidium, 68
 developmental stages, 68
Bethell process, 417, 418
Bioconversion, 450-451
Biodecolorization, 451
Biodegradation agents, 85
Biodeterioration, 13, *See also* decays, stains
 saprogen, 14
 substratum, 14
Biological control, 269, 448, 449
 agents, 85
 stain fungi, 449
Biological protection, *See* Biological control
Biomodification, 13
 pulp and paper, 451
Biopulping, wood, 450

Biotechnology
  chemical waste management, 451–452
  pulp and paper, 450
Bjerkandera adusta, 65, 66, 300, 315, 320
Blue stain, *See* Sapstain
Boron, 423
Bound water, 159
Brown rots, 45, 171, 176, 179, 250, 254, 256, 355
Brown rot, cubical, 56, 171
Brown rot fungi
  chemical changes, major features, 202
  wood components, 197, 199
Brown rots, ultrastructural changes, wood, 232

Cambium, 137
Camponotus pennsylvanicus, 39
Carbon dioxide, fungal growth relationships, 102
Carbon dioxide levels, tree stems, 101
Carpenter ants, 39
  biology, 39
  damage, 31, 39
Carpenter bees, 39
Catabolism, 116
CCA, *See* Preservatives, chromated copper arsenate
Cell wall
  bordered pits, 143
  chemical constituents, 145
  components distribution, 153, 154
  layers, 143
  pittings, 143
  ultrastructure, 141
Cell wall layer
  S1, 143
  S2, 141
  S3, 143
Cell wall layers
  primary, 141, 143
  secondary, 141, 143
Cell wall penetration, 184, 187
Cell wall zones
  crystalline, 157
  gross capillary, 157
  transient capillary, 157
Cellobiose: quinone oxidoreductase, 205
Cellulase, 61, 124, 204
Cellulose
  chemical structure, 146
  degree of polymerization, 146
  glycosidic linkage, 147
  structural configuration, 147
  unit cell, 147
Cellulose decomposition, 204
  brown rot fungi, 207
  brown rot fungi, non-enzymatic oxidizing agent, 207
  C_1–C_x concept, 204
  soft rot fungi, 208
  white rot fungi, enzymatic cleavage sites, 206
  white rot fungi, pathways, 205
  white rot fungi, regulation reactions, 207
Cellulose microfibril, 146
  molecular arrangement, 148
Cerambycidae, 37
Ceratocystis coerulescens, 336
Chaetomium globosum, 76, 178, 184, 203
Chaetomium thermophile, 104, 320
Chemical decomposition, wood, 26
  acid attack, 27
  alkali attack, 27
  strength reduction, 28
Chitin, 61
Chitosan, 61
Chlamydospores, 62
Chloropicrin, 424
Chromated zinc chloride, 423
Citric acid cycle, 127
Clamp connections, 57, 69
  developmental stages, 59
  medallion, 59
  multiple, 59
  types, 59
Clavariaceae, 80
Cleistothecium, 67
Climax community, 267
Co-metabolism, 269
Coating discolorations, *See* paint mildew
Coenzyme, 118
Coleoptera, 34–38, *See also* Beetles
  families of wood-destroying beetles, 35
  post-harvest beetles, 35
  pre-harvest beetles, 34
Colonization, sequences in wood, 272
  dying and dead trees, 272
  ground contact, treated, 274
  ground contact, untreated, 273
  seasoning lumber, 273
  standing trees, 272
Colonization strategies
  combatants, 271
  opportunists, 271, 301

stem decay fungi, 300
stress resistors, 271
Colonization, wood-inhabiting fungi, 184
Columbian timber beetle, 35, 36
Combustion, *See* Thermal decomposition
Commensalism, 268
Community, 267
Compartmentalization, stem decay, 287, 288
Competition, 266
Compression wood, 157
Coniophora puteana, 101
Copper naphthenate, 421
Copper-8-quinolinolate, 422
Cu-8, *See* Copper-8-quinolinolate
Cytoplasmic membrane, 60

Dacrymyces stillatus, 79
Dacrymycetales, 79
Daldinia concentrica, 76
Death-watch beetles, 36
Decay, 13
  anatomical features, 183
  color changes, 174
  effects on acoustic properties, wood, 261
  effects on caloric values, wood, 258, 259
  effects on electrical properties, 260
  effects on permeability, wood, 259
  effects on pulping and paper properties, 257
  effects on wood density loss, 250
  effects on wood hygroscopicity, 251
  effects on wood strength, 251
  effects on wood weight loss, 249
Decay, internal, 365
  macroscopic features, 170, 175
  microscopic evidences, 176
  microscopic features, 177
  patterns, 170
  shrinkage patterns, 174
  visual evidences, 173
  wood property changes, 172
Decay anatomy, historical, 182
Decay control
  growth requisites, adversely affected, 130, 131, 132, 133
  new approaches, 446
  principle, 94, 360, 361
Decay detection
  chemical indicators, 375
  culturing, laboratory, 374
  difficulties, 365
  infra-red spectroscopy, 375

lectins, 375
locations, sampling, 366
methods, 372
  acoustic, acoustic emissions, 372
  acoustic, stress wave timers, 373
  acoustic, wave form analysis, 373
  electrical, 371
  electrical, moisture meters, 371
  electrical, Shigometer, 371
  mechanical, 368
  mechanical, compression tests, 368
  mechanical, extensiometer, 369
  mechanical, penetration resistance, 369
  mechanical, splinter test, 369
  mechanical, torque release, drilling, 369
  mechanical, vibration, 370
  physical, 367
  physical, boring, 368
  physical, increment cores, visual examination, 368
  physical, sounding, 367
microscopy, 375
serological tests, 375
tomography, 372
X-ray, 372
Decay development, sequential, 188
Decay fungi, wood products
  buildings, 348
  conifers, prevalence, 347
  hardwoods, prevalence, 347
  types, 347
Decay hazard rankings, 346
Decay
  in bridge timbers, 358
  in buildings, 348-353
    hazard zones, 350
    moisture accumulation indicators, 351
    moisture sources, 349-353
    prevention, 353
  in cooling towers, 359
  in marine piling, 357
  in pallets and boxes, 360
  in panel products, 360
  in railroad ties, 358
  in utility poles, 353-357
    controls, 356
    external decay, 354
    internal decay, checking, 354
    locations and patterns, 356
    remedial treatments, 356
      brown rot, 355
      soft rot, 355
      white rot, 355

Decay (cont.)
  in wooden boats, 359–360
Decay losses, 6, 173
  reduction, 6
  wood products, 344
Decay rate, affected by rot type, substrate, and time, 250
Decay resistance
  commercial wood species, 401
  evaluations, 408
  factors affecting, 402
  heartwood extractives, 405
  stem position, 402
  variations, 406–407
Decay stages, 169
Decay types, 171, 179
  cell-wall destruction, modes, 189
Deterioration, 12
Deuteromycotina, 55,77
Dichomitus squalens, 213
Dictyosomes, 62
Digestion, 122
Discoloration, 13
Disease, 12
  host, 14
  pathogen, 14
  signs, 13
  symptoms, 14
Dolipore septum, 57,58

Earlywood, 138
Echinodontium tinctorium, 286, 293, 297, 301
Electron carriers, 117
Electron microscopy, 225
  scanning, 227
  transmission, 226
  wood
    detection of wood preservatives, 239
    quantification of wood preservatives, 239
Electron-transport chain, 128
Endo-1-4-β-glucanase, 124, 204, 205, 207, 208, 209
Endohydrolase, 123
Endoplasmic reticulum, 60
Energy dispersive x-ray analysis, 227
Enzymatic stains, 327–330
  kiln brown, 327, 328, 329
    causes, 328
    controls, 328
    oxidative, hardwoods, 330

Enzyme, 118
  action model, 120
  cofactors, 121
  function, 120, 121
  inhibitors, 130
  structure, 118
  types, 121–122
Enzyme locations, decayed cell walls
  immunocytological techniques, 232
  lignin peroxidase, 233
Equilibrium moisture content, 249
Erwinia nimipressuralis, 83
Eukaryons, 53
Eumycota, 74
Exo-1-4-β-glucanases, 123, 124, 204, 205, 208
Exohydrolase, 123

Facultative anaerobes, 96
Fermentation, 96, 129
  wood wastes, 448
Fiber saturation point, 5, 94, 159
Fire retardent chemicals, 26
Flavenoids, 405, 406
Fomes fomentarius, 299
Fomitopsis officinalis, 298
Forest pathology, 7
  historical aspects, 284
Forest Products Laboratory, Madison, 9
Formicidae, 39
Frass, 29
Fruiting body, 67
FSP, *See* Fiber saturation point
Fumigants, 424
Fungi
  characteristics, 53
  classification, 73
  cultural characteristics, 64
  growth requirements, 72–73
  life cycles, 69
  macroscopic features, 55
  microscopic features, 57
  parasexual cycle, 71
  reproduction, 66
  reproductive capacity, 71
  reproductive potential, 85
  roles in ecosystem, 55, 60, 84, 85
  ultrastructural features, 228
  variability, 72
    aneuploidy, 72
    crossing over, 72
    mutations, 72
    plasmids, 72

Ganoderma applanatum, 71, 174, 198, 235, 299
Gasteromycetes, 68, 78
Gelatinous fibers, 158
Gloeophyllum sepiarium, 56, 65, 66, 105, 106, 315, 347
Gloeophyllum trabeum, 105, 106, 199, 253, 347
Glucan, 61
Glucomannan, 148
Glucose oxidase, 207, 214
β-glucosidases, 204, 207, 208
1,4-β-glucosidase, 124, 205
Glycolysis, 125
Glyoxal oxidase, 214
Golgi apparatus, 62
Graphium rigidum, 336
Gross capillary system, 159
Growth requisites, fungi, 90

Haustoria, 62
Heart rots, 181, 285
Heart rot fungi, oxygen sensitivity, 99, 101
Heartwood, 138
  extractives, 406
Hemicellulose
  chemical structures, 149
  detection in cell walls, 228
  decomposition
    mannans, 209
    xylans, 208, 209
Hemicelluloses, 124, 146
Heterobasidion annosum, 179
Heterotroph, 117
Hirschioporus abietinus, 273, 300
Hirschioporus pargamenus, 315, 320
Hormoconis resinae, 77, 274, 332, 356
Horntail wasps, 38
Hydnaceae, 80
Hydrogen bonding, 162
Hydrogen ion concentration, 108, See also pH, 108
Hydrolases, 91, 121, 122
Hydrolytic reactions, 123
Hylotrupes bajulus, 37
Hymenium, 67
Hymenomycete, life cycle, typical, 69–71
Hymenomycetes, 68,78,79
Hymenoptera, 38
Hyphae, 55, 57
Hyphae
  autolysis, 60
  coenocytic, 57
  dikaryon stage, 57
  "hyphal system," 85
  microscopic features, identifications, 65
  polymers in cell walls, 61
  septations, 57
  specialized forms, 62
  types in basidiomata, 64
  ultrastructural features, 63
  wall structure, 58–60
Hyphal sheaths, 62, 230, 232, 233
Hysteresis, 160

Incipient decay, discolorations, 332
Incising, 425
Inonotus glomeratus, 290, 299
Inonotus obliquus, 285, 299
Insect damage, wood, 29
Insects, life cycles, 29
Iron stain, 331
Irpex lacteus, 179, 184, 209, 271, 300
Isopods, 43, See also Limnoria
Isoptera, 29–34, See also Termites

Karyogamy, 69
Kerfing, 425
Kickback, 416
Kindling temperature, 26
Klason lignin, 27

Laccase, 205, 213
Latewood, 138
Laurilia (Stereum) taxodii, 300
Lentinus lepideus, See Neolentinus
Lenzites betulina, 272
Light, fungal growth, 110
Lignification, 151, 156, 402
Lignin, 124
  cell wall distribution, 156
  cell wall functions, 150
  degradation products, 150
  detection in cell walls, 228
  monomer linkages, 151
  precursors, 150
  structure, 150
  structural model, 151, 152
Lignin decomposition, 209
  white rot fungi
    chemical modifications, 211
    cleavage of C_α-C_β bonds in side chains, 214

Lignin decomposition *(cont.)*
  degradation products, 213
  laccase roles, 214
Lignin degradation
  brown rot fungi, 215
  products, 150
  soft rot fungi, 215
Lignin determinations, 210
Lignin peroxidase (LIP), 212
Ligninase, 212
Limnoria, 43
  biology, 43
  damage control, 44
Log and bolt storage, 308
  decay penetrations, 309–310
  general control practices, 310
    biological, 313
    chemical, 312
    water spraying, 312
    water storage, 311
Lowry process, 417, 418
Lumen, 141
Lyctidae, 37

Manganese peroxidase, 212, 213
Mannan, 61
β-mannosidase, 209
Margo, 144
Marine borer damage, 40
Mastigomycotina, 74,75
Mechanical wear, wood, 28
Meruliporia incrassata, 56, 101, 347, 352, 353
Mesophilic fungi, 104
Metabolism, 116
Microbial ecology, 266
Microbial, interactions, 268
  in decay, 239
Microbiology, applied, 7
Microfibrils, 161
Microtubules, 62
Middle lamella, 141
Mildew, 55
  paint, 385
Mildewcides, 392
  evaluations, 392–393
Mineral stain, 152, 330, 331
Minor metals, fungal growth, 110
Mitochondria, 62
Moisture content, *See* Water content
Mold, discolorations, 332
Molds, 44, 55

allergy problems, 395
growth, inert materials, 395
growth problems, homes, 395
preservative detoxification, 332
Monera, 53
Mushroom production, wood, 447
Mutualism, 268
Mycangia, 35
Mycelial fans, 55,56, 174
Mycelium, 55
  biomass estimation, 60
Mycology, 7
Mycopathogens, 269, 271

NAD, 118, 119
Neolentinus lepideus, 79, 106, 199, 347, 449
Niche, 267
Nicotinic adenine dinucleotide, *See* NAD
Nitrogen, fungal growth, 108–110
Nitrogen fixation, 109
Nitrogen levels, tree stems, 110
Nuclear condition
  dikaryon, 69
  diploid, 69
  heterokaryon, 71
  monokaryon, 69
Nuclei, 62

Obligate anaerobes, 96
Old-house borer, 37
Oomycetes, 57
Ophiostoma ips, 336
Ophiostoma picea, 273, 336
Ophiostoma piliferum, 336
Oven-dry weight, 159, 249
Oxidase reaction, 64
Oxidation, 117
Oxidative phosphorylation, 128
Oxidoreductases, 121, 122
Oxine copper, *See* Copper-8-quinolinolate
Oxygen, fungal growth, 96
Oxygen concentrations
  carbon dioxide, growth effects, 100
  fungal growth, 100
  maximum levels for decay, 99
  minimum levels for decay, 98
  optimum levels for decay, 99
  survival of fungi, 101
Oxygen levels, tree stems, 101

Paint mildew
  control practices, 392
  factors affecting development, 390
Paint
  biodeterioration, types, 384-385
  microflora, 386
  types and compositions, 385
Parasitism, 269
Pathogens, 55
Penetration pegs, 185, 187
Penicillium dupontii, 104
Pentose shunt, 128
Peptidoglycan, 54
Periodic acid, thiocarbohydrazide silver proteinate, 228
Perithecium, 67
Peroxidase, 124
pH, fungal growth, 108
Phaeolus schweinitzii, 298
Phanerochaete chrysosporium, 205, 212, 230, 233, 234, 320, 360, 450
Phanerochaete gigantea, 271, 315
Phellinus pini, 170, 186, 188, 285, 286, 290, 292, 293
Phellinus igniarius, 171, 298
Phellinus robiniae, 300
Phenol oxidizing enzymes, 212
Phialophora hoffmannii, 238
Phialophora mutabilis, 237, 238
Pholads, 43
Phytopathology, 7
Pilodyne, 369
Pinosylvin, 405, 407
Pits, 138
  bordered pit membrane, 144
  simple, 145
  semibordered, 145
Plasmogamy, 69
Poles and piling, storage, 313-315
Polyphenols, 405
Polyporaceae, 81
Population, 267
Postia amara, 171, 290, 300
Postia placenta, 176, 177, 183, 184, 197, 199, 209, 234, 235, 251, 253, 273, 347
Powder post beetles, 36-37
Predation, 269
Preservative protection, treatments, 413
Preservative resistance, 434
Preservative testing, 427
  decay cellars, 429
  field stakes, 429, 430
  marine borers, 430

Petri plate test, 427
pole test, 432, 433
soil block test, 427
wood assemblies, 433
Preservative threshold values, 428
Preservative tolerance, *See* Resistance
Preservative treatment
  empty cell process, 417
  full cell process, 416
  nonpressure, 415
  pressure, 416
  remedial, 424
  thermal process, 416
Preservative treatment cycles, 418
Preservatives
  ammoniacal copper arsenate, 423
  chemical structures, 420
  chromated copper arsenate, 422
    decomposition, 436
  copper naphthenate, 420
  copper-8-quinolinolate, 420
  creosote, 420
    decomposition, 434
  environmental considerations, 426
  early, 412-413
  miscellaneous compounds, 423
  oilborne, 419
  pentachlorophenol, 420
    decomposition, 435
  tributyltinoxide, 420
  waterborne, 422
Pressure treating cylinders, 417
Pretreatment decay, 313
  poles, 354
Proboscis hyphae, 238
Products pathology, 7
Prokaryons, 53
Protista, 53
Psychrophilic fungi, 103
Pulpwood chip pile, heat sources, 320
Pulpwood chip storage, 317-321
  colonization, microorganisms, 319
  deterioration controls, 321
  pile temperatures, factors affecting, 317
  principal defects, 317
Pulpwood storage
  losses, 315-316
  decay controls, 316
  losses by decay type, 316
  principal decay fungi, 315
  weight losses by species, locations, 315
Pyruvate, 125

Radial drilling, 425
Reaction wood, 157
Redox, 117
Reduction, 117
Respiration quotient, 97
Respiration
  aerobic, 96
  anaerobic, 96
Respirometry, 97
Reticulitermes, 30
Rueping process, 417, 418
Rhinocladiella atrovirens, 274
Rhizomorphs, 55, 56
Ribosomes, 60
Root rot pathogens, 286

Sap rots, 286
Sap stainers, 180
Saprobes, 55
Saprobic decays, 181
Sapstain, 333–336
  anatomical features, 333, 335
  eastern white pine, 334
  causal fungi, major, 336
  control practices, 339
  effects, wood properties, 336–337
  susceptible species, 333
  wood-inhabiting insects, relationship, 333
Sapstain controls, 337–339
  chemical treatments, 337
  piling procedures, 338
  rapid drying, 337
Sapwood, 138
Scavengers, 180
Schizophyllum commune, 59, 64, 77, 106, 347, 360
Scytalidium lignicola, 77
Serpula lacrimans, 352, 353
Shigometer, 260
Shipworms, 41
  biology, 41
  control, 43
  damage, 42
  Teredo, 41
  Bankia, 41
Siricidae, 38
Sistotrema brinkmannii, 273
Slash rots, 181
Slime formation
  controls, 394
  pulp and paper manufacture, 394–395
Slimicides, 394

Soft rot, 8, 10, 178, 179, 183, 355
  cavities, 236
  damage, 358
Soft rot fungi
  chemical changes
    wood components, 201, 202, 203
    major features, 202
  longitudinal bore holes, 186
  oxygen sensitivity, 101
Soft rots, 45
  Type 1, 178, 256
    damage, 186–187
  Type 2, 179, 256
    damage, 186
  ultrastructural changes, wood, 235
Spike kill, 358
Spores, 63
  asexual, 66
  conidial origins
    blastic, 66
    enteroblastic, 66
    holoblastic, 66
  sexual
    ascospores, basidiospores, 66
Stains, 45
Stains and discolorations, types, 327
Standing timber decays, 284, *See also* Stem decays, 284
Starch, 149
Stem canker pathogens, 286
Stem decay fungi, 294–296
  listing by timber species, 294–296
Stem decays
  development rates, 289
  origins, 285
  recognition, 290, 291
  types, 284
Stem wounds
  decay origins, 286
  tissue reactions, 287
Stilbenes, 405, 406
Strength of wood
  compression tests, 253
  early decay stage, 254
  toughness, 253
  weight loss relationships, 256
Stroma, 55
Substrates, fungal growth, 107
Succession
  stem decay, 287
  wood decay, 274, 275
Symbionts, 55
Symbiosis, 267
Synergism, 269

Subject Index / 475

Talaromyces dupontii, 320
Tall oil, 406
Tannins, 407
Taxifolin, 401, 407
TBTO, *See* Tributytinoxide
Temperature, fungal growth, 102
  cardinal levels, 103, 104
Temperature groupings, decay fungi, 104
Temperature, lethal, fungi, 106
  time-relationships, 106–107
Tension wood, 157
Termites
  biology, 30
  control, 34
  damage, 31
  dampwood, 33
  distribution, 32
  drywood, 33
  Formosan, 33
  major families, 32
  prevention, 33
  subterranean, 30
  tests, 429
Terpenoids, 405
Thelephoraceae, 80
Thermal decomposition
  wood, 24
  wood, combustion, 26
  wood, fire-retardent chemicals, 26
  wood, property changes, 25
  wood, pyrolysis, 26
  wood, strength losses, 25
Thermophilic fungi, 104, 320
Through-boring, 425
α-Thujaplicin, 407
Torus, 144
Toughness testing machine, 253
Trametes versicolor, 98, 99, 100, 170, 171,
  174, 176, 177, 179, 182, 183, 197, 198,
  212, 230, 251, 252, 315, 320, 347, 450
Transient capillary zone, 92
Treatability
  improvements, 425
  variability, 425
Tributyltinoxide, 422
Trichoderma viride, 259, 313, 316
Tropolones, 406
Tylosis, 145

Ustulina deusta, 76, 299

Vapam, 424

Vitamins, fungal growth, 110
Vorlex, 424

Water, roles in decay, 91
Water activity, 93
Water content
  wood, cardinal levels for decay, 95
  wood, maximum levels for decay, 94
  wood, minimum levels for decay, 93
  wood, optimum levels for decay, 95
  wood, survival of fungi, 95
Water potential, 93
  matrix potential, 93
  osmotic potential, 93
Weathering, wood, 23
  anatomical features, 24
  causal agents, 23
  loss rates, 24
  preventive treatments, 24
Wet wood, 101, 130
White rots, 45, 176, 250, 254, 256, 355
  lignin distribution, cell wall after decay,
    231
  lignin distribution, cell wall before decay,
    231
  pocket, 171
  sequential, 179
  simultaneous, 179
White rot fungi
  chemical changes, major features, 200
  chemical changes, wood components,
    197
  selective wood element removal, 229
  ultrastructural changes, wood, 229
Wood
  ash content, 151
  chemical compositions, 154, 155
  disadvantages, 5
  extractives, 151
  functions, 13
  gross features, 137
  growth patterns, 137
  inhabiting fungi, anatomical, chemical
    features, 181
  macroscopic features, 139
  microscopic features, 142
    cross section, 139–140
    radial section, 139
    tangential section, 139
  potential uses, 4
  principal cell types, 138, 140, 141
  ultrastructural features, 227
  values and uses, 4

Wood decay
  chemical changes, decay type, 19
  chemical mechanisms, 203
  control principle, 163
  microbial ecology, 270
  model of major events, 217
Wood deterioration, major types, 22, 23, 46
Wood durability, See Decay resistance
Wood microbiology, 6
  concepts and terminology, 12
  major research contributions, 10
  research history, 8
Wood pathology, 6, See also Wood microbiology
Wood preservation, 7
Wood preservatives
  detoxification, fungal, 354
  tolerance, 274
Wood storage, biodeterioration
  general control practices, 307–308
  types of loss, 307
Wood supplies, 446
Wood variability, 163
Wood–water relationships, 158, 160
  adsorption, 159
  desorption, 159
  dimensional changes, 161
  equilibrium moisture content, 159
  sorption characteristics, 161
  sorption isotherms, 160

Xylan, 148
Xylaria polymorpha, 175
β-xylosidases, 209

Zone lines, 170, 174
Zygomycotina, 57, 74, 75